Gribov-85 Memorial Volume
Exploring Quantum Field Theory

The Memorial Workshop Devoted to the
85th Birthday of V N Gribov

Gribov-85 Memorial Volume
Exploring Quantum Field Theory

The Memorial Workshop Devoted to the
85th Birthday of V N Gribov

L. D. Landau Institute for Theoretical Physics
Chernogolovka, Russia

17 – 20 June 2015

Editors

Yuri L Dokshitzer
University of Paris VI, France

Peter Levai
MTA Wigner RCP, Hungary

Julia Nyiri
MTA Wigner RCP, Hungary

World Scientific

NEW JERSEY • LONDON • SINGAPORE • BEIJING • SHANGHAI • HONG KONG • TAIPEI • CHENNAI • TOKYO

Published by

World Scientific Publishing Co. Pte. Ltd.
5 Toh Tuck Link, Singapore 596224
USA office: 27 Warren Street, Suite 401-402, Hackensack, NJ 07601
UK office: 57 Shelton Street, Covent Garden, London WC2H 9HE

Library of Congress Cataloging-in-Publication Data
Names: Gribov Memorial Workshop (2015 : Budapest, Hungary) | Dokshitzer, Yuri, editor. | Lévai, P. (Péter), 1962– editor. | Nyiri, J. (Julia), 1939– editor. | Gribov, V. N. (Vladimir N.), honouree.
Title: Gribov-85 memorial volume : exploring quantum field theory : proceedings of the Memorial Workshop devoted to the 85th birthday of V.N. Gribov, Budapest, Hungary, 17–20 June 2015 / editors, Yu L. Dokshitzer (University of Paris VI, France), P. Levai and J. Nyiri (Hungarian Academy of Sciences, Hungary).
Other titles: Exploring quantum field theory
Description: Hackensack, NJ : World Scientific, [2016] | Includes bibliographical references and index.
Identifiers: LCCN 2016035673| ISBN 9789813141698 (hardcover ; alk. paper) | ISBN 9813141697 (hardcover ; alk. paper)
Subjects: LCSH: Quantum field theory--Congresses.
Classification: LCC QC174.45.A1 G75 2015 | DDC 530.14/3--dc23
LC record available at https://lccn.loc.gov/2016035673

British Library Cataloguing-in-Publication Data
A catalogue record for this book is available from the British Library.

Desk Editor: Christopher Teo

Typeset by Stallion Press
Email: enquiries@stallionpress.com

Printed in Singapore

PREFACE

A memorial workshop devoted to the 85th anniversary of Vladimir Naumovich Gribov took place at the Landau Institute for Theoretical Physics in June 2015.

Physicists from different parts of Russia as well as from various corners of the world came to Chernogolovka, Moscow region, for the meeting. This book accumulates the talks presented at the workshop as well as contributions sent from all over the world by those who could not come to the meeting but were willing to pay tribute to a visionary man and a great physicist who laid the foundation of modern theoretical high energy physics.

Most contributions present new research results and reviews of various branches of theoretical physics. A separate chapter is devoted to biographical notes and short personal recollections by Volodya's colleagues and friends.

This book demonstrates that Gribov's ideas and approaches to various challenging problems of particle physics, such as the theoretical construction of a consistent picture of high energy hadron interactions (Gribov reggeon field theory), quantum anomalies, color confinement, to name a few, are being actively used and developed in the 21th century.

The 2015 Nobel Prize in physics was awarded for the discovery of neutrino oscillations, the theoretical basis for which was laid in Gribov and Pontecorvo's 1969 article "Neutrino Astronomy and Lepton Charge." The 2015 High Energy and Particle Physics Prize of the European Physical Society was awarded in 2015 "for developing a probabilistic field theory framework for the dynamics of quarks and gluons, enabling a quantitative understanding of high-energy collisions involving hadrons," a concept pioneered by Gribov and Lipatov in the pre-QCD epoch of early 1970s. These prizes demonstrate that Gribov's legacy continues to shine today.

We are grateful to Alexander Belavin for his initiative to choose Chernogolovka as workshop venue, to the Landau Institute, to Nikolai Nikolaev for organizing the conference, and to the Hungarian OTKA grant No. NK106119 for partial financial support.

We are thankful to Gergely Gábor Barnaföldi and Árpád Lukács for their help in preparing the book.

Yuri Dokshitzer, Péter Lévai, and Júlia Nyiri

4th V.N. Gribov Memorial Workshop, L.D. Landau Institute of Theoretical Physics, Chernogolovka, 2015.

Yu.L. Dokshitzer, V.N. Gribov and P. Nyiri, Paris, 1997.

L.N.Lipatov and G. Eliashberg, Chernogolovka, 2015.

I.M. Khalatnikov, Chernogolovka, 2015.

B.L. Ioffe, Chernogolovka, 2015.

CONTENTS

REMEMBERING VOLODYA

HE SEEMED PERMANENTLY INSPIRED

YURI F. ORLOV
Laboratory of Nuclear Studies, Cornell University Ithaca, New York, 14853, USA
E-mail: yfo@cornell.edu

I attended several of Gribov's legendary seminars. He seemed permanently inspired. It was a deep pleasure to follow him in his arguments.

My most vivid memory of him in private was when we went to Lev Okun's apartment for tea in 1977, after Berestetsky's funeral. Conversation turned to the achievements of the various groups: the Leningrad group, the ITEP group, and so on. At that point I had been without work for three years because, as a participant in the democratic movement, I had been fired from my physics job, just as I had been fired from my job in ITEP in 1956 for a pro-democracy speech. So I remarked that I felt I had personally underachieved as a scientist. Gribov, who knew my history perfectly well, said very firmly: 'What you have already achieved in physics, plus beyond physics, is completely sufficient.'

Volodya lived as if outside of politics. But he definitely had a political outlook, the outlook of a civilized and decent person, and he had the generosity of spirit to express it at a moment when it meant a lot to me.

Yu.F. Orlov and V.N. Gribov, Minneapolis, in front of the Cathedral of St. Paul, 1995.

REMEMBERING GRIBOV

ANDREW SZENT-GYÖRGYI
Brandeis University, Waltham/Boston, Massachusetts 02254-9110, USA
E-mail: szentgyo@brandeis.edu

It was always a pleasure to remember your husband Gribov, and how much it meant to spend time with him. Although I am not a physicist, it is clear what a great scientist Volodya was, innovative and deep. At the same time he was an extremely nice person.

He was not only a great scientist, but a person who, apart from his great contributions to physics, was also able to clearly see what was right and what was wrong in life. His presence, his opinion about people and science was always a delightful experience. Albert was the only other person I have known well who also clearly knew what was right and what was wrong, and was willing to stand up for his principles. We are lucky to have known and lived with such great individuals.

VLADIMIR NAUMOVICH GRIBOV (1930–1997)*

ANDOR FRENKEL

Wigner RCP of the H.A.S, P.O. Box 49, H-1525, Budapest, Hungary
E-mail: frenkel.andor@wigner.mta.hu

V.N. Gribov was incapable of sparing himself. All his life he has acted as passionately, as intensively as he did in his youth when he worked with Landau and Pomeranchuk and led the Theoretical Physics Department of the Ioffe Institute and later of the Nuclear Physics Institute in Leningrad. For him "leading" meant just hiring the most talented students to join the institute and then engaging in merciless, endless but fruitful discussions with them on the problems they worked on. He did not care about who was right, he cared only about the right answer. Incidentally, only in physics did he push tirelessly to reach a decision. In conversations about literature or politics as a rule he argued mildly and kindly. Not that he did not have thoughtful and firm opinions about many questions, but because he seemed to realize that in these matters different points of view were arguable.

In the beginning of the eighties when in addition to his position at the Landau Institute in Moscow he was appointed as a part-time scientific adviser at the Institute for Particle and Nuclear Physics in Budapest, Gribov was already known world-wide as one of the leading scientists in the theory of particle physics and in quantum field theory. Together with Bruno Pontecorvo he had formulated the theory of neutrino oscillation; he provided basic theorems of the dispersion relations and of the non-Abelian gauge theory; he worked out the Gribov Regge calculus and made other far-reaching contributions to theoretical physics. In the capacity of a scientific adviser he could, indeed, choose to give only advices. Instead, he began then, at the age of fifty, to work on the solution of one of the most exciting problems of particle physics, namely on the theory of quark confinement. He considered this to be the most important of the questions he had ever dealt with.

---

*Adapted from the funeral commemoration in August 1997, published in the Hungarian journal "Fizikai Szemle" (Review of Physics) 1997/9

Gribov belonged to those few who speak openly about their maturing ideas and about their unfinished work. This offered a glimpse into the workshop secrets of the Landau school. First a loose sketch is prepared, the ingredients of which may come from various fields of physics. In his theory of quark confinement Gribov relied on the theory of supercharged nuclei, on the theory of superconductivity and on a modified Dirac sea. Then came the mathematical elaboration, in the course of which the original sketch may undergo modifications.

Gribov always tried to prove himself wrong, to show that the whole line of thoughts or some parts of it had to be changed, and indeed he did so several times. The work took almost twenty years and only near the end of his life did he feel that the time for summing up had come. He was still able to finish the first part, published under the title "QCD at short and large distances". The second part, "The theory of quark confinement", has been assembled by his wife Julia Nyiri and by Yuri Dokshitzer on the basis of Gribov's manuscripts and notes. If the ideas exposed in these papers are right, Gribov will be among those who decoded the mystery of quark confinement.

The fact that Gribov got a position at the Ioffe Institute only in 1954, two years after he finished his university studies, was probably related to the anti-semitic course of the Soviet leadership at that time, which lasted until the death of Stalin in 1953. Later the official promotions of Gribov were hampered by his refusal to make compromises he considered unacceptable. Although Landau proposed to elect the 26-year old Gribov a member of the Academy of Sciences of the Soviet Union already in 1962, ha was admitted as a corresponding member only in 1972.

In 1962 Gribov was permitted to participate in the International High Energy Physics Conference in Geneva, but after that he was not allowed to travel for the next 15 years even to socialist countries. In 1977 Livia Jenik, one of the organizers of the EPS Conference in Budapest, had the lucky idea to invite Gribov as a tourist. To our surprise he was able to come and he gave a talk that was not included into the program of the conference. Nor dared we to publish the talk in the proceedings, fearing that he would not be allowed to come even as a tourist.

I got acquainted with Vladimir Naumovich during this conference. Above forty years of age friendships are rarely formed, but it happened to us. I got acquainted also with Lilya, Volodya's first wife and with their son Lyonya who lost his life in 1984 at the age of 27 in a tragic accident. I keep with piety the text of those Vysotsky songs I heard from them, written down for me by Lyonya.

In the life of Volodya a much more important event happened during the conference. It was there and then that he and Julia Nyiri found each other and in 1980 they married. This became a very happy marriage. It always seemed to me that they looked at each other with the same joy as in the beginning. With Pali, Julia's son from her first marriage, they formed a nice family.

If Volodya would have known how to spare himself, perhaps he could have lived longer, but I do not think that he would have a happier and wholer life.

VLADIMIR NAUMOVICH GRIBOV: PIECES OF BIOGRAPHY

YA. I. AZIMOV

Petersburg Nuclear Physics Institute,
National Research Center "Kurchatov Institute",
Gatchina, 188300, Russia
E-mail: azimov@thd.pnpi.spb.ru

The talk presents the main lines of biography of the prominent physicist and bright personality. Also given is a necessarily brief description of Gribov's scientific work.

This talk I would like to begin with some personal recollections. I was lucky to work with V.N. Gribov for 20 years. It was not *under* Gribov, but just *with*. We have several common papers, several of my papers were made according to Gribov's suggestions. But most of my work were formally independent of Gribov. Nevertheless, all my work of those 20 years, in more or less detail, went through discussions with Gribov.

Of course, physics was not the only interest of Gribov, but, undoubtedly, it was his main interest producing very strong emotions. One could discuss with Gribov various physics problems, even rather far from his own studies (a couple of most bright examples will be mentioned below). But those discussions were not easy. Gribov was hearing very attentively and was ready to immediately jump into battle if something seemed incorrect to him. He had a rare feature: his arguments were very hard to reject even when his position appeared unjust after all (I think nobody can be just in all cases). The same situation was both in private discussions and at seminars. That is why talks at Gribov's seminar were very difficult for speakers. As a result, some people feared to speak at his seminar. However, if a person were sufficiently brave, the talk appeared very useful for the speaker. After such a trial the author himself began to better understand his own work and results.

Gribov's name and his papers (at least some of them) are well known today. But his biography is less known. Here I present its main lines.

Vladimir (Russian nickname Volodya) Gribov was born on March 25, 1930 in Leningrad (St. Petersburg, before 1914 and now). His father died in 1938, at the time of the Great Terror in the Soviet Union. "Luckily", his death was the result of a disease, and not because of repression (in which case the whole family might have been repressed as well). But the situation was not simple: the mother stayed alone with two small children, Volodya and his younger sister.

Gribov's mother worked in one of Leningrad theaters (not as an actress). When Germany attacked the Soviet Union in 1941, the family was evacuated from Leningrad together with the theater. They were moving with the theater over Siberia, Far East, Urals. Nevertheless, Volodya was continuing his school education, started in 1937, without any delay. Only in summer 1945 the family was able to return to Leningrad (to enter the city after the blockade, every person needed to have a special permit). In 1947 Volodya finished here the school course. The natural question arose: what to do after that?

Volodya was growing up in theatrical environment, and he dreamed to become an actor, best of all, a cinema actor. However, when being in senior school classes, he had a possibility to expose himself to filming. As appeared, under camera he became "frozen" and lost his natural mobility. After that, one of professional actors advised him to choose some other specialization. At school, Volodya was quite successful in physics and mathematics. He preferred the former.

In 1947 Gribov enrolled in Physical Faculty of the Leningrad University. He was a student together with D.V. Volkov (later, the full member of the Ukrainian Academy of Sciences) and G.M. Eliashberg (now the full member of the Russian Academy of Sciences, chief researcher of the Landau Institute for Theoretical Physics). In 1950 Volkov was included into the special group to study nuclear physics in more detail, and in 1951 was transferred to the Kharkov University. Gribov also wished to be in that group, but was rejected.

In 1952 Gribov presented his diploma thesis considering interaction of two electrons in Quantum Electrodynamics (under the supervision of Yu.V. Novozhilov). It was evaluated very high. According to decision of the examination commission, Bulletin of the Leningrad University published the paper based on this diploma work. It was the first publication signed by V.N. Gribov. Thus, in summer 1952 he was graduated from the University with *diploma cum laude*.

The year 1952 in the Soviet Union was not favorable for persons of Jewish origin, one of which was Vladimir Naumovich Gribov. It was the last year for the trial of the Jewish Antifascist Committee (after its end, more than 100 persons were sentenced, more than 20 of them were shot). It was also the preparatory year for the "case of physicians" ("killers in white coats") which was officially opened in January 1953. In such situation the only job available for Gribov, a young physicist with *diploma cum laude*, was the position of a physics teacher in an evening school, organized for working people who could not, by any reason, complete their school education in the childhood. The salary was low, and year later Gribov found additional part-time work.

Nevertheless, Volodya wished and was continuing to do science. He was able to contact Professor L.E. Gurevich. To the beginning of 1954, they together prepared two papers, on properties of matter in external fields, electric and gravitation. The papers were submitted to Journal of Experimental and Theoretical Physics and accepted for publication. In addition, Gribov participated in the theoretical seminar of the Leningrad Physico-Technical Institute (now the A.F. Ioffe Physico-Technical Institute, further PTI) headed by I.M. Shmushkevich and K.A. Ter-Martirosyan.

At last, in May 1954 (after Stalin's death and extinction of the "case of physicians"), Gribov was able to be employed by PTI as the senior laboratory assistant in the group for nuclear theory. It was headed by I.M. Shmushkevich, who was simultaneously an informal head of the Theoretical Department as a whole. Now Gribov's progress was very fast. A year later he received the higher position of the junior scientist. In March 1956 he presented his dissertation for the degree of Candidate of Sciences (analog of PhD). It was concerned with interactions of neutrons with nuclei. The problem was suggested by K.A. Ter-Martirosyan, but at the presentation of the work he emphasized that methods for calculations were invented by Gribov himself.

After the defense of the dissertation, Shmushkevich and Ter-Martirosyan organized contacts of Gribov with L.D. Landau and I.Ya. Pomeranchuk. Gribov began to go regularly to Moscow for participating in Landau's seminar. Initially, Landau was skeptical about Gribov (he said: "I know one Gribov, the theatrical actor, and that is enough"). But he rapidly changed his mind. In 1958, when the Scientific Council of PTI discussed the next higher position for Gribov, the senior scientist, it received the very favorable recommendation from Landau. Later, Gribov always considered Landau as his main Teacher in theoretical physics.

In 1957, Shmushkevich invited Gribov to give lectures for students of the Leningrad Polytechnical Institute, where he was the Chair of the Theoretical Physics Department. Later, Gribov began to lecture in his *alma mater*, in the Leningrad University, and in 1968 he became the Professor of the University (the highest scientific degree, Doctor of Sciences, necessary for this, Gribov received in 1964). In parallel, Gribov presented lectures at various Physics Schools, both in the Soviet Union and (later) abroad.

Gribov's scientific work after receiving the Candidate degree became also very active and self-reliant. More and more often he appeared a source of ideas for his colleagues. For instance, in 1958 he investigated three-pion decays of the K-meson. Pair-energy distributions in the decay were shown to depend on pion-pion scattering length. For several years after that, the series of papers were published by members of Shmushkevich's group (including Gribov himself) on various inelastic near-threshold reactions. They could allow to obtain information on hadron interactions (*e.g.*, pion-pion ones) unreachable in conventional ways. Regretfully, corresponding experimental attempts, as appeared, could not give definite results at those times because of insufficient quality of experimental technique.

One of his further directions of interest (probably, under influence of Pomeranchuk) was high-energy behavior of strong interactions. It was generally assumed for a long time to be similar to classical diffraction of light on black screen. Gribov showed that such behavior would be inconsistent with analytic properties of strong-interaction amplitudes. This work of 1961 initiated international interest to his activity. Trying to overcome this diffraction difficulty, Gribov, partly in collaboration with Pomeranchuk, developed reggeology, method of Regge poles and, later, cuts as well. This direction was also actively supported by his colleagues in PTI. He became one of leaders of reggeology not only in the Soviet Union, but in the whole world as well.

At the same time, Gribov tried to use the known quantum field theories as test-grounds for studying their high-energy properties. In Quantum Electrodynamics, as he showed, a special role play the so-called "double-logarithmic terms". That is why the group of enthusiasts of this approach (G.V. Frolov, V.G. Gorshkov, and L.N. Lipatov first of all) was informally called "the double-logarithmic academy". Their attempts were crowned by papers of Gribov and Lipatov on summation of those "double-logarithmic terms", very famous now (they became a basis for studying the evolution of partons).

Gribov could efficiently discuss even such problems in which he was not previously active. For instance, he did not work himself with weak interactions and, in particular, with neutrinos, but when encountered with Pontecorvo's idea of neutrino oscillations Gribov immediately began to construct the corresponding formalism. This resulted in the joint paper of V.N. Gribov and B.M. Pontecorvo, considered now to be classic in the neutrino physics.

There was one more, less known example. In his talk at one of Gribov's seminars, Ya.B. Zeldovich explained that a charged rotating black hole should lose energy by radiation, so its rotation should slow down. When it stops, radiation, according Zeldovich, would stop as well. Here Gribov interrupted him by the statement that this is not correct, radiation would continue. Zeldovich waved away this statement, and the seminar talk was continued without discussion of Gribov's suggestion (it seems, however, that their discussions on this problem went on later, but Zeldovich stayed rigid). At that seminar, I restored for myself the lines of Gribov's thought as follows. Gribov, as a physicist, had grown up mainly on quantum physics (in difference with Zeldovich), and he was aware quite well about the Schwinger effect: strong enough electric field, even homogeneous and static (*i.e.*, large gradient of electromagnetic potential), generates electron-positron pairs due to quantum tunneling. Near a black-hole horizon, there is very large gradient of the gravitational potential, which should analogously produce particle-antiparticle pairs.

Later, Zeldovich again talked at Gribov's seminar, now about the famous paper of Hawking. In particular, Zeldovich said: "Volodya Gribov had tried to assure me that radiation of a black hole should continue, but I did not believe". That is how he lost the interesting and important result.

Gribov always tried to organize his whole knowledge into some consistent picture. As a rule, such an approach is very useful. But not always. It may be not quite adequate if the picture needs strong changes. For Gribov, such was the case of quarks. The idea of quarks was publicized by M. Gell-Mann and G. Zweig in 1964 on the base of resonance spectroscopy. It was further supported by the discovery of scaling in deep-inelastic scattering (in 1968) and, especially, by the unexpected discovery of J/ψ (in 1974). However, Gribov did not believe yet in quarks and, correspondingly, was skeptical in respect to Quantum Chromodynamics (QCD), which arose in 1971–1972. It was not a question of taste; for his position Gribov had definite rational arguments. At that time, I asked him once on reasons of his disbelieving in quarks. He answered: "The quarks should be strong-interacting objects; therefore, each of them should be surrounded by a pion cloud. Where is it?" (I should confess, that a clear consistent answer to this question is still absent. Note also that such negative relation to quarks did not prevent Gribov from suggesting to E.M. Levin and L.L. Frankfurt in early days of quarks to compare meson and baryon cross sections in terms of quarks.)

However, in 1976, after the discovery of charmed particles, predicted by the quark picture, Gribov changed his mind. He began to study quarks and QCD very intensively. And in 1977 he was able to find a new feature of QCD, known now as the Gribov horizon or Gribov copies. It is interesting to note that many–many people worked with QCD to that moment, but existence of the horizon stayed unnoticed. Initially, Gribov hoped that it is just the horizon which determines the origin of confinement of quarks and gluons. But soon he came to conclusion of its insufficiency. Since then he worked hard trying various ways to understand and describe a still unknown mechanism of confinement.

The administrative career of Gribov may look successful. To 1962 the Gatchina site of PTI had the working nuclear reactor and the proton accelerator under construction. There appeared necessity to have there a separate Theoretical Department. Gribov was suggested to organize it. Later, in 1969, after the death of I.M. Shmushkevich, Gribov was returned to the central part of PTI and became the head of its Theoretical Department. In 1971 the Gatchina site of PTI was transformed to be a new institute, Leningrad (now Petersburg) Nuclear Physics Institute (LNPI, now PNPI). All the activity on atomic nuclei and elementary particles was transferred from PTI to the new Institute. Its Theoretical Department was headed, of course, by Gribov. He was a rather good administrator, but, in my opinion, he disliked administrative duties which interfered with his scientific work. Meanwhile, those duties increased along with the Department becoming more populous. In 1980 Gribov moved from LNPI to L.D. Landau Institute of Theoretical Physics in Chernogolovka (near Moscow). There were several reasons for this step, and one of them, as I think, was his wish to diminish necessary administrative duties.

As a physicist, V.N. Gribov was recognized world-wide. He was a speaker at various conferences and schools in the Soviet Union and, sometimes, even abroad. Many foreign physicists, being in the Soviet Union, were eager to visit LNPI for discussions with Gribov. For his scientific work Gribov won various prizes, but he was most strongly proud to be the first recipient of the L.D. Landau Prize established in 1971 by the Academy of Sciences of the USSR, which honored his Teacher. In the same year, 1971, he became a member of the American Academy of Arts and Sciences. In the next year, 1972, he was elected (after several unsuccessful attempts) to be a corresponding member of the Academy of Sciences of the USSR. However, official position in respect to him was demonstrated by the fact that Gribov has never been elected to be a full member of the Soviet Academy.

After 1980, Gribov obtained the possibility to be part-time in Budapest, where his second wife, Júlia Nyíri, worked in Physics Institute. This made somewhat easier his contacts with Western physicists. And after 1990, longer abroad trips became possible, to be a visiting Professor of various Institutes and Universities, both in Europe and in the US.

Active and intensive work of Gribov was unexpectedly interrupted by the acute stroke in 1997 during one of scientific conferences. He was taken to a hospital and, after stabilization of his state, was transported to another hospital, in Budapest. Even there he tried to continue investigations of confinement. The medical treatment looked successful, and the physicians planned that Gribov would be able soon to go home. However, on August 13, 1997 he passed away. His grave in Budapest is marked by the memorial that reminds a beautiful fading flower, with the simple epitaph

VLADIMIR GRIBOV
FIZIKUS
1930–1997

on the basement.

After Gribov's death, his works have not been forgotten. Just opposite, many of his papers are republished. His lectures. which were mainly written up in Russian, but not published, are now collected, translated into English and published as books. Therefore, his results become available and well-known even for younger generations of physicists. If judging by references, the most famous and operative of those results seem to be the Gribov copies (they are especially essential now in lattice calculations) and DGLAP equations for evolution of partons (here G stays just for Gribov). There established are various stipends and prizes called by Gribov's name.

And yet there are Gribov's last papers or even notes. They are mainly not completed or, at least, not quite understood by the world community. Meanwhile, they tried various ways to solve the problem of confinement, one of the hottest problems in strong interactions. In some sense the situation is similar to the fate of Einstein's last ideas. During his life, they looked to be out of mainstream, but now many of them feed new theoretical approaches. Such a future is not excluded also for Gribov's last ideas. Look forward...

Acknowledgments

This work is supported by Russian Science Foundation (No.14-22-00281).

HOW A MEETING IN 1970 WITH VOLODYA TURNED INTO A VERY CLOSE FRIENDSHIP

URI MAOR
Tel-Aviv University, School of Physics and Astronomy, Ramat Aviv, Tel-Aviv 69978, Israel
E-mail: maor@post.tau.ac.il

The close friendship between Volodya Gribov and myself dates back to the 1970 Rochester High Energy Conference, which took place in Kiev. At that time I was a young Visiting Associate Professor in Berkeley, CA, active, like most of G.F. Chew group members, in both theory and phenomenology studies of S-Matrix Theory of Regge Poles. The Berkeley school approach was initiated by Chew and Frautschi in their 1961 paper, which led to the bootstrap theory. Less known in the West were Gribov's late 1950s and early 1960s papers which led to the definition of the Pomeron, the Reggeon field theory and the Pomeron partonic decomposition. In the early 1970 I was invited to give a talk at the Kiev Rochester Conference. I was very naive thinking that being invited would get me a visa to the Soviet Union, even though it had broken its diplomatic relations with Israel. A short visit to the Soviet Consulate in San Francisco ended with a rude and very clear statement that I did not have a chance to get the desired visa. Luckily, quite a few senior colleagues, who were also invited, convinced the Soviet authorities to grant a small number of Israelis, myself included, the desired visa, provided that we apply in Europe. I had to fly from San Francisco to Helsinki to get my visa and proceed via Leningrad and Moscow on my way to Kiev. Arriving in Kiev I was confronted by a very friendly person, introducing himself as Volodya Gribov. I knew that he was one of the Soviet stars of particle physics, but not much more. Volodya suggested that we should spend our free time in Kiev parks, where we could talk freely. Volodya's suggestion was too exotic to be declined, and we ended spending lots of hours, night after night, in Kiev parks, changing our route each night. Most of the time we were alone, joined by Lev Okun a couple of times. Our conversations centred, at the beginning, on Israeli politics, the Palestinian conflict with Israel and other obvious subjects. The issue of massive immigration from the Soviet Union to Israel following the collapse of the Soviet Union seemed too exotic to even be discussed. Volodya kept asking me about the quality of physics teaching and research back home, but, obviously, this subject was secondary on his priority list. What surprised me was how quickly we started to discuss also very personal issues. In general, I refrain from personal discussions on very private matters. But there was something touching in Volodya's wish to talk. A week later, when the Kiev Conference ended, both of us realised that we have become very close friends. I met Volodya again a year after his son Lenya died in a mountaineering accident, shortly after completing his Ph.D. It was a very sad meeting and in my opinion Volodya partially recovered thanks to Júlia. Years later, I still miss him.

U. Maor (right) with N.N. Nikolaev, Chernogolovka, 2015.

SNAPSHOTS

NOEMI SMORODINSKAYA

Institute for Theoretical and Experimental Physics
25 ul. B. Cheremushkinskaya, 117218 Moscow, Russia
E-mail: naya@itep.ru

It may sound too trivial, but it is very important that life consists of details. It becomes still more important when you are trying to get together the recollections using the details you remember. Snapshots render memorable moments in imperfect images. This is exactly what is presented below viewed by an external observer.

I saw Volodya Gribov for the first time in 1965 at ITEP when he had a heated discussion with Karen Ter-Martirosyan in the large hall near the conference room. The traditional Monday seminar was scheduled to begin in a few minutes. Professor Shapiro, who was my diploma supervisor, stopped near me and said: "You are young. Always do your best to distinguish and remember outstanding physicists." This is Gribov. It makes no sense for Karen to mess with him because Gribov will never let go. He is known around the world not only for his talent but also, and mostly, for his ingenious intuition in science. Karen will obviously lose in this argument, I am sure.

In 1965 I was a fifth year student, and Volodya Gribov was already famous for his talent. I could understand only some general points of what Volodya was talking about, but when he just started his talk, from his first words at the blackboard, anyone would immediately realize how gifted he was.

Many years later, in summer 1978, when I came to see my close friend Júlia Nyíri in her Dubna apartment, I met there a person whom I by no means expected to meet. It was Volodya Gribov. Pali, Julia's son who was six years old at that time, presented Volodya to me. He said: This is my friend Volodya. Later it turned out that Volodya drove alone overnight from what then was Leningrad (now St. Petersburg) to come to Dubna in the early morning. It was not an easy drive because the roads in the USSR were never kept in good shape and, besides, did not have any lighting. A glance was enough to understand why he came. He wanted to see Júlia as soon as possible, and I was witnessing a big love.

Later Volodya left Leningrad and moved to Moscow. Júlia, Volodya, and Pali had an apartment not far from the former store Leipzig where Volodya and Pali enjoyed buying new toys. Volodya, like a child, was enthusiastic about any new toy he would see in the store. Pali had a big toy railway with many accessories and the whole family often played together. Pali and Volodya loved each other. Pali was an inquisitive boy and never missed a chance to ask a question. I always admired Volodya, who with permanent patience went into all details of Pali's problem.

Volodya used to work all day long, day after day. He worked at home and Júlia was doing her best to organize his life the best possible way. Sometimes Volodya left his desk and entered the living room to relax. Then it was a real pleasure to watch Júlia and Volodya talking to each other. Actually, they even did not need to talk. They could understand each other without words, and their glances did not need translation.

N. Smorodinskaya, Chernogolovka, 2015.

GRIBOV'S ROLE

VLADIMIR V. ANISOVICH

Petersburg Nuclear Physics Institute,
National Research Center "Kurchatov Institute",
Gatchina, 188300, Russia
E-mail: anisovich@thd.pnpi.spb.ru

Time puts things in their places, and as time passes, Volodya's role as a universal physicist with an exceptional intuition becomes more and more significant.

HE HAS BEEN WITH US...

E.G. DRUKAREV

National Research Center "Kurchatov Institute"
B.P. Konstantinov Petersburg Nuclear Physics Institute
Gatchina, St. Petersburg 188300, Russia
E-mail: drukarev@thd.pnpi.spb.ru

1. Bus seminar

The first thing which surprised me in the theoretical department was the seminar. Before graduating from the physical faculty of the Petersburg (Leningrad at that time) university I participated in a number of the faculty seminars. They did not differ much from the lectures. One hour talk was followed by several questions which were put in a very quiet and calm way. They were answered in the same manner. Sometimes there was a short discussion.

The seminar at the theoretical department of our institute was quite different. It lasted at least for three hours. Note that the technology which nowadays enables one to exhibit the illustrative material did not exist yet, and the speaker just wrote the equations on the blackboard. The usual scenario of the seminar looked like this. The speaker was interrupted by the participants. They put questions and made comments. Vladimir Naumovich Gribov (the colleagues called him BH, following the Latin transcription of the Russian initials) went to the blackboard more and more often. During the last half an hour it was BH who talked to the participants. He explained what was the main idea of the author, what have been done properly, and what are the weak points. It is important to note that often BH did not think beforehand over the problem which was the subject of the seminar.

In the fall of 1971 and later on the theoretical department staff had to go to Gatchina one day a week (it was Thursday). We were given a bus which started from the Physical Technical Institute. After crossing the city of Leningrad, it arrived to Gatchina in about an hour and a half after the departure. The younger people used this time to discuss the physical problems with their senior colleagues. I studied some atomic physics problems at that time and found a specific quasifree mechanism for the double photoionization. It was somewhat unusual for the atomic physics, and the experts in the field did not want to accept it. Also, it was not easy to find a partner in the department for discussing the point. The discovery of neutral weak currents, experiments for detection of the neutrino mass, the problem of heavy neutrino, where the atomic physics effects are strongly involved were still ahead. "Atomic physics is trivial since the Lagrangian is known and everything can be calculated" was a rather common statement. Anyway, BH did not share this point of view. Once Victor Gorshkov (I was his postgraduate) told him about these findings. "I agree—said BH turning to me—there is such a mechanism. But in your case it will not contribute much. It cannot be detected in present experiments. Thus the community will not be interested. You should try to find a process where this mechanism dominates". After some time it appeared that BH was right. The mechanism was confirmed by the experiments 35 years later.

The bus left for Leningrad at 6 p.m. It reached the city in forty minutes, and this was the most interesting part of the day. It was some kind of informal seminar with conversations about the frontiers of particle physics. In the seventies it developed very rapidly. The Weinberg–Salam theory of electroweak interactions, quantum chromodynamics... It was a qualitative discussion without equations and without the blackboard. Usually BH gave short comments when the bus entered the city and left the bus at the first metro station.

I was not involved in particle physics directly at that time, and it was not always easy to follow the discussion. However it was very useful to see the general picture and the way of thinking. The bus seminars gave me not less than the regular seminars at the institute with formulas and with the quantitative results. Unfortunately the trips of the special bus for the theoretical department were canceled in the early eighties.

2. Working with Gribov

In 1972 BH gave lectures on quantum electrodynamics for the last year students of University. He did not do this when I was a student. I attended the course. Two hours, twice a week. It appeared to be very interesting to listen to the course, having already some experience of working in the field. Also, in contrast to the student years, I did not have to worry about future exams. The lectures were very artistic. The intonations and gestures were as important as the physical contents itself. Sometimes the scenario of the lectures reminded of detective stories. Making a short pause after putting a question BH wrote the answer on the blackboard. After that BH gave detailed explanation of the intermediate steps which enabled to obtain the result.

Soon I had a possibility to work with BH directly. One of the leading experimentalists of the institute, A.A. Vorobiev, asked BH to look into the problem of the energy loss of a charged particle in dense matter. There was a theory developed by *N.* and the experiments which supported the theory. However the results obtained by Vorobiev differed from those ones. BH suggested me to study the problem. In his analysis *N.* employed the language which differed from the one used in our department. This made my task more complicated.

That was the time when the discussions on the confinement of quarks became hot, and BH was very busy. Thus he usually fixed the time of our work for the second half of the day. But during twenty or thirty minutes of conversations with me he "switched off" all other problems. Very often he answered my questions just rephrasing them a little. The answer to the question put in such a new form could be found very easily. Once BH said "Everything became clear. Thus it is not interesting any more, *N.* was wrong. Do not calculate this term. Try to find another one which cancels it". Unfortunately, BH did not agree to put his name on the author list.

We finished the work in February 1980. A month later BH moved to Moscow.

3. Lenya

One day in July 1984 I met Lenya Gribov on my way to the institute. We both planned to receive the vacation pay. Lenya was the son of BH. He was also our young employee. Several months earlier he received his PhD.

At that time Lenya and I worked with our senior colleague E. Levin on the nucleon structure functions in limiting cases. We discussed some physics on the way to the institute and agreed to continue after we came back from our vacations.

Lenya worked in our department for several years, and progressed rapidly. He new a lot of other things besides physics. During our walk he told me about Andrew Stackenschneider, the famous architect who was born on the mill near our institute. Lenya was very polite and gentle, without any trace of haughtiness which one could expect from the son of the head of the department.

Would he become a bright physicist, as bright as his father? The question remained unanswered. Lenya never returned from his vacations. That summer he died in an accident during hiking in the mountains.

BH was one of the brightest people whom I met during my life.

"The life of the dead is placed in the memory of the living"

– Marcus Tullius Cicero

A FEW LINES ABOUT VOLODYA

DOMINIQUE SCHIFF

Laboratoire de Physique Theorique d'Orsay, L'Universit Paris-Sud Orsay, France
E-mail: Dominique.Schiff@th.u-psud.fr

I would like to write a few words, in memory of Volodya, for the occasion of this workshop which is organized under his name and will start by publicly remembering his essential presence in our theoretical physics world. We had in Orsay (LPT, theoretical physics laboratory) the incredible luck to meet him in 1992 for the first time when we could finally invite this famous and celebrated figure of modern particle physics. He gave a series of lectures this year: Orsay lectures on confinement—in which he mainly developed the picture of confinement based on light quarks that led to many discussions which contributed to open the road of the search actually still going on... The way he was giving his talks nobody will forget. He always started by describing the field he would talk about in a very passionate and extraordinary way: he would say: "I have a picture" which would force even the most far away spectators to participate in an actively engaged vision of the problem he was talking about. He was building with enthusiasm the theoretical image which led to the result he wanted to show. He will remain in our memory as a rare model of intellectual passion. This led him to formulate in a unique way precious theoretical results. Thank you Volodya...

STRONG INTERACTIONS AT HIGH ENERGIES

GRIBOV UNIVERSALITY OF HADRON CROSS SECTIONS AT ULTRAHIGH ENERGIES

V.V. ANISOVICH

National Research Centre 'Kurchatov Institute',
Petersburg Nuclear Physics Institute, Gatchina, 188300, Russia

The current status of ultrahigh energy diffractive collisions is given. Profile function and K-matrix function techniques in the impact parameter space are presented, and the Gribov universality of the hadron total cross sections is discussed.

PACS: 13.85.Lg 13.75.Cs 14.20.Dh

More then forty years ago, in 1973, Gribov predicted[1] the universality of all hadron cross sections:

$$\sigma_{tot}(pp) : \sigma_{tot}(\pi p) : \sigma_{tot}(\pi\pi) = 1 : 1 : 1 \quad \text{at} \quad \sqrt{s} \to \infty\,. \tag{1}$$

At that time the data for 50-500 GeV[2,3] led to a different result: $\sigma_{tot}(pp) : \sigma_{tot}(\pi p) : \sigma_{tot}(\pi\pi) \simeq 9 : 6 : 4$ (Levin-Frankfurt relation[4]). The situation changed later with the appearance of the ISR data[5] which indicate definitely a growth of total cross sections with energy increase. The data imply parton clouds with gluonic (or, glueballic) content as an origin of the total cross section increase. The recent LHC and cosmic ray data[6–10] support this point.

The sources of parton clouds are valence quarks that give the total cross sections at moderately high energies to be proportional to the number of colliding quarks[4]. With energy increase the parton densities are increasing as well as their size. At LHC energies the parton clouds overlap to an extent that the black spot appears in the profile function, see Fig. 1 for illustration. As a result, not only the total cross sections should undergo changes — the reorganization of parton clouds with energy growth should result in transformations of particle production, the phenomena for spectra in fragmentation region were discussed in[11,12].

The appearance of a black spot at LHC energy was observed and commented in a number of papers devoted to ultrahigh energy total cross section growth, see[13–19] and references therein. The black disk model corresponds to the non-coherent interaction of the partons of the colliding hadrons.

1. Black disk asymptotic regime

If the asymptotic regime is realized in the black disk mode the ultrahigh energy behavior of hadron cross sections can be predicted on the basis of unitarity and analyticity using the data at pre-LHC energies. Such a model was suggested in Ref. 20, the Dakhno-Nikonov model takes into account the quark structure of colliding hadrons, the gluon origin of the t-channel interactions and and color screening effects in collisions. The model can be considered as a realization of the Good-Walker eikonal approach[21] for a continuous number of channels. The extension of the results of Ref. 20 to the asymptotic energy region was performed in Ref. 17, 18.

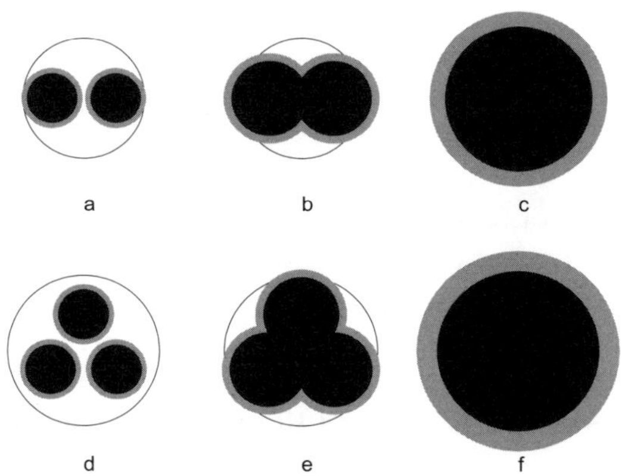

Fig. 1. Meson and proton pictures of profile functions in the impact parameter space and their transformations with increasing energy. a,d) Parton densities at moderately high energies (Serpukhov[2]-FNAL[3]) with constituent quark structure realized; b,e) the parton densities at LHC energies where overlapping of the quark cloud starts and creation of the black spot occurs; c,f) picture at cosmic ray energy region where the universal black disks are formed.

The total cross section growth in the black disk mode is demonstrated on Fig. 2a for $\sigma_{tot}(pp)$, $\sigma_{tot}(\pi p)$, $\sigma_{tot}(\pi\pi)$. It is seen that, the ratios of Eq. (1) are fulfilled within 15% accuracy at $\sqrt{s} \sim 10^{15}$ GeV. For comparison Fig. 2b demonstrates proton-proton asymptotic characteristics which for

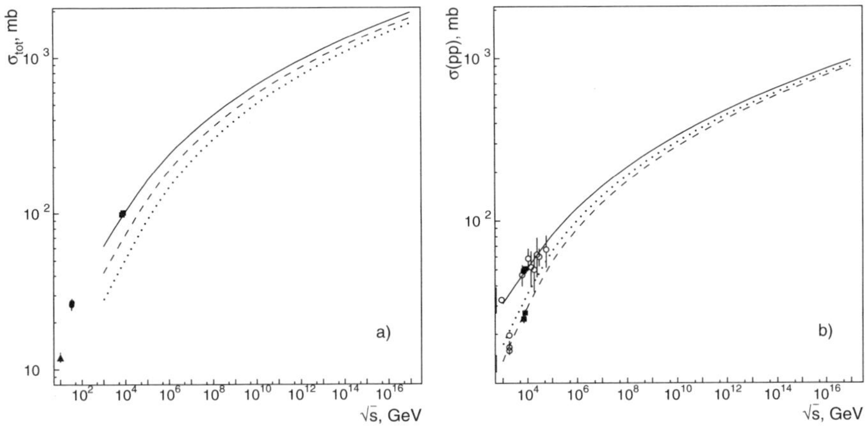

Fig. 2. a) Total cross sections: $\sigma_{tot}(pp)$ (solid line), $\sigma_{tot}(\pi p)$ (dashed line) and $\sigma_{tot}(\pi\pi)$ (dotted line) in the black disk mode. Squares (pp) are from Ref. 6, circles (πp) are from Ref. 22, triangles ($\pi\pi$) are from Ref. 23. b) Proton-proton collisions: $\frac{1}{2}\sigma_{tot}(pp)$ (solid line), $\sigma_{el}(pp) + 2\sigma_{D(p)}(pp)$ (dotted line) and $\sigma_{el}(pp)$ (dashed line). Data[5,6,10] stand for $\frac{1}{2}\sigma_{tot}(pp)$ and $\sigma_{el}(pp)$. The calculated curves are from Refs. 17, 18.

the black disk mode reads $\frac{1}{2}\sigma_{tot}(pp) : \sigma_{el}(pp) = 1 : 1$, which is observed at lower energies with the same accuracy, $\sqrt{s} \sim 10^7$ GeV, that means, that the universality of cross sections sets in the energy region where the parton cloud size exceeds the static hadron one.

A steady increase of the parton disk is determined by parameters of the leading t-channel singularity that are the pomeron intercept $\alpha(0) = 1 + \Delta$, where $\Delta \simeq 0.25 - 0.30$, and the pomeron trajectory slope α'_P. The leading pomeron is formed by effective gluons (G) which are massive, $m_G \sim 700 - 1000$ MeV.[24,25] The pomeron slope parameter is small $\alpha'_P \simeq (0.13 - 0.25)$ GeV^{-2} $\sim \frac{1}{4m_G^2}$, that correspond to hard pomeron[26]. The characteristics of the leading pomeron provide the value of the black disk radius $R_{black\ disk} \simeq 2\sqrt{\Delta\alpha'_P}\ \ln s$ with $2\sqrt{\Delta\alpha'_P} \simeq 0.08$ fm.[17,18]

The s-channel unitarization of the scattering amplitude damps the strong pomeron pole singularity, transforming it into a multipomeron one. Therefore, we face an intersection of problems of the gluon content of t-channel states at ultrahigh energies and the physics of gluonic states, glueballs. At present the glueball states are subjects of discussions, see for example Refs. 27–29 and references therein. Studies of glueballs and multigluon states are enlightening the problem of the confinement singularity - see the discussion in Ref. 30. The large value of mass of the soft effective gluon (and the corresponding value the low-lying glueballs) and the slow rate of the black disk increase appear to be related phenomena.

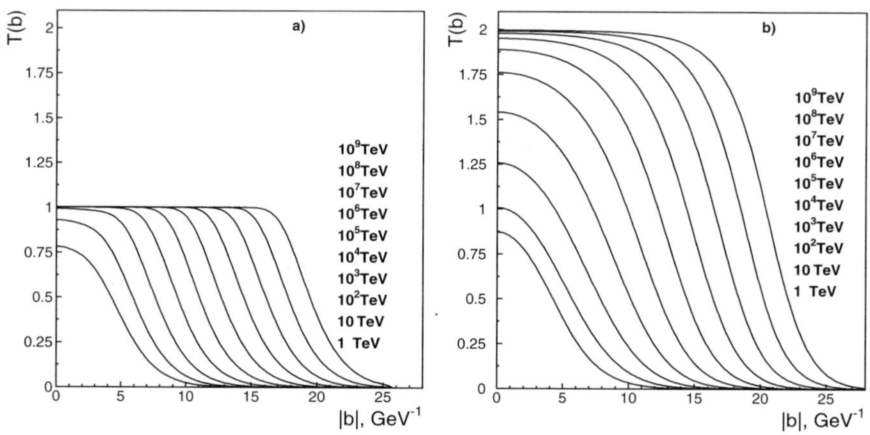

Fig. 3. a) Profile functions, $T(b)$, at $\sqrt{s} = 1$, 10, $10^2, ...10^9$ TeV for the black disk regime with $T(b) \to 1^{18}$ and b) resonant disk regime, $T(b) \to 2^{33}$. At $\sqrt{s} = 1 - 10$ TeV the profile functions in both modes are nearly the same.

2. Profile function for scattering amplitude in the impact parameter space and the K-matrix representation

The black disk mode is usually discussed in terms of the profile function and optical density, for discussion of asymptotics beyond this mode the K-matrix representation is useful. In the impact parameter space the profile

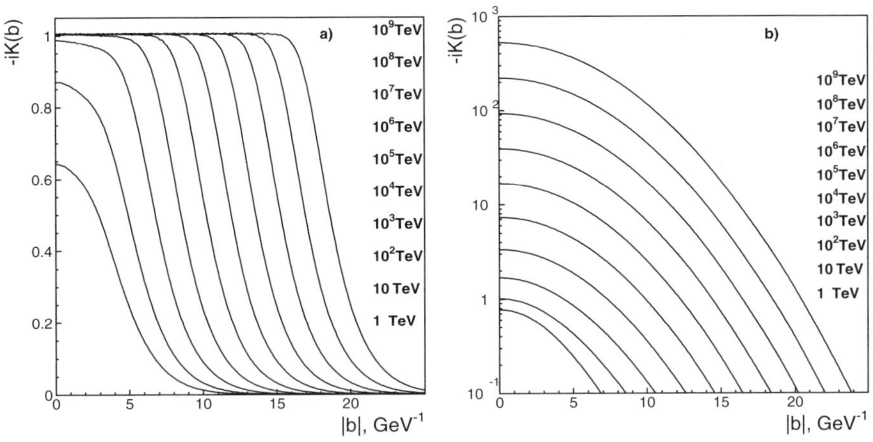

Fig. 4. The K-matrix functions, $-iK(b)$, for a) the black disk mode ($[-iK(b)]_{\xi\to\infty} \to 1$ at $b < R_0\xi$) and b) the resonant disk mode ($[-iK(b)]_{\xi\to\infty} \to \infty$ at $b < R_0\xi$). The calculated curves are from Ref. 33.

function $T(b)$ is determined at high energies as:

$$\sigma_{tot} = 2\int d^2b\, T(b)\,, \tag{2}$$

$$A_{el}(\mathbf{q}_\perp^2) = i\int d^2b e^{i\mathbf{b}\mathbf{q}_\perp} T(b)\,, \quad 4\pi\frac{d\sigma_{el}}{d\mathbf{q}_\perp^2} = |A_{el}(\mathbf{q}_\perp^2)|^2,$$

$$T(b) = 1 - \eta(b)\, e^{2i\delta(b)} = 1 - e^{-\frac{1}{2}\chi(b)} = \frac{-2iK(b)}{1 - iK(b)},$$

here $A_{el}(\mathbf{q}_\perp^2)$ is the elastic scattering amplitude. The profile function can be presented either in the standard form using the inelasticity parameter $\eta(b)$ and the phase shift $\delta(b)$ or in terms of the optical density $\chi(b)$ and the K-matrix function $K(b)$. The K-matrix approach is based on the separation of the elastic rescatterings in the intermediate states: the function $K(b)$ includes only the multiparticle states thus being complex valued. The small value of the $\mathrm{Re}\, A_{el}/\, \mathrm{Im}\, A_{el}$ tells that $K(b)$ is dominantly imaginary.

The profile functions $T(b)$ and the K-matrix functions $-iK(b)$ for the black disk mode are shown in Figs. 3a, 4a. The K-matrix function in the pre-LHC region increases with energy being mainly concentrated at $b < 1$ fm. The black disk regime for the K-matrix function means its 'freezing', $-iK(b) \to 1$, in the disk area. But the black disk mode does not provide maximal total cross section, the maximal values correspond to $-iK(b) \to \infty$ and $T(b) \to 2$.[31–33]

3. Resonant disk asymptotics

If the growth of the $-iK(b)$ continues with increasing energy, the interaction area turns into a resonant disk. The profile functions $T(b)$ and the K-matrix functions $-iK(b)$ for the resonant disk mode are shown on Figs. 3b, 4b. In this case asymptotically $\sigma_{tot}(pp) \sim \ln^2 s$, $\sigma_{el}(pp) \sim \ln^2 s$ with $[\sigma_{el}(pp)/\sigma_{tot}(pp)]_{s\to\infty} \to 1$, see Fig. 5. The resonant disk area is surrounded by a black border band, see Fig. 6, that provides a more slow increase of other cross sections: $\sigma_{inel}(pp) \sim \ln s$, $\sigma_D(pp) \sim \ln s$, $\sigma_{DD}(pp) \sim \ln s$.

It is definitely seen, Fig. 5, that the data at $\sqrt{s} \sim 1 - 10$ TeV are not sensitive to the versions of the disk: the initial stages are similar in both modes. Distinctions are seen at $\sqrt{s} \sim 10^3 - 10^4$ TeV. Apparently, the study and interpretation of the cosmic ray data at such energies are the problems on the agenda.

At the energy $\sqrt{s} \sim 10$ TeV the black cloud fills out the proper hadron domain, the region ≤ 1 fm, and that happens in both modes. It is demonstrated in Figs. 3, 4: the profile functions $T(b)$ coincide practically in both modes as well as the K-functions $-iK(b)$. The energy behavior of σ_{tot}, σ_{el} and σ_{inel} coincide also at $\sqrt{s} \sim 1 - 100$ TeV in both modes, Fig. 5.

Differences appear at $\sqrt{s} \sim 1000$ TeV: here $T(b) \simeq 1.5$ at $b \lesssim 0.5$ fm and the black zone has shifted to $b \simeq 1.0 - 1.5$ fm, Fig. 3b. With further energy increase the radius of the black band increases as $2\sqrt{\Delta\alpha'}\ln s$. The parton picture in the resonant disk mode is illustrated by Fig. 6.

The rate of growth in both modes is determined by the leading singularity and the fit of the data in the region $\sqrt{s} \sim 1-10$ TeV gives approximately the same values of Δ and α' for both cases thus providing the same rate of the size growth $R_{black\ disk} \simeq R_{resonant\ disk}$.

4. The Gribov universality of cross sections

The interaction of soft gluons determines the physics of hadrons. The effective gluons are massive and their mass is of the order of 1 GeV that is seen directly in radiative decays of heavy quarkonia,[24,25] $\psi \to \gamma + hadrons$ and $\Upsilon \to \gamma + hadrons$. The effective gluon mass is determinative both for low energy physics, making possible to introduce the notion of the constituent quark, and for high energy physics, dictating the rate of the growth of the

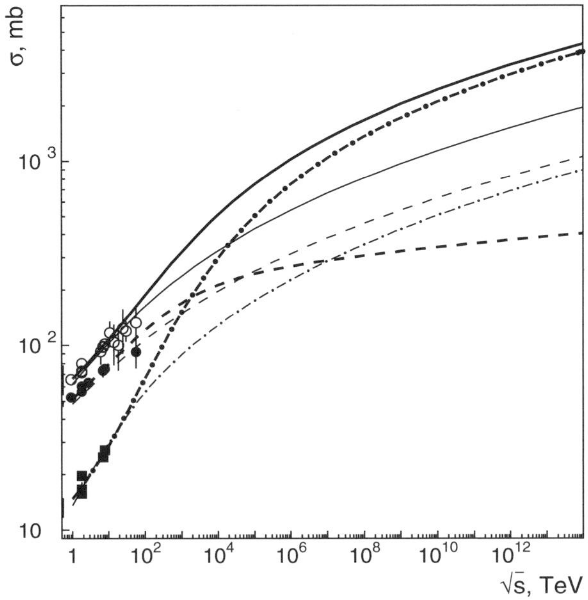

Fig. 5. Total, elastic and inelastic cross sections for resonant disk (thick lines) and black disk (thin lines) modes: open circles and solid lines are σ_{tot}, squares and dot-dashed lines are σ_{el}, filled circles and dashed lines are σ_{inel}.

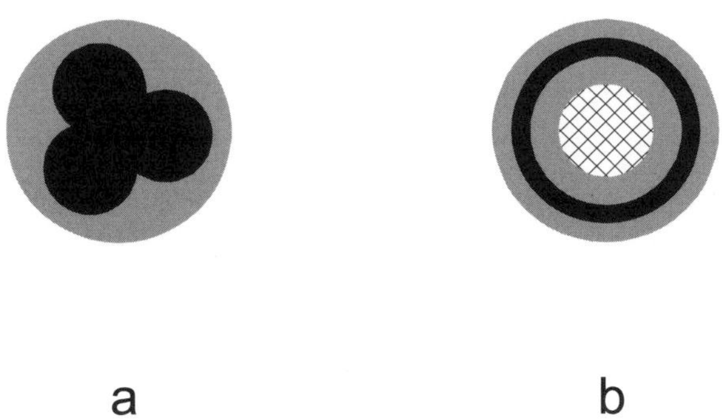

Fig. 6. Resonant disk mode, the picture of proton in the impact parameter space: a) black spot appears at the LHC energies and b) transforms with energy increase to black band around the area with maximal scattering amplitude (the resonant disk).

interaction radius. High energy physics is the physics of large logarithms, $\ln s/s_0 \gg 1$, and the value $\sqrt{s_0} \sim m_{effective\,gluon}$ corresponds to a start of the asymptotic regime at $\sqrt{s} \sim 1$ TeV. However, the initial increments of the measured characteristics such as σ_{tot} , σ_{el} and σ_{inel} are visually similar for different versions of asymptotics, and therefore their behavior in the region of LHC energy does not distinguish between different modes. A real discrimination of the modes can appear when cross section data are discussed at much larger energies, $\sqrt{s} \sim 10^3 - 10^4$ TeV.

Parton disks determine the ultrahigh energy physics, they are universal for all types of hadrons. It results in the universality of the cross sections: for the total cross sections, σ_{tot}, and the diffractive ones σ_{el}, σ_D and σ_{DD}. Though the mode of the parton disk is not deciphered yet but within any mode the universality of cross sections remains true.

Cosmic ray data probably can provide information to fix the asymptotic mode. The diffractive inelastic processes which differ strongly for different modes[34,35] can also beat a path to the problem.

Acknowledgment

I thank Y.I. Azimov, M.A. Matveev, V.A. Nikonov, J. Nyiri, M.G. Ryskin for useful discussons. The work was supported by grants RFBR-13-02-00425 and RSGSS-4801.2012.2.

References

1. V.N. Gribov, *Yad. Fiz.* **17**, 603 (1973), [*Sov. J. Nucl. Phys.* **17**, 313 (1973)].
2. S. P. Denisov *et al.*, *Phys. Lett.* **B36**, 415 (1971).
3. A.S. Carroll *et al.*, *Phys. Lett.* **B61**, 303 (1976); **B80**, 423 (1978).
4. E.M. Levin and L.L. Frankfurt, *Pis'ma v ZhETF*, **2**, 105 (1965), [*JETP Lett.* **2**, 65 (1965)].
5. UA1 Collaboration, *Phys. Lett.* **B128**, 336 (1982);
   UA4 Collaboration, *Phys. Lett.* **B147** , 385 (1984);
   E710 Collaboration, *Phys. Lett.* **B247**, 127 (1990);
   UA4/2 Collaboration, *Phys. Lett.* **B316**, 448 (1993);
   CDF Collaboration, *Phys. Rev.* **D50**, 5518 (1994).
6. G. Latino for the TOTEM collaboration, *Summary of Physics Results from the TOTEM Experiment*, arXiv:1302.2098(2013) [hep-ph].
7. ATLAS collaboration, G. Aad *et al.*, *Eur. Phys. J.* **C72** , 1926 (2012), arXiv:1201.2808 [hep-ph].

8. CMS collaboration, *Measurement of diffraction dissociation cross sections at $\sqrt{s} = 7TeV$ at the LHC, CMS-PAS-FSQ-12-005*, 2013.
9. ALICE collaboration, B. Abelev *et al.*, *Eur. Phys. J.* **C73** , 2456 (2013), arXiv:1208.4968 [hep-ph].
10. Pierre Auger Collaboration, P. Abreu *et al.*, *Phys. Rev. Lett.* **109**, 062002 (2012).
11. V.V. Anisovich, V.M. Shekhter, *Yad. Fiz.* **28**, 1079 (1978), [*Sov. J. Nucl. Phys.* **28**, 561 (1978)].
12. V.V. Anisovich, E.M. Levin, M.G. Ryskin, *Yad. Fiz.* **29**, 1311 (1979), [*Sov. J. Nucl. Phys.* **29**, 674 (1979)].
13. M.M. Block and F. Halzen, *Phys. Rev.* **D 86**, 051504 (2012).
14. V. Uzhinsky and A. Galoyan, arXiv:1111.4984v5 (2012) [hep-ph].
15. V.A. Schegelsky, M.G. Ryskin, *Phys. Rev.* **D85**, 094024 (2012).
16. E. Martynov, *Phys. Rev.* **D87**, 114018 (2013).
17. V.V. Anisovich, K.V. Nikonov, and V.A. Nikonov, *Phys. Rev.* **D 88**, 014039 (2013).
18. V.V. Anisovich, V.A. Nikonov, and J. Nyiri, *Phys. Rev.* **D 88**, 094015 (2013).
19. I.M. Dremin, *Usp. Fiz. Nauk.* **183**, 3 (2013), [*Phys. Usp.* **83**, 3 (2013)].
20. L.G. Dakhno and V.A. Nikonov, *Eur. Phys. J.* **A 5**, 209 (1999).
21. M.L. Good, W.D. Walker, *Phys. Rev.* **120**, 1857 (1960).
22. U. Dersch *et al.* [SELEX Collaboration], *Nucl. Phys.* **B579**, 277 (2000) arXiv:hep-ex/9910052.
23. H. Abramowicz, M. Gorski, G. Sinapius, A. Wroblewski, A. Zieminski, H. J. Lubatti, K. Moriyasu and C. D. Rees *et al.*, *Nucl. Phys.* **B166**, 62 (1980).
24. G. Parisi and R. Petronzio, *Phys. Lett.* **B94**, 51 (1980).
25. M. Consoli, J.H. Field, *Phys. Rev.* **D49**, 1293 (1994).
26. V.N. Gribov, *Nucl. Phys.* **B106**, 189 (1976).
27. V.V. Anisovich, *Usp. Fiz. Nauk.* **47**, 45 (2004), [*Phys. Usp.* **47** (2004) 45].
28. E. Klempt and A. Zaitsev, *Phys. Rept.* **454**, 1 (2007).
29. W. Ochs, *J. Phys.* **40**, 043001 (2013).
30. A.V. Anisovich, V.A. Nikonov, A.V. Sarantsev, V.V. Anisovich, M.A. Matveev, T.O. Vulfs, K.V. Nikonov, and J. Nyiri, *Phys. Rev.* **D84**, 076001 (2011).
31. S.M. Troshin and N.E. Tyurin, *Int. J. Mod. Phys.* **A 22**, 4437 (2007) arXiv:1408.2650 [hep-ph].
32. I.M. Dremin, arXiv:1406.2153 [hep-ph].

33. V.V. Anisovich, V.A. Nikonov and J. Nyiri, *Phys. Rev.* **D90**, 074005 (2014); arXiv:1408.0692 [hep-ph].
34. V.V. Anisovich, M.A. Matveev, V.A. Nikonov, *Int. J. Mod. Phys.* **A 30**, 155054 (2015) arXiv:1407.4588 [hep-ph].
35. V.V. Anisovich, M.A. Matveev, V.A. Nikonov, and J. Nyiri. *Int. J. Mod. Phys.* **A 30**, 1450176 (2014) arXiv:1407.4588 [hep-ph].

EULER-LAGRANGE EQUATIONS FOR THE GRIBOV REGGEON CALCULUS IN QCD AND IN GRAVITY*

L.N. LIPATOV

St. Petersburg State University, Petersburg Nuclear Physics Institute, Gatchina, Orlova Roshcha, 188300, St. Petersburg, Russia
E-mail: lipatov@thd.pnpi.spb.ru

The theory of the high energy scattering in QCD and gravity is based on the reggeization of gluons and gravitons, respectively. We discuss the corresponding effective actions for reggeized particle interactions. The Euler-Lagrange equations in these theories are constructed with a variational approach for the effective actions and by using their invariance under the gauge and general coordinate transformations.

1. Introduction

The hadron-hadron scattering amplitudes $A(s,t)$ at high energies $\sqrt{s}$ and fixed momentum transfers $q = \sqrt{-t}$

$$s = (p_1 + p_2)^2 \gg -t = -(p_1 - p_1')^2 \tag{1}$$

are described usually in the framework of the Regge model.[1] They can be presented as sums of the contributions A^p with definite signatures $p = \pm 1$ in the form of the Mellin integral

$$A^p(s,t) = \int_{a-i\infty}^{a+i\infty} \frac{dj}{2\pi} \left((-s)^j + p\, s^j\right) f_j^p(t)\,. \tag{2}$$

It is assumed that the partial waves $f_j^p(t)$ have the poles with their positions depending on t

$$f_j^p(t) \sim \frac{1}{j - j(t)}\,, \tag{3}$$

which leads to the Regge behavior of the scattering amplitudes

$$A^p(s,t) \sim (-s)^{j(t)} + p s^{j(t)}\,. \tag{4}$$

*The investigation is supported by the St.Petersburg State University grant and RFBI grant 14-2200281.

There is a special Regge pole with $p = 1$ and $j(0) \approx 1$, which is responsible for an approximately constant behavior of total cross sections σ_t at large energies and for the fulfillment of the Pomeranchuk theorem claiming that the particle-particle and particle-anti-particle cross sections are equal. This Regge pole has vacuum quantum numbers and is called Pomeron.

The exchange of two or more reggeons leads to the appearance of more complicated singularities of $f_j(t)$ in the j-plane.[2] These so-called Mandelstam cuts can be obtained also from an analytic continuation of multi-particle unitarity conditions in the t-channel.[3] To take into account contributions of all possible Mandelstam cuts for Pomerons V.N. Gribov constructed the Reggeon calculus based on the 2+1 field theory of a complex scalar field.[4] In this model the particle rapidity takes the role of time.

In the Quantum Chromodynamics (QCD) the scattering amplitudes with the gluon quantum numbers in the t-channel have the Regge form in the so-called leading logarithmic approximation.[5] Pomeron as a colorless object is a composite state of the reggeized gluons. Its wave function in QCD and in other field models with the gauge group $SU(N_c)$ satisfies the BFKL equation.[5] It is remarkable that equations for the composite states of reggeized gluons with the color singlet and octet quantum numbers are integrable at large N_c.[6,7] Thus, it looks natural to reformulate QCD and other gauge models in terms of the effective degrees of freedom corresponding to the reggeized gluons. The gluon Regge trajectory and various reggeon couplings in upper orders of perturbation theory can be calculated from the effective action presented in Ref. 8, 9. In this paper we consider the Euler-Lagrange equation for this action and for its generalization in gravity.

2. High energy effective action in QCD

The effective action in QCD is written for a cluster of usual quarks and gluons interacting with reggeized gluons and having their rapidities y in a small interval around the central value y_0

$$y = \frac{1}{2} \ln \frac{\epsilon_k + |k|}{\epsilon_k - |k|}, \; |y - y_0| < \eta, \; \eta \ll \ln s \,. \tag{5}$$

Apart from the anti-Hermitian matrix $N_c * N_c$ of the gauge field v_μ for usual gluons we introduce also the fields $A_\pm$ describing the production and annihilation of the reggeized gluons

$$v_\mu(x) = -iT^a v_\mu^a(x) \,, \; A_\pm(x) = -iT^a A_\pm^a(x) \,, \; [T^a, T^b] = if^{abc}T_c \,, \tag{6}$$

where v_μ^a and $A_\pm^a$ belong to the adjoint representation of the $SU(N_c)$-group.

The fields $A_\pm = A^\mp$ are invariant,

$$\delta A^\mp = 0\,, \tag{7}$$

under the local gauge transformations of the gluon field,

$$\delta v_\mu = \frac{1}{g}\,[D_\mu, \chi(x)]\,,\;\; D_\mu = \partial_\mu + g v_\mu\,, \tag{8}$$

with parameters $\chi(x)$ vanishing at $x \to \infty$. They satisfy the kinematical constraints,

$$\partial_- A^- = \partial_+ A^+ = 0\,,\;\; \partial_\pm = \frac{\partial}{\partial x^0} \pm \frac{\partial}{\partial x^3}\,, \tag{9}$$

corresponding to the fact that the Sudakov components α_i, β_i of the cluster momenta k_i,

$$k_i = \beta_i p_1 + \alpha_i p_2 + k_i^\perp\,, \tag{10}$$

are strongly ordered $\beta_i \gg \beta_{i+1}$, $\alpha_i \ll \alpha_{i+1}$ in the multi-Regge kinematics.

The effective action for the cluster of real and virtual particles with approximately equal rapidities has the form[8]

$$S_{eff} = \int d^4x\, L_{QCD} + S_{ind}\,,\;\; S_{ind} = Tr \int d^4x\, (V_+ \partial_\mu^2 A^+ + V_- \partial_\mu^2 A^-)\,, \tag{11}$$

where L_{QCD} is the usual QCD Lagrangian,

$$L_{QCD} = i\bar{\psi}\hat{\partial}\psi + \frac{1}{2} Tr\, G_{\mu\nu}^2\,,\;\; G_{\mu\nu} = \frac{1}{g}\,[D_\mu, D_\nu]\,. \tag{12}$$

The anti-Hermitian operators $V_\pm$ in the induced action S_{ind} are expressed in terms of eikonal amplitudes for the massless particle scattering off the external gluon field[8]

$$V_\pm = -\frac{1}{g}\,\partial_\pm \frac{1}{D_\pm}\,\partial_\pm \equiv -\frac{\partial_\pm}{g} + v_\pm - v_\pm \frac{g}{\partial_\pm} v_\pm + v_\pm \frac{g}{\partial_\pm} v_\pm \frac{g}{\partial_\pm} v_\pm + \ldots . \tag{13}$$

The term $-\partial_\pm/g$ is negligible in S_{int}, because the fields $A_\mp$ do not depend on $x^\pm$. The operators $1/\partial_\pm$ correspond to the propagators of the energetic particles emitting the gluons in the given interval of rapidity. Their virtualities $k^2 \sim |k_\pm|$ are large and it looks natural to define the action of the operator $1/\partial_\pm$ as follows

$$\begin{aligned} \frac{1}{\partial_\pm} f(x^\pm) &= \frac{1}{2}\left(\int_{-\infty}^{x^\pm} d\widetilde{x}^\pm f(\widetilde{x}^\pm) - \int_{x^\pm}^{\infty} d\widetilde{x}^\pm f(\widetilde{x}^\pm)\right) \\ &= \frac{1}{2}\int_{-\infty}^{\infty} d\widetilde{x}^\pm \epsilon(x^\pm - \widetilde{x}^\pm)\, f(\widetilde{x}^\pm), \end{aligned} \tag{14}$$

which is compatible with the anti-Hermiticity property of $V_\pm$ and leads to the principal value prescription for the corresponding poles $1/k_\pm$ in the momentum space. But it is possible to consider different boundary conditions for the integral operator $1/D_\pm$ at $x^\pm \to \infty$.

The invariance of the action under the gauge group is a consequence of the relation

$$V'_\pm = -\frac{1}{g}\,\partial_\pm\, e^{-\chi}\,\frac{1}{D_\pm}\,e^{\chi}\,\partial_\pm \to V_\pm\,, \tag{15}$$

which is valid after its integration over $x^\pm$, providing that the gauge parameters χ vanish at $x \to \pm\infty$.

There are simple operator representations of $1/D_\pm$

$$\begin{aligned}\frac{1}{D_\pm} &= P\,e^{-\frac{g}{2}\int_{-\infty}^{x\pm} dx'^{\pm} v_\pm}\,\frac{1}{\partial_\pm}\,\bar{P} e^{\frac{g}{2}\int_{-\infty}^{x\pm} dx'^{\pm} v_\pm} \\ &= \bar{P}\,e^{\frac{g}{2}\int_{x\pm}^{\infty} dx'^{\pm} v_\pm}\,\frac{1}{\partial_\pm}\,P\,e^{-\frac{g}{2}\int_{x\pm}^{\infty} dx'^{\pm} v_\pm}\,.\end{aligned} \tag{16}$$

The induced action after the integration over $x^\pm$ can be written as follows

$$\begin{aligned}S_{ind} = -\frac{Tr}{2g}\int d^2x_\perp\Big(&\int_{-\infty}^{\infty} dx^-\,\partial_\mu^2 A^+\,\frac{1}{D_+}\,\partial_+|_{x^+=-\infty}^{x^+=\infty} \\ &+\int_{-\infty}^{\infty} dx^+\,\partial_\mu^2 A^-\,\frac{1}{D_-}\,\partial_-|_{x^-=-\infty}^{x^-=\infty}\Big)\,,\end{aligned} \tag{17}$$

where

$$\frac{1}{D_\pm}\partial_\pm \equiv 1 - \frac{g}{\partial_\pm}v_\pm + \frac{g}{\partial_\pm}v_\pm\frac{g}{\partial_\pm}v_\pm - \frac{g}{\partial_\pm}v_\pm\frac{g}{\partial_\pm}v_\pm\frac{g}{\partial_\pm}v_\pm + \ldots \tag{18}$$

In accordance with the Wilsonian ideas the induced action S_{ind} can be derived as a coherent effect of the functional integration over the color fields for strongly virtual particles in arbitrary color representations. It means, that with our accuracy the principal value prescription for the operator $1/\partial_\pm$ is not unique. Below we present $D_\pm^{-1}\partial_\pm$ in terms of the P-exponent[8]

$$\frac{1}{D_\pm}\partial_\pm = P\exp\left(-\frac{g}{2}\int_{-\infty}^{x^\pm} dx'^{\pm} v_\pm\right),$$

multiplied on a constant unitary matrix.

Note that the action should contain also an additional term bilinear in the reggeon fields

$$\Delta S = 2\,Tr\int d^4x\,\partial_\mu^\perp A^-\partial_\mu^\perp A^+\,, \tag{19}$$

where the fields A^- and A^+ describe the reggeized gluons with different rapidities $\Delta y > \eta$.

3. Classical equation for the effective action in QCD

The variation of the effective action S_{ind} in $v_\pm$ can be found with the use of the relation

$$\delta \frac{1}{D_\pm} = -\frac{1}{D_\pm}\, g\, \delta v_\pm \,\frac{1}{D_\pm}\,, \tag{20}$$

where $\delta v_\pm$ is an infinitesimal gluon field. Thus, with the integration by parts we obtain

$$\begin{aligned}\delta S_{ind} = Tr \int d^4x \Bigg(&\left(\frac{1}{D_+}\partial_+\right)^+ \delta v_+ \,\frac{1}{D_+}\,\partial_+\,\partial_\mu^2 A^+ \\ &+ \left(\frac{1}{D_-}\partial_-\right)^+ \delta v_- \,\frac{1}{D_-}\,\partial_-\,\partial_\mu^2 A^- \Bigg)\,,\end{aligned} \tag{21}$$

where O^+ means the Hermitian conjugation of the operator O. There is an ambiguity in the integrand, because we have at least two different representations for the corresponding quantity

$$\begin{aligned}\left(\frac{1}{D_\pm}\partial_\pm\right)^+ \delta v_\pm \,\frac{1}{D_\pm}\,\partial_\pm =& \bar{P}\,\exp\left(\frac{g}{2}\int_{-\infty}^{x^\pm} v_\pm(x')dx'^\pm\right)\delta v_\pm \\ &\times P\exp\left(-\frac{g}{2}\int_{-\infty}^{x^\pm} v_\pm(x'^\pm)dx'^\pm\right)\end{aligned}$$

or

$$\begin{aligned}\left(\frac{1}{D_\pm}\partial_\pm\right)^+ \delta v_\pm \,\frac{1}{D_\pm}\,\partial_\pm =& P\,\exp\left(-\frac{g}{2}\int_{x^\pm}^{\infty} v_\pm(x')dx'^\pm\right)\delta v_\pm \\ &\times \bar{P}\exp\left(\frac{g}{2}\int_{x^\pm}^{\infty} v_\pm(x')dx'^\pm\right).\end{aligned}$$

These expressions are related by a global similarity transformation in accordance with different boundary conditions for the operator $1/D_\pm$. Below we remove this freedom by choosing the constraint $D_\pm^{-1}\partial_\pm = 1$ at the central point $x^\pm = 0$. It leads to the following result for the variation of S_{ind},

$$\begin{aligned}\delta S_{ind} = Tr \int d^4x \sum_{r=+,-} &\delta v_r \exp\left(-\frac{g}{2}\int_0^{x^r} v_r(x'^r)dx'^r\right) \\ &\times \partial_\mu^2 A^r \exp\left(\frac{g}{2}\int_0^{x^r} v_r(x'^r)dx'^r\right),\end{aligned} \tag{22}$$

where

$$\exp\left(\frac{g}{2}\int_0^{x^\pm} v_\pm dx'^\pm\right) \equiv \theta(x^\pm)P\exp\left(\frac{g}{2}\int_0^{x^\pm} v_\pm dx'^\pm\right) + \theta(-x^\pm)P\exp\left(-\frac{g}{2}\int_{x^\pm}^{0} v_\pm dx'^r\right). \tag{23}$$

The Euler-Lagrange equations for v_σ following from the effective action can be written in the following form (for the pure glue-dynamics),

$$[D_\mu, G^\perp_{\mu\sigma}] = 0\,,\ [D_\mu, G_{\mu\pm}] = j^{ind}_\pm\,, \tag{24}$$

with the current $j^{ind}_\pm$ given below

$$j^{ind}_\pm = U(v_\mp)(\partial^2_\sigma A_\pm)U^+(v_\mp)\,. \tag{25}$$

The matrix $U(v_\mp)$ with our constraint $U(v_\mp) = 1$ at $x^\pm = 0$ is given below

$$U(v_\mp) = \exp\left(-\frac{g}{2}\int_0^{x^\mp} dx'^\mp v_\mp\right), \tag{26}$$

where the ordering of non-commuting operators was discussed above. The unitary matrix $U(v_\mp)$ describes a gauge transformation of an arbitrary gluon field v_μ,

$$v'_\mu = U^+(v_\mp)\left(\frac{\partial_\mu}{g} + v_\mu\right)U(v_\mp)\,, \tag{27}$$

to the new field v'_μ satisfying the light cone condition

$$v'_\mp = 0\,. \tag{28}$$

From the last equation we derive the gauge transformation law for the matrix $U(v_\mp)$,

$$U(v_\mp) \to e^{\chi(x)}U(v_\mp)\,,\ v_\mu \to e^{\chi(x)}\left(\frac{\partial_\mu}{g} + v_\mu\right)e^{-\chi(x)}\,, \tag{29}$$

which corresponds to the gauge symmetry of the current $j^{ind}_\pm$,

$$j^{ind}_\pm \to e^{\chi(x)} j^{ind}_\pm e^{-\chi(x)}\,. \tag{30}$$

The Euler-Lagrange equations are self-consistent due to the current conservation,

$$[D_\mp, j^{ind}_\pm] = 0\,. \tag{31}$$

It is possible to use this equality to derive a recurrent relation for $j^{ind}_\pm$.[8,9]

We can solve the above Euler-Lagrange equations in the quasi-elastic kinematics when there is only one induced contribution $\sim \partial_\mu^2 A_+$ or $\sim \partial_\mu^2 A_-$ in the current j^{ind}. Indeed, in the corresponding light-cone gauge the solution is $\widetilde{v'}_\sigma = \delta_{\sigma\mp} A_\mp$ which coincides with a superposition of known shock wave solutions. For other gauges one should perform its gauge transformation,

$$\widetilde{v}_\sigma = V\left(\delta_{\sigma\pm} A_\mp + g^{-1}\partial_\sigma\right) V^{-1}\,, \tag{32}$$

with an unitary matrix V depending on x. These fields are solutions of the Euler-Lagrange equation with the gauge covariant current

$$j_\pm^{ind} = U(v_\mp)\partial_\mu^2 A_\pm U^{-1}(v_\mp)\,, \tag{33}$$

where $U(v_\mp)$ is the above P-exponent. Our approach allows to derive $j_\pm^{ind}$ for the quasi-elastic case without using the effective action. For a general case of the cluster produced in the central rapidity region the Euler-Lagrange equations are obtained assuming, that j_σ^{ind} is a sum of two derived terms $\delta_\sigma^+ j_+^{ind}$ and $\delta_\sigma^- j_-^{ind}$. But in this region the classical equations can be solved only in the framework of perturbation theory with the use of the Feynman rules constructed in Ref. 9. Expanding the gluon fields $v_\mu = \widetilde{v}_\mu + \delta v_\mu$ near the classical solutions $\widetilde{v}_\mu$ we can integrate over the quantum fluctuations δv_μ to calculate effective vertices for the gluon production and the loop corrections for the multi-reggeon vertices.[10–12] Thus, the reggeon fields $A^\pm(x)$ can be considered as soliton-type solutions of the corresponding classical equations in the quasi-elastic kinematics. The effective action at large N_c allows to find the interaction of these solitons with the use of integrability of an effective spin chain model.[6,7]

4. Effective action for the high energy gravity

It is well known, that the $N = 4$-super-symmetric gauge theory is equivalent to the 10-dimensional super-string model living on the 10-dimensional Anti-de-Sitter space.[13–15] It means, that the BFKL Pomeron in this theory is equivalent to the reggeized graviton (see, for example, Ref. 16). Therefore, to construct the Gribov calculus for Pomerons in the Maldacena model, it is needed to study the reggeized graviton interactions. The graviton Regge trajectory and its various couplings were calculated many years ago.[17]

The generally covariant effective action in gravity is obtained for a cluster of gravitons and reggeized gravitons having their rapidities in an interval around their central value. Apart from the usual Einstein-Hilbert term in pure gravity it contains the induced contribution S_{ind},[18]

$$S = -\frac{1}{2\kappa^2} \int d^4x \sqrt{-g}\, R + S_{ind}\,,$$
$$S_{ind} = -\frac{1}{2\kappa^2} \int d^4x \left(\frac{j_{++}}{2}\, \partial_\sigma^2 A^{++} + \frac{j_{--}}{2}\, \partial_\sigma^2 A^{++} \right), \tag{34}$$

with the currents

$$j_{\pm\pm} = j_{\alpha\beta}\, g_\pm^\alpha\, g_\pm^\beta \tag{35}$$

being functionals of the metric tensor $g_{\mu\nu}$. The reggeon fields A^{++} and A^{--} describe the production and annihilation of the reggeized gravitons. They are invariant $\delta A^{\pm\pm} = 0$ under the general coordinate transformations and satisfy the kinematical constraints

$$\partial_+ A^{++} = \partial_- A^{--} = 0\,, \tag{36}$$

corresponding to the strong ordering of the Sudakov components for momenta of the produced clusters in the multi-Regge kinematics. Note, that there is an additional term in the lagrangian for the reggeon fields $\Delta L \sim \partial_\sigma A_{++} \partial_\sigma A_{--}$. The gravity fields $h_{\mu\nu}$ are introduced as usual

$$g_{\mu\nu} = \eta_{\mu\nu} + h_{\mu\nu}\,, \tag{37}$$

where the Minkowski metric tensor $\eta_{\mu\nu}$ has the diagonal structure $(1,-1,-1,-1)$. One can expand the Hilbert-Einstein Lagrangian in h after its integration by parts as follows

$$\sqrt{-g}\, R = \frac{\partial_\sigma h_{\mu\sigma}}{2}\, (\partial_\mu h_{\rho\rho} - \partial_\rho h_{\mu\rho}) + \frac{(\partial_\sigma h_{\mu\nu})^2 - (\partial_\sigma h_{\mu\mu})^2}{4} + O(h^3)\,. \tag{38}$$

As a result, a solution of the corresponding Euler-Lagrange equation at small $A_{\pm\pm}$ has the form

$$h_{\pm\pm} = A_{\pm\pm} + O(A^2)\,, \tag{39}$$

if the currents $j_{\pm\pm}$ have the following expansion

$$j_{\pm\pm} = h_{\pm\pm} + O(h^2)\,. \tag{40}$$

To construct $j_{\pm\pm}$ as a functional of the metric tensor from the condition of general covariance of the effective action it is convenient to consider initially the quasi-elastic kinematics, where one of the fields $A_{\pm\pm}$ is absent. For this case in the coordinate systems with the light-cone global times $x'^+ = 0$ and $x'^- = 0$, corresponding to the constraints $g'^{+\sigma} = \eta^{+\sigma}$ and $g'^{-\sigma} = \eta^{-\sigma}$, the right hand side of classical equations does not contain any dependence from $g^{\mu\nu}$. In this case the solution of the Euler-Lagrange equations is

$$g^{\mu\nu} = \eta^{\mu\nu} + g^{\mu}_{\mp} g^{\nu}_{\mp} \, A^{\mp\mp} \tag{41}$$

in an accordance with the condition, that in the light-cone systems the reggeon fields should coincide with classical values for the corresponding components of $g^{\mu\nu}$. With a corresponding coordinate transformation we obtain a generally covariant energy-momentum tensor for this kinematics.

On the other hand, the dependence of $j_{\pm\pm}$ from $g_{\mu\nu}$ is also fixed by general covariance. To show it we introduce another current $j^{\mp}$ with the definition

$$j_{\pm\pm} = \partial_{\pm} j^{\mp} \,. \tag{42}$$

Then this current satisfies the non-linear equation[18]

$$g^{\sigma\mp} \, \partial_{\sigma} j^{\mp} = g^{\mp\mp} + \frac{g^{\rho\sigma}}{4} \, (\partial_{\rho} j^{\mp})(\partial_{\sigma} j^{\mp}) \,. \tag{43}$$

One can look for its solution in the form

$$j^{\mp} = 2x^{\mp} - \omega^{\mp} \,. \tag{44}$$

The function $\omega^{\mp}$ obeys the generally covariant Hamilton-Jacobi equation,[18]

$$g^{\rho\sigma} \, (\partial_{\rho} \omega^{\mp})(\partial_{\sigma} \omega^{\mp}) = 0 \,, \tag{45}$$

with the initial condition

$$\lim_{\tau \to -\infty} \, \omega^{\mp} = 2x^{\mp} + const \,, \tag{46}$$

where τ is the proper time. The solution $\omega^{\mp}$ describes the light-front shock wave moving in the gravitational field.

We look for a solution of the Hamilton-Jacobi equation in the form

$$\omega^{\mp} = 2x^{\mp} - 2x^{\mp}(x^{\pm}, x^{\perp}) \,. \tag{47}$$

Due to the HJ equation $\omega^{\mp}/2$ can be considered as a component of the coordinate transformation $x'(x)$ to the system with the global light-cone time $x'^{\mp}$ corresponding to the metric constraint

$$g'^{\mp\mp} = 0 \,. \tag{48}$$

Note, that $j^{\mp}$ in the effective action can be substituted by $-\omega^{\mp}$ because the term $2x^{\mp}$ gives a vanishing contribution. After its integration over $x^{\pm}$

the induced action can be expressed only in terms of the Hamilton-Jacobi functions $\omega^{\mp}$ at $x^{\pm}=\infty$ and is generally invariant

$$S_{ind}=-\frac{1}{2\kappa^2}\int d^2x^{\perp}\Big(-\int_{-\infty}^{\infty}\frac{dx^-}{4}\,\omega^-\,\partial^2_{\sigma\perp}A^{++}\,|^{x^+=\infty}_{x^+=-\infty}$$
$$-\int_{-\infty}^{\infty}\frac{dx^+}{4}\,\omega^+\,\partial^2_{\sigma\perp}A^{--}|^{x^-=\infty}_{x^-=-\infty}\Big)\,.$$

It is obvious, that in the coordinate system, where $g^{\mp\mp}=0$, we have $j_{\pm\pm}=j^{\mp}=0$, which can be verified by an iteration of the HJ equation in $g^{\mp\mp}$. This fact does not mean, that in this system the interaction with the corresponding reggeon field $A^{\pm\pm}$ disappears, because the induced energy-momentum tensor is proportional to the variational derivative of S_{ind} over $g^{\mu\nu}$ being non-zero even for $g^{\mp\mp}=0$. The variation of S_{ind} contains the factor $\sqrt{-g}$ as a result of transformation to new coordinates in an accordance with its general covariance. The induced energy-momentum tensor $\theta^{\mu\nu}$ in the quasi-elastic kinematics, where, for example, $A^{++}=0$, is proportional to $\partial^2_{\mu}A^{--}$ in the light cone coordinate system corresponding to $g'^{++}=0$.

5. Classical equation for the effective gravity

Similar to the case of the induced current j^{ind} in QCD we shall attempt to find the induced energy-momentum tensor $\theta_{\mu\nu}$ in the Einstein equation for the effective high energy theory,

$$R_{\mu\nu}-\frac{1}{2}g_{\mu\nu}R=-\theta_{\mu\nu}\,, \tag{49}$$

from considerations related to the general covariance without using the effective action. To begin with, let us consider the quasi-elastic kinematics, where $A^{++}=0$. Initially it is convenient to consider here the coordinate system with the metric tensor obeying the light-cone constraint,

$$g'_{-\sigma}=\eta_{-\sigma}\,, \tag{50}$$

for $\sigma=(+,-,1,2)$. In such a coordinate system the tensor component of the Euler-Lagrange equations containing in the right hand side the contribution proportional to A^{--} has the form

$$R^{--}-\frac{g^{--}}{2}R=-T^{--}\,,\ T^{--}=\partial^2_{\tau}A^{--}\,. \tag{51}$$

One can find an explicit solution of these classical equations for the quasi-elastic kinematics,

$$g'^{\rho\sigma} = \eta^{\rho\sigma} + \delta^{\rho}_{-}\delta^{\sigma}_{-}A^{--}\,, \tag{52}$$

which is a superposition of the plane-wave solutions of Aichelburg and Sexl $g'^{--}_{AS}(x^+, x^\perp)$ with the gravitation centers at $x^+ = z^+$, $x^\perp = z^\perp$ distributed with the weight $\partial^{\perp 2}_{\sigma} A^{--}(z^+, z^\perp)$,

$$g'^{\rho\sigma} = \eta^{\rho\sigma} - \delta^{\rho}_{-}\delta^{\sigma}_{-}\int \frac{d^2 z}{4\pi}\,\ln(|x-z|^2)\,\partial^{\perp 2}_{r}\,A^{--}(z^+, z^\perp)\,. \tag{53}$$

Indeed, the operator $\partial^{\perp 2}_{r}$ applied to g'^{--}_{AS} is proportional to the product of δ-functions,

$$\partial^{\perp 2}_{\tau}\, g'^{--}_{AS}(x^+ - z^+, x^\perp - z^\perp) \sim \delta^2(x^\perp - z^\perp)\delta(x^+ - z^+)\,, \tag{54}$$

which corresponds to the energy-momentum tensor for a massless point-like particle.

To derive the Euler-Lagrange equation valid for all coordinate systems we should construct the generally covariant energy-momentum tensor $\theta_{\mu\nu}$ which can be reduced to the above AS tensor $T_{\mu\nu}$ in the light-cone coordinate system. The coordinate transformation $x' = x'(x)$ to this light cone system satisfies the set of equations

$$g_{\mu\nu}\,\frac{\partial x^\mu}{\partial x'^-}\,\frac{\partial x^\mu}{\partial x'^\sigma} = \eta_{-\sigma} \tag{55}$$

for $\sigma = (\pm, 1, 2)$.

Its solution x'^ρ allows to construct the generally covariant energy-momentum tensor $\theta^{\mu\nu}$ in the quasi-elastic kinematics for a general coordinate system by inverting the relations

$$\theta_{\mu\nu}\,\frac{\partial x^\mu}{\partial x'^\rho}\,\frac{\partial x^\nu}{\partial x'^\sigma} = \delta^+_\rho\delta^+_\sigma\,T^{--}(x')\,. \tag{56}$$

Note, that we could calculate $\theta^{\mu\nu}$ also by transforming the Einstein tensor $G_{\mu\nu}$ to an arbitrary coordinate system from light-cone one.

The condition $g'^{++} = 0$ for the global light-cone time leads to the Hamilton-Jacobi equation,

$$g^{\mu\nu}\,\partial_\mu x'^+\partial_\nu x'^+ = 0\,, \tag{57}$$

for the corresponding coordinate transformation $x'^+ = x'^+(x)$. Its solution gives a possibility to construct the covariant energy-momentum tensor for our kinematics,

$$\theta_{\mu\nu} = \partial_\mu x'^+\partial_\nu x'^+\partial^2_\sigma A^{--}\,, \tag{58}$$

where new coordinates x' should be expressed in terms of initial coordinates x. This expression for $\theta_{\mu\nu}$ is reduced to the Aichelburg-Sexl energy-momentum tensor in the light-cone system with $g^{+\sigma} = \eta^{+\sigma}$. It can be obtained also from the effective action with taking into account, that in fact in L^{ind} the integration is performed over the coordinate x', which should be transformed to the coordinate x in the variation $\delta g^{\mu\nu}$ with the use of the corresponding Jacobian $\sqrt{-g}$.

One can construct the energy-momentum tensor also for the general multi-Regge kinematics,

$$\theta_{\mu\nu} = \partial_\mu x'^- \partial_\nu x'^- \partial_\sigma^2 A^{++}(x') + \partial_\mu x'^+ \partial_\nu x'^+ \partial_\sigma^2 A^{--}(x') \,. \tag{59}$$

Its trace is equal to zero,

$$g^{\mu\nu}\, \theta_{\mu\nu} = 0 \,, \tag{60}$$

due to the Hamilton-Jacobi equations for $\omega^\pm$, describing the massless particle in an external gravitation field. Further, $\theta_{\mu\nu}$ is conserved,

$$D_\mu\, \theta^{\mu\nu} = 0 \,, \tag{61}$$

due to its covariance under the general coordinate transformation $x \to \widetilde{x}$,

$$\widetilde{\theta}_{\rho\sigma} = \frac{\partial x^\mu}{\partial \widetilde{x}^\rho} \frac{\partial x^\nu}{\partial \widetilde{x}^\sigma}\, \theta_{\mu\nu} \,, \tag{62}$$

and the kinematical constraints $\partial_\pm A^{\pm\pm} = 0$ for the reggeon fields.

The solution of the Euler-Lagrange equations with the above energy-momentum tensor is simplified in one of two light-cone coordinate systems, where $g^{++} = 0$ or $g^{--} = 0$. In these cases the corresponding contribution $\sim A^{--}$ or $\sim A^{++}$ in $\theta_{\mu\nu}$ does not depend on $g^{\mu\nu}$. The second contribution can be simplified, because the effective currents $j^\pm$ in the induced action S_{ind} are known for the particle in the gravitational shock wave of Aichelburg and Sexl,[18]

$$\begin{gathered} j^- = -a\,\mu \left(\ln |\overrightarrow{x}| + \ln f(z) - \frac{1}{4}\, \frac{z}{f^2(z)} \right) , \\ f(z) = \frac{1}{2} + \frac{\sqrt{1+2z}}{2} \,, \quad z = a\, \frac{x^+}{|\overrightarrow{x}|^2} \,, \end{gathered} \tag{63}$$

where a is proportional the mass density of the shock wave.

Thus, we have constructed the generally covariant Euler-Lagrange equations for the effective action in gravity. The solution of these equations for the metric tensor in the quasi-elastic kinematics coincides with the reggeized graviton field $A^{\pm\pm}$ in the corresponding light-cone coordinate system. For the general multi-Regge kinematics their solutions can be found in perturbation theory, which allows to calculate various reggeon vertices in the tree approximation in an agreement with the known results.[17] The integration over fluctuations around these solutions gives a possibility to find corrections to reggeized graviton interactions. In particular, using the graviton Regge trajectory obtained in Refs. 17, 18 one can construct the scattering amplitudes of high energy gravitons in the double-logarithmic approximation.[19] The generalization of our approach to the super-gravity in the 10-dimensional Anti-de-Sitter space will be interesting for the construction of the Gribov Reggeon calculus for Pomerons in N=4 SUSY.

References

1. V.N. Gribov, *Sov. Phys. JETP* **14**, 478 (1962).
2. S. Mandelstam, *Nuovo Cim.* **30**, 1148 (1963).
3. V.N. Gribov, I.Ya. Pomeranchuk and K.A. Ter-Martirosyan, *Phys. Rev.* **B 139**, 184 (1965).
4. V.N. Gribov, *Sov. Phys. JETP* **26**, 414 (1968).
5. L.N. Lipatov, *Sov. J. Nucl. Phys.* **23**, 338 (1976); V.S. Fadin, E.A. Kuraev, L.N. Lipatov, *Phys. Lett.* **B 60**, 50 (1975); E.A. Kuraev, L.N. Lipatov, V.S. Fadin, Sov. Phys. JETP **44**, 443 (1976);Ya.Ya. Balitsky, L.N. Lipatov, *Sov. J. Nucl. Phys.* **28** (1978) 822.
6. L.N. Lipatov *High energy asymptotics of multi-colour QCD and exactly solvable lattice models*, arXiv:hep-th/9311037, unpublished.
7. L.N. Lipatov, *J. Phys.* **A 42**:304020 (2009).
8. L.N. Lipatov, *Nucl. Phys.* **B 452**, 369 (1995); *Phys. Rept.* **286**, 131 (1997).
9. E.N. Antonov, L.N. Lipatov, E.A. Kuraev, I.O. Cherednikov, *Nucl. Phys.* **B 721**, 111 (2005).
10. V.S. Fadin, L.N. Lipatov, *Phys. Lett.* **B 429**, 127 (1998); M. Ciafaloni and G. Camici, *Phys. Lett.* **B 430**, 349 (1998).
11. A.V. Kotikov, L.N. Lipatov, *Nucl. Phys.* **B 582**, 19 (2000).
12. A.V. Kotikov, L.N. Lipatov, *Nucl. Phys.* **B 661**, 19 (2003).
13. J.M. Maldacena, *Adv. Theor. Math. Phys.* **2**, 231 (1998).
14. S.S. Gubser, I.R. Klebanov, A.M. Polyakov, *Phys. Lett.* **B 428**, 105 (1998).
15. E. Witten, *Adv. Theor. Math. Phys.* **2**, 253 (1998).
16. A.V. Kotikov, L.N. Lipatov, Nucl. Phys. **B 874**, 889 (2013).
17. L.N. Lipatov, Phys. Lett. **116**, 411 (1982).

18. L.N. Lipatov, *Phys. Part. Nucl.* **44** 391 (2013), arXiv:1105.31277 [hep-ph].
19. J. Bartels, L.N. Lipatov, A. Sabio Vera, *J. High Energy Phys.* **07**, 056 (2014).

L.N. Lipatov (middle) with V. Fadin (left), N.V. Abramovskaya (right), N. Smorodinskaya and J. Nyiri (back row), Chernogolovka, 2015.

PERTURBATIVE POMERON VERTICES AND THE CONFORMAL BOOTSTRAP

CARLO EWERZ

Institut für Theoretische Physik, Universität Heidelberg
Philosophenweg 16, D-69120 Heidelberg, Germany

High energy scattering processes in QCD can be described in terms of interacting pomerons. Vertices for the interaction of pomerons have been derived in perturbative QCD. Their properties hint at the existence of a 2+1-dimensional conformal field theory of interacting pomerons. Here we study in particular the 1-to-2 and the 1-to-3 pomeron vertex and find them to be related to each other by a conformal bootstrap relation.

1. High energy scattering and a theory of interacting pomerons

The scattering of hadrons at very high energy has since many years attracted great interest. The main question to be answered is very simple: how does the cross section depend on the energy? But to date it remains an open problem to answer that questions and to derive a full theoretical understanding of high energy scattering from first principles, that is from Quantum Chromodynamics (QCD). While a full answer to this problem has not yet been found, many important steps have been made. Regge theory, developed long before QCD was discovered as the microscopic theory of the strong interaction, incorporates many fundamental properties of high energy scattering. In Regge theory, the leading contribution to the total cross section is attributed to the exchange of a pomeron. More precisely, the total cross section is via the optical theorem related to the imaginary part of the forward elastic amplitude. The latter, in turn, is given by the exchange of a colorless object between the scattering hadrons — which at high energy is the pomeron. The pomeron therefore acts as the main building block in the theoretical description of high energy scattering.

Building a full theory of high energy scattering then requires to also include the interaction of many pomerons. A beautiful construction of such a theory of pomeron interactions in the Regge framework has been devised by Vladimir Gribov.[1] The ultimate goal would be to derive a similar picture from the interaction of quarks and gluons in QCD. This turns out to be extremely difficult for a number of reasons. One of them is that most hadronic scattering processes at high energy involve a large contribution from low momenta, necessitating a nonperturbative treatment. Another difficulty arises due to the large phase space for parton emissions at high energy, resulting in high parton densities. The dynamics of such a dense system of partons is obviously nonlinear, and hence very challenging to describe. At least the first of these two problems can be circumvented by considering scattering processes involving hard momentum scales provided by the scattering particles, hence making a perturbative description possible. An example is the scattering of highly virtual photons which turn into quark–antiquark pairs that then interact with each other by exchanging gluons. Such processes are in general rare and more difficult to study experimentally, but they should provide a good starting point for the theoretical investigation of high-energy QCD. The challenge of describing the highly nonlinear evolution of a dense system of partons, however, remains. It is exactly the emergence of multiple parton cascades in the scattering hadrons and of their recombination which gives rise to the interactions of pomerons.

2. Pomerons and their interactions from QCD

The classical approach to studying the pomeron in perturbative QCD is based on the resummation of logarithms of the energy $\sqrt{s}$ which in perturbation theory emerge in certain types of diagrams. At high energy, such logarithms are large and can compensate the smallness of the coupling α_s. Therefore one collects, in the leading logarithmic approximation[2,3] (LLA), all terms in the perturbative series that contain factors of the form $(\alpha_s \log s)^n$. The result of this procedure is the BFKL equation, a linear evolution equation for the exchange of two interacting gluons. Diagrammatically, it corresponds to the exchange of a gluon ladder in the t-channel with the rungs representing the interaction between the gluons via the kernel of the BFKL integral equation. While the interaction of the two t-channel gluons is due to real gluons, there is also a contribution to the BFKL equation representing virtual corrections. These give rise to the reggeization of the gluon at high energy,[4] a phenomenon due to which the t-channel gluons can in a sense be regarded as composite objects of (again) reggeized gluons. In other words, the reggeized gluon is a collective excitation of the gluon field. At higher and higher energies the LLA turns out to be insufficient at least from a theoretical point of view as it eventually violates unitarity. The perturbative diagrams that are crucial for the restoration of unitarity in the perturbative series can be identified as those with more reggeized gluons in the t-channel. In this picture, the nonlinear effects mentioned above correspond to fluctuations in the number of reggeized gluons in the t-channel evolution due to splittings and recombinations. To include such effects naturally requires an approximation scheme beyond the LLA.

Various perturbative approaches and techniques have been developed in order to extend the LLA by including nonlinearities. Among the most prominent directions are the dipole picture,[5] the color glass condensate approach,[6] the operator expansion of Wilson lines,[7] and the generalization of the resummation of large logarithms to higher numbers of gluons. Our present work is based on results obtained in the latter approach, and we therefore describe it in more detail momentarily. Before we do that we should mention that at least in cases that allow the direct comparison between the different approaches equivalent results are obtained. For example, all these approaches reproduce the BFKL equation at low parton densities.

The resummation technique allows one to go beyond the LLA in a straightforward way by analyzing diagrams with more than two reggeized gluons in the t-channel. For that one considers a given number n of gluons in the t-channel, and then resums all perturbative contributions containing the maximally possible number of logarithms of the energy for that given number. The corresponding approximation scheme is called the generalized leading logarithmic approximation (GLLA). There are two different ways of performing this resummation. In the simpler scheme one always keeps the number of gluons in the t-channel fixed.[8–10] As a consequence one obtains a separate integral equation for each n describing the evolution of the corresponding n-gluon compound state. The kernel of the integral equation is similar to that of the BFKL equation, now representing all pairwise interactions of the n gluons. One is hence led to consider the quantum mechanics of n-gluon states which turns out to exhibit a very rich structure including complete integrability in the large-N_c limit.[11,12]

The more advanced version of the GLLA in addition collects diagrams in which the number of reggeized gluons can fluctuate during the t-channel evolution.[13,14] This extended scheme of the GLLA is called EGLLA. Here, the n-gluon amplitudes are described by a hierarchy of coupled integral equations. It is worth pointing out that some essential features of the amplitudes of the EGLLA (as compared to the simpler version of the GLLA with fixed numbers of gluons) emerge from formally subleading contributions in N_c. It is therefore crucial that the EGLLA does not rely on the large-N_c limit.

The EGLLA is most conveniently formulated in terms of amplitudes for the production of n reggeized gluons in the t-channel starting from a given impact factor defined by the external particles in the scattering process, for example by virtual photons. It has been shown that a suitable arrangement of these amplitudes brings them into a form corresponding to a field theory in which different compound states with fixed numbers of gluons interact via effective vertices.[14–16] This field theory lives in 2+1 dimensions. The two dimensions are the spatial directions transverse to the direction of the colliding external particles. The additional dimension is given by rapidity $Y \sim \log s$ which acts as a time-like parameter in the t-channel evolution of the system of reggeized gluons. The effective transition vertices coupling states with different numbers of gluons to each other are local in rapidity. The reggeization of the gluon at high energies[4,17] mentioned above plays an essential role in the emergence of this field theory structure.

If one considers the charge-parity even channel of the scattering process — which is relevant for the elastic amplitude and hence for the high energy behavior of the total cross section — one finds that only states with even numbers of gluons occur in this effective field theory, and only the corresponding vertices are relevant here. (States with odd numbers of gluons and their vertices would contribute in the charge-parity odd channel.) So far, the explicit forms of the two-to-four[16] and of the two-to-six gluon vertex[14] have been calculated. Going to higher vertices with the same techniques appears challenging. It would therefore be very valuable to understand the structure of the effective field theory on a deeper level, in particular its underlying symmetries.

Starting from the effective field theory of reggeized gluons with its structure of transition vertices one can make an additional approximation by considering only pomeron states, that is two-gluon BFKL ladders, and products of such pomeron states. This reduces the complexity of the problem considerably. However, this additional approximation is not necessarily a good one as can be seen already in the simple case of four gluons. A general state of four interacting reggeized gluons has obviously a different spectrum than can be obtained from product states of color-singlet two-gluon states only. In general, one can even argue that a general $2m$-gluon state leads to a more rapid increase of the cross section at high energy than an

m-pomeron state. It is therefore very likely that, even when including arbitrarily many pomerons in the t-channel, the approximation due to only considering pomerons leads to a different behavior than the summation over general n-gluon states. Despite this caveat it turns out to very interesting to study the theory of interacting pomerons for several reasons. Depending on the scattering process under consideration, the pomeron diagrams can be the leading terms in N_c. Consequently, in some processes the restriction on pomeron states could actually be a good approximation, like for example in scattering processes on large nuclei. Moreover, the resulting theory of interacting pomerons has a very rich structure and can give us important information about various mechanisms of QCD at high energy. It should also be the QCD equivalent of the pomeron calculus traditionally considered in Regge theory.

Technically, the reduction from general n-gluon states to pomerons is done by projecting the transition vertices of the effective theory of interacting gluons onto two-gluon BFKL pomeron states. It is worth emphasizing that this does not only involve the projection of the gluons onto pairwise color singlets, but also a projection onto eigenstates of the BFKL equation with their particular dependence on the coordinates in transverse space. In this way one obtains the 1-to-2 pomeron vertex (or triple-pomeron vertex) $V_{3\mathbb{P}}$ from the 2-to-4 reggeized gluon vertex,[18,19] and a 1-to-3 pomeron vertex (or 4-pomeron vertex) $V_{4\mathbb{P}}$ from the 2-to-6 reggeized gluon vertex.[20]

In the theory of interacting pomerons one can make a further approximation, valid at low parton densities, by restricting the pomeron interactions to only the splitting of one into two pomerons, that is the triple-pomeron vertex, and its iteration. This gives rise to so-called fan diagrams and contains the first possible nonlinear term. That approximation is encoded in the BK equation.[7,21] While that equation is useful for the phenomenological analysis of various processes, many interesting properties of the effective field theory of interacting pomerons can be understood only if one considers also higher pomeron vertices. The findings that we present in section 4 below are an example for such a property that characterizes the underlying structure of the pomeron field theory.

3. Conformal symmetry in high energy QCD

A very interesting property of the BFKL pomeron is its conformal invariance in two-dimensional impact parameter space.[22] Identifying position vectors in impact parameter space with complex numbers, $\rho = \rho_x + i\rho_y$

and $\rho^* = \rho_x - i\rho_y$, the BFKL integral kernel is invariant under conformal $SL(2,\mathbb{C})$ transformations

$$\rho \longrightarrow \frac{a\rho + b}{c\rho + d} \qquad \text{with} \quad ad - bc = 1\,. \tag{1}$$

As a consequence of conformal symmetry one can find the eigenfunctions of the BFKL equation, given by

$$E^h(\rho_{10}, \rho_{20}) = \left(\frac{\rho_{12}}{\rho_{10}\rho_{20}}\right)^h \left(\frac{\rho_{12}^*}{\rho_{10}^*\rho_{20}^*}\right)^{\bar{h}} \tag{2}$$

with the notation $\rho_{ij} = \rho_i - \rho_j$, and with the conformal weights

$$h = \frac{1+n}{2} + i\nu\,, \qquad \bar{h} = \frac{1-n}{2} + i\nu = 1 - h^*\,. \tag{3}$$

The parameter $n \in \mathbb{Z}$ is the conformal spin, and $\nu \in \mathbb{R}$ is the scaling dimension. Hence n and ν are the quantum numbers of the perturbative pomeron.

Also the transition vertices of the effective theory of reggeized gluons have the remarkable property of being conformally invariant.[23,24] That is, they do not change when a simultaneous transformation of the type (1) is applied to the transverse coordinates of all gluons. The same holds for the n-gluon compound states. Hence, all known elements of the effective 2+1-dimensional theory are conformally invariant in the two space-like dimensions, for each given value of the additional time-like parameter Y. Based on the expectation that also higher vertices of reggeized gluons are conformally invariant, together with some other necessary conditions like IR-finiteness and symmetry under permutations, it was even possible to make a conjecture about the explicit form of all 2-to-$2l$ gluon vertices $V_{2\to 2l}$ in the effective theory.[24] The observation of conformal symmetry raises hope that the effective theory of reggeized gluons can be cast into the form of a conformal field theory (CFT). Conformal symmetry in two dimensions is a particularly powerful symmetry which strongly constrains the coordinate dependence as well as further properties of correlation functions. To identify this CFT would therefore be a significant step in the direction of understanding the high energy limit of QCD.

Also after the restriction to two-gluon pomeron states, the resulting effective theory of interacting pomerons exhibits conformal invariance in two-dimensional impact parameter space, and rapidity Y again acts as a time-like parameter. The pomeron vertices $V_{3\mathbb{P}}$ and $V_{4\mathbb{P}}$ depend on the coordinates ρ_i and conformal weights h_i of the pomerons. This dependence is found to have exactly the form of correlation functions of quasiprimary fields in a CFT.[18–20] The pomeron vertices exhibit also a further important property required in a CFT, as we will discuss now.

4. Conformal bootstrap for perturbative pomeron vertices

In conformal field theories, correlation functions are in general related to each other by conformal bootstrap relations. They are consequences of the associativity of the conformal operator algebra. In the following I describe the finding that the perturbative pomeron vertices $V_{3\mathbb{P}}$ and $V_{4\mathbb{P}}$ satisfy such a conformal bootstrap relation. This is a result I have found in collaboration with T. Bittig, G. P. Korchemsky, and S. Wallon.[25]

The three-pomeron vertex, as derived from the EGLLA,[18,19] consists of two contributions (coming with different powers of N_c),

$$V_{3\mathbb{P}}(\rho_i, h_i) = g^4 C_1 \left[V^{(0)}(\rho_i, h_i) + \frac{C_2}{N_c^2} V^{(1)}(\rho_i, h_i) \right] \tag{4}$$

with $C_{1,2}$ being independent of the positions ρ_i. The two functions $V^{(0)}$ and $V^{(1)}$ are obtained as certain integrals over BFKL eigenfunctions E^h, see (2). For example,

$$V^{(1)}(\rho_\mu, h_\mu) = \int \frac{d^2\rho_1 d^2\rho_2}{|\rho_{12}|^4} E^{h_\alpha}(\rho_{1\alpha}, \rho_{2\alpha}) E^{h_\beta}(\rho_{1\beta}, \rho_{2\beta}) E^{h_\gamma}(\rho_{1\gamma}, \rho_{2\gamma}) , \tag{5}$$

and $V^{(0)}$ is given by an integral of a similar type.

The four-pomeron vertex $V_{4\mathbb{P}}$ consists of several contributions[20] that all come with the same power of N_c. For illustrating our main result it will not be necessary to display it in full detail here. Suffice it so say that it can be expressed[25] in terms of three different functions, $\Phi(\rho_j, h_j)$, $\Theta(\rho_j, h_j)$ and $\Pi(\rho_j, h_j)$, which depend on the four pomeron coordinates ρ_j and the corresponding conformal weights h_j ($j = a, b, c, d$). Explicit integral representations for these functions in terms of BFKL eigenfunctions are known.* For example,

$$\Phi(\rho_i, h_i) = \int \frac{d^2\rho_1 d^2\rho_2}{|\rho_{12}|^4} E^{h_a^*}(\rho_{1a}, \rho_{2a}) E^{h_b^*}(\rho_{1b}, \rho_{2b}) E^{h_c^*}(\rho_{1c}, \rho_{2c}) \times E^{h_d}(\rho_{1d}, \rho_{2d}) , \tag{6}$$

*Strictly speaking, the full derivation[20] has been performed for vanishing conformal spins, but the generalization to $n_j \neq 0$ is straightforward.

and the other two functions Θ and Π are given by integrals of a similar type.

Using a completeness relation for BFKL pomeron states one can then derive conformal bootstrap relations for the functions $\Phi(\rho_j, h_j)$, $\Theta(\rho_j, h_j)$ and $\Pi(\rho_j, h_j)$ in the four-pomeron vertex. As a result, these functions are expressed in terms products of the functions $V^{(0)}$ and $V^{(1)}$ in the three-pomeron vertex. For the function Φ shown above, for example, the bootstrap relation reads

$$\Phi(\rho_i, h_i) = \sum_{n_k=-\infty}^{+\infty} \int_{-\infty}^{+\infty} d\nu_k \, f(h_k) \int d^2\rho_k \, V^{(1)}(\rho_a^*, \rho_b^*, \rho_k, h_a^*, h_b^*, h_k) \times V^{(1)}(\rho_k^*, \rho_c^*, \rho_d, h_k^*, h_c^*, h_d) \qquad (7)$$

with the weight factor $f(h_k) = |h_k - 1/2|^2/\pi^4$. Note that the sum and integral are over a full set of intermediate BFKL states labeled by the index k. The other two functions Θ and Π in the four-pomeron vertex can be written in a similar way as products of $V^{(0)}$ and $V^{(1)}$, where the difference between them originates from crossed arguments of the BFKL eigenfunctions in the integral.

Interestingly, the analogous product of $V^{(0)}$ with itself is not part of the four-pomeron vertex derived from the EGLLA. This can be understood as a consequence of the resummation procedure used in the EGLLA. However, one could try to devise an improved version of the EGLLA which would include diagrams that lead to an additional contribution to the four-pomeron vertex, representing the product of $V^{(0)}$ with itself.

The bootstrap relations described here strongly support the expectation that the effective theory of interacting pomerons is indeed a conformal field theory. We hope that further insight into the structure of this CFT can be gained by a more detailed analysis of how exactly the different contributions to the three-pomeron vertex and the four-pomeron vertex occur in the bootstrap relations. In this way it might even be possible to derive strong constraints on the yet unknown higher n-pomeron vertices from these conformal bootstrap relations.

Finally, we would like to point out that the above considerations are not restricted to QCD. The same effective transition vertices between different n-gluon compound states and consequently the same pomeron vertices are found when the same approximations are applied to $\mathcal{N} = 4$ supersymmetric Yang-Mills theory.[26]

5. Summary

The high energy limit of QCD can be studied by resumming large logarithms of the energy to all orders in the perturbative series. Various approximation schemes can be used in this context. We have concentrated on the EGLLA scheme, in which a 2+1-dimensional effective theory of reggeized gluons is obtained. Here, the 2 dimensions refer to impact parameter space, and the additional dimension is given by rapidity which acts as a time-like parameter. In a further approximation step the n-gluon states and vertices of that effective theory can be projected onto two-gluon pomeron states. This leads to an effective 2+1-dimensional theory of interacting pomerons exhibiting various properties which are typical for conformal field theories. We have found in particular that the three-pomeron vertex and the four-pomeron vertex are related to each other by a conformal bootstrap relation. We hope that this discovery is a further useful step towards deriving a full Regge calculus of interacting pomerons from QCD, possibly in the form of a conformal field theory.

Acknowledgments

I am grateful to Julia Nyiri for the kind invitation to contribute to this volume. I would like to thank Thomas Bittig, Gregory Korchemsky, and Samuel Wallon for enjoyable collaboration.

References

1. V.N. Gribov, *Sov. Phys. JETP* **26**, 414 (1968) [*Zh. Eksp. Teor. Fiz.* **53**, 654 (1967)].
2. E.A. Kuraev, L.N. Lipatov and V.S. Fadin, *Sov. Phys. JETP* **45**, 199 (1977) [*Zh. Eksp. Teor. Fiz.* **72**, 377 (1977)].
3. I.I. Balitsky and L.N. Lipatov, *Sov. J. Nucl. Phys.* **28**, 822 (1978) [*Yad. Fiz.* **28**, 1597 (1978)].
4. L.N. Lipatov, *Sov. J. Nucl. Phys.* **23**, 338 (1976) [*Yad. Fiz.* **23**, 642 (1976)].
5. A.H. Mueller, *Nucl. Phys.* **B415**, 373 (1994).
6. E. Iancu and R. Venugopalan, in *Quark-Gluon Plasma 3*, Eds. R.C. Hwa and X.-N. Wang, World Scientific (2003), arXiv:hep-ph/0303204.
7. I. Balitsky, *Nucl. Phys.* **B463**, 99 (1996) arXiv:hep-ph/9509348.
8. J. Bartels, *Nucl. Phys.* **B151**, 293 (1979).

9. J. Bartels, *Nucl. Phys.* **B175**, 365 (1980).
10. J. Kwieciński and M. Praszałowicz, *Phys. Lett.* **B94**, 413 (1980).
11. L.N. Lipatov, *JETP Lett.* **59**, 596 (1994) [*Pisma Zh. Eksp. Teor. Fiz.* **59**, 571 (1994)] arXiv:hep-th/9311037.
12. L.D. Faddeev and G.P. Korchemsky, *Phys. Lett.* **B342**, 311 (1995) arXiv:hep-th/9404173.
13. J. Bartels, DESY 91-074 (unpublished)
14. J. Bartels and C. Ewerz, *JHEP* **9909**, 026 (1999) arXiv:hep-ph/9908454.
15. J. Bartels, *Z. Phys.* **C60**, 471 (1993).
16. J. Bartels and M. Wüsthoff, *Z. Phys.* **C66**, 157 (1995).
17. C. Ewerz, *JHEP* **0104**, 031 (2001) arXiv:hep-ph/0103260.
18. H. Lotter, Ph. D. Thesis, Hamburg University 1996, arXiv:hep-ph/9705288.
19. G.P. Korchemsky, *Nucl. Phys.* **B550**, 397 (1999) arXiv:hep-ph/9711277.
20. C. Ewerz and V. Schatz, *Nucl. Phys.* **A736**, 371 (2004) arXiv:hep-ph/0308056.
21. Y.V. Kovchegov, *Phys. Rev.* **D60**, 034008 (1999) arXiv:hep-ph/9901281.
22. L.N. Lipatov, *Sov. Phys. JETP* **63**, 904 (1986) [*Zh. Eksp. Teor. Fiz.* **90**, 1536 (1986)].
23. J. Bartels, L.N. Lipatov and M. Wüsthoff, *Nucl. Phys.* **B464**, 298 (1996) arXiv:hep-ph/9509303.
24. C. Ewerz, *Phys. Lett.* **B512**, 239 (2001) arXiv:hep-ph/0105181.
25. T. Bittig, C. Ewerz, G. P. Korchemsky, S. Wallon, in preparation
26. J. Bartels, C. Ewerz, M. Hentschinski and A.-M. Mischler, *JHEP* **1005**, 018 (2010) arXiv:0912.4759 [hep-th].

MULTIPOMERON THEORY IN THE GRIBOV APPROACH

V.A. ABRAMOVSKY, N.V. ABRAMOVSKAYA, N.V. EVSTIGNEEVA
Novgorod State University
E-mail: Victor.Abramovsky@novsu.ru

Pomeron and non vacuum reggeon parameters which reasonably describe total and elastic cross sections are obtained. The ratio of the real and imaginary part of the forward elastic scattering amplitude is calculated with these parameters and it coincides with the experimental data in the whole energy region. The shower enhancement coefficient in the quasi-eikonal approach appears to be less than unity which corresponds to the definite distribution of parton showers in incident hadrons.

Before we begin to talk about science, we V.A.A., as the student of Vladimir Naumovich, and N.V.A., as the student of V.A.A. (so she is Gribov's "scientific granddaughter"*) would like to say that the scientific school of Gribov and its traditions will exist in his students and "scientific grandchildren". V.A.A. really misses his advice, his critics, his approval.

1. Introduction

The status of hadronic physics has changed with the advent of QCD. It is believed that QCD is the microtheory of strong interactions. However, fundamental problems such as the description the large-scale structure of the vacuum or the confinement of color are still not solved within the framework of QCD. So finding regularities of such a complex process with many degrees of freedom as multiple production of hadrons from the first principles QCD is not yet possible. Therefore, we must turn to more phenomenological and model approaches.

In this paper we consider the Gribov multipomeron theory.[1]

*This is the definition of Julia Nyiri and A. Zichichi

2. Elastic scattering amplitude

The contribution of multipomeron exchange to elastic scattering amplitude has the form

$$\begin{aligned}\frac{A^v(s,Q^2)}{s} &= (i+\tan\varphi)g^2e^{\Delta Y}\sum_{n=1}^{\infty}\frac{(-1)^{n-1}}{n\cdot n!}\left(\frac{z}{2\cos\varphi\sqrt{1+\beta^2}}\right)^{n-1} \\ &\times \exp\left(-(\alpha' Y+R^2)\frac{Q^2}{n}\right)\exp\left(i\left(\theta(n-1)+\frac{Q^2}{n}\frac{\alpha'\pi}{2}\right)\right),\end{aligned} \tag{1}$$

where

$$z=\frac{g^2e^{\Delta Y}c}{4\pi(\alpha' Y+R^2)}, \tag{2}$$

$Y = \ln s/s_0$, $s_0 = 1$ GeV, $\varphi = \frac{\pi\Delta}{2}$, $\Delta = \alpha_P(0)-1$, $\beta = \frac{\pi\alpha'}{2(\alpha' Y+R^2)}$, $\theta = \arctan\beta-\varphi$, $t=-Q^2$ is transverse momentum, α' is slope of Pomeron trajectory, and c is shower enhancement coefficient, which historically was defined as

$$\frac{\sigma_{el}+\sigma_{diff}}{\sigma_{el}} = c>1, \qquad c\approx 1.3\div 1.5. \tag{3}$$

We use the approximation for the signature factor of Pomeron,

$$\frac{i\exp\left(-i\frac{\pi}{2}(\Delta-\alpha' Q^2)\right)}{\cos\left(\frac{\pi}{2}(\Delta-\alpha' Q^2)\right)} \approx \frac{i\exp\left(-i\frac{\pi}{2}(\Delta-\alpha' Q^2)\right)}{\cos\left(\frac{\pi}{2}\Delta\right)}. \tag{4}$$

Formula (1) can be rewritten in the form

$$\begin{aligned}\frac{A^v(s,Q^2)}{s} &= (i+\beta)\frac{8\pi(\alpha' Y+R^2)}{c}\sum_{n=1}^{\infty}\frac{(-1)^{n-1}}{n\cdot n!}\left(\frac{ze^{i\theta}}{2\cos\varphi\sqrt{1+\beta^2}}\right)^{n} \\ &\times \exp\left(-(\alpha' Y+R^2)\frac{Q^2}{n}\right)\exp\left(\frac{i\alpha'\pi}{2}\frac{Q^2}{n}\right).\end{aligned} \tag{5}$$

For $Q^2=0$ the series in (5) can be convoluted

$$\begin{aligned}\sum_{n=1}^{\infty}\frac{(-1)^{n-1}}{n\cdot n!}\left(\frac{ze^{i\theta}}{2\cos\varphi\sqrt{1+\beta^2}}\right)^{n} &= \mathrm{E}_1\left(\frac{ze^{i\theta}}{2\cos\varphi\sqrt{1+\beta^2}}\right) \\ &+\gamma+\ln\left(\frac{ze^{i\theta}}{2\cos\varphi\sqrt{1+\beta^2}}\right)\end{aligned} \tag{6}$$

where γ is Euler's constant and $\mathrm{E}_1(\tau)$ is the exponential integral function,

$$\mathrm{E}_1(\tau)=\int_{\tau}^{\infty}\frac{\mathrm{d}t}{t}e^{-t}.$$

From (1) and (3) we get real and imaginary parts of elastic scattering. For scattering with $Q^2 = 0$ we have

$$\sigma_{tot} = \frac{\operatorname{Im} A^v(s, Q^2 = 0)}{s} = \frac{8\pi(\alpha' Y + R^2)}{c}\left[\operatorname{Re}\mathrm{E}_1\left(\frac{z}{2}e^{i\theta}\right) + \gamma + \right.$$
$$\left. + \ln\left(\frac{z}{2\cos\varphi\sqrt{1+\beta^2}}\right) + \beta\left[\operatorname{Im}\mathrm{E}_1\left(\frac{z}{2}e^{i\theta}\right) + \theta\right]\right], \tag{7}$$

and

$$\frac{\operatorname{Re} A^v(s, Q^2 = 0)}{s} = \frac{8\pi(\alpha' Y + R^2)}{c}\left[\beta\left[\operatorname{Re}\mathrm{E}_1\left(\frac{z}{2}e^{i\theta}\right) + \gamma + \right.\right.$$
$$\left.\left. + \ln\left(\frac{z}{2\cos\varphi\sqrt{1+\beta^2}}\right)\right] - \left[\operatorname{Im}\mathrm{E}_1\left(\frac{z}{2}e^{i\theta}\right) - \theta\right]\right]. \tag{8}$$

For the pure imaginary Pomeron with $\beta = 0$, $\varphi = 0$ we have

$$\frac{\operatorname{Im} A^v(s, Q^2 = 0)}{s} = \frac{8\pi(\alpha' Y + R^2)}{c}\left[\mathrm{E}_1\left(\frac{z}{2}\right) + \gamma + \ln\left(\frac{z}{2}\right)\right].$$

Non vacuum exchange can be written as follows:

$$\frac{\operatorname{Im} A_{pp(\bar{p})}(s, Q^2 = 0)}{s} = g_f^2 s^{-\Delta_f} \mp g_R^2 s^{-\Delta_R}, \tag{9}$$

and

$$\frac{\operatorname{Re} A_{pp(\bar{p})}(s, Q^2 = 0)}{s} = -\tan\left(\frac{\pi}{2}\Delta_f\right) g_f^2 s^{-\Delta_f} \pm \cot\left(\frac{\pi}{2}\Delta_R\right) g_R^2 s^{-\Delta_R}, \tag{10}$$

where g_f, g_R are the interaction vertices for C-even and C-odd reggeons. Furthermore $\Delta_f = 1 - \alpha_f$, $\Delta_R = 1 - \alpha_R$. The vacuum contribution to the elastic cross section is given by

$$\sigma_{el}^v = \frac{1}{16\pi}\int\left|\frac{A^v(s, Q^2)}{s}\right|^2 = \frac{4\pi(\alpha' Y + R^2)}{c^2}\left[\gamma - \mathrm{E}_1\left(\frac{z\cos\theta}{\sqrt{1+\beta^2}\cos\varphi}\right) + \right.$$
$$\left. + \mathrm{E}_1\left(\frac{z e^{i\theta}}{2\sqrt{1+\beta^2}\cos\varphi}\right) + \mathrm{E}_1\left(\frac{z e^{-i\theta}}{2\sqrt{1+\beta^2}\cos\varphi}\right) + \ln\left(\frac{z\sqrt{1+\beta^2}\cos\varphi}{4}\right)\right]. \tag{11}$$

3. Results

Our aim was to obtain parameters of this theory by a simultaneous fit of experimental data on total and elastic cross sections of proton–proton and proton–antiproton scattering; the energy range is from $\sqrt{s} = 3.5$ GeV to $\sqrt{s} = 8$ TeV. The data were taken from the Particle Data Group.[2] The result of the fit is shown in Figure 1 for the total cross section and in Figure 2 for elastic cross sections, and the values of parameters are listed in Table 1. The overall ratio of χ^2 to number of degrees of freedom for the fit equals to 1.5.

Table 1. Parameters of simultaneous fit

Parameter	Value
Vacuum parameters	
g	21.0 ± 0.3 $\mathrm{mb}^{1/2}$
Δ	0.10 ± 0.01
c	0.8 ± 0.2
α'	0.24 ± 0.01 GeV^{-2}
R^2	2.8 ± 0.1 GeV^{-2}
Non vacuum parameters	
g_f	70.6 ± 1.2 $\mathrm{mb}^{1/2}$
g_R	44.6 ± 1.8 $\mathrm{mb}^{1/2}$
Δ_f	0.38 ± 0.01
Δ_R	0.61 ± 0.01

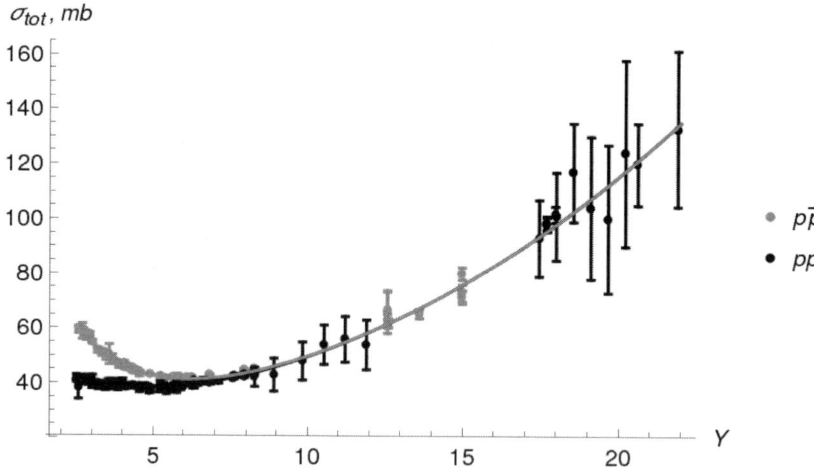

Fig. 1. Fitting of total cross sections of pp and $p\bar{p}$ collisions

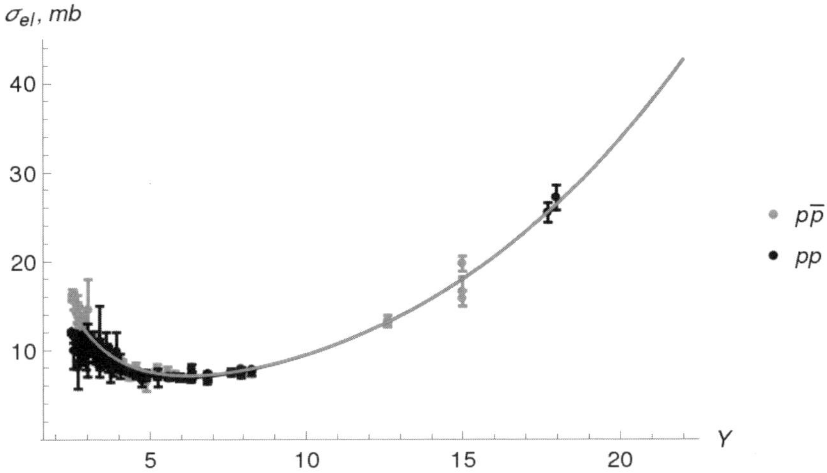

Fig. 2. Fitting of elastic cross sections of pp and $p\bar{p}$ collisions

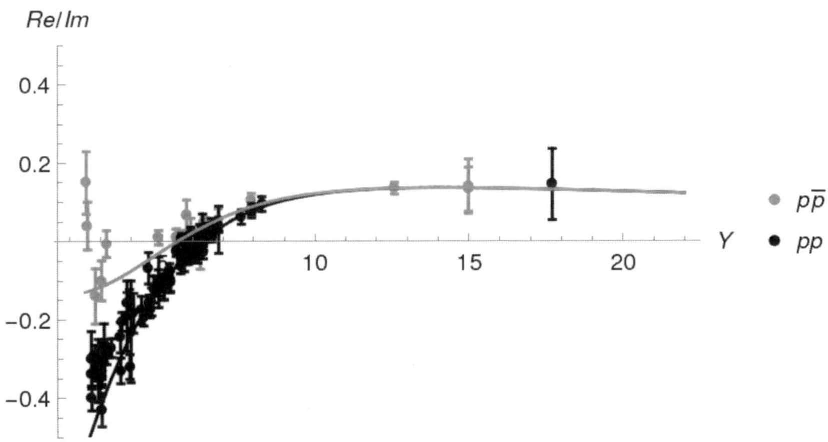

Fig. 3. Ratio of real to imaginary parts of elastic scattering amplitude of pp and $p\bar{p}$ scattering with calculated curve collisions

4. Conclusion

We have shown one of the possible sets of Regge parameters which simultaneously describe the total and elastic cross section. This set does not describe differential elastic cross sections. Such a problem requires further investigation.

The value of the shower enhancement coefficient is an interesting result. In contrast with its definition by formula (3), where $c > 1$, we have obtained in our fit the value $c = 0.8 \pm 0.2$. We can define the parameter c in different ways. In paper[3] the distribution of number of parton chains was derived

$$\begin{aligned} w_m &= \frac{1}{c}\exp(-\lambda c)\frac{(-\lambda c)^m}{m!}, \quad m > 0 \\ w_0 &= \frac{1}{c}\left[\exp(-\lambda c) - 1 + c\right] \end{aligned} \tag{12}$$

where $\lambda = g^2 e^{\Delta Y}/\sigma_0$. Here σ_0 is the absorption cross section of a parton chain by a target in a laboratory reference frame. In this way the distribution is characterized by two parameters, $\langle m \rangle = \lambda$, which means the number of parton chains and $\frac{\langle m(m-1)\rangle}{\langle m \rangle^2} = c$. With this definition it is possible to have c less than unity.

This work is supported under RSCF grant No.15-11-10023.

References

1. V.N. Gribov, *JETP* **53**, 654 (1967).
2. K.A. Olive *et al.* (Particle Data Group), *Chin. Phys.* **C 38**, 090001 (2014)
3. V.A. Abramovsky, E.V. Gedalin, E.G. Gurvich, and O.V. Kancheli, *"Inelastic interactions at high energies and QCD"*. Metzniereba, Tbilisi, 1986, pp. 111–112.

V. Abramovsky and N.V. Abramovskaya, Chernogolovka, 2015.

THE GRIBOV LEGACY, GAUGE THEORIES AND THE PHYSICAL S-MATRIX

ALAN R. WHITE
Argonne National Laboratory, Il 60439, USA.

Reggeon unitarity and non-Abelian gauge field copies are focused on as two Gribov discoveries that, it is suggested, may ultimately be seen as the most significant and that could, in the far distant future, form the cornerstones of his legacy. The crucial role played by the Gribov ambiguity in the construction of gauge theory bound-state amplitudes via reggeon unitarity is described. It is suggested that the existence of a physical, unitary, S-Matrix in a gauge theory is a major requirement that could even determine the theory.

1. Introduction

Volodya Gribov was a brilliant theoretical physicist, ranking amongst the greatest of the twentieth century. The list of his discoveries and achievements is very long and is generally well-known. In this paper* I will focus on just two discoveries that, from my own work, I think are the most significant and that I think could, in the far distant future, form the cornerstones of his legacy. They are reggeon unitarity and the Gribov ambiguity due to non-Abelian gauge field copies. These discoveries were made well over 10 years apart and superficially involve very different physics. Both may be even more profound than is generally perceived and I will link them by elaborating the crucial role played by the Gribov ambiguity in my construction of gauge theory bound-state amplitudes via reggeon unitarity. I will suggest that the existence of a physical, unitary, S-Matrix in a gauge theory could be a highly non-trivial constraint that may even determine the theory.

*I was very sorry to miss the Gribov-85 conference, primarily because I could not obtain a visa. I am very happy to have this opportunity to express my deep admiration for Gribov's work.

2. Reggeon Unitarity

This was the outcome of a particularly intense collaboration between two equally brilliant physicists, Gribov and Pomeranchuk, during a period of several years shortly before Pomeranchuk's death. It was a stunning leap forward. They took the low-order, already elaborate, t-channel unitarity calculations of Feynman diagrams by Mandelstam and built a beautiful multiparticle complex angular momentum formalism[1] that led to the existence of a unitarity condition for all multiple Regge pole states. I will describe what I think are the most dramatic implications, not all of which are generally appreciated, but first I will discuss some early developments that followed the initial discovery.

2.1. *Analyticity Problems*

The GPT analysis[1] made fundamental technical assumptions about the complex angular momentum and helicity continuations of multiparticle amplitudes. Subsequently, these assumptions seemed to be invalidated when the analyticity properties of the simplest (relevant) Feynman production amplitudes were investigated.[2] Unfortunately, perhaps, this convinced Gribov that because of the complicated Landau singularity structure of multiparticle momentum space amplitudes, the complex angular momentum continuation of multiparticle t-channel unitarity could not work as assumed by GPT. As a result, he turned to field theory based models and formulated the Reggeon Calculus,[3] which was effectively an explicit solution of reggeon unitarity. Regrettably, the apparent model dependence of the formulation allowed the theory community at large to believe that the physical relevance of the reggeon calculus, and the equivalent reggeon field theory,[4] was an open question.

2.2. *Simplified Asymptotic Singularity Structure*

Gribov remained convinced, nevertheless, that even though the complex angular momentum and helicity continuations assumed might not exist, the GPT analysis was valid near the Regge cuts (angular momentum plane branch-points) associated with multiple Regge pole exchanges. He was unable to justify this but from my own work, in particular my collaboration with Henry Stapp on asymptotic dispersion relations,[5] I had understood that multiparticle complex angular momentum theory should be based on the simplified asymptotic analytic structure of many particle amplitudes in multi-Regge regions. The asymptotic dispersion relations break up amplitudes into component parts that each have distinct asymptotic cut structures (as determined by the Steinmann relations or, in essence, according to the time-ordering of interactions). The component amplitudes then have distinct complex angular momentum and helicity continuations[6] that can be used in a more elaborate version of the GPT analysis. Ultimately, this analysis showed that Gribov's faith in the GPT discontinuity formulae was justified.

2.3. *Implications of Reggeon Unitarity*

The reggeon unitarity discontinuity formulae have an analogous structure to the momentum space unitarity equations and, similarly, are only well-defined when the basic complex angular momentum singularities are Regge poles with factorizing residues. Extensive t-channel analytic continuation of the unitarity equations can then be used to show[7] that Regge poles and Regge cuts are the only singularities remaining (on the physical sheet) near $t = 0$ and so they control the s-channel high-energy limit. Most importantly, the complex angular momentum and helicity formalism based on asymptotic dispersion relations shows[6] that the reggeon unitarity equations extend to all multi-Regge regions of multiparticle amplitudes. Consequently, if the basic angular momentum plane singularities of physical amplitudes are Regge poles then a complete asymptotic solution of multiparticle t-channel unitarity can be obtained. In particular, reggeon unitarity can be applied to the study of bound-state amplitudes via multi-Regge theory.

Basic singularities that are not Regge poles are analogous to massless particles in momentum space in that they do not give well-defined multiple singularity contributions. As a result they are unable to provide an angular momentum plane solution of multiparticle t-channel unitarity. I believe that, eventually, the asymptotic dispersion relation based multi-Regge theory that I have formulated will attract substantial general interest, and will undergo the exposure and further development that is probably necessary for reggeon unitarity to become part of the established basis of particle physics theory. If this happens, the implications will be dramatic and will include the following.

2.3.1. *The Physical Pomeron*

Reggeon unitarity implies that the leading singularity in the vacuum channel must be a Regge pole(s) together with associated Regge cuts. If total cross-sections do not fall asymptotically, there must be a vacuum Regge pole with unit intercept and indeed, experimentally, there is very strong evidence[8,9] that the pomeron is approximately a single Regge pole. In this case, all the associated Regge cuts will also have unit intercept and the only known solution of reggeon unitarity is the Critical Pomeron.

2.3.2. *Gauge Theories*

In a general non-Abelian gauge theory, and QCD in particular, the conventional assumption is that perturbative parton model calculations, at large transverse momentum, can be extended into the Regge region. It being assumed that confinement is a small transverse momentum phenomenon that can be ignored at sufficiently large transverse momentum. In leading order this leads to a two (gluon) reggeon cut which, after iteration of interactions (via a scale-invariant kernel) produces the BFKL pomeron as the leading singularity. As is widely known, it is an angular momentum cut[11] which has no possibility to satisfy reggeon unitarity. Consequently, if the physical S-Matrix satisfies reggeon unitarity, it can not contain the BFKL pomeron as a physical singularity.

2.3.3. *Quantized Gravity?*

If the graviton is a Regge pole, because it has spin two the Regge pole intercept also has to be two. Reggeon unitarity then requires a two graviton Regge cut with intercept three. In addition, multiple graviton Regge cuts must also exist. They have analogous intercepts that are increasing integers and that will produce corresponding power increases of the total cross-section. Since reggeon unitarity is an exact S-Matrix property, it does not allow Regge cut contributions to be removed by summation, for example, into an eikonal amplitude. Unavoidably, if the graviton is a Regge pole, the forward amplitude is not polynomially bounded. Consequently, gravity can not be quantized and described by an S-Matrix with conventional analyticity properties, that satisfies reggeon unitarity. Most likely, from my viewpoint, gravity is not quantized.

2.4. *Massive Gauge Theory Reggeons*

In a spontaneously-broken gauge theory, with a suitably chosen symmetry breaking scalar sector, all the elementary gauge bosons and fermions acquire masses, while also becoming Regge poles (reggeizing) in low-order perturbation theory. Therefore, reggeon unitarity should be satisfied and, indeed, as higher-orders are constructed (at high-energy), the multi-reggeon diagrams appear[10] that are predicted. Using the broad applicability of reggeon unitarity discussed above, the structure of reggeon diagrams in general multiparticle multi-Regge limits can then be predicted. In principle, the contributions of arbitrarily high order Feynman diagrams are included. In the next Section, I will discuss the massless limit for reggeon amplitudes and the pattern of infra-red divergences that occur. Because the Regge region includes low transverse momenta we could expect to discuss possible bound-states that are a consequence of the divergences. Moreover, because of the range of multi-Regge limits that can be considered, bound-state scattering amplitudes should also be accessible. As I will discuss in the next Section, the Gribov ambiguity, implied by the existence of gauge field copies, plays a crucial role in the massless limit and in the formation of bound-state scattering amplitudes.

3. The Gribov Ambiguity

When Gribov first described the phenomenon of non-Abelian gauge field copies to a western audience (at the 1977 EPS conference in Budapest), it caused much excitement. Most western physicists were, of course, unaware that Gribov had previously announced[12] his discovery at the Leningrad Winter School. At first the copies seemed only to require a straightforward modification[13] of the functional space in a gauge theory functional integral. Moreover, as Gribov himself suggested,[13] it seemed that this might produce the confinement of color, via the growth of the coupling, that the theory community was desperate to discover.

Unfortunately, perhaps, in the nearly forty years since Gribov's discovery, no satisfactory explicit proposal for how the functional integral is to be handled has appeared. Moreover, there is no acceptance of the idea that eliminating the copies will actually produce the conventionally desired confinement. Instead, the copies may actually imply that there is no well-defined non-perturbative definition of a functional integral giving correlation functions for a non-Abelian gauge theory. In reality, as is well-known, the infinite volume divergence of the integral is already a big threat to it's definitive existence. However, without a well-defined function space, it seems that it can not provide even a formal vehicle for the non-perturbative definition of the theory.

I will, instead, focus on Gribov's simple statement[12] of the basic problem,

"in distinction to electrodynamics, in non-Abelian theories it is not possible to uniquely introduce three-dimensional transverse fields (in particular, transverse fields can be pure gauge)."

The vectors in massive reggeon diagrams contain three-dimensional transverse components and so if we can take the massless limit in a well-defined way we might obtain a unique result that satisfies reggeon unitarity. There are a variety of non-trivial issues involved in the taking of this limit and the crucial role of the Gribov ambiguity emerged (in my work) as a necessary component of a gauge theory origin for the Critical Pomeron. However, I will initially give a more general description of the phenomena involved.

3.1. *Partial Gauge Symmetry and Reggeon Anomaly Vertices*

With a transverse momentum cut-off, the general expectation would be that the reggeon massless limit will produce a cancellation of infra-red divergences in color zero channels and exponentiation (reggeization) of divergences in color non-zero channels. However, if massless fermions are present and we first **restore only** an SU(2) sub-group of a larger gauge symmetry for the vector reggeons, a new phenomenon emerges that has far-reaching ramifications.

In addition to the cancellation of reggeization divergences, an overall infra-red scaling divergence appears in SU(2) color zero amplitudes that contain combinations of "anomalous wee gluon reggeons" coupled via **anomaly pole reggeon vertices**. The wee gluon reggeon combinations are multi-reggeon (non-local) generalizations of the well-known anomaly current. They are massless, with overall SU(2) color zero, anomalous color parity ($\neq$ signature), and all carrying zero transverse momentum. The divergence selects physical reggeon amplitudes. Because it is an overall divergence, the states and interactions appear simultaneously and so vacuum-like "universal wee partons" are necessarily present in all states and interactions.

Anomaly poles[14] are produced by chirality transitions in massless fermion triangle diagrams. They appear in triangle diagram reggeon vertices that contribute in non-planar multi-Regge limits where bound-state amplitudes appear. The vector and axial vector couplings involved have two sources. Firstly, when multiple massless vector (gluon) reggeons are present in each of three distinct (non-planar) rapidity channels and couple into the same fermion triangle vertex, orthogonal γ-matrices accumulate to produce a γ_5. Secondly, on-shell **massive** vector exchange between a massless fermion/antifermion reggeon pair produces an effective vector vertex that couples to triangle anomalies. A pseudoscalar anomaly pole can couple, therefore, to a reggeon state composed of a fermion/antifermion pair and anomalous wee gluons.

As the full gauge symmetry is restored (in the specific cases we discuss below) there is an exploitation of the Gribov ambiguity. The zero momentum transverse component of the on-shell massive vector exchanged between the fermion/antifermion pair, in pseudoscalar anomaly poles, does not decouple as it becomes massless. These anomaly poles become the pseudoscalar Goldstone bosons associated with chiral symmetry breaking. They are also essential components of the fermion bound-states. Therefore, the Gribov ambiguity is fundamental in the formation of the spectrum of physical states that appear in the unbroken gauge theory.

3.2. *Full Gauge Symmetry With Only Massless Fermions*

The transverse momentum cut-off has to be removed first if the full massless limit is to be well-defined. This can be done smoothly if the scalar sector to be decoupled is asymptotically free. To achieve this, many fermions are needed to slow down the evolution of the gauge coupling - the simplest example being QCD with the asymptotic freedom constraint saturated. An even more restrictive constraint is that the anomaly pole coupled infra-red divergence has to be preserved in all orders of perturbation theory. This requires that the fermions are massless and produce an infra-red fixed point.

3.2.1. *Massless $QCD_S \equiv 6$ triplet quarks + 2 sextet quarks*

This[14–16] is the only, physically relevant, possibility to saturate QCD with massless fermions. A pair of color sextets (that can be associated with electroweak symmetry breaking) is added to the physical triplet sector. We initially consider color superconducting QCD_S - with the gauge symmetry broken to SU(2) by an (asymptotically free) color triplet scalar field. As implied above, when the overall divergence is subtracted to define physical amplitudes we obtain an SU(2) color confining "parton model" in which anomalous wee gluons provide a vacuum-like "wee parton condensate" within states and interactions. Anomaly poles produce chiral "pion" poles (containing both fermion pairs and fermion/antifermion pairs) and also couple wee gluons present in distinct reggeon channels. There is both confinement and chiral symmetry breaking!

The exchanged pomeron interaction is a massive, SU(2) singlet, gluon reggeon accompanied by the wee gluons. It is exchange degenerate with the gluon reggeon and can be identified with the supercritical pomeron. Consequently, the massless limit restoring SU(3) color is described[14–16] by the Critical Pomeron. Many experimentally desirable features appear.

- *Bound-states are pseudoscalar mesons and baryons containing triplet and sextet quarks, but with no hybrid sextet/triplet states,*
- *Anomaly color factors give larger masses and x-sections for sextet states.*
- *There are no glueballs, no BFKL pomeron, and no odderon.*
- *The Critical Pomeron gives maximal parton model factorization.*

The infra-red fixed-point implies off-shell amplitudes have to be scale invariant and so can not contain massive states. Therefore, if a bound-state S-Matrix containing massive particles exists, there must be **no off-shell correlation functions!** Indeed, the existence of off-shell amplitudes seems very unlikely to be consistent with the anomaly pole dynamics. Nevertheless, a huge chiral symmetry implies the S-Matrix necessarily has many massless Goldstones that would pose a very serious, if not insoluble, infrared problem.

At this point, it seems questionable that the Critical Pomeron can appear in a physical S-Matrix that has only massive states (as in the real world!) Fortunately, the electroweak interaction provides a way out of this dilemma. A remarkable result emerges when we consider combining the electroweak interaction with QCD_S. We discover a unique theory[14,16–21] that is again massless, asymptotically free, and saturated with fermions, but has left-handed couplings to all fermions.

3.2.2. *QUD*

Asking for asymptotic freedom and no short-distance anomaly, massless QCD_S and the electroweak interaction together embed uniquely in

QUD ≡ SU(5) gauge theory with left-handed massless fermions in the $5 \oplus 15 \oplus 40 \oplus 45^*$ representation.

The asymptotic freedom constraint is again saturated and there is an infrared fixed-point. Now, however, there are no exact chiral symmetries and so all bound-states can acquire S-Matrix masses via interactions. Amazingly,

QUD has the additional structure needed to generate a bound-state S-Matrix that produces all the physics of the Standard Model via massless fermion anomaly dynamics.

The only "beyond the Standard Model" elements are a dark matter sector and neutrino masses, both of which are extremely welcome.

There are three "generations" of both elementary leptons and elementary triplet quarks, and the theory is vector-like with respect to SU(3)xU(1)$_{em}$. The SU(2)xU(1) quantum numbers are not quite right but, in the S-Matrix constructed[9,15,19] via multi-Regge theory, all elementary fermions are confined and only Standard Model interactions and states emerge. The color sextet sector provides "sextet pions" that produce electroweak symmetry breaking and sextet baryons - with the sextet neutron and antineutron providing stable, massive, **dark matter** particles. There is also a color octet sector that is responsible, via large $k_\perp$ anomalies, for the generation structure of the physical states. We can briefly summarize how we obtain QUD bound-state amplitudes from reggeon diagrams as follows.

3.2.3. *QUD Reggeon Diagrams*

QUD reggeon diagrams are initially defined with massive reggeons. Fermion masses are removed first, then the gauge boson global symmetries are restored through a sequence of fundamental representation scalar decouplings

$$\to SU(2)_C \ , \to SU(4), \ \lambda_\perp \to \infty, \ \to SU(5)$$

The last scalar is asymptotically free and so the $\lambda_\perp \to \infty$ limit can be taken between the SU(4) and SU(5) limits. The SU(2)$_C$ symmetry restoration produces anomaly poles, but with a difference. As in QCD$_S$, the chirality transitions left by the fermion mass removal do not conflict with the **vector** gauge symmetry but, crucially, they do break the **non-vector** part of the gauge symmetry. The SU(5) symmetry is broken to SU(3)$_C\otimes$U(1)$_{em}$, but **in the reggeon anomaly vertices only.** The final "universal wee partons" are combinations of vector coupling anomalous wee gauge bosons in the adjoint SU(5) representation. The global SU(5) symmetry is restored within the reggeon interactions between anomaly vertices. The **Gribov ambiguity is essential** for the breaking, **via the anomaly vertices,** of the SU(5) symmetry in the bound-state S-Matrix and, in particular, for the electroweak vector boson masses that are generated by mixing with the sextet pions.

3.2.4. *QUD Interactions and States*

As I have described in my papers, much of the Standard Model is clearly present in the QUD bound-state amplitudes extracted from the reggeon diagrams. Indeed, most likely, it is all there but much remains to be better understood. Standard Model vector interactions between bound-states are

- *Critical Pomeron ≈ SU(3) gluon reggeon + wee gauge bosons.*
- *Photon ≈ a* $U(1)_{em}$ *gauge boson + wee gauge bosons*
- *Weak Interaction ≈ l-handed gauge bosons mixed with sextet pions +wee gauge bosons.*

The elementary QUD coupling stays very small because of the infra-red fixed-point. The Standard Model couplings are enhanced by anomaly color factors that imply

$$\alpha_{QCD} > \alpha_{em} \gg \alpha_{QUD} \sim 1/120\,.$$

Bound-states contain elementary fermions that combine with color octet "pions" (present at infinite light-cone momentum as an anomaly contribution) to form SU(5) singlets. Standard Model generations of physical hadrons and leptons appear. Anomaly vertex mixing, combined with fermion and wee parton color factors, produces a wide range of mass scale, with the very small QUD coupling, presumably, the origin of small neutrino masses. In general, anomaly color factors imply

$$M_{hadrons} \gg M_{leptons} \gg M_{\nu' s} \sim \alpha_{QUD}\,.$$

There are three generations of lepton bound states. Each physical lepton contains three elementary leptons, with two of them originating from an anomaly pole. The electron is almost elementary since, in effect, the anomaly pole disturbs the Dirac sea minimally. The muon has the same constituents, but in a different dynamical configuration that will obviously generate a significant mass.

We can summarize hadronic sector results, very briefly, as follows.

- *Two triplet generations of Standard Model hadrons that mix.*
- *The physical b quark is a mixture of all three QUD generations.*
- *Sextet pion/vector boson mixing → electroweak symmetry breaking.*
- *Sextet neutrons are stable and provide dark matter.*
- *l-handed "top quark" mixes with exotic quarks, no low mass states.*
- *Triplet/sextet mixing gives mixed-parity* η_3 *and* η_6 *scalars.*
- η_6 *gives Standard Model "top physics" with sextet mass scale.*
- η_3 *could be the "Higgs boson".*

More details of these results can be found in my papers. Theoretical virtues for the QUD origin of the Standard Model are also listed in several of the papers. A brief list is

- *Strong/electromagnetic/weak interaction parities explained.*
- *Confinement/chiral symmetry breaking/the parton model/the Critical Pomeron all appear in QCD.*
- *The massless photon partners the "massless" Critical* $I\!P$.
- *Anomaly vertices/wee parton color factors give a wide range of scales and masses, with neutrino masses* $\sim$ *very small* α_{QUD}.
- *The only new physics is a high mass strong interaction giving electroweak symmetry breaking and dark matter.*
- *Particles and fields are truly distinct. Hadrons and leptons have equal status.*
- *Symmetries and masses are dynamical S-Matrix properties. There are no off-shell amplitudes and there is no Higgs field.*
- *Einstein gravity is induced with zero cosmological constant. Gravity is not quantized !! Particle Physics is described by an S-Matrix.*

Obviously, it would be incredible if the Standard Model has the underlying simplicity of QUD. It is important to emphasize again that there is no freedom for variation. It is an "all or nothing" explanation of the origin of the Standard Model which predicts the, apparently observed, "nightmare scenario" of a "Higgs boson" produced at the LHC without other new short-distance physics! As is also described in my papers, there is much suggestive cosmic ray and accelerator experimental evidence for the electroweak scale strong (not short-distance) QCD interaction of the sextet quark sector. Particularly striking are[22] the latest AMS-02 results and the most recent observation of dark matter strong self-interaction properties.

4. Conclusions and Outlook

I was led to the QUD bound-state S-Matrix by looking for a gauge theory origin for the Critical Pomeron - an S-Matrix phenomenon formulated far from the framework of quantum field theory. It could well be, as I have suggested,[9] that the Gribov copies are a major reason why the path-integral fails to provide a non-perturbative formulation of a non-Abelian gauge theory. In this case, the field theory can be well-defined perturbatively and yet the particle S-Matrix could be the only well-defined non-perturbative element that exists.

I have argued that the Critical Pomeron high-energy S-Matrix is uniquely associated with QUD. I have emphasized that QUD is self-contained and must reproduce the established physics of the Standard Model, or else it is simply wrong ! The potential scientific and aesthetic importance of QUD, as an underlying unifying massless field theory, is overwhelming. So can the necessity for the underlying theory to be QUD be seen without the Critical Pomeron?

The infra-red fixed-point is required to enhance infra-red fermion anomaly interactions and for the color-superconductivity starting point that provides the exploitation of the Gribov ambiguity. The vector gauge group has to be as large as SU(3) to produce a universal wee gluon distribution, as well as an infinite momentum "parton model" giving an ultra-violet finite S-Matrix. If the vector gauge group is larger than SU(3), the universality of the wee gluon distribution is lost. Left-handed interactions, that acquire a mass via infra-red anomalies, and also generate bound state masses, can be added. Asking that all bound-states acquire masses, plus no short-distance anomaly leads, perhaps uniquely, to QUD. So, it could be that

The Standard Model is reproducing the unique, unitary, S-Matrix ??

If it is eventually established that there is indeed a unique S-Matrix and the Gribov discoveries of reggeon unitarity and gauge field copies are the fundamental elements that determine it's existence then it will be a dramatic addition to the legacy of Volodya Gribov.†

5. Epilogue

As is surely evident, there is an enormous amount of work needed to fully develop my multi-Regge construction of the QUD S-matrix into a practical formalism. Far more than I could ever do on my own.‡ I have given[21] a partial list of the multi-Regge developments that are needed. Once the diagrams have been fully characterized, there is still the introduction of scales, including the regularization of the anomaly vertices and the description of the Critical Pomeron limit. Given the disappearance of the intense scientific environment that produced Gribov (a bizarre consequence, mainly, of

†Even though it is very doubtful that he would have actually endorsed my argumentation.
‡Despairing over the situation I wrote to Bj, *"What I really need and, of course, have zero chance of getting, is a "young Bjorken" to work with me!* Bj replied *"Much better would be a young Gribov"* I replied *"If only !!!!!!"*

Soviet repression), and the current focus of the field on inventive intellectual forays that are far away from unsolved hard core physics problems, it is difficult to imagine where and when the effort will be made to develop all the details of the physical picture I have outlined.

The most serious obstacle is, of course, that a major change of theory paradigm is involved. I am saying that quantum fields exist only as short-distance, gauge-dependent, fluctuations for which the only manifest physical reality is the particle S-Matrix. This is in stark contrast with the concepts and philosophy of the vast majority of today's "Beyond the Standard Model" research. Nevertheless, if it is reality, it will eventually emerge and, I at least, can imagine that with his scientific background, Gribov would not have been entirely unsympathetic!

References

1. V.N. Gribov, I.Ya. Pomeranchuk and K. A. Ter-Martirosyan, *Phys. Rev.* **139B**, 184 (1965).
2. A.A. Anselm, Y.I. Azimov, G.S. Danilov, I.T. Dyatlov, V.N. Gribov, *Annals of Phys.* **37**, 227 (1966).
3. V.N. Gribov, *Sov. Phys.* **JETP 26**, 414 (1968), [*Zh. Eksp. Teor. Fiz.* **53**, 654 (1967)].
4. H.D. I. Abarbanel, J.B. Bronzan, R.L. Sugar, A.R. White, *Phys. Rept.*, **21**, 119 (1975).
5. H.P. Stapp "Many-particle Dispersion Relations". In R. Balian and D. Iagolnitzer (eds.), *Structural Analysis of Collision Amplitudes*, pp 191-274. North Holland (1976), H.P . Stapp H and A.R. White *Phys. Rev.* **D26**, 2145 (1982).
6. A.R. White "The Analytic Foundations of Regge Theory." In R. Balian R and D. Iagolnitzer (eds.), *Structural Analysis of Collision Amplitudes*, pp 427-503. North Holland (1976), *Int. J. Mod. Phys.* **A11**, 1859 (1991), *Phys. Rev.* **D58**, 074008 (1998).
7. Y.A. Simonov, *Zh. Exp. Teor. Fiz.* **48**, 242 (1965) [*Sov. Phys. JETP* **21**, 160 (1965)].
8. TOTEM Collaboration, G. Antchev et al. "Measurement of proton-proton elastic scattering and total cross-section at $\sqrt{s} = 7$ TeV" CERN-PH-EP-2012-239, *Europhys. Lett.* **101** (2013) 21002.
9. A.R. White, "The Nightmare Scenario and the Origin of the Standard Model. "We Got it Wrong ...How did we misread the signals? ... What to Do?"", arXiv:1406.1441 [hep-ph].

10. J.B. Bronzan and R.L. Sugar, *Phys. Rev.* **D17**, 585 (1978).
11. V.S. Fadin, E.A. Kuraev, L.N. Lipatov, *Phys. Lett.* **B60**, 50 (1975).
12. V.N. Gribov, Lecture at the 12th Winter School of the Leningrad Nuclear Physics Institute, translated by J. Bartels and W. Nahm *CERN preprint*, H.D.I. Abarbanel *SLAC-TRANS-176.*
13. V.N. Gribov, *Nucl. Phys.* **B139**, 1 (1978).
14. A.R. White, "Could a Weak Coupling Massless SU(5) Theory Underly the Standard Model S-Matrix", 2011qcb..conf..496W. (2010)
Anomaly poles are described in the appendix to this paper published in the archive version arXiv:1009.4850 [hep-ph].
15. A.R. White, *Phys. Rev.* **D72**, (036007 (2005), arXiv:hep-ph/0412062.
16. A.R. White, "New Physics at the Tevatron and the LHC May Relate to Dark Matter Visible in UHE Cosmic Rays", arXiv:1106.5662 [hep-ph] (2011).
17. K. Kang and A.R. White, *Int. J. Mod. Phys.* **A2**, 409 (1987).
18. A.R. White, "The Potential Importance of Low Luminosity and High Energy at the LHC", arXiv:1310.6649 [hep-ph] (2013).
19. A.R. White, "A Higgs-Like Scalar With Mixed-Parity and Standard-Model-Like Couplings Could be Further Evidence for Underlying Massless SU(5) Unification", arXiv:1301.5628 [hep-ph] (2013).
20. A.R. White, "A Pseudoscalar Resonance That Could Resemble the Higgs", arXiv:1206.0192 [hep-ph] (2012).
21. A.R. White, "Past, Present, and Future Multi-Regge Theory", arXiv:1403.4454 [hep-ph] (2014).
22. A.R. White, "AMS-02, Strongly Self-Interacting Dark Matter, and QUD", arXiv:1409.6250 [hep-ph] (2014).

COLOR FLUCTUATIONS IN HIGH ENERGY HADRON- AND PHOTON-NUCLEUS COLLISIONS

LEONID FRANKFURT

Particle Physics Department, School of Physics and Astronomy, Tel Aviv University, 69978 Tel Aviv, Israel

MARK STRIKMAN

Department of Physics, the Pennsylvania State University, State College, PA 16802, USA

We explain that coherence of high energy QED and QCD processes implies existence of new kind of phenomena which are beyond a framework based on Regge poles (cuts). New phenomena emerge as the consequence of compositeness of the bound states and the Lorentz slowing down of interaction. We focus on the color fluctuations phenomena predicted earlier for pA collisions within QCD and recent evidence for this phenomenon from pA LHC run, significant modification of nuclear shadowing phenomenon in the diffractive photoproduction of vector mesons observed recently in the ultra peripheral collisions at LHC. We outlined briefly general properties of color fluctuations phenomena and perspectives of future studies of this phenomenon in electron (photon) collisions with nuclei.

1. Introduction

We explain in this paper that coherence of high energy processes leads to serious modifications of semiclassical approximation, to a variety of new phenomena in QCD and QED. To visualize physics of coherent phenomena we consider two phenomena observed recently at LHC.

The Glauber model[1] for the hadron-nucleus interactions was first suggested within the non-relativistic Schrödinger equation framework with hadron-nucleon interactions described by a potential. The model provided a high precision description of total and elastic cross sections of hadron-nucleus scattering at intermediate energies ($\sim$ 1 GeV).

Applicability of the model for higher energies looks problematic. First, the interactions at high enough energies are predominantly inelastic and cannot be described by a potential. Second, a substitution of the potential exchange by a high energy hN amplitude requires considering of a planar

Feynman diagrams. However it was demonstrated in Refs. 2, 3 that the contribution of planar diagrams corresponds to the amplitudes decreasing with energy as $1/s$. The physical reason for this is that a sufficiently energetic hadron being a composite particle can be considered as the superposition of configurations of constituents described by wave function of a hadron. Due to the Lorentz slowing down of interaction such configurations are frozen during the coherent length/time interval,

$$l_{coh} = \frac{2P_h}{m^2_{conf} - m^2_h}\,, \tag{1}$$

when coherence length is much larger that the nucleus size. Thus at high energies interaction of the projectile hadron with a target is given by the sum over the interactions of frozen configurations of constituents of the projectile. This is the origin of coherence phenomena specific for high energy processes.

Note here that at low energies inter-hadron interactions involve mostly their surfaces. As a result a theory of low energy dynamics may ignore inner structure of hadrons. This is qualitatively different from the high energy scattering where inelastic collisions dominate and as the result of coherence the inelastic interaction depends significantly on the 3D structure of the colliding hadrons. Therefore approximations which ignore inner structure of hadrons such as concepts of pions as Goldstone bosons, Regge pole factorization, string tension which is the same for all quark-gluon configurations are insufficient for the theoretical description of the high energy processes.

First step to account for coherent contribution of inelastic diffractive processes in the intermediate states has been made by Gribov before the advent of QCD. He demonstrated[4] how the contribution of the diffractive nonplanar diagrams which survives at high energies can be expressed for the case of scattering off the deuteron through the cross section of the diffractive processes at $t \approx 0$. For the total cross section of hadron-nucleus scattering the Gribov result for double scattering differs from the Glauber shadowing correction by the factor

$$1 + \left.\frac{\dfrac{d\sigma(h+p \to X+p)}{dt}}{\dfrac{d\sigma(h+p \to h+p)}{dt}}\right|_{t=0} \equiv 1 + \omega_\sigma, \tag{2}$$

where a sum over diffractively produced states X (which differ from the elastic one), including the triple Pomeron contribution is implied.

To calculate the contribution of the diagrams involving interactions with $\nu > 2$ nucleons one needs to know diffractive production of inelastic states off several nucleons which cannot be extracted from the data on hN interaction.

Exploring QCD allows alternative derivation of the Gribov result in terms of quark-gluon wave function of sufficiently energetic projectile hadron and generalizing it to the scattering off any nuclear target. The starting point is the observation that a projectile hadron scatters off the nucleus in different configurations of constituents, each of which is frozen during the passing through the nucleus. Hence the strength of its interactions with nucleons along its path should remain constant and it is legitimate to introduce probability distribution for the hadron to be in configuration with the interaction strength σ – $P_h(\sigma)$.[a] The sum rules

$$\int P_h(\sigma)d\sigma = 1\,, \quad \int P_h(\sigma)\sigma d\sigma = \langle\sigma\rangle \equiv \sigma_{tot}, \tag{3}$$

follow from probability conservation and from the definition of the configuration-averaged cross section. The variance of the distribution is given by

$$\left\langle\sigma^2\right\rangle / \left\langle\sigma\right\rangle^2 - 1 = \omega_\sigma. \tag{4}$$

Eq. (4) follows directly from the optical theorem and the definition of $P_h(\sigma)$, and was first introduced in Ref. 5 by reinterpreting formulae of quasieikonal approximation in the spirit of the approach of Ref. 6.

Within QED more general formulae have been derived for the probability of the propagation of sufficiently energetic positronium through a thin foil of length[7] L which accounts for the distribution over strengths of interaction in terms of square of the wave function of a bound state. The account of coherence of high energy processes leads to the substitution in the case of $l_{coh} > L$ of the textbook expression for the survival probability $\exp(-\sigma\rho L)$ by a function which behaves as $\propto 1/(\sigma\rho L)$ for $\sigma\rho L \gg 1$. These formulae can be directly generalized to QCD by exploring the quark model of hadrons.

Several considerations constrain the shape of $P_h(\sigma)$.[8] For values $\sigma \sim \langle\sigma\rangle$, $P_h(\sigma)$ is expected to have the Gaussian like form which follow from the account of small fluctuations in the number of constituents as well as

[a]The notion of $P_h(\sigma)$ was first introduced in Ref. 5 for pp scattering before advent of QCD. For small σ it was expressed in QCD in terms of the wave function of projectile.[8] This allows to derive QCD dynamics constraints in Ref. 8, see discussion below.

the transverse area occupied by constituents. For $\sigma \ll \langle\sigma\rangle$, configurations with a small number constituents, n_q, should dominate. Account of the properties of the wave function of the projectile hadron suggested in the theoretical analysis of the measured form factors of hadrons allows to predict behavior of $P_h(\sigma)$ for small σ:

$$P_h(\sigma) \propto \sigma^{n_q-2}. \tag{5}$$

The resulting form of $P_h(\sigma)$ represents a smooth interpolation between small and large σ regimes with parameters chosen to reproduce the first three moments of the distribution as given by the data. The analysis[8] of the measurements of coherent diffraction in proton–deuteron collisions with $E_p = 400$ GeV has shown[9] that $P_N(\sigma_{tot})$ is approximately symmetric around $\sigma = \sigma_{tot}$,[8] in agreement with the expected Gaussian form.

Based on these considerations the parametrization of $P_h(\sigma)$ for the proton projectile was suggested in Ref. 8 which we will use below (though many results are mostly sensitive to the value of the variance and the value of the total cross section):

$$P_N(\sigma_{tot}) = \frac{\rho}{\sigma_0}\left(\frac{\sigma_{tot}}{\sigma_{tot}+\sigma_0}\right)\exp\left\{-\frac{(\sigma_{tot}/\sigma_0-1)^2}{\Omega^2}\right\}. \tag{6}$$

Note that Eq. (6) is qualitatively different from $P_N(\sigma)$ suggested in Ref. 5 to describe pN scattering in terms of the pre-QCD idea that only wee partons are involved in the high-energy hadron-hadron interaction. In particular the model of Ref. 5 contained a contribution $\propto \delta(\sigma)$.

The first application of this approach was the calculation of the inclusive cross section of coherent diffraction of nuclei which agrees with the data[10] and study of fluctuations in the heavy ion collisions at fixed target energies.[11]

Below we will review several recent applications of our approach which we refer to as color fluctuation approach since fluctuations of the transverse area occupied by color provide one of the natural sources of the fluctuations of the strength of the interaction.

2. Color fluctuation effects in pA collisions

In the Glauber model the cross section of inelastic interaction of projectile with exactly ν nucleons can be calculated.[12] Formulae account for the finite radius of hN interactions and NN correlations in nuclei. A compact expression for $\sigma_{in}(pA)$ can be written, if internucleon correlations in the nucleus and the finite radius of the NN interaction are neglected[12]:

$$\sigma_{in}^{hA} = \sum_{\nu=1}^{A} \sigma_\nu, \qquad \sigma_\nu = \frac{A!}{(A-\nu)!\,\nu!} \int d^2\vec{b}\, x(b)^\nu \left[1 - x(b)\right]^{A-\nu}, \tag{7}$$

where $x(b) = \sigma_{in} T(b)/A$ and the normalization is $\int d^2\vec{b}\, T(b) = A$.

Fluctuations of the strength of hN interactions obviously result in an increase of fluctuations in the number of wounded nucleons. To account for this effect one needs the distribution over inelastic cross sections, $P_{in}(\sigma_{in})$. This distribution is calculable in terms of $P_N(\sigma)$ since all above discussed restrictions on $P_{in}(\sigma_{in})$ and $P_N(\sigma)$ are the same except for the normalization. Moreover, since experimentally the fraction $1-\lambda$ of the total cross section due to elastic scattering is a rather weak function of the incident energy, it is natural to assume that this is also true for each individual proton fluctuation (This assumption also ensures that for fluctuations close to average interaction at small impact parameters interaction is close to black as it is for the observed cross section.). Thus, the variances of $P_{in}(\sigma_{in})$ and $P_N(\sigma)$ are equal and one can restore $P_{in}(\sigma_{in})$ using the following relation:

$$P_{in}(\sigma_{in}) = \sigma \frac{\sigma_{in}}{\sigma_{tot}} = \frac{\sigma_{in}}{\sigma_{tot}} P_N(\sigma)\,. \tag{8}$$

Using $P_{in}(\sigma_{in})$ we obtain

$$\sigma_\nu = \int d\sigma P_{in}(\sigma) \frac{A!}{(A-\nu)!\,N\nu!} \int d^2\vec{b}\, x(b)^\nu \left[1 - x(b)\right]^{A-\nu}. \tag{9}$$

This formulae resembles formulae of the Gribov Reggeon calculus, where the factor of $x(b)^\nu$ corresponds to ν cut Pomeron exchanges and the factor of $[1-x(b)]^{A-\nu}$ to $A-\nu$ uncut Pomeron exchanges while the integral over σ with weight $P_{in}(\sigma)$ describes the ν-Pomeron-projectile coupling. The important difference is in the account of the distribution over strengths of interactions in Eq. (9).

Eq. (9) is essentially probabilistic, reflecting the semiclassical picture of high energy inelastic interactions with nuclei. The Monte Carlo (MC) which includes accurately both geometry of the NN interactions and nuclear correlations was presented in Ref. 13. The calculation confirms much broader distribution over ν in the color fluctuation approach, see Fig. 1. In the calculation we use parametrization Eq. (6) with $\omega_\sigma = 0.1$ which corresponds to our estimate of ω_σ at the LHC energies using Eq. (2) and limited LHC data on diffraction. To illustrate sensitivity to the value of ω_σ we present also results for $\omega_\sigma = 0.2$.

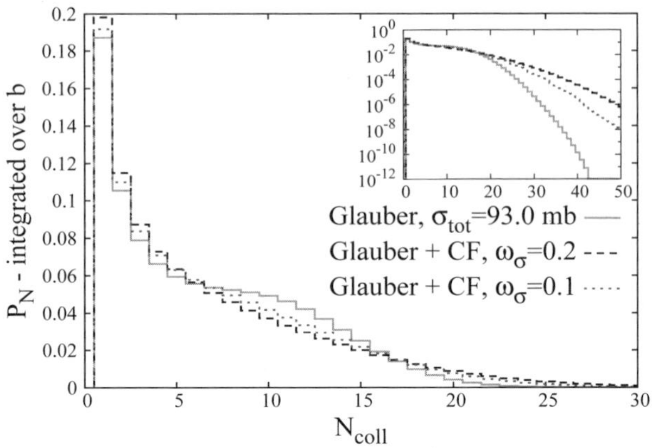

Fig. 1. The probabilities P_N of having ν wounded nucleons, averaged over the global impact parameter b, as a function of ν for the Glauber model ($\omega_\sigma = 0$) and in the CF model with $\omega_\sigma = 0.1$ (our base value for analysis of the LHC data) and $\omega_\sigma = 0.2$. The inset is in the log scale.

The predicted broader distribution over ν is consistent with the analysis of ATLAS which studied dependence of the multiplicity at the central rapidities as a function of transverse energy, $\sum \Sigma_T$ at -4.9 $< \eta <$ - 3.2 in the nucleus-going direction in pA have collisions at $\sqrt{s} = 5.02$ TeV.[14]

An interesting question is how the strength of the interaction for fluctuations correlates with its parton structure. In models of hadrons where few quarks in the rest frame interact with each other through a potential which is Coulomb-like at short distances the size of a given configuration decreases with increasing quark momentum. Thus quarks with large x in the fast frame arise from configurations with large relative momenta in the rest frame and, thus, a smaller size. The density of the gluon field in these configurations is necessarily reduced. Additionally, in QCD the main contribution to the parton density in configurations with $x \gg \langle x \rangle$ is from configurations with the minimal number of partons, leading to the quark counting rules.[15] Hence for such configurations, the nucleons $q\bar{q}$ cloud is suppressed, leading to a reduction of the soft cross-section even on the nonperturbative scale.[16] In the pion case these expectations are corroborated by the observation of color transparency in the coherent diffractive production of two jets at FNAL and color transparency effect in exclusive electroproduction of pions at TJNAF.

To test the presence of the correlation between presence of large-x in the hadron and its strength of interaction with a nucleon it was suggested to study processes proton-nucleus collisions with a hard trigger in the nucleon fragmentation region and study correlation between value of x for the parton in the nucleon, x_p, and number of wounded nucleons.[16] Such studies became possible in pA collisions at the LHC.

Forward jet production in pA collisions was studied by Jet production measurements ATLAS and CMS.[17,18] Studies of inclusive pA collisions at the LHC confirmed the pQCD expectations for the total production rate. At the same time the jet production rate, σ^{hard}, was compared to the Glauber model expectation through the ratio

$$R_{\nu}^{hard} = \left(\sigma_{\nu}^{hard}/\sigma_{\nu}\right) / \left(\nu \cdot \sigma_{NN}^{hard} / \langle\sigma_{in}\rangle\right). \tag{10}$$

Here σ_{NN}^{hard} is the rate of the hard process in NN collisions. Large deviations from the expected $R_{\nu}^{hard} = 1$ were observed for jets produced along the proton-going direction: namely, an enhancement for peripheral (small-ν) collisions and a suppression for central (large-ν) collisions, which compensate each other in the inclusive cross section. These findings are not sensitive neither to finite size effects[19] in the Glauber modeling nor, as explained above, to energy-conservation effects. In the ATLAS data,[17] R_{ν}^{hard} is presented as a function of the fraction of the energy of the proton carried by the jet $z = E_{jet}/E_p$ which, in the forward kinematics of interest, practically coincides with the x_p of the parton in the proton involved in the hard interaction.[b] The ATLAS data demonstrate that for fixed energy release ΣE_T in the nuclear hemisphere, R_{ν}^{hard} is predominantly a function of z and not of the jet p_t or rapidity alone. For large z, R_{ν}^{hard} was observed to decrease strongly for the central collisions, why the opposite behavior is observed for peripheral collisions. Note here that the total transverse energy, ΣE_T, near the nuclear fragmentation region is not expected to be influenced by energy conservation effects (due to the approximate Feynman scaling in this region), or to be strongly correlated with the activity in the rapidity-separated central and forward regions. This expectation is validated by a recent measurement of ΣE_T as a function of hard scattering kinematics in pp collisions.[20]

Qualitatively the observed pattern is consistent with the decrease of the effective cross section of interaction for the projectiles which contain partons with large x. Smaller $\langle\sigma(x)\rangle$ leads to decrease of the number of collisions

[b]Studying the correlation between the initial parton–parton kinematics and the final state kinematics of the produced jets in MC event generators confirms that events with a forward jet with energy E have a distribution of x_p values narrowly peaked around $x_p = 2E/\sqrt{s}$.

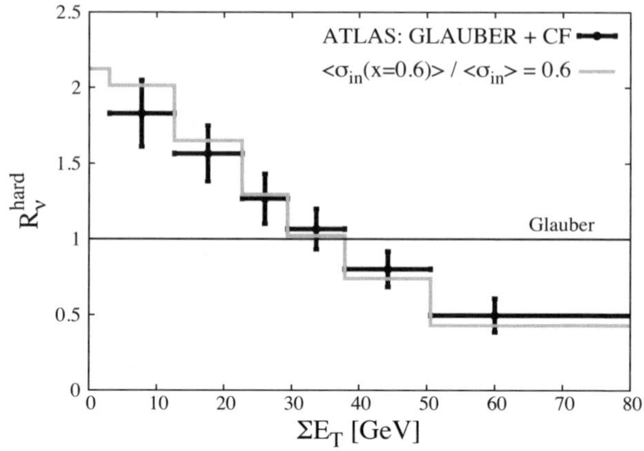

Fig. 2. R_ν^{hard} for pA collisions with $x = E_{jet}/E_p = 0.6$, and $\lambda = 0.6$ for centrality bins extracted from the ATLAS data[17] and using ν distributions given by the CF model.[14] Errors are combined statistical and systematic errors. The $R_\nu^{hard} = 1$ line is the Glauber model expectation.

and hence to an enhancement of the fraction of the collisions which look like peripheral collisions.

In our quantitative analysis,[21] we combine the ATLAS model for the ν-dependence of ΣE_T with the distribution of ν given by a CF Monte Carlo approach.[13,19] The strength of the interaction σ_{in} at given impact parameter b is generated with the measure $P_h(\sigma)$ with inelastic profile function $\Gamma_{in}(b)$ evaluated using the optical theorem. The transverse spread of partons in the colliding nucleons were generated using generalized parton densities which take into account a much stronger localization of hard interactions relative to soft ones as well as the spatial NN correlations in the nucleus.[22] (For a detailed description of the calculation of the hard collision rate as a function of ν see Ref. 19.)

We focused on the data at the highest values of $z \sim 0.6$. A priori fluctuations of σ_{in} for configurations with given z maybe different from those for the bulk. However we did check that for the intervals of centrality studied by ATLAS sensitivity to the value of ω_σ is small (see Fig. 2 in Ref. 21). Hence it was taken the same as for the bulk. We found that $\lambda = \sigma_{in}(x)/\sigma_{in} = 0.6$ gives a good description of the dependence of R_{hard} on ΣE_T (that is effectively on the centrality of the collisions). The results of the calculation are presented in Fig. 2.

The configurations of smaller size are expected to interact with strength which grows with energy faster than than that for the average configurations. So we expect that $\lambda(x)_{LHC} > \lambda(x)_{RHIC}$ which can be tested using deuteron-nucleus (proton-nucleus) collisions at RHIC.

A weaker interaction strength for configurations with $x \geq 0.5$ has implications for our understanding of the EMC effect. This follows from the analysis of Ref. 16, in which the Schrödinger equation for the bound state of the nucleus included a potential term which depends on the internal coordinates of the nucleons. In this potential, the overall attractive nature of the NN interaction results in a smaller binding energy for nucleons in small configurations. Thus by the variational principle, the probability of such configurations is suppressed. The magnitude of this suppression is comparable to the strength of the EMC effect at $x \geq 0.5$, for the recent discussion see Ref. 23.

3. Color fluctuation effects in diffractive photoproduction of ρ-meson in ultra peripheral processes

Vector meson photoproduction of ρ meson in the $\gamma A \to \rho A$ process is usually considered as due to transition of $\gamma \to \rho$ and consequent interaction of the ρ meson with a target A. Such a description (vector dominance model-VDM) which correctly accounts for the space-time evolution[24] reasonably describes data in the cases of nucleon,nuclear targets at the range of photon energies in the target rest frame in the ten GeV energy range.[25] Recently this process was studied at significantly larger energies in ultra peripheral collisions (UPC) of heavy ions at the LHCs.[26] A comparison of the data with a "classical" VDM in which one assumes that $\sigma_{tot}(\rho N) = \sigma_{tot}(\pi N)$ and takes into account for elastic shadowing significantly overestimates cross section of the coherent ρ-meson production in $\gamma Pb \to \rho Pb$ at $W \sim 50 GeV$.

First let us briefly consider the data on the ρ-meson photoproduction off a proton. It is convenient to consider this process in the quark-gluon basis even though the process is dominated by soft physics since the reduction of the strength of the interaction with reduction of overall area occupied by color appears to be present even in the nonperturbative domain. To visualize this phenomenon it is convenient to explore completeness over produced diffractive states. As a result one comes to the quark -gluon basis for the description of the process. Since photon is a point- like particle nonperturbative quark-gluon wave function of photon rather weakly depends on transverse momenta of constituents. On the contrary, the wave function of a light vector meson which is a bound state rapidly decreases with increase of transverse momenta of constituents. Hence the dominant contribution to the matrix element for the transition $\gamma \to \rho$ in the interaction with a nucleon: $\langle\psi_\gamma|\,\hat{\sigma}\,|\psi_\rho\rangle$ originates from somewhat smaller transverse size configurations than the matrix element for elastic ρN scattering: $\langle\psi_\rho|\,\hat{\sigma}\,|\psi_\rho\rangle$ Such a squeezing of of hadronic component of wave function of photon as compared to wave function of ρ meson leads to a relatively small reduction of the nonperturbative interaction of hadronic component of photon wave function with a target and hence to a reduction of the cross section of the $\gamma + p \to \rho p$ process at $t = 0$.[27] One can easily see the phenomenon of squeezing within the dipole model for photoproduction processes though the model a priori is applicable only for perturbative region. To interpret this reduction using a hadronic basis for the photon wave function one would need to introduce a contribution of the higher mass states which would reduce the ρ - photoproduction cross section.

A fit to the existing high energy data was performed in Ref. 27. It was found that to fit the data one needs to reduce the prediction of the VDM with $\sigma_{tot}(\rho - N) = \sigma_{tot}(\pi - N)$ by the factor

$$\left.\frac{d\sigma(\gamma p \to \rho p)/dt}{d\sigma(\gamma p \to \rho p)_{VDM}/dt}\right|_{t=0} = 0.7, \tag{11}$$

in agreement with the above expectations. To account for this effect we define the effective ρ–nucleon cross section $\hat{\sigma}_{\rho N}$:

$$\hat{\sigma}_{\rho N}(W_{\gamma p}) = \frac{f_\rho}{e}\sqrt{16\pi\frac{d\sigma^{\exp}_{\gamma p \to \rho p}(t=0)}{dt}}\,. \tag{12}$$

We refer to this model as the modified vector meson dominance (mVMD) model.[c]

Calculation of the nuclear cross section within the Glauber model explains the bulk of the reduction of the cross section as compared to the impulse approximation, though a significant discrepancy remains which tends to grow with the increase of the collision energy, see Fig. 3 from Ref. 27.

Another effect becomes important with increase of collision energy is the contribution of inelastic diffraction in the intermediate states with masses m_{int}. This is because with increase of collision energy minimal momentum transferred to the nuclear target $t_{min} = -m_N^2(m_{int})^4/s^2$ is rapidly decreasing with increase of collision energy s. Thus allowed range of masses

[c] The observed suppression of the photoproduction cross section is larger than suggested by a fit[28] to the early HERA data in which the VDM result was renormalized by the factor 0.84.

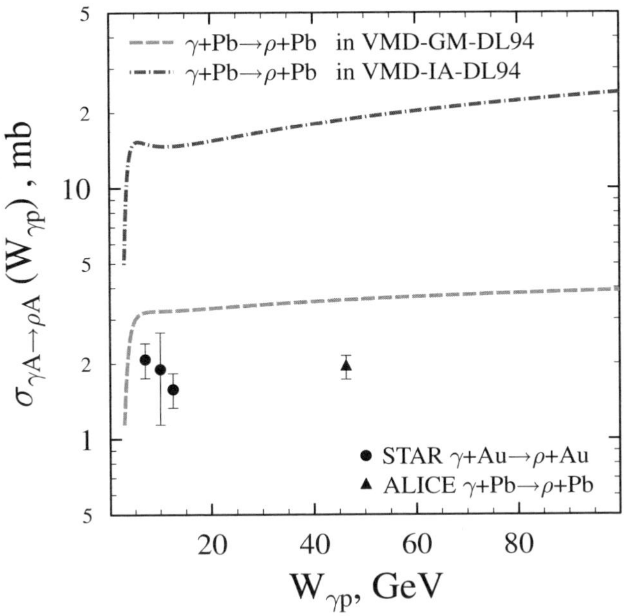

Fig. 3. The $\gamma A \to \rho A$ cross section as a function of $W_{\gamma p}$. The VMD-GM (Glauber model) (red dashed curve) and VMD-IA (blue dot-dashed line) predictions for a ^{208}Pb target based on the DL94 parametrization of the ρN cross section[29] are compared to the experimental values extracted from the STAR and ALICE UPC measurements.

of diffractively produced inelastic states in the intermediate states is rapidly increasing. This property leads to emergence of inelastic nuclear shadowing due to diffractive processes[4] (for the deuteron case shadowing is enhanced by the factor $1+\omega_\sigma$). In the case of coherent production inelastic shadowing contribution is enhanced as compared to the total cross section since the elastic cross section is proportional to the square of the total cross section.

For the case of the elastic proton-nucleus scattering effect of inelastic shadowing can be calculated using $P_N(\sigma)$ introduced in Section 2:

$$\sigma_{pA\to pA} = \int d^2\vec{b} \left| \int d\sigma P_N(\sigma) \left(1 - e^{-\frac{\sigma}{2}T_A(b)}\right) \right|^2 , \tag{13}$$

In the case of ρ-meson we need to take into account both color fluctuations due to transition to large masses and reduction of the effective cross section leading to

$$\sigma_{\gamma A\to\rho A}^{\mathrm{mVMD-GGM}} = \left(\frac{e}{f_\rho}\right)^2 \int d^2\vec{b} \left| \int d\sigma P_\rho(\sigma)\left(1 - e^{-\frac{\sigma}{2}T_A(b)}\right)\right|^2, \tag{14}$$

In the case of proton scattering the inelastic shadowing is known to be a small correction. It is primarily due to the large value of $\sigma_{tot}(NN)$ which results in interaction been nearly black for a large range of impact parameters and not sensitive to inelastic shadowing contribution.

With a decrease of σ_{hN} effect of inelastic shadowing increases because the interaction remains grey for a large range of the impact parameters. Also the ratio of the inelastic and elastic diffractive cross sections at $t = 0$ grows with decrease of $\sigma_{tot}(hN)$. In particular in a wide energy range where Regge factorization for diffraction is applicable one expects

$$\frac{\omega_{h_1}}{\omega_{h_2}} \approx \frac{\sigma_{tot}(h_2N)}{\sigma_{tot}(h_1N)}. \tag{15}$$

For the highest energy reached in UPC we estimate $\omega_\sigma(\rho N) \sim 0.5 \div 0.6$ with the large mass diffraction giving a very important contribution and increasing with energy contribution to the inelastic diffraction cross section, while in the popular now dipole models no such contribution is present.

Based on the similarity between the pion and ρ meson wave functions suggested by the additive quark model, it is natural to assume that $P_\rho(\sigma)$ for the ρN interaction should be similar to the pion $P_\pi(\sigma)$ with a small enhancement at small σ due to the enhanced contribution of small σ in the ρN interaction:

$$P(\sigma) = C\frac{1}{1+(\sigma/\sigma_0)^2}e^{-(\sigma/\sigma_0-1)^2/\Omega^2}. \tag{16}$$

The parameterization of Eq. (16) satisfies the basic QCD constraint of $P_\rho(\sigma = 0) \neq 0$ and also $P(\sigma \to \infty) \to 0$. The free parameters C, σ_0 and Ω are found from the constraints similar to those for $P_N(\sigma)$ which we discussed in Section 2.

Account of inelastic shadowing and color fluctuations allows to explain existing data on diffractive ρ meson production in ultra peripheral processes, see Fig. 4 from Ref. 27.

Evidently the same reasoning is applicable to the theoretical description of the processes of photoproduction of ϕ and J/ψ mesons also. Relative role of elastic, inelastic shadowing and color fluctuations depends on flavor of constituents. Elastic contribution decreases with increase of the mass of flavor since $\sigma(\rho N) > \sigma(\phi N) > \sigma(J/\psi N)$. Inelastic shadowing becomes

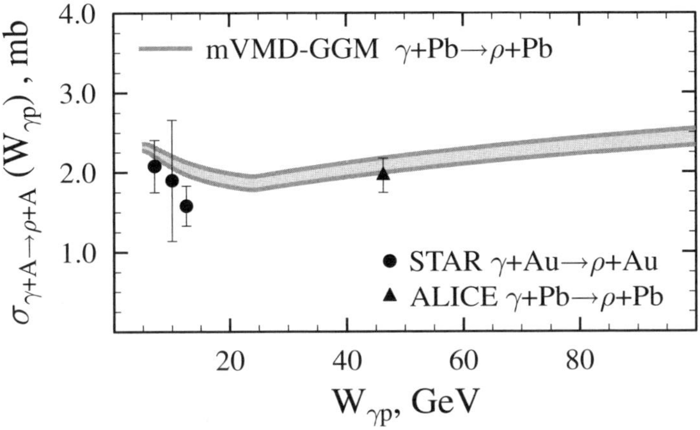

Fig. 4. The $\sigma_{\gamma A\to\rho A}$ cross section as a function of $W_{\gamma p}$. The theoretical predictions using the mVMD model for the $\gamma p\to\rho p$ cross section and the Gribov–Glauber model with cross section fluctuations for the $\gamma A\to\rho A$ amplitude are compared to the STAR (circle) and ALICE (triangle) data. The shaded area reflects the theoretical uncertainty in the estimate of the strength of cross section fluctuations (see Ref. 27).

dominant in J/ψ meson photoproduction processes. Indeed, a naive expectation is that shadowing effects should be small (essentially a higher twist effect) in the case of coherent J/ψ photoproduction. However the nuclear shadowing for the amplitude of the coherent J/ψ photoproduction off lead, S, turns out to be pretty large: $S(x\approx 0.001)=0.61^{+0.05}_{-0.04}$. It agrees well with prediction of the theory of leading twist shadowing, see Ref. 30 and references therein.

4. Strengthonometry in inelastic γA collisions at collider energies

The experimental studies of the jet production in pA described in Section 2 indicate that in the case when average configurations in nucleon are selected by the trigger, distribution in ΣE_T gives an accurate information about *average* strength of interaction of nucleon with the nucleons of the nucleus. Conversely, one can extract from the distribution in ΣE_T at rapidities far enough from the projectile average strength of interaction for a particular trigger.

It was demonstrated in Ref. 31 that in the ultraperipheral ion collisions at the LHC it is feasible to study photon-nucleus interactions up to $W_{\gamma N} \sim 0.5 TeV$. This opens a way to exploring the high energy photon wave function in great detail along the lines we described in Section 2 for pA scattering.

In fact fluctuation effects should be larger in this case. Indeed for the component corresponding to the VDM contribution we expect $\omega_\sigma \sim 0.5 \div 0.6$ like for the pion-nucleon interaction. A large fraction of the non VDM contribution originates from aligned jet configurations (large mass, low transverse momentum $q\bar{q}$ configurations) which interact with a strength comparable to that of pion, cf. discussion of the aligned jet model in Ref. 32 and likely to have similar color fluctuations. There is also a point like contribution effectively corresponding to $\sigma = 0$.

Distribution over ν for γA scattering can be calculated using Eq. (9). The production of hadrons at large negative rapidities is expected to be very similar for proton and photon projectiles for the same ν. Hence by studying distribution in ΣE_T for different forward triggers it would be possible to determine both $\langle\sigma\rangle$ and variance of the distribution for selected configuration in the photon. For example, one can estimate roughly $\langle\sigma\rangle$ for photon-nucleon interactions as

$$\frac{\Sigma E_T(\gamma A)}{\Sigma E_T(A)} = \frac{\langle\sigma_{\gamma N}\rangle}{\sigma_{in}(NN)}. \tag{17}$$

A starting point would be to investigate the direct photon dijet production for $x_A \geq 0.01$ for which we expect $\nu \approx 1$. In this case the distribution over ΣE_T should be similar to that in γp scattering. With a decrease of x_γ distribution over ΣE_T should become broader and approach for small x_γ distribution for generic γA collisions. Our first numerical estimates using $P_\gamma(\sigma)$ given by Eq. (16) indicate that the distribution over ν should be very broad with a significant fraction of events with $\nu \geq 5$.

Other interesting triggers are production of leading charm, strangeness, etc. Of separate interest are direct photon interactions for $x_A < 10^{-2}$. In this case leading twist shadowing is present resulting in a rather broad distribution over ν, see discussion in Sections 6.3, 6.4 of Ref. 33.

5. Conclusions

Analysis of high energy processes performed in the paper found significant probability of color fluctuations in pA collisions investigated at LHC, color fluctuations in the elastic photoproduction of vector mesons off nuclear targets studied in the ultra peripheral processes at LHC. We predict also color coherent phenomena at the future eA collider and ultraperipheral collisions at the LHC.

From the theoretical side observation of significant color coherent phenomena demonstrates necessity to account for the coherence of high energy processes and composite structure of hadrons. Such phenomena is a challenge for the theoretical approaches which describe non perturbative amplitudes of high energy processes in terms of exchanges by mathematical objects whose inner structure is universal – independent of the projectile (like reggeons (cuts), string tension, etc). One of the reasons for the difference is that color fluctuations approach accounts for both soft and hard QCD dynamics.

Acknowledgements: We thank our collaborators in the studies described above: M. Alvioli, G. Baym, B. Blaettel, B. Cole, V. Guzey, H. Heiselberg, D. Perepelitsa, M. Zhalov for numerous discussions.

References

1. R.J. Glauber, In *Lectures in Theoretical Physics*, ed. W.E. Brittin and L.G. Dunham, v.1, p. 315. New York: Interscience, 1959.
2. S. Mandelstam, *Nuovo Cim.* **30**, 1148 (1963).
3. V.N. Gribov, I.Y. Pomeranchuk and K.A. Ter-Martirosian, *Phys. Rev.* **139**, B184 (1965).
4. V.N. Gribov, *Sov. Phys. JETP* **29**, 483 (1969).
5. H.I. Miettinen and J. Pumplin, *Phys. Rev.* **D 18**, 1696 (1978).
6. M.L. Good and W.D. Walker, *Phys. Rev.* **120**, 1857 (1960).
7. L. Frankfurt and M. Strikman, *Prog. Part. Nucl. Phys.* **27**, 135 (1991).
8. B. Blaettel, G. Baym, L.L. Frankfurt and M. Strikman, *Phys. Rev. Lett.* **70**, 896 (1993); *Phys. Rev.* **D47**, 2761 (1993).
9. Y. Akimov, L. Golovanov, S. Mukhin, V. Tsarev, E. Malamud, R. Yamada, P. Zimmerman and R. Cool *et al.*, *Phys. Rev.* **D14**, 3148 (1976).
10. L. Frankfurt, G.A. Miller and M. Strikman, *Phys. Rev. Lett.* **71**, 2859 (1993) arXiv:hep-ph/9309285.
11. H. Heiselberg, G. Baym, B. Blaettel, L.L. Frankfurt and M. Strikman, *Phys. Rev. Lett.* **67**, 2946 (1991); *Phys. Rev.* **C52**, 1604 (1995).
12. L. Bertocchi and D. Treleani, *J. Phys. G* **3**, 147 (1977).
13. M. Alvioli and M. Strikman, *Phys. Lett.* **B722**, 347 (2013).
14. ATLAS Collaboration, Conference note ATLAS-CONF-2013-096, https://cds.cern.ch/record/1599773.
15. S.J. Brodsky and G.R. Farrar, *Phys. Rev. Lett.* **31**, 1153 (1973).

16. L.L. Frankfurt and M.I. Strikman, *Nucl. Phys.* **B 250**, 143 (1985).
17. G. Aad *et al.* [ATLAS Collaboration], *Phys. Lett.* **B 748**, 392 (2015) arXiv:1412.4092 [hep-ex].
18. CMS Collaboration, *Eur. Phys. J. C* **74**, 2951 (2014) arXiv:1401.4433 [nucl-ex].
19. M. Alvioli, L. Frankfurt, V. Guzey and M. Strikman, *Phys. Rev.* **C 90**, 034914 (2014) arXiv:1402.2868 [hep-ph].
20. ATLAS Collaboration, Conference note ATLAS-CONF-2015-019, http://cdsweb.cern.ch/record/2029360.
21. M. Alvioli, B. A. Cole, L. Frankfurt, D. V. Perepelitsa and M. Strikman, arXiv:1409.7381 [hep-ph].
22. M. Alvioli, H.-J. Drescher and M. Strikman, *Phys. Lett.* **B 680**, 225 (2009).
23. A.J. Freese, M.M. Sargsian and M.I. Strikman, arXiv:1411.6605 [hep-ph].
24. R.P. Feynman, *Photon-hadron interactions*, Benjamin, Reading 1972, p. 282.
25. T.H. Bauer, *et al.*, *Rev. Mod. Phys.* **50**, 261 (1978) [Erratum-ibid. **51**, 407 (1979)].
26. J. Adam *et al.* [ALICE Collaboration], arXiv:1503.09177 [nucl-ex].
27. L. Frankfurt, V. Guzey, M. Strikman and M. Zhalov, arXiv:1506.07150 [hep-ph].
28. A. Donnachie and P.V. Landshoff, *Phys. Lett.* **B478**, 146 (2000) arXiv:hep-ph/9912312.
29. A. Donnachie and P.V. Landshoff, *Phys. Lett.* **B348**, 213 (1995) arXiv:hep-ph/9411368.
30. V. Guzey, E. Kryshen, M. Strikman and M. Zhalov, *Phys. Lett.* **B726**, 290 (2013) arXiv:1305.1724 [hep-ph].
31. A.J. Baltz, G. Baur, D. d'Enterria, L. Frankfurt, F. Gelis, V. Guzey, K. Hencken, and Y. Kharlov *et al.*, *Phys. Rept.* **458**, 1 (2008) arXiv:0706.3356 [nucl-ex].
32. L.L. Frankfurt and M.I. Strikman, *Phys. Rept.* **160**, 235 (1988).
33. L. Frankfurt, V. Guzey and M. Strikman, *Phys. Rept.* **512**, 255 (2012) arXiv:1106.2091 [hep-ph].

TIME EVOLUTION OF THE ANISOTROPIES OF THE HYDRODYNAMICALLY EXPANDING SQGP

A. BAGOLY, M. CSANÁD*

Department of Atomic Physics, Eötvös University,
1/a Pázmány P. s., Budapest, H-1117, Hungary
**E-mail: csanad@elte.hu*

In high energy heavy ion collisions of RHIC and LHC, a strongly interacting quark gluon plasma (sQGP) is created. This medium undergoes a hydrodynamic evolution, before it freezes out to form a hadronic matter. The initial state of the sQGP is determined by the initial distribution of the participating nucleons and their interactions. Due to the finite number of nucleons, the initial distribution fluctuates on an event-by-event basis. The transverse plane anisotropy of the initial state can be translated into a series of anisotropy coefficients or eccentricities: second, third, fourth-order anisotropy etc. These anisotropies then evolve in time, and result in measurable momentum-space anisotropies, to be measured with respect to their respective symmetry planes. In this paper we investigate the time evolution of the anisotropies. With a numerical hydrodynamic code, we analyze how the speed of sound and viscosity influence this evolution.

Keywords: heavy ion collisions, numerical hydrodynamics, equation of state, azimuthal anisotropy

1. Hydrodynamics in ultra-relativistic heavy ion collisions

The medium created in ultra-relativistic heavy ion collisions is successfully modeled by perfect fluid hydrodynamics, i.e. the observables of soft hadrons, photons and leptons can described by hydro models. Exact solutions are particularly interesting, since one gains an analytic insight on the connection between the initial and final state of the hydro evolution.

These solutions fulfill a simple set of laws, governing many systems on many scales. Hydrodynamics only assumes the local conservation of energy and momentum, and, depending on the circumstances, of entropy or some conserved charges. The equations of non-relativistic hydrodynamics are:

$$\frac{\partial \rho}{\partial t} + \boldsymbol{\nabla} \rho \boldsymbol{v} = 0 \tag{1}$$

$$\rho \left(\frac{\partial \boldsymbol{v}}{\partial t} + (\boldsymbol{v}\boldsymbol{\nabla})\boldsymbol{v} \right) = -\boldsymbol{\nabla} p + \mu \Delta \boldsymbol{v} + \left(\zeta + \frac{\mu}{3} \right) \boldsymbol{\nabla}(\boldsymbol{\nabla} \boldsymbol{v}) \tag{2}$$

$$\frac{\partial \varepsilon}{\partial t} + \boldsymbol{\nabla} \varepsilon \boldsymbol{v} = -p \boldsymbol{\nabla} \boldsymbol{v} + \boldsymbol{\nabla}(\sigma \boldsymbol{v}) \tag{3}$$

where ρ is matter density, $\boldsymbol{v}$ is the velocity field, ε is the (internal) energy density, p is pressure, σ is the viscous stress tensor of the homogeneous and isotropic medium, μ is the dynamic viscosity coefficient, λ the bulk viscosity coefficient, while t represents time and $\boldsymbol{\nabla} = (\partial_x, \partial_y, \partial_z)$ the spatial derivative vector. If we however would like to utilize hydrodynamics in a relativistically expanding medium, we have to use the equations of relativistic hydrodynamics. In case of perfect hydro, these take the form of

$$T^{\mu\nu} = (\varepsilon + p) \frac{u^\mu u^\nu}{c^2} - p g^{\mu\nu}, \quad \partial_\mu T^{\mu\nu} = 0. \tag{4}$$

To close this set of equations, we need to specify a relation between energy density ε and pressure p. This we obtain using the below Equation of State:

$$\varepsilon = \kappa(T) p, \tag{5}$$

where $\kappa = c_s^{-2}$ is the inverse square of speed of sound. It's temperature dependence can be taken from lattice QCD calculations, but often κ is simply kept constant.

2. Initial conditions and anisotropies

There are many solutions that fulfill the above equations, we however need to start from an initial state that corresponds to the energy distribution deposited by the nucleons participating in the collision. In nucleus-nucleus collisions the initially formed medium can be spatially approximated by an ellipsoid, which can be described by the contour levels of a scale variable s

$$s = \frac{x^2}{X^2} + \frac{y^2}{Y^2} + \frac{z^2}{Z^2} \tag{6}$$

where X, Y, Z are the axes of the ellipsoid described by the equation $s = 1$. Multiple exact solutions of hydrodynamics were found[1,2] that resemble this symmetry, i.e. where the thermodynamical quantities are constant on s =const. curves. It is important to see however, that nuclei contain a finite number of nucleons, and thus the participants of a nucleus-nucleus collision form an event-by-event fluctuating shape (see Fig. 1).

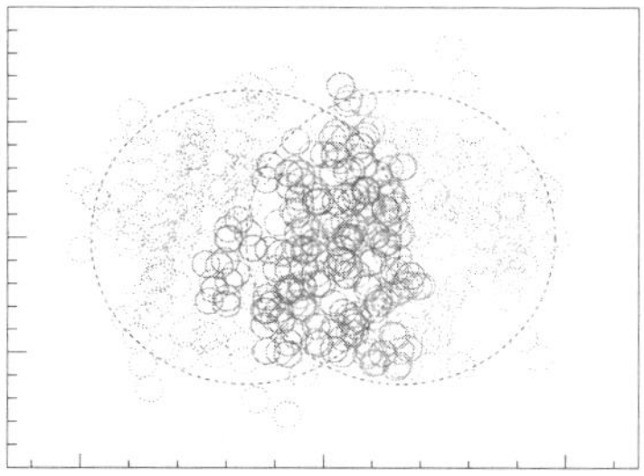

Figure 1. Simulated participant nucleon distribution (figure from Ref. 3)

Thus a simple ellipsoidal symmetry is not sufficient to describe the hydrodynamically expanding medium, not to mention that it does not describe higher order flow coefficients either. To overcome this difficulty, one can assume higher order anisotropies in the scaling variable, by redefining it as

$$s = \frac{r^2}{R^2}\left(1 + \varepsilon_2 \cos(2\phi)\right) + \frac{z^2}{Z^2} \tag{7}$$

with r and ϕ being the polar coordinates in the transverse plane, while R and ε_2 are in a direct relation to X and Y. This can then be generalized to

$$s = \frac{r^2}{R^2}\left(1 + \sum_{n=2,3,\ldots} \varepsilon_n \cos n\varphi\right) + \frac{z^2}{Z^2} \tag{8}$$

in the $\cos n\varphi$ expression a Ψ_n n-th order event plane can also be introduced. Example distributions generated with this scaling variable are shown in Fig. 2. Solutions utilizing scaling variables like the above s were presented in Ref. 4. However, these solutions, as well as the ones in Ref. 1 are valid only in case of a 3D Hubble-flow, which corresponds to the lack of pressure gradients. This also means that the anisotropies will remain constant in time, which may not be the case throughout the whole evolution. In our analysis, we thus apply numerical calculations to include pressure gradients in the initial conditions, and analyze the time evolution of the anisotropies.

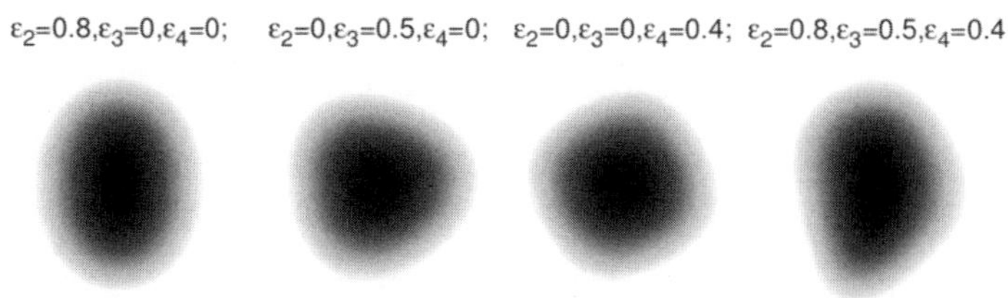

Figure 2. Initial distribution with the outlined s scaling variable, for different ε_n values.

Thus we will start the hydrodynamic evolution from the above described set of initial conditions. We would like to extract the anisotropy coefficients at later stages of the evolution, for which we need a definition. We introduced the anisotropies of density, energy density and velocity fields as follows:

$$\epsilon_n = \langle \cos(n\varphi) \rangle_{\rho/\varepsilon/\mathrm{w}}, \tag{9}$$

where $w = \exp\left(-v_x^2 - v_y^2\right)$ is defined to calculate the asymmetry of velocity field. To understand the relation of the asymmetry parameters introduced by 9 to the scale variable anisotropy coefficients, ε_n's, we checked their relation. As a first order approximation (in ε_n), we obtained

$$\epsilon_n = -\varepsilon_n/2 \tag{10}$$

however, in a more accurate approximation the result is:

$$\epsilon_1 \approx \frac{(\varepsilon_2+\varepsilon_4)\varepsilon_3}{2+\sum_n \varepsilon_n^2} = \frac{\epsilon_3(\epsilon_2+\epsilon_4)}{2} + \mathcal{O}(\epsilon_n^4) \tag{11}$$

$$\epsilon_2 \approx \frac{-\varepsilon_2+\varepsilon_2\varepsilon_4}{2+\sum_n \varepsilon_n^2} = -\frac{\epsilon_2}{2} + \frac{\epsilon_2\epsilon_4}{2} + \frac{1}{4}\epsilon_2 \sum_n \epsilon_n^2 + \mathcal{O}(\epsilon_n^4) \tag{12}$$

$$\epsilon_3 \approx \frac{-\varepsilon_3}{2+\sum_n \varepsilon_n^2} = -\frac{\epsilon_3}{2} + \frac{1}{4}\epsilon_3 \sum_n \epsilon_n^2 + \mathcal{O}(\epsilon_n^4) \tag{13}$$

$$\epsilon_4 \approx \frac{-\varepsilon_4+\frac{1}{2}\varepsilon_2^2}{2+\sum_n \varepsilon_n^2} = -\frac{\epsilon_4}{2} + \frac{\epsilon_2^2}{4} - \frac{1}{4}\epsilon_4 \sum_n \epsilon_n^2 + \mathcal{O}(\epsilon_n^4) \tag{14}$$

this means that even if there is no ε_1 coefficient, we will still get an ϵ_1. It is also important to note that $\varepsilon_2 \neq 0$ and $\varepsilon_{n\neq 2} = 0$ will result in a nonzero $\epsilon_4 = \varepsilon_2^2/4$. See an example with multiple nonzero ε_n coefficients in Table 1, where we give an example how a set of ε_n values result in a different set of ϵ_n values. The actual results can be accurately approximated with Eqs. (12)-(14).

Table 1. Example for the $\epsilon_n - \varepsilon_n$ relation.

n	ε_n	ϵ_n direct calculation	approximation	first order estimate
1	0	0.00278	0.00298	0
2	0.1	-0.04927	-0.04869	-0.05
3	0.05	-0.02533	-0.02484	-0.025
4	0.02	-0.00769	-0.00745	-0.01

3. Numerical method

Let us now discuss how we evolve the previously defined initial conditions in time. In order to do that, we first transform the equations of (1+2 dimensional) hydrodynamics to a convective form as usual:

$$\partial_t \mathbf{Q} + \partial_x \mathbf{F}(\mathbf{Q}) + \partial_y \mathbf{G}(\mathbf{Q}) = 0 \tag{15}$$

where $\mathbf{Q}$ is calculated from the hydrodynamical fields, while $\mathbf{F}$ and $\mathbf{G}$ are functions of these, transformed from the original hydro equations. The finite volume method discussed below and introduced in Ref. 5 works for many different systems of partial differential equations (PDE's), one just needs to determine the given functions to transform the PDE's to the above form. This can be done in a straightforward manner, see for example Refs.[5–8].

The basis of finite volume methods is to discretize the $\mathbf{Q}$ vector, by defining it on a space-time grid, and obtaining the discrete values by averaging in each cell. Then, we have to evaluate fluxes $\mathbf{F}$ and $\mathbf{G}$ between grid points (these are then called intercell fluxes). This is difficult, since the fluxes depend on the values of $\mathbf{Q}$, but these are defined on the grid points. Our method, outlined in Ref. 5 and used for example in Ref. 7, is based on an accurate multi-stage (MUSTA) predictor-corrector estimation of the intercell fluxes. We discuss this method briefly in 1+1 dimensions (and with a single-component Q only) – it is straightforward to generalize it to multiple spatial dimensions via operator splitting (we used the Lie type of splitting method here). In case of viscous hydrodynamics, we also used operator splitting to separate ideal and viscous fluxes.

To estimate the intercell fluxes between cells i and $i+1$, we start from the Q_i^n and Q_{i+1}^n values, where n represents the current time-step. Let us call these values $Q_{i,i+1}^{(0)}$, indicating that these represent a "zeroth order" estimate of the intercell values. We will then make a series of estimates in the given cell, the lth estimate being indicated with $Q_{i,i+1}^{(l)}$. The flux in the cell centers is then $F_{i,i+1}^{(l)} \equiv F\left(Q_{i,i+1}^{(l)}\right)$. We define an intermediate Q value and flux F as follows:

$$Q_{i+\frac{1}{2}}^{(l)} = \frac{1}{2}\left[Q_i^{(l)} + Q_{i+1}^{(l)}\right] - \frac{1}{2}\frac{\Delta t}{\Delta x}\left[F_{i+1}^{(l)} - F_i^{(l)}\right], \quad F_M^{(l)} \equiv F\left(Q_{i+\frac{1}{2}}^{(l)}\right) \tag{16}$$

Then our intercell flux estimate is

$$F_{i+\frac{1}{2}}^{(l)} = \frac{1}{4}\left[F_{i+1}^{(l)} + 2F_M^{(l)} + F_i^{(l)} - \frac{\Delta x}{\Delta t}\left(Q_{i+1}^{(l)} - Q_i^{(l)}\right)\right] \tag{17}$$

This is the so called FORCE flux approximation. The essence of our the applied MUSTA algorithm is that now we make a new prediction, $Q_i^{(l+1)}$

and with that we repeat the steps and we get a better flux approximation, and with that we can make better prediction, and again better flux approximation. To make a new prediction simply we use the equations we want to solve:

$$Q_i^{(l+1)} = Q_i^{(l)} - \frac{\Delta t}{\Delta x}\left[F_{i+\frac{1}{2}}^{(l)} - F_i^{(l)}\right] \tag{18}$$

$$Q_{i+1}^{(l+1)} = Q_{i+1}^{(l)} - \frac{\Delta t}{\Delta x}\left[F_{i+1}^{(l)} - F_{i+\frac{1}{2}}^{(l)}\right] \tag{19}$$

We may then stop at a given l value, by setting $F^n_{i+\frac{1}{2}} = F^{(l)}_{i+\frac{1}{2}}$. With this, we simply calculate the next time-step as

$$Q_i^{n+1} = Q_i^n - \frac{\Delta x}{\Delta t}\left(F^n_{i+\frac{1}{2}} - F^n_{i-\frac{1}{2}}\right). \tag{20}$$

where $F^n_{i-\frac{1}{2}}$ can also be calculated by the procedure outlined above, if we start from cells $i-1$ and i.

In order to complete the method, let us mention that we used the initial conditions as described above, we used non-reflective boundary conditions, and the timestep was determined adaptively via the Courant-Friedrichs-Lewy condition. We tested our method with the usual Sod shock tube and known hydrodynamic solutions as well.

4. Results

With the above outlined method, we solved the equations of nonrelativistic hydrodynamics with viscosity, as well as the equations of relativistic perfect hydrodynamics. Our goal was to investigate how various speed of

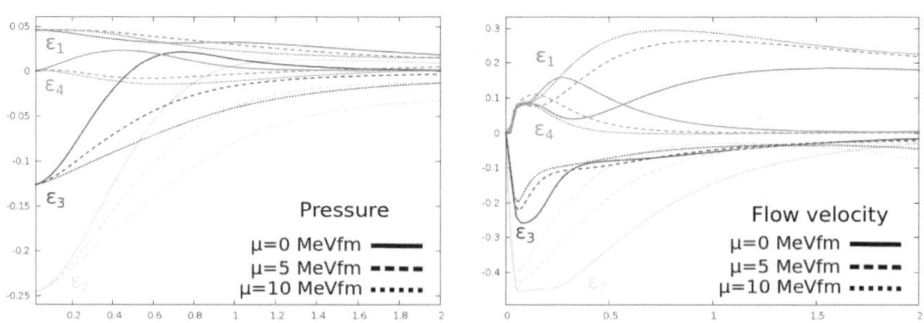

Figure 3. Time evolution of pressure anisotropies is shown in the left panel, while anisotropies in the flow field are shown in the right panel.

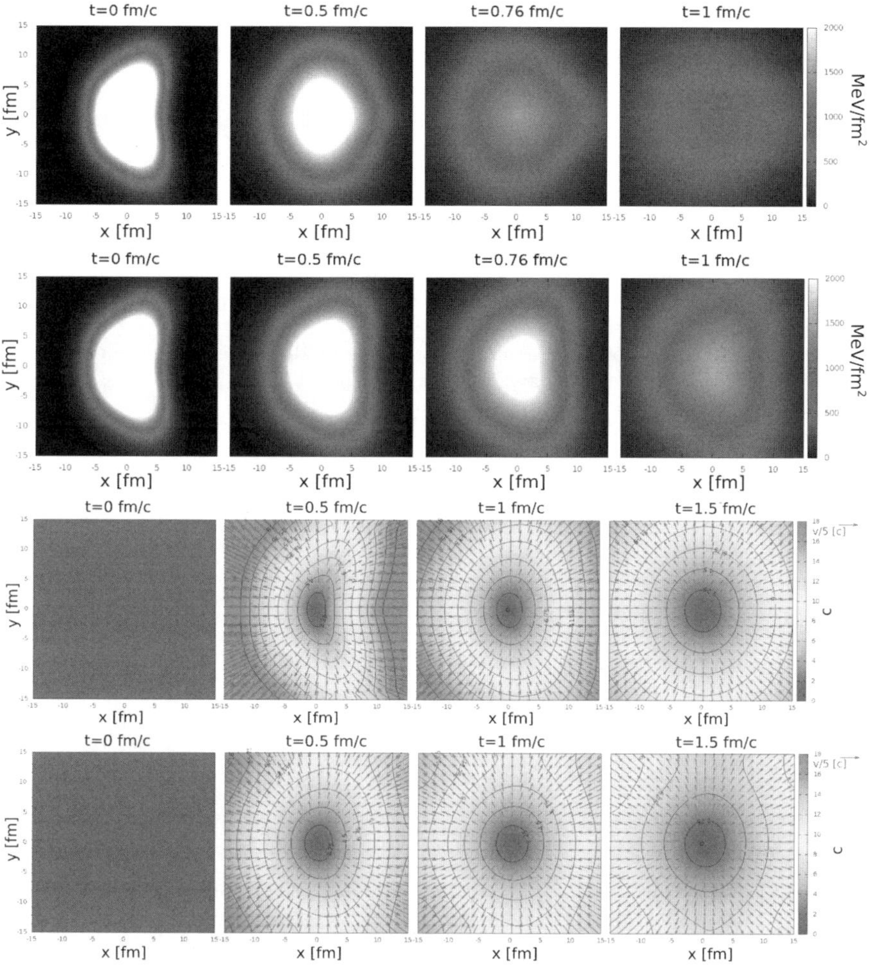

Figure 4. Time evolution of the energy density (first two rows) and of the flow field (last two rows). The first and third rows show the viscosity free case, the second and fourth rows show the viscous case with $\mu = 10\mathrm{MeV} \cdot \mathrm{fm}$.

sound values and various viscosity values affect the time evolution of the asymmetries.

Let us first discuss the effects of viscosity (in nonrelativistic hydro). As shown in Fig. 3 viscosity makes the disappearance of pressure anisotropies slower, and the same is in the number density. In other words, in case of nonzero viscosity the pressure asymmetries remain in the system for a longer time. This may be due to the fact that viscosity makes the flow slower, as illustrated in the first two plots of Fig. 4. As shown in the

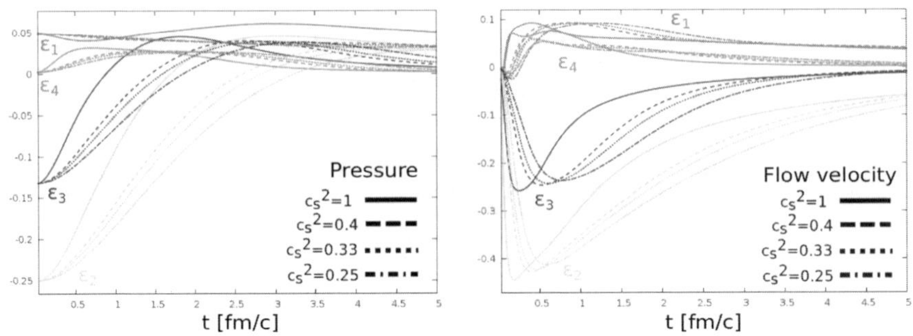

Figure 5. Time evolution of pressure anisotropies is shown in the left panel, while anisotropies in the flow field are shown in the right panel.

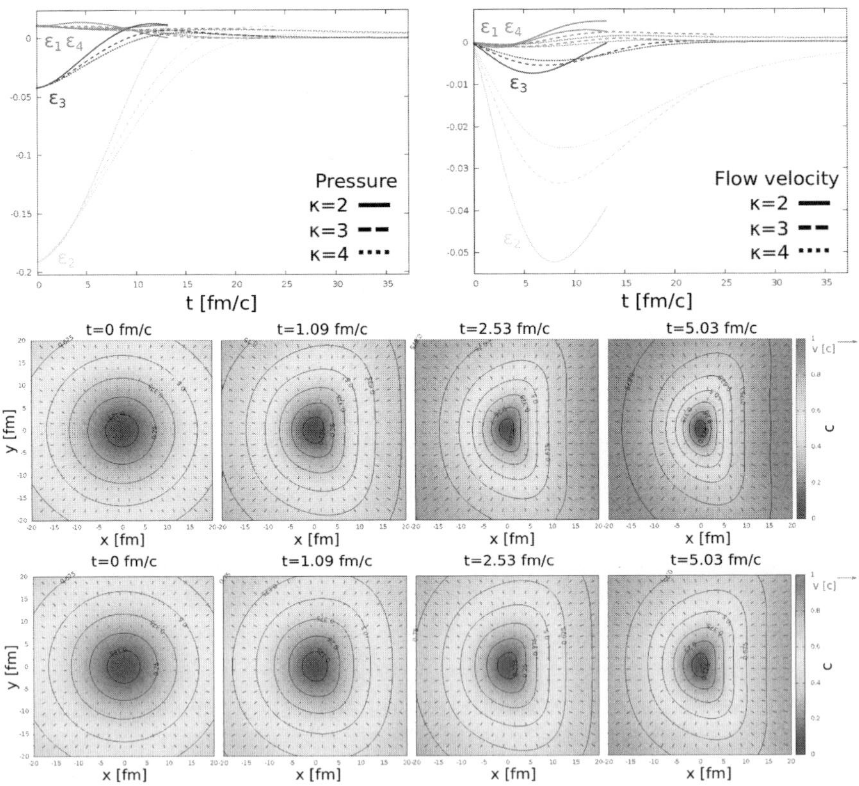

Figure 6. The time evolution of velocity field. The top row shows $\kappa = 2$ case, the bottom row the $\kappa = 4$ case. Time evolution of pressure anisotropies is shown in the left panel, while anisotropies in the flow field are shown in the right panel.

bottom plot of Fig. 3 and the third and fourth plots in Fig. 4, the effect of viscosity on the flow field is however the opposite: viscosity makes the flow anisotropies disappear faster. The reason of this may be that the viscous force contains the second spatial derivatives, thus the fluid elements with larger asymmetries feel a stronger viscous force, which means that viscosity "washes out" the anisotropies.

We also investigated the effects of a change in the speed of sound. As shown in Fig. 5, the reduction of speed of sound makes the anisotropies disappear slower. The reason for this is that pressure waves travel with the speed of sound, so the equalization of pressure is slower in case of smaller speed of sound values. This is confirmed by relativistic calculations as well. As shown in the top panels of Fig. 6, in case of a larger κ (i.e. a smaller c_s^2 value) the anisotropies disappear slower, and also the cooling is slower. This also means that of course the system will freeze out later. The time evolution of the flow field is shown in the bottom panels of Fig. 6.

5. Summary

In this paper we analyzed the time evolution of the asymmetries of the QCD matter created in high energy heavy ion collisions. We utilized numerical hydrodynamics instead of exact solutions to investigate situations not describable by the presently known hydrodynamic solutions. We however did not start from the most realistic initial conditions, but started instead from an initial condition that is very similar to one described by known analytic solutions – except in pressure, where we used a pressure profile similar to the density profile given in these models. We analyzed both non-relativistic and relativistic hydrodynamics, and arrived at similar conclusions. It turns out that a stiffer equation of state, i.e. a larger speed of sound makes the pressure anisotropies disappear faster. The appearance of viscosity also makes the flow anisotropies disappear faster, however, pressure anisotropies remain in the system longer, due to the slower flow of the viscous system.

Bibliography

1. T. Csörgő, L.P. Csernai, Y. Hama, and T. Kodama, *Heavy Ion Phys.* **A21**, 73 (2004).
2. T. Csörgő, S.V. Akkelin, Y. Hama, B. Lukács, and Y.M. Sinyukov, *Phys. Rev.* **C67**, 034904 (2003).
3. C. Loizides, J. Nagle, and P. Steinberg, *SoftwareX* **1-2**, 13-18 (2015).
4. M. Csanád and A. Szabó, *Phys. Rev.* **C90**, 054911 (2014).

5. E.F. Toro and V.A. Titarev, *J. Comp. Phys.* **216**, 403 (2006).
6. H. Nishikawa and K. Kitamura, *J. Comp. Phys.* **227**, 2560 (2008).
7. M. Takamoto and S.-I. Inutsuka, *J. Comput. Phys.* **230**, 7002 (2011).
8. I. Karpenko, P. Huovinen, and M. Bleicher, *Comput. Phys. Commun.* **185**, 3016 (2014).

AN ANALYTIC HYDRODYNAMICAL MODEL OF ROTATING 3D EXPANSION IN HEAVY-ION COLLISIONS

M. I. NAGY

Dept. of Atomic Physics, Eötvös University,
H-1117 Pázmány Péter sétány 1/A, Budapest, Hungary

T. CSÖRGŐ

HAS Wigner RCP RMI, H-1525 Budapest 114, POB 49, Hungary

A new exact and analytic solution of non-relativistic fireball hydrodynamics is presented. It describes an expanding three-axis ellipsoid that rotates along one of its principal axes. The observables are calculated using simple analytic formulas. Following earlier works, it is pointed out that azimuthal oscillation of the off-diagonal Bertsch-Pratt radii of Bose-Einstein correlations as well as rapidity dependent directed and third flow measurements provide means to determine the magnitude of the rotation of the fireball. It is argued that observing this rotation and its dependence on collision energy may lead to new information on the equation of state of the strongly interacting quark gluon plasma produced in high energy heavy ion collisions.

1. Introduction

The applications of hydrodynamics (both numerical and exact analytic solutions) in the description of the "soft" particle production in heavy-ion collisions has a good tradition. Here we deal with an analytic model. (For a recent work on numerical hydrodynamics, see e.g. Ref. 1; the many references therein provide a good list of recent developments in numerical hydrodynamical modeling). Exact solutions require some "luck" to find, however, they also have a good line of success: the early days of high energy physics saw the development of the Landau-Khalatnikov solution[2] and the Hwa-Bjorken solution,[3] both invaluably useful for high energy collision phenomenology. Since then, many physically realistic hydrodynamical solutions have been found, relativistic as well as non-relativistic ones.

The rotation of the expanding hot and dense matter produced in non-central heavy ion collisions drew much theoretical attention recently. Many numerical models try to grasp the effects of rotation on the observables (see e.g. Refs. 4, 5). In this work we present an exact solution of the non-relativistic perfect fluid hydrodynamical equations that describes the desired rotating expansion of a three-axis ellipsoid-like fireball, just as one imagines the space-time picture of a non-central heavy-ion collision. We also calculate the observables using simple analytic formulas and point out those features of them that would reveal the magnitude of rotation.

We mention some of our works of which this step is a natural continuation: the discovery of accelerating relativistic solutions[6,7] as well as the first rotating relativistic[8] and non-relativistic[9] exact solutions. There are also exact solutions available for arbitrary equation of state in the relativistic[10] domain (even ones containing multipole asymmetries[11]) as well as in the non-relativistic domain[12] with realistic, ellipsoid-like symmetry. The solution and the calculation of the observables presented below is thus a next example of usefulness of exact solutions in heavy-ion phenomenology.

2. A new rotating solution of hydrodynamics

The equations of non-relativistic hydrodynamics are the Euler equation, the energy conservation equation and the particle number conservation equation. They are recited here in the form suitable for heavy-ion physics phenomenology:

$$nm_0\left(\partial_t+\mathbf{v}\nabla\right)\mathbf{v} = -\nabla p, \tag{1}$$

$$\left(\partial_t+\mathbf{v}\nabla\right)\varepsilon = -\left(\varepsilon+p\right)\nabla\mathbf{v}, \tag{2}$$

$$\left(\partial_t+\mathbf{v}\nabla\right)n = -n\nabla\mathbf{v}. \tag{3}$$

Here n stands for the particle number density (thus nm_0 is the mass density), T for temperature, p for pressure, $\mathbf{v}$ for the velocity field, ε for the energy density. The equations above need to be supplemented with a suitable Equation of State (EoS) which we customarily choose as

$$\varepsilon = \kappa\left(T\right)p, \tag{4}$$

$$p = nT. \tag{5}$$

In non-central heavy-ion collisions, an almond-shaped region of hot and dense matter forms which has non-zero angular momentum. The almond shape is approximated with a three-axis ellipsoid, and the following solution of hydrodynamics takes rotation and expansion into account. Let the x axis point in the direction of the impact parameter and the z axis in the direction of the colliding beams; thus the x–z plane is the event plane, and the rotation is around the y axis. We call the lab frame the K frame, to be distinguished from the K' frame, which co-rotates with the principal axes of the expanding ellipsoid. The solution is described in terms of the time-dependent angle of rotation $\vartheta(t)$, and the three principal axes of the rotated ellipsoid $X(t)$, $Y(t)$, $Z(t)$. The scaling variable, whose level surfaces correspond to self-similar ellipsoids is*

$$s(\mathbf{r},t) \equiv \frac{r'^2_x}{X^2} + \frac{r'^2_y}{Y^2} + \frac{r'^2_z}{Z^2}, \tag{6}$$

$$\begin{aligned} r'_x &= r_x \cos\vartheta - r_z \sin\vartheta, \\ r'_y &= r_y, \\ r'_z &= r_x \sin\vartheta + r_z \sin\vartheta. \end{aligned} \tag{7}$$

The velocity field is taken to be a linear, rotating Hubble-like one, for which the above s variable is a proper "scaling variable", i.e. its co-moving derivative vanishes. We introduce the $V = XYZ$ notation. One can directly check that for constant κ the above and the following formulas provide the desired rotating solution to the (1)–(3) hydrodynamical equations:

$$\mathbf{v}'(\mathbf{r},t) = \begin{pmatrix} \frac{\dot{X}}{X} r'_x + g\frac{X}{Z} r'_z \\ \frac{\dot{Y}}{Y} r'_y \\ \frac{\dot{Z}}{Z} r'_z - g\frac{Z}{X} r'_z \end{pmatrix}, \tag{8}$$

$$\dot{\vartheta}(t) = g(t) = \frac{\omega_0}{2}\left(\frac{X_0+Z_0}{X+Z}\right)^2, \tag{9}$$

$$n = n_0 \frac{V_0}{V} \nu(s), \tag{10}$$

$$T = T_0 \left(\frac{V_0}{V}\right)^{\frac{1}{\kappa}} \tau(s), \tag{11}$$

$$\nu(s) = \frac{1}{\tau(s)} e^{-\frac{1}{2}\int_0^s \frac{ds'}{\tau(s')}}. \tag{12}$$

For non-constant κ, the above formulas can be modified to form a valid solution. A solution is found with spatially homogeneous T and Gaussian-like n as

$$n = n_0 \frac{V_0}{V} e^{-s/2}, \tag{13}$$

*In the following, we do not explicitly write up the time dependence of functions which were introduced as time-dependent ones.

$$T \equiv T(t), \quad \frac{V_0}{V} = \exp\left\{\int_{T_0}^{T} \mathrm{d}\beta \left(\frac{\mathrm{d}\kappa}{\mathrm{d}\beta} + \frac{\kappa}{\beta}\right)\right\}. \tag{14}$$

Furthermore, the time development of the X, Y, Z principal axes (and thus that of $\dot{\vartheta}$ and g) is governed by a Lagrangian for a point-like particle of mass m_0, which we write up only in the $\kappa = const$ case:

$$L = \frac{m_0}{2}\left(\dot{X}^2 + \dot{Y}^2 + \dot{Z}^2\right) - \frac{1}{\kappa}\left(\frac{V_0}{V}\right)^{\frac{1}{\kappa}} - \frac{m_0\omega_0^2}{4}\frac{(X_0+Z_0)^4}{(X+Z)^2}. \tag{15}$$

Note that the velocity profile is written up in the co-rotating K' frame. The free parameter ω_0 is related to the total angular momentum M_y of the flow (which points in the y direction); for example, in the $\nu(s) = e^{-s/2}$ Gaussian case

$$M_y = m_0 n_0 V_0 (2\pi)^{3/2} \frac{\omega_0}{2} (X_0 + Z_0)^2 . \tag{16}$$

The solution thus specified is a natural generalization of the earlier results for 3d expanding non-rotating ellipsoids[12,13] (indeed, our formulation resembles very much to those ones), as well as the rotating exact solution presented in Ref. 9.

3. Observables from the new solution

One can calculate the hadronic observables from the above solution by specifying the emission function (source function) and a suitable freeze-out condition. Following Ref. 12, for simplicity (and because the solution for arbitrary $\kappa(T)$ is then available), we take a Gaussian ansatz for the density profile (thus the temperature is spatially homogeneous), as well as take the freeze-out to happen at a constant time t_f (which in this case corresponds to a given T_f freeze-out temperature). We assume that at the freeze-out, particles with mass m are created. The source function is then taken to be the non-relativistic Boltzmann distribution for a particle with mass m:

$$S(\mathbf{r},\mathbf{p}) = \frac{n(t_f,\mathbf{r})}{(2\pi m T_f)^{3/2}} \exp\left\{-\frac{(\mathbf{p} - m\mathbf{v}(t_f,\mathbf{r}))^2}{2mT(t_f)}\right\}. \tag{17}$$

In the following, the $_f$ index means the value taken at the freeze-out time t_f, but it is mostly dropped: all quantities are to be understood as their value at t_f. The single-particle spectrum and the two-particle Bose-Einstein correlation function are calculated as

$$N_1(\mathbf{p}) = \int \mathrm{d}^3\mathbf{r}\, S(\mathbf{r},\mathbf{p}), \tag{18}$$

$$C\left(\mathbf{K},\mathbf{q}\right)=1+\lambda\frac{\left|\tilde{S}\left(\mathbf{K},\mathbf{q}\right)\right|^{2}}{\left|\tilde{S}\left(\mathbf{K},\mathbf{0}\right)\right|^{2}}, \tag{19}$$

where $\mathbf{K}$ and $\mathbf{q}$ are the average pair momentum and the relative momentum, respectively, and λ is the effective intercept parameter of the correlation function. For the mentioned Gaussian density case, all the integrals can be performed analytically, yielding simple results. We introduce the following quantities for convenience:

$$T_{xx}^{*}\equiv T_{f}+m\dot{X}^{2}+\frac{m\omega_{0}^{2}}{4}\frac{\left(X_{0}+Z_{0}\right)^{4}}{\left(X+Z\right)^{2}}\equiv T_{f}+m\dot{X}^{2}+m\omega^{2}R^{2}, \tag{20}$$

$$T_{zz}^{*}\equiv T_{f}+m\dot{Z}^{2}+\frac{m\omega_{0}^{2}}{4}\frac{\left(X_{0}+Z_{0}\right)^{4}}{\left(X+Z\right)^{2}}\equiv T_{f}+m\dot{Z}^{2}+m\omega^{2}R^{2}, \tag{21}$$

$$T_{xz}^{*}\equiv\frac{m\omega_{0}}{2}\frac{\left(X_{0}+Z_{0}\right)^{2}}{X+Z}\left(\dot{X}-\dot{Z}\right)\equiv m\omega R\left(\dot{X}-\dot{Z}\right), \tag{22}$$

where we introduced another simplifying notation: the "average" radius of the ellipsoid R, and the "average" angular velocity $\omega\left(t\right)$ as

$$R\equiv\frac{X+Z}{2},\qquad R_{0}\equiv\frac{X_{0}+Z_{0}}{2},\qquad \omega\left(t\right)\equiv\omega_{0}\frac{R_{0}^{2}}{R^{2}\left(t\right)}. \tag{23}$$

Note that these definitions can be straightforwardly specialized to the case when the two principal axes X and Z are equal (equal to R), and thus one deals with the expansion of a rotating spheroid.

The single particle spectrum can be then written as

$$\frac{\mathrm{d}n}{\mathrm{d}^{3}\mathbf{p}}\propto\exp\left(-\frac{p_{x}^{2}}{2mT_{x}}-\frac{p_{y}^{2}}{2mT_{y}}-\frac{p_{z}^{2}}{2mT_{z}}-\frac{\beta_{xz}}{m}p_{x}p_{z}\right)=$$

$$=\exp\left(-\frac{p'^{2}_{x}}{2mT'_{x}}-\frac{p'^{2}_{y}}{2mT'_{y}}-\frac{p'^{2}_{z}}{2mT'_{z}}-\frac{\beta'_{xz}}{m}p'_{x}p'_{z}\right), \tag{24}$$

where T_x, T_y, T_z, and β_{xz}, as well as T'_x, T'_y, T'_z and β'_{xz} characterize the slope parameters (and the cross-terms) in the K (laboratory) frame and the K' frame (the eigen-frame of the rotated ellipsoid), respectively. The expression of these parameters are given as

$$T'_x = T^*_{xx} - \frac{T^{*2}_{xz}}{T^*_{zz}}, \tag{25}$$

$$T'_z = T^*_{zz} - \frac{T^{*2}_{xz}}{T^*_{xx}}, \tag{26}$$

$$\beta'_{xz} = \frac{T^*_{xz}}{T^*_{xx}T^*_{zz} - T^{*2}_{xz}}, \tag{27}$$

$$\frac{1}{T_x} = \frac{\cos^2\vartheta}{T'_x} + \frac{\sin^2\vartheta}{T'_z} + \beta'_{xz}\sin(2\vartheta), \tag{28}$$

$$\frac{1}{T_z} = \frac{\sin^2\vartheta}{T'_x} + \frac{\cos^2\vartheta}{T'_z} - \beta'_{xz}\sin(2\vartheta), \tag{29}$$

$$\beta_{xz} = \frac{\sin(2\vartheta)}{2}\left(\frac{1}{T'_z} - \frac{1}{T'_x}\right) + \beta'_{xz}\cos(2\vartheta), \tag{30}$$

$$T'_y = T'_y = T'_{0y} = T + m\dot{Y}^2. \tag{31}$$

We see that the formulas relating the parameters in the K and K' frames are simply the ones describing a rotation by a fixed θ angle (the value of the tilt angle at freeze-out). The fact that a cross term (i.e. nonzero β'_{xz}) appears (this is a new feature compared to Ref. 12) signals that the single-particle spectrum is *not* diagonal in the K' frame: the tilt of the coordinate-space ellipsoid is not the same as that of the ellipsoid determined from the single-particle spectrum.

The expression of the azimuthal angle-averaged p_T spectrum and the anisotropy (v_n) parameters follow the footsteps of Ref. 12: one can introduce the scaling variables ν and w and the effective temperature $T_{\rm eff}$ as

$$\nu \equiv -\frac{\beta_{xz}}{m}p_z p_T, \quad w \equiv \frac{p_T^2}{4m}\left(\frac{1}{T_y} - \frac{1}{T_x}\right), \quad \frac{1}{T_{\rm eff}} = \frac{1}{2}\left(\frac{1}{T_y} + \frac{1}{T_x}\right). \tag{32}$$

Using the $I_n \equiv I_n(w) = \frac{1}{\pi}\int_0^\pi \cos(n\varphi)\, e^{w\cos\varphi}\mathrm{d}\varphi$ modified Bessel functions of w, up to second order in v (and thus in the y rapidity) one has

$$E\frac{\mathrm{d}n}{\mathrm{d}^3\mathbf{p}} = \frac{1}{2\pi p_T}\frac{\mathrm{d}n}{\mathrm{d}p_T\mathrm{d}y}\left\{1 + 2\sum_{n=1}^{\infty} v_n\cos\left[n\left(\varphi - \Psi_n\right)\right]\right\} \quad \Rightarrow \tag{33}$$

$$\frac{1}{2\pi p_T}\frac{\mathrm{d}n}{\mathrm{d}p_T\mathrm{d}y} \propto \exp\left\{-\frac{p_z^2}{2mT_z} - \frac{p_T^2}{2mT_{\rm eff}}\right\} \times \left[I_0(w) + \frac{I_0(w) + I_1(w)}{4}\nu^2\right], \tag{34}$$

$$v_1 = \frac{\nu}{2}\frac{I_0(w) + I_1(w)}{I_0(w)}, \tag{35}$$

$$v_2 = \frac{I_1(w)}{I_0(w)} + \frac{\nu^2}{8}\left[1 + \frac{I_2(w)}{I_0(w)} - 2\frac{I_1^2(w)}{I_0^2(w)}\right], \tag{36}$$

$$v_3 = \frac{\nu}{2}\frac{I_1(w) + I_2(w)}{I_0(w)}. \tag{37}$$

Here φ denotes the azimuthal angle, Ψ_n is the nth order event plane (in our model, all the event planes coincide, i.e. fluctuation effects are not taken into account). Fig. 1 presents the characteristic rapidity dependence of v_1, v_2 and v_3. The two-particle Bose-Einstein correlation also turns out

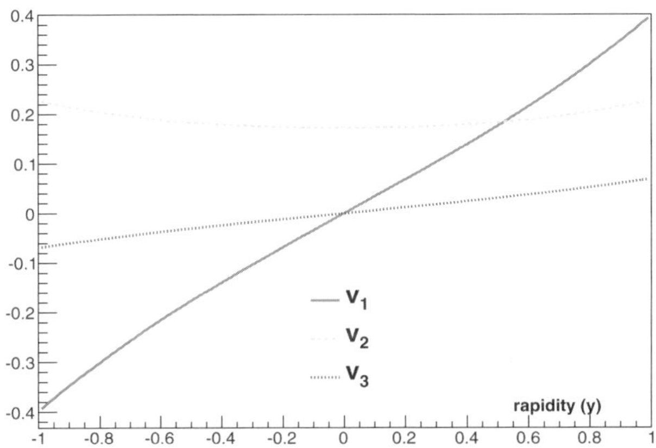

Fig. 1. Dependence of the azimuthal anisotropy parameters v_1, v_2 and v_3 on rapidity y, for a reasonable choice of the freeze-out parameters (namely: $\vartheta_f = \pi/5$, $\beta'_{xz} = 5 \cdot 10^{-4}$ 1/MeV, $p_T = 300$ MeV/c, $m = 140$ MeV, $T_x = 270$ MeV, $T_y = 170$ MeV, $T_z = 600$ MeV). The dependence of v_1 and v_3 is characteristic to a system tilted by some ϑ_f angle.

to be a Gaussian, and can either be expressed in terms of the $\mathbf{q}'$ of the $\mathbf{q}$ relative momentum (in the K and K' frames, respectively), or using the Bertsch-Pratt parametrization. Simple calculation yields

$$C(\mathbf{K}, \mathbf{q}) = 1 + \lambda \exp\left\{-\sum_{ij} q'_i q'_j R'^2_{ij}\right\} = 1 + \lambda \exp\left\{-\sum_{ij} q_i q_j R^2_{ij}\right\}, \tag{38}$$

$$R'^2_x = X^2 \frac{T_f}{T'_x}\,, \tag{39}$$

$$R'^2_z = Z^2 \frac{T_f}{T'_z}\,, \tag{40}$$

$$R'^2_{xz} = XZT_f\beta'_{xz}\,, \tag{41}$$

where we utilized the already calculated values of the T'_x, T'_z, β'_{xz} slope parameters. These radii are valid in the K' frame; transformation into the K frame (lab frame) is straightforward:

$$R^2_x = R'^2_x \cos^2\vartheta + R'^2_z \sin^2\vartheta + R'^2_{xz} \sin(2\vartheta)\,, \tag{42}$$

$$R^2_z = R'^2_x \sin^2\vartheta + R'^2_z \cos^2\vartheta - R'^2_{xz} \sin(2\vartheta)\,, \tag{43}$$

$$R^2_{xz} = \frac{\sin(2\vartheta)}{2}\left(R'^2_x - R'^2_z\right) + \cos(2\vartheta)\, R'^2_{xz}\,. \tag{44}$$

The y direction, as before, is much simpler:

$$R'^2_y = R^2_y = \frac{Y^2 T_f}{T_y}\,. \tag{45}$$

Customarily introducing the o, s, l (out, side, long) Bertsch-Pratt components of $\mathbf{q}$ in the longitudinally co-moving system (LCMS), one has

$$C(\mathbf{K},\mathbf{q}) = 1 + \lambda \exp\left\{ - \sum_{a,b=o,s,l} q_a q_b R^2_{ab} \right\}\,, \tag{46}$$

and the radii as functions of the pair momentum azimuthal angle φ are

$$\begin{aligned}
R'^2_{oo} &\equiv R^2_o = R^2_x \cos^2\varphi + R^2_y \sin^2\varphi,\\
R'^2_{ss} &\equiv R^2_s = R^2_x \sin^2\varphi + R^2_y \cos^2\varphi,\\
R'^2_{oo} &\equiv R^2_l = R^2_z,\\
R'^2_{os} &= \left(R^2_y - R^2_x\right) \sin\varphi \cos\varphi,\\
R'^2_{ol} &= R^2_{xz} \cos\varphi,\\
R'^2_{sl} &= R^2_{xz} \sin\varphi.
\end{aligned} \tag{47}$$

Fig. 2 shows the Bertsch-Pratt radii as a function of pair momentum azimuthal angle. The 2nd order oscillating ($\cos^2\varphi$, $\sin^2\varphi$ containing) BP-radii, also the cross-terms, were recently extensively measured by the STAR collaboration,[14] however, as we see, it is not these oscillations that signal rotational expansion.

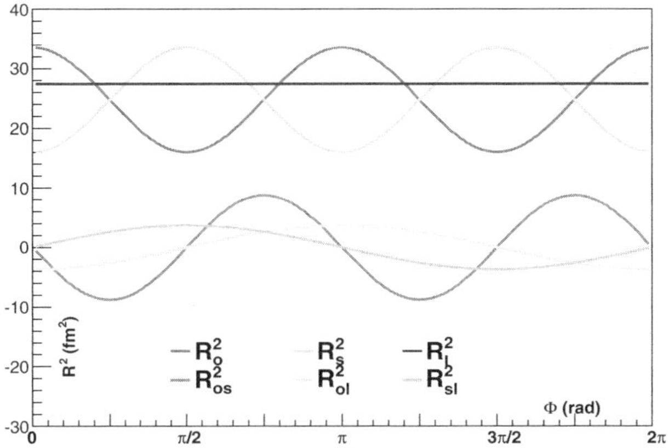

Fig. 2. φ dependence of the Bertsch-Pratt radii for a reasonable choice of the freeze-out parameters (namely: $\vartheta_f = \pi/5$, $R'^2_x = 25$ fm^2, $R'^2_y = 16$ fm^2, $R'^2_z = 36$ fm^2, $R'^2_{xz} = 5$ fm^2). The oscillation of R_s, R_o and R_{os} with a π period is characteristic to an ellipsoid-shaped source. The 2π period oscillation of R_{ol} and R_{sl} are characteristic to a tilted (and rotating) source.

4. Discussion and outlook

Most of the formulae of the observables derived here show a very close similarity (even identity) to those calculated in Ref. 12, where a *stationary, but tilted* ellipsoidal source was assumed. The universal scaling of the elliptic flow and other anisotropies (a feature of experimental data that was successfully explained in terms of the Buda-Lund model[15]) is preserved here, just the scaling variables are expressed in a bit more involved way, see Eq. (32). The oscillation of the HBT radii characteristic to rotation also shows up in other models of tilted ellipsoidal sources (just as in Ref. 12; a simpler model was introduced earlier in Ref. 16).

The new result in the present work is two-fold: first, an actual (and in some sense, unique) rotating and expanding hydrodynamical solution is found which naturally leads to tilted sources from a non-tilted initial condition (and it can be applied to follow the time-evolution of rotation, for any given $\kappa(T)$ EoS). Second, the result for the observables are refined in a way that takes not only the tilt but also the rotational flow into account.

The most striking consequence of tilted (and also rotating) expansion are seen in the rapidity dependence of the anisotropy parameters (Fig. 1) as well as the *directed flow-like* (i.e. azimuthally 2π-periodic) oscillations of the R_{sl} and R_{ol} HBT radius parameters (the appearance of these cross-terms is purely due to rotation, see Eq. (47)). A natural proposal is thus to measure these observables to infer the rotational angle ϑ_f and possibly the angular velocity. It is worthwhile to mention that to do these measurements, one needs experimental information on not only the usual (2nd order) reaction plane but also on the first order reaction plane; this may be done precisely by utilizing the y-dependence of the v_1 anisotropy parameter. It is also worth to mention that in the case of models with spheroidal expansion (that was discussed e.g. in Refs. 4, 9), the "angle of rotation" ϑ_f becomes ill-defined, so the experimental signatures of rotation become much harder to be identified.

The detection of rotation has some far-going promises. A softer equation of state (caused e.g. by the presence of a critical endpoint on the QCD phase diagram) would mean that the matter expands less violently, thus the angle of rotation will be greater. In this manner, measuring the rotation angle as a function of collision energy can be of great importance in mapping out the critical endpoint and the location of quark-hadron phase transition. So we look forward to see whether our presented model is applicable to new collision-energy dependent measurements of HBT radii oscillations as well as rapidity dependent anisotropy parameters to infer the rotation of the expanding hot and dense matter. It would lead to new knowledge of the strongly interacting quark-gluon-plasma produced in heavy-ion collisions.

This work was supported by the Hungarian OTKA grant NK101438. M. N. was supported by the TÁMOP 4.2.4. A/1-11-1-2012-0001 "National Excellence Program", financed by the European Union and the State of Hungary, co-financed by the European Social Fund.

References

1. I. Karpenko, P. Huovinen and M. Bleicher, *Comput. Phys. Commun.* **185**, 3016 (2014) arXiv:1312.4160 [nucl-th].
2. L.D. Landau, *Izv. Akad. Nauk Ser. Fiz.* **17**, 51 (1953),
   I.M. Khalatnikov, *Zhur. Eksp. Teor. Fiz.* **27**, 529 (1954),
   S.Z. Belenkij, L.D. Landau, *Nuovo Cim. Suppl.* **3S10**, 15 (1956).
3. R.C. Hwa, *Phys. Rev.* **D10**, 2260 (1974),
   J.D. Bjorken, *Phys. Rev.* **D27**, 140 (1983).
4. L.P. Csernai, S. Velle and D.J. Wang, *Phys. Rev.* **C89**, 034916 (2014) arXiv:1305.0396 [nucl-th].
5. L.P. Csernai, D.J. Wang and T. Csörgő, *Phys. Rev.* **C90**, 024901 (2014) arXiv:1406.1017 [hep-ph].

6. T. Csörgő, M.I. Nagy and M. Csanád, Phys. Lett. B **663**, 306 (2008) arXiv:nucl-th/0605070.
7. M.I. Nagy, T. Csörgő and M. Csanád, *Phys. Rev.* **C77**, 024908 (2008) arXiv:0709.3677 [nucl-th].
8. M.I. Nagy, *Phys. Rev.* **C83**, 054901 (2011) arXiv:0909.4285 [nucl-th].
9. T. Csörgő and M.I. Nagy, *Phys. Rev.* **C89**, no. 4, 044901 (2014) arXiv:1309.4390 [nucl-th].
10. M. Csanád, M.I. Nagy and S. Lökös, *Eur. Phys. J.* **A48**, 173 (2012) arXiv:1205.5965 [nucl-th].
11. M. Csanád and A. Szabó, *Phys. Rev.* **C90**, 054911 (2014) arXiv:1405.3877 [nucl-th].
12. T. Csörgő, S.V. Akkelin, Y. Hama, B. Lukacs and Y.M. Sinyukov, *Phys. Rev.* **C67**, 034904 (2003) arXiv:hep-ph/0108067.
13. T. Csorgo, *Acta Phys. Polon.* **B37**, 483 (2006) arXiv:hep-ph/0111139.
14. L. Adamczyk *et al.* [STAR Collaboration], arXiv:1403.4972 [nucl-ex].
15. M. Csanád, B. Tomasik and T. Csörgő, *Eur. Phys. J.* **A37**, 111 (2008) arXiv:0801.4434 [nucl-th].
16. M.A. Lisa, U.W. Heinz and U.A. Wiedemann, *Phys. Lett.* **B489**, 287 (2000) arXiv:nucl-th/0003022.

DIFFRACTION AND SHADOWING PHENOMENA

DIFFRACTION AND UNITARITY

I.M. DREMIN

Lebedev Physical Institute, Moscow 119991, Russia
and
National Research Nuclear University MEPhI, Moscow 115409, Russia

I begin with a tribute to V.N. Gribov and then come to a particular problem which would be of interest for him. His first paper on reggeology was devoted to elastic scatterings of hadrons. Here, using the unitarity relation in combination with experimental data about the elastic scattering in the diffraction cone, I show how the shape and the darkness of the interaction region of colliding protons change with the increase of their energies. In particular, the collisions become fully absorptive at small impact parameters at LHC energies that results in some special features of inelastic processes as well. The possible evolution with increasing energy of the shape from the dark core at the LHC to the fully transparent one at higher energies is discussed. It implies that the terminology of the black disk would be replaced by the black torus.

1. Gribov's tribute

The life of Gribov in science was completely framed in the second half of the XXth century. I describe my impressions about this period specially emphasizing the events where our interests went closely.

1953 - I was already a student when I first heard about the new paper of Landau which dealt with the hydrodynamic theory of multiparticle production.

Л.Д. Ландау, О множественном образовании частиц при столкновениях быстрых частиц. Изв. АН СССР. Сер. физ., **17**, 51 (1953). [L.D. Landau, Multiparticle production in fast particle collisions, Izvestia AS USSR (1953)].

I tried to understand the details but it was quite difficult for me at that time. Later, I learned that Landau himself said that this work was the most hard for him among all his papers.

1956 - At the end of my fourth grade I.Ya. Pomeranchuk became my tutor and I started to visit regularly the Landau seminars on Thursdays at the Kapitsa institute and the Pomeranchuk seminars on Wednesdays at ITEP. In parallel to the student duties, I passed the famous Landau exams.

Once it happened that no talk was scheduled at the Pomeranchuk seminar and Landau proposed to discuss the physics schools in the USSR. His opinions impressed me very much and strongly influenced my own estimates.

1957 - It was at the very end of 1957 when I first heard the name Gribov. At one of Landau seminars, Pomeranchuk said: "Dau, there is a very talented physicist Gribov in Leningrad." Landau reacted immediately: "I know a talented actor Gribov but not a physicist Gribov." Actually, the actor Gribov from Moscow Art Academic Theater was very famous at that time.

1958 - In about a year, Gribov gave a talk at the Landau seminar on Regge papers of 1957 and commented how Regge's results could be used for high energy scattering of hadrons.

1958 - By the recommendation of Pomeranchuk, I started my PhD studies at the Lebedev Physical Institute. The cosmic ray school was very strong there and at some seminars I heard that beside central collisions of hadrons described by the Landau theory there are observed events where peripheral interactions play a crucial role. The peripheral region of a proton was usually treated as a cloud occupied by virtual pions with a single one at the very periphery. Therefore, in December 1958 I derived the formula for truly peripheral collisions of protons where both protons get excited due to the exchange by a single pion and emit pions. That was the very first proposal of the one-pion exchange model.

1959 - There were numerous discussions of this proposal at LPI where I.E. Tamm invited leading specialists. With D.S. Chernavsky we compared the results to recent data from the Dubna accelerator. The short talk was presented at the Kiev mini-"Rochester".

1960 - The complete publication in JETP only appeared in January 1960.

И.М. Дремин, Д.С. Чернавский, Периферические взаимодействия нуклонов при энергии 9 БэВ. ЖЭТФ **38**(1), 229 (1960). [I.M. Dremin, D.S. Chernavsky, Peripheral interactions of nucleons at 9 BeV, *Sov. Phys. JETP* **11**(1), 187 (1960)].

1961 - At that time, Gribov worked on implementation of his ideas to elastic scattering of hadrons and published the two papers.

В.Н. Грибов, О возможном асимптотическом поведении упругого рассеяния, ЖЭТФ, **41**(2), 667 (1961) [V.N. Gribov, Possible Asymptotic Behavior of Elastic Scattering, *Sov. Phys. JETP* **14**(2), 478 (1962)].

В.Н. Грибов, Парциальные волны с комплексными орбитальными моментами и асимптотическое поведение амплитуды рассеяния,

ЖЭТФ **41**(6), 1962 (1961). [V.N. Gribov, Partial Waves with Complex Orbital Angular Momenta and the Asymptotic Behavior of the Scattering Amplitude, *Sov. Phys. JETP* **14**(6), 1395 (1962)].

1962 - Our activities seemed to be not connected directly until the following paper was published:

D. Amati, A. Stanghellini, and S. Fubini, Theory of High Energy Scattering and Multiple Production, *Nuovo Cim.* **24**, 896 (1962).

It is shown there that the multiperipheral model generalizing the one-pion exchange approach leads directly to the Regge-Gribov formulae for elastic scattering when applying the unitarity condition.

60s-70s-80s - An explosive flow of publications on multiperipheral and multireggeon models was observed at these years. At the very beginning, Gribov did not accept the idea that they really can be used for the underlying mechanism of elastic processes considered as a shadow of inelastic ones. We had numerous hard discussions! However, his own pupils persuaded him to believe that this is true. He invited me to be a referee of dissertations of E.M. Levin (with the famous misprint!) and M.G. Ryskin where this relationship was discussed. The influence of K.A. Ter-Martirosyan and A.B. Kaidalov working on multireggeon models was also very important.

90s - Gribov's 60th birthday was celebrated at Orsay. He gave a talk on confinement at the seminar there. I remember that somehow people left quite fast after the seminar but two of us still stayed at the blackboard and discussed the talk. I mentioned that the Keldysh diagram technique could possibly be used for this purpose in place of Feynman technique and Gribov said that he thought about it but did not get enough time to progress there. Nowadays there is some progress in using that technique for nuclei collisions. The title of the talk was close to the title of his last paper,

V.N. Gribov, The theory of quark confinement, *Eur. Phys. J.* **C 10** (1), 91 (1999), hep-ph/9902279.

2. Elastic and inelastic processes united by unitarity

I have mentioned already that elastic scattering is a shadow of inelastic processes. They are directly related by the unitarity condition which is nothing else as an indubitable statement that the total probability of all processes is equal to 1. In this talk, I show that the shape and the opacity of the inelastic interaction region of two colliding protons is strictly determined by the behavior of their elastic scattering amplitude at high energies.

Let me stress that all conclusions are obtained just in the framework of the indisputable unitarity condition using experimental data about the elastic scattering of protons in the diffraction cone without any reference to quantum chromodynamics (QCD) or phenomenological approaches.

The only information about elastic scatterings at a fixed energy s comes from the measurement of the differential cross section as a function of the transferred momentum t at its experimentally available values and of the ratio of the real and imaginary parts $\rho(s,t) = \mathrm{Re}f(s,t)/\mathrm{Im}f(s,t)$ of the elastic scattering amplitude $f(s,t)$ just in forward direction $t = 0$ $\rho(s,0) = \rho_0$ but not at any other values of t. The latter one is obtained from studies of the interference between the nuclear and Coulomb contributions to the amplitude f which becomes practically noticeable only in the near-forward direction.

The differential cross section is related to the scattering amplitude as

$$\frac{d\sigma}{dt} = |f(s,t)|^2. \tag{1}$$

Thus, from measurements of the differential cross section at any energy of colliding particles we get the knowledge only about the modulus of the amplitude at the experimentally available values of t. As a first approximation at present energies, it can be described at comparatively small transferred momenta by the exponential shape with the slope B such that

$$\frac{d\sigma}{dt} = \frac{\sigma_t^2}{16\pi}\exp(-B|t|). \tag{2}$$

This region is called the diffraction peak. It becomes higher and its width shrinks with increasing energy because both the total cross section σ_t and the slope B increase with energy.

The unitarity of the S-matrix SS^+=1 imposes definite requirements on the amplitude of elastic scatterings $f(s,t)$ and amplitudes of inelastic processes M_i. In the s-channel it looks[1,2] like

$$\mathrm{Im}f(p,\theta) = I_2(p,\theta) + g(p,\theta) =$$

$$\frac{s}{8\pi^{3/2}}\int\int d\theta_1 d\theta_2 \frac{\sin\theta_1 \sin\theta_2 f(p,\theta_1) f^*(p,\theta_2)}{\sqrt{[\cos\theta - \cos(\theta_1+\theta_2)][\cos(\theta_1-\theta_2) - \cos\theta]}} + g(p,\theta). \tag{3}$$

For the region of integration in Eq. (3) see Refs. 1, 2. The non-linear integral term represents the two-particle intermediate states of the incoming particles. The term

$$g(p,\theta) \propto \sum_i \int d\Phi_i M_i M_i^*(\theta) \tag{4}$$

represents the shadowing contribution of inelastic processes to the imaginary part of the elastic scattering amplitude (the overlap function[3]).

At $t = 0$ it leads to the optical theorem,

$$\mathrm{Im} f(s,0) = \sigma_t / 4\sqrt{\pi}\,, \tag{5}$$

and, after being integrated over b, to the general statement that the total cross section is the sum of cross sections of elastic and inelastic processes,

$$\sigma_t = \sigma_{el} + \sigma_{in}\,. \tag{6}$$

To define the geometry of the collision we must express all characteristics presented by the angle θ and the transferred momentum t in terms of the transverse distance between the centers of the colliding protons called the impact parameter b. It is easily done by the Fourier – Bessel transform of the amplitude f which retranslates the momentum data to the transverse space features and is written as

$$i\Gamma(s,b) = \frac{1}{2\sqrt{\pi}} \int_0^\infty d|t| f(s,t) J_0(b\sqrt{|t|})\,. \tag{7}$$

The unitarity condition in the b-representation reads

$$G(s,b) = 2\mathrm{Re}\Gamma(s,b) - |\Gamma(s,b)|^2\,. \tag{8}$$

The left-hand side (the overlap function in b-representation) describes the transverse impact-parameter profile of inelastic collisions of protons. It is just the Fourier – Bessel transform of the overlap function g. It satisfies the inequalities $0 \leq G(s,b) \leq 1$ and determines how absorptive is the interaction region depending on the impact parameter (with $G = 1$ for the full absorption and $G = 0$ for the complete transparency). The profile of elastic processes is determined by the subtrahend in Eq. (8).

Even though the impact parameter can not be directly measured, the geometric picture is instructive and closely related to such experimentally found characteristics as the ratio of the diffraction cone slope to the total cross section that provides immediate guides to its energy evolution.

The diffraction cone contributes mostly to the Fourier - Bessel transform of the amplitude. Using the above formulae, one can write the dimensionless Γ as

$$i\Gamma(s,b) = \frac{\sigma_t}{8\pi} \int_0^\infty d|t| \exp(-B|t|/2)(i + \rho(s,t)) J_0(b\sqrt{|t|})\,. \tag{9}$$

Here, the diffraction cone approximation (2) is inserted. Then one calculates

$$\mathrm{Re}\Gamma(s,b) = \frac{1}{Z} \exp(-\frac{b^2}{2B})\,, \tag{10}$$

where we introduce the dimensionless ratio of the cone slope (or σ_{el}) to the total cross section,

$$Z = \frac{4\pi B}{\sigma_t} = \frac{\sigma_t}{4\sigma_{el}} \, . \tag{11}$$

The ratio $\rho(s,t)$ is small at $t = 0$ and, at the beginning, we neglect it and get the robust formula describing well the experimental shape of G in Fig. 1,

$$G(s,b) = \frac{2}{Z} \exp(-\frac{b^2}{2B}) - \frac{1}{Z^2} \exp(-\frac{b^2}{B}) \, . \tag{12}$$

For central collisions with $b = 0$ it gives

$$G(s, b=0) = \frac{2Z-1}{Z^2} \, . \tag{13}$$

Thus, the darkness at the very center is fully determined by the parameter Z, i.e. by the ratio of experimentally measured characteristics - the slope B (or σ_{el}) to the total cross section. Their energy evolution defines the evolution of the absorption value. It becomes completely absorptive in the center only at $Z = 1$ and diminishes for other values of Z. At ISR energies $Z \approx 1.5$, then it decreases systematically and at LHC energies becomes equal to 1 within the accuracy of the measurements of B and σ_t (see the Table 1).

Table 1. The energy behavior of Z and $G(s,0)$ obtained from experiment.

$\sqrt{s}$, GeV	2.70	4.11	4.74	7.62	13.8	62.5	546	1800	7000
Z	0.64	1.02	1.09	1.34	1.45	1.50	1.20	1.08	1.00
$G(s,0)$	0.68	1.00	0.993	0.94	0.904	0.89	0.97	0.995	1.00

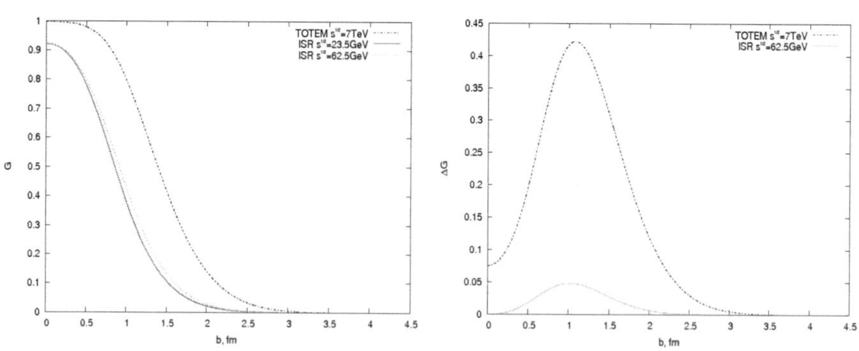

Figure 1. (left) The overlap function $G(s,b)$ at 7 TeV (upper curve)[4] compared to those at ISR energies 23.5 GeV and 62.5 GeV. (right) Their differences (computed using the fit of experimental data by the model[5]).

The energy evolution of the inelastic profile $G(s,b)$ is shown in Fig. 1. The central core is very absorptive at LHC. It is responsible for jets.[6] It is possible[7] to use this information from elastic processes for analysis of experimental data of CMS collaboration at 7 TeV about high multiplicity inelastic collisions triggered by the hadron jet production.[8] Its important conclusion is that all Monte Carlo models fail to describe this subclass of the processes and ask for some theory improvement at high multiplicities.

The positivity of $G(s,b)$, i.e., of $\sigma_{inel}(s,b)$, imposes some limits on the relative role of B and σ_t. It follows from Eq. (13) that

$$2Z = \frac{8\pi B}{\sigma_t} = \frac{\sigma_t}{2\sigma_{el}} \geq 1\,, \tag{14}$$

i.e., the slope B should increase asymptotically at least as strong as the total cross section σ_t. The values $Z < 0.5$ are unphysical since they lead to negative $\sigma_{inel}(s,b)$. However, they were discussed at the name of the antishadow or reflective effect in Ref. 9 and of the resonant disk mode in Ref. 10.

It is usually stated that the equality $Z = 0.5$ corresponds to the black disk limit with equal elastic and inelastic cross sections $\sigma_{el} = \sigma_{in} = 0.5\sigma_t$. One can see from Eq. (13) that $G(s,b=0) = 0$ at $Z = 0.5$, i.e. no inelastic interactions take place in central collisions. Surely, one should not call it as a black disk. This paradox is resolved[11] if we write the inelastic profile of the interaction region using Eq. (12). At $Z = 0.5$ it looks like

$$G(s,b) = 4[\exp(-\frac{b^2}{2B}) - \exp(-\frac{b^2}{B})]. \tag{15}$$

We see that one must rename the black disk as a black torus (or a black ring in the two-dimensional projection) with full absorption $G(s,b_m) = 1$ at the impact parameter $b_m = \sqrt{2B\ln 2}$, complete transparency at $b = 0$ and rather large half-width. Thus, the evolution to values of Z smaller than 1 at higher energies (this can happen if the decreasing tendency of Z with energy from ISR to LHC persists as, e.g., is claimed in some models[12]) would imply a quite special transition from the critical two-scale regime at the LHC to the concave torus-like configurations of the interaction region if the exponential shape of the diffraction cone, described by Eq. (2), persists. It looks as if the protons penetrate through one another at central collisions, just scattering elastically, while peripheral collisions become responsible for inelastic processes. They begin to prevail over elastic scattering only at the periphery $b > b_m$.

A signature of transition to such a regime at 7 TeV at the level 10^{-4} in a single point was claimed in Ref. 13 while the accuracy of measuring Z is about 10^{-3}. No such signature is seen in Ref. 4. Different results of Refs. 4, 13 using the fit by the same model[5] prevent from any (even preliminary) statements. Here we just come to the close connection of properties of inelastic collisions to the results obtained with the help of the unitarity relation from experimental data on elastic scattering.

The maximum absorption at central collisions at LHC energies must reveal itself in some special features of inelastic collisions in such a critical regime with $Z \approx 1$. One of such features is the abundant production of jets. Below, we demonstrate that they are created mostly at the very central collisions.

The diffraction cone contributes mostly to $G(s,b)$. The large-$|t|$ elastic scattering can not serve as an effective trigger of the black core. One of the typical features of the high energy inelastic processes is the production of high energy jets, i.e. collimated groups of particles. The inelastic exclusive processes can be more effectively used for the analysis of the central black core. One needs such triggers which enhance its contribution. Following the suggestions of Refs. 6, 14, it becomes possible[7] to study the details of the central core using the experimental data of CMS collaboration at 7 TeV about inelastic collisions with high multiplicity triggered by the hadron jet production[8] as well as some other related data. Triggers (charged particles or jets) with large transverse momenta are produced in central collisions. Therefore, the black plateau in the central part of the interaction region with $b < 0.4 - 0.5$ fm should result in the corresponding plateau of the charged particle density in the transverse region $60^\circ < |\Delta\phi| < 120^\circ$ defined as follows:

$$\mu_{tr} = \frac{N_{ch}^{tr}}{\Delta\eta\Delta(\Delta\phi)}, \tag{16}$$

where N_{ch}^{tr} is the charged particle multiplicity in the transverse region, $\Delta\eta$ is the pseudorapidity range studied, $\Delta(\Delta\phi)$ is the azimuthal width of the transverse region. This is really the case as shown in Fig. 2.

Let us explain it. Starting from large transverse momenta of triggers on the right-hand side of Fig. 2 and going to the left, we somehow feel at the beginning the central region from $b = 0$ to the end of the plateau. Then the density of accompanying particles in the transverse region should not change until we approach the end of it. Only then the decrease of the

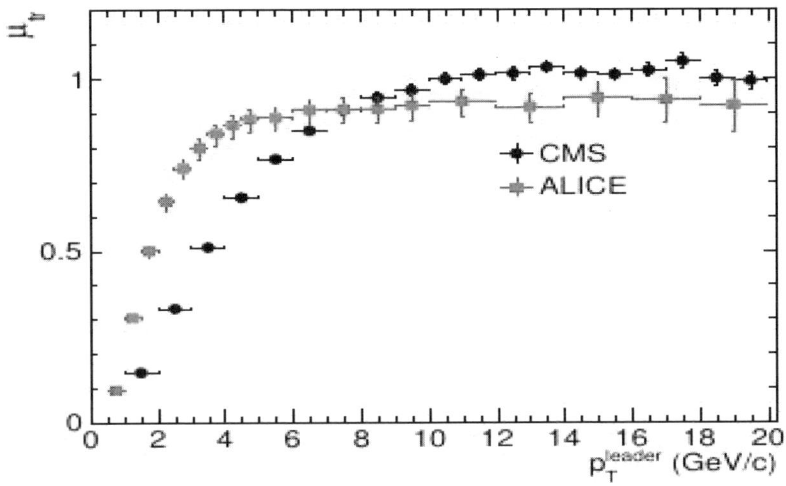

Figure 2. Charged-particle density in the transverse region as a function of p_T of leading object[7] (CMS - charged-particle jet, ALICE - charged particle). CMS analyses particles with $p_T > 0.5$ GeV/c and $|\eta| < 2.4$, ALICE - with $p_T > 0.5$ GeV/c and $|\eta| < 0.8$.

distribution of the accompanying particles should start. The difference in the positions of such a decrease in the two plots is defined by the difference in the choice of the leading trigger used by the two collaborations. The flat dependence of μ_{tr} on p_T shows that the activity in the transverse region is independent of the hard process scale, provided that the scale is hard enough that all proton-proton interactions are central.

Many other characteristics of such inelastic processes considered in Ref. 7 support this conclusion. Using very high multiplicity events in combination with jet properties is crucial. In particular, the most important observation is that the significant reduction of the jet rate at very high multiplicities compared to MC predictions asks for new inputs in the models. Separating the core contribution with the help of these triggers, one comes to the important conclusion that the simple increase of the geometrical overlap area of the colliding protons does not account for properties of jet production at very high multiplicities. It looks as if the parton (gluon) density must strongly increase in central collisions and rare configurations (fluctuations) of the partonic structure of protons are involved. Up to now, the paper[7] is the first attempt to explain the distinctive features of jets and underlying event properties at LHC energies. The correlation studies of jets (see, e.g., Ref. 15) can be used for further femtoscopy of the fine structured system. Surely, some further proposals to learn the critical regime at 7 - 8 TeV will be proposed.

At the same time, implications of energy evolution of Z (if it will be observed!) for inelastic processes are of great interest. The relative roles of elastic and inelastic cross sections will start approaching one another in the case if values of Z would diminish below 1 with increase of energy. The mean multiplicity will, probably, decrease because of the more peripheral origin of newly created particles. The decreasing role of central interactions can lead to some changes in the shape of the multiplicity distributions (lower tails?) and diminished share of jets which will acquire new features. Jets will become produced at the periphery in distinction to the situation described above. That would imply that they will have to penetrate larger distances in transverse direction compared to forward directions. It would give rise to their stronger depletion in the transverse plane and, therefore, to the azimuthal asymmetries which were looked for in Ref. 15. Surely, there will be found other criteria of transition to the concave shape of the interaction region of inelastic processes.

Hardly, we will be able to reach the regime with extremely small values of $Z < 0.5$ when inelastic processes would play negligibly small role compared to elastic ones.

At the very end, let me briefly mention the problem of the ratio of real and imaginary parts outside the diffraction cone. No phenomenological model has been able to describe the behavior of the elastic differential cross section in this region. It was shown long ago[1,16] that this can be done with the help of the unitarity condition. The imaginary part of the amplitude f outside the diffraction cone can be derived from the general unitarity condition, Eq. (3), which is reduced there to the inhomogeneous linear integral equation. Its solution prescribes the exponential (the exponent is predicted in the analytical form!) decrease with $\sqrt{|t|}$ plus imposed on it damped oscillations responsible for the dip of the differential cross section and no zeros while all phenomenological models ask for its zero there. This problem has not been resolved up to now.

The only unknown ingredient in the predicted exponent is the ratio of real to imaginary parts of the amplitude outside the diffraction cone. Replacing it by its average value there one is able to fit experimental data. The great surprise of the fit of TOTEM data at 7 TeV[17] was necessity to use there the large (in modulus) negative value of this ratio $\rho_l \approx -2.1$ (if $\rho_0 = 0.14$) not predicted by phenomenology that poses another problem. It is required by the increased (compared to lower energies) slope in this region.

I am grateful for support by the RFBR grants 12-02-91504-CERN-a, 14-02-00099 and the RAS-CERN program.

Bibliography

1. I.V. Andreev, I.M. Dremin, *Pis'ma Zh. Eksp. Teor. Fiz.* **6**, 810 (1967).
2. I.M. Dremin, *UFN* **183**, 3 (2013); *Physics-Uspekhi* **56**, 3 (2013).
3. L. Van Hove, *Nuovo Cimento* **28**, 798 (1963).
4. I.M. Dremin, V.A. Nechitailo, *Nucl. Phys.* **A916**, 241 (2013).
5. U. Amaldi, K.R. Schubert, *Nucl. Phys.* **B166** 301 (1980).
6. L. Frankfurt, M. Strikman, C. Weiss, *Phys. Rev.* **D83**, 054012 (2004).
7. M.Yu. Azarkin, I.M. Dremin, M. Strikman, *Phys. Lett.* **B735**, 244 (2014).
8. CMS Collaboration, *Eur. Phys. J.* **C73**, 2674 (2013).
9. S.M. Troshin, N.E. Tyurin, *Phys. Lett.* **B316**, 175 (1993).
10. V.V. Anisovich, V.A. Nikonov, J. Nyiri, arXiv:1408.0692 [hep-ph].
11. I.M. Dremin, *JETP Lett.* **99**, 243 (2014).
12. D.A. Fagundes, M.J. Menon, *Nucl. Phys.* **A880**, 1 (2012); arXiv:1112.5115 [hep-ph].
13. A. Alkin, E. Martynov, O. Kovalenko, S.M. Troshin, *Phys. Rev.* **D89** 091501(R) (2014).
14. L. Frankfurt, M. Strikman, C. Weiss, *Phys. Rev.* **D 69**, 114010 (2004).
15. CMS Collaboration PAS FSQ-13-005 (2013).
16. I.V. Andreev, I.M. Dremin, *Sov. J. Nucl. Phys.* **8**, 473 (1969).
17. I.M. Dremin, V.A. Nechitailo, *Phys. Rev.* **D85**, 074009 (2012).

HADRON DIFFRACTIVE PRODUCTION AT ULTRAHIGH ENERGIES AND SHADOW EFFECTS

V.V. ANISOVICH, M.A. MATVEEV

National Research Centre "Kurchatov Institute",
Petersburg Nuclear Physics Institute, Gatchina, 188300, Russia

V.A. NIKONOV

National Research Centre "Kurchatov Institute",
Petersburg Nuclear Physics Institute, Gatchina, 188300, Russia
and
Helmholtz-Institut für Strahlen- und Kernphysik, Universität Bonn, Germany

Shadow effects at collisions of hadrons with light nuclei at high energies were subject of scientific interest of V.N. Gribov, first, we mean his study of the hadron-deuteron scattering, see *Sov. Phys. JETP* **29**, 483 (1969) [*Zh. Eksp. Teor. Fiz.* **56**, 892 (1969)] and discovery of the reinforcement of shadowing due to inelastic diffractive rescatterings. It turns out that the similar effect exists on hadron level though at ultrahigh energies.

Diffractive production is considered in the ultrahigh energy region where pomeron exchange amplitudes are transformed into black disk ones due to rescattering corrections. The corresponding corrections in hadron reactions $h_1 + h_3 \to h_1 + h_2 + h_3$ with small momenta transferred ($q^2_{1\to 1} \sim m^2/\ln^2 s$, $q^2_{3\to 3} \sim m^2/\ln^2 s$) are calculated in terms of the K-matrix technique modified for ultrahigh energies. Small values of the momenta transferred are crucial for introducing equations for amplitudes. The three-body equation for hadron diffractive production reaction $h_1 + h_3 \to h_1 + h_2 + h_3$ is written and solved precisely in the eikonal approach. In the black disk regime final state scattering processes do not change the shapes of amplitudes principally but dump amplitudes by a factor $\sim \frac{1}{4}$; initial state rescatterings result in additional factor $\sim \frac{1}{2}$. In the resonant disk regime initial and final state scatterings damp strongly the production amplitude that corresponds to $\sigma_{inel}/\sigma_{tot} \to 0$ at $\sqrt{s} \to \infty$ in this mode.

1. Introduction

Recent data for diffractive production of hadrons[1,2] demonstrate a remarkable phenomenon: an appearance of a black spot in the impact parameter presentation of the pp-scattering amplitude. So, one may suppose that the black disk picture starts at $\sqrt{s} \sim 10 - 100$ TeV. The key point is the steady growth of total and elastic cross sections up to the region $\sqrt{s} \sim 5-50$ TeV, for the pre-LHC data see Refs. 3–8. The phenomenon of the increase of high energy cross sections was discussed for a long time. First, the power growth, s^α with $\alpha > 1$, was suggested[9–11] on the basis of the reggeon exchange notion. Then, it was shown in Refs. 12–14 that the power-type growth of scattering amplitudes with energy is dumped to $\ln^2 s$-type within the s-channel unitarization. The black disk picture at ultrahigh energies is realized in the Dakhno-Nikonov model[15] for πp and $p^\pm p$ collisions. The model, being QCD-motivated, takes into account the quark structure of colliding hadrons, the gluon origin of the input pomeron and the color screening effects in collisions. The model can be considered as a realization of the Good-Walker eikonal approach[16] for a continuous set of channels.

An appropriate way for the description of the diffractive scattering data at ultrahigh energies seems to be the use of the profile function in a version of the Good-Walker approach with the Froissart bound[17] (though exceeding the Froissart bound does not violate necessarily the general constraints[18]). Examples of such descriptions can be found in Refs. 19–23 and references therein.

The description of the pre-LHC and LHC data in terms of the Dakhno-Nikonov model[23] and the extension of results into the ultrahigh energy region[24] tells that the 5-50 TeV region turns out to be that where the asymptotic behavior starts. The asymptotic regime should reveal itself definitely at $10^2 - 10^4$ TeV; a short summary of predictions is given in Ref. 25.

For the ultrahigh energy limit the black disk picture predicts a $(\ln^2 s)$-growth for total and elastic hadron-hadron cross sections: $\sigma_{tot} \sim \ln^2 s$ with $[\sigma_{el}/\sigma_{tot}]_{\ln s\to\infty} \to 1/2$. Further, the differential elastic cross sections depend asymptotically on transverse momenta with a relation for τ-scaling: $d\sigma_{el}(\tau)/d\tau = D(\tau)$ with $\int_0^\infty d\tau D(\tau) = \sigma_{el}(s)$ and $\tau = \mathbf{q}_\perp^2 \sigma_{tot} \propto \mathbf{q}_\perp^2 \ln^2 s$. The black disk picture with a universal growth of radius realizes the Gribov universality[26]: the asymptotic equality $\sigma_{tot}(pp)/\sigma_{tot}(\pi p) \to 1$ and $\sigma_{tot}(\pi p)/\sigma_{tot}(\pi\pi) \to 1$ at $\ln s \to \infty$. The universal behavior of all total and elastic cross sections is the consequence of the universality of the colliding disk structure, or the structure of parton clouds at ultrahigh energy. The diffractive dissociation processes are increasing at asymptotic energies ($\sigma_D \propto \ln s$, $\sigma_{DD} \propto \ln s$) but their relative contribution tends to zero ($\sigma_D/\sigma_{tot} \to 0$, $\sigma_{DD}/\sigma_{tot} \to 0$).[24]

In the $(\ln s, \mathbf{b})$-space ($\mathbf{b}$ is impact parameter) disks form solitons[27] which are running from small $|\mathbf{b}|$ to larger values with energy increase.

The universal character of the black disks and the τ-scaling phenomenon open a path for consideration of hadron productions in diffractive collisions. The simplest process of this type is the diffractive production of hadron shower, $p+p \to M_X+p$, the cross section of this process at moderately high energies is usually modeled by three-pomeron diagrams. More complicated for consideration is the process of diffractive scattering with the production of a third particle, $pp \to php$, with large pair energies ($s_{ph} \sim s_{hp} \sim m\sqrt{s}$) and small momenta transferred to protons ($\mathbf{q}^2_{p\perp} \sim m^2/\ln^2 s$). This process is the subject of our studies in this paper (see also Fig. 1).

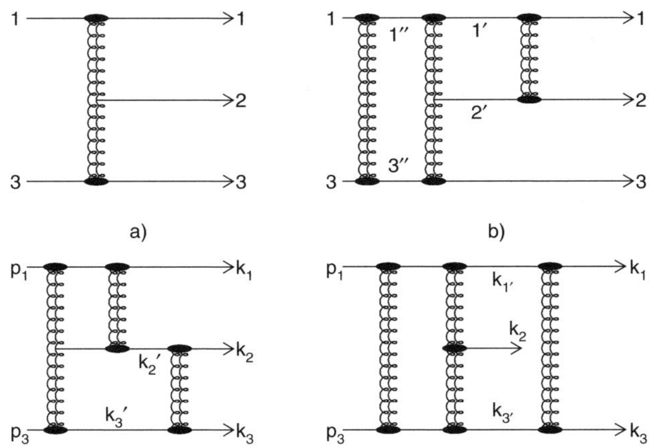

Fig. 1. Input diagram for diffractive production (a) and diagrams with rescatterings in initial and final states: the last interactions in 12-channel (b), in 23-channel (c) and 13-channel (d).

Using the K-matrix method, or the dispersion relation N/D-approach, we get an appropriate way for the consideration of three-particle production processes.[28] Therefore, the first problem we face here is the extension of these techniques to ultrahigh energies. In Section 2 we consider examples of diagrams for the production of three hadrons with very small momenta transferred, $q^2 \sim m^2/\ln^2 s$. The examples demonstrate us specific features of the K-matrix technique at such energies. In Section 3, in the framework of the developed technique we write in the impact parameter space a system of equations which determines the diffractive production amplitude in hadron-hadron collisions at ultrahigh energies; we solve the equation. The hypothesis about the black disk structure of the two-particle interaction amplitude allows an easy calculation of screening effects inherent to ultrahigh energies.

We concentrate our attention on study of the three particle production amplitude within the black disk scenario for ultra high energy collisions - in this case we have definite predictions. Going out the black disk mode we consider extreme case when $-iK(b) \to \infty$ at $b < R_{disk}$. It leads to the resonant disk mode which is a subject of recent discussions.[29–33] We demonstrate that within this scenario diffractive production amplitude is strongly damped with energy increase.

2. Three particle production amplitude and initial/final state interactions

At ultrahigh energies the initial state and final state rescatterings are to be taken into account: the growth of total and elastic cross sections definitely tells us that the effect of rescatterings is not small. Here we consider examples of such processes using the impact parameter representation. But first we recall the corresponding presentation for the elastic scattering amplitude.

For the two-particle scattering amplitude the standard determination of the profile function $T(b,\xi)$ in the impact parameter space, $\mathbf{b}$, can be written as:

$$4\pi \frac{d\sigma_{el}}{dq_\perp^2} = A^2(q_\perp^2,\xi), \quad A(q_\perp^2,\xi) = \int d^2b\, e^{i\mathbf{b}\mathbf{q}_\perp} T(b,\xi)\,,$$
$$T(b,\xi) = 1 - e^{-\frac{1}{2}\chi(b,\xi)} = 1 - \eta(b,\xi)\, e^{2i\delta(b,\xi)} = \frac{-2iK(b,\xi)}{1 - iK(b,\xi)},$$
$$b = |\mathbf{b}|, \quad \xi = \ln s, \tag{1}$$

where the profile function is presented in terms of the optical density $\chi(b,\xi)$, the inelasticity parameter and the phase shift, $\eta(b,\xi)$ and $\delta(b,\xi)$, and using the K-matrix approach (the function $K(b,\xi)$ for the multichannel case is complex value).

Below we calculate explicitly examples of diagrams for the production of three particles using the K-matrix technique in the $\mathbf{b}$-space.

2.1. *Feynman diagram technique and eikonal approach*

Let us consider the amplitude of the Fig. 1(c),(d) type (last interaction in 23-channel). The Feynman integral for the loop related to the intermediate

$2'3'$-state reads:

$$\int \frac{d^4k_{2'}}{(2\pi)^4 i} \frac{A_{2\to 3}(k_1, k_{2'}, k_{3'})\, A_{2\to 2}(k_{2'}, k_{3'})}{(m^2 - k_{2'}^2 - i0)(m^2 - k_{3'}^2 - i0)}. \tag{2}$$

The key point is that interactions at ultrahigh energies turn out to be effectively instantaneous. This is the result of shrinking of the diffractive cones (the effect of the τ-scaling): the substantial regions of integration over momenta transferred are small, of the order of $\mathbf{q}^2 \sim m^2/\xi^2$ with $\xi \equiv \ln s >> 1$.

It is convenient to consider three-particle production processes in the cm-system where the initial particle momenta are determined as $p_1 = (p_0, \mathbf{p}_\perp, p_z) \simeq (p + m^2/2p, 0, p)$ and $p_3 \simeq (p + m^2/2p, 0, -p)$. Therefore, in this system we have the following relations:

$$\begin{aligned}
&\mathbf{k}_{1\perp} + \mathbf{k}_{2'\perp} + \mathbf{k}_{3'\perp} = 0, \quad \mathbf{k}_{1\perp} + \mathbf{k}_{2\perp} + \mathbf{k}_{3\perp} = 0,\\
&m_{2'\perp}^2 = m^2 + \mathbf{k}_{2'\perp}^2, \quad q_1^2 = (p_1 - k_1)^2 \simeq -\mathbf{k}_{1\perp}^2,\\
&q_{3'}^2 = (p_3 - k_{3'})^2 \simeq -\mathbf{k}_{3'\perp}^2,\\
&q_{3'3}^2 = (k_{3'} - k_3)^2 \simeq -(\mathbf{k}_{3\perp} - \mathbf{k}_{3'\perp})^2.
\end{aligned} \tag{3}$$

The integral (2) within this kinematics is written as:

$$\int \frac{dk_{2'}^{(+)} dk_{2'}^{(-)} d^2k_{2'\perp}\, A_{2\to 3}(k_1, k_{2'}, k_{3'})\, A_{2\to 2}(k_{2'}, k_{3'})}{2i(2\pi)^4 \left(k_{2'}^{(+)}k_{2'}^{(-)} - (m^2 + \mathbf{k}_{2'\perp}^2) + i0\right)\left(k_{3'}^{(+)}k_{3'}^{(-)} - (m^2 + \mathbf{k}_{3'\perp}^2) + i0\right)}. \tag{4}$$

where $k_{2'}^{(\pm)} = k_{2'0} \pm k_{2'z}$ and $k_{3'}^{(\pm)} = k_{3'0} \pm k_{3'z}$. The eikonal approach corresponds to the mass-on-shell calculation of loop diagrams. It can be seen when considering the K-matrix elements. The K-matrix function $(-i)K(b,\xi)$ of a scattering amplitude is real for the black disk regime. This means that the imaginary parts in loop diagrams are dominant. For the rescattering diagrams of the type given in Fig. 1, it is realized by the replacement:

$$\begin{aligned}
\left[(m^2 - k_{2'}^2 - i0)(m^2 - k_{3'}^2 - i0)\right]^{-1} &\to -2\pi^2\delta(m^2 - k_{2'}^2)\delta(m^2 - k_{3'}^2)\\
&= -2\pi^2\delta\left(k_{2'}^{(+)}k_{2'}^{(-)} - (m^2 + \mathbf{k}_{2'\perp}^2)\right)\\
&\quad \times\delta\left(k_{3'}^{(+)}k_{3'}^{(-)} - (m^2 + \mathbf{k}_{3'\perp}^2)\right).
\end{aligned} \tag{5}$$

Then the amplitude (2) reads (below we skip index $\perp$):

$$\int \frac{d^2k_{2'}}{(2\pi)^2} \frac{i}{4s_{23}} \left[A_{2\to 3}(k_1, k_{2'}, k_{3'})\, A_{2\to 2}(k_{2'}, k_{3'})\right]_{k_{3'}^2 = m^2}^{k_{2'}^2 = m^2}. \tag{6}$$

For mass-on-shell amplitudes we introduce the notation:

$$\left[A_{2\to3}\left(k_1,k_{2'},k_{3'}\right)\right]^{k^2_{2'}=m^2}_{k^2_{3'}=m^2} \equiv A^{(23)}\left(\mathbf{k}_1^2,\xi_{12};\mathbf{k}^2_{3'},\xi_{23}\right),$$
$$\frac{1}{4s_{23}}\left[A_{2\to2}\left(k_{2'},k_{3'}\right)\right]^{k^2_{2'}=m^2}_{k^2_{3'}=m^2} \equiv K\left((\mathbf{k}_{3'}-\mathbf{k}_3)^2,\xi_{23}\right), \tag{7}$$

where $K\left((\mathbf{k}_{3'}-\mathbf{k}_3)^2,\xi_{23}\right)$ is the K-matrix function in momentum representation, $\xi_{23} = \ln s_{23}$ and $s_{23} = (k_2+k_3)^2$. After changing integration $d^2\mathbf{k}_{2'} \to d^2\mathbf{k}_{3'}$, we, finally, write for Eq. (2):

$$\int \frac{d^2k_{3'}}{(2\pi)^2} A^{(23)}\left(\mathbf{k}_1^2,\xi_{12};\ \mathbf{k}^2_{3'},\xi_{23}\right)\ iK\left((\mathbf{k}_{3'}-\mathbf{k}_3)^2,\xi_{23}\right). \tag{8}$$

The procedure of calculating mass-on-shell rescatterings given here corresponds exactly to the eikonal approach used in Refs. 15, 23, 24. The Fourier transform of Eq. (8), the operator reads as

$$\int \frac{d^2k_1}{(2\pi)^2}\exp\left(-i\mathbf{k}_1\mathbf{b}_1\right)\int \frac{d^2k_3}{(2\pi)^2}\exp\left(-i\mathbf{k}_3\mathbf{b}_3\right), \tag{9}$$

gives us the amplitude in the impact parameter space.

2.2. *Amplitudes of initial and final state rescatterings in the impact parameter space*

We continue to consider examples of diagrams with initial and final state screenings but in terms of the K-matrix. The best way for that is to use the impact parameter representation. For the scattering amplitude we write $T(b,\xi) = (-2iK(b,\xi))\,/\,(1-iK(b,\xi))$, for production amplitudes we should use the corresponding Fourier transforms.

2.2.1. *Initial state rescatterings*

The bare amplitude for the production of three particles and its Fourier transform (see Fig. 1a) are written as:

$$\phi_0(\mathbf{k}_1^2,\xi_{12};\ \mathbf{k}_3^2,\xi_{23}) = \int d^2b_1d^2b_3 f_0(b_1,\xi_{12}\ ;b_3\ ,\xi_{23})$$
$$\times\ \exp\left(i\mathbf{k}_1\mathbf{b}_1 + i\mathbf{k}_3\mathbf{b}_3\right). \tag{10}$$

We use the same notations for the bare amplitude and its Fourier transform supposing that do not lead to misunderstanding.

The rescattering in the initial state gives an additional factor in the impact parameter space

$$\phi_1(\mathbf{k}_1^2,\xi_{12};\ \mathbf{k}_3^2,\xi_{23}) = \int d^2b_1 d^2b_3\ iK(b,\xi)$$
$$\times\ f_0(b_1,\xi_{12}\ ;b_3\ ,\xi_{23})\exp\left(i\mathbf{k}_1\mathbf{b}_1 + i\mathbf{k}_3\mathbf{b}_3\right),$$
$$\xi = \xi_{12}+\xi_{23}, \quad \mathbf{b} = \mathbf{b}_1 + \mathbf{b}_3, \tag{11}$$

two rescatterings result in $[iK(b,\xi)]^2$ and so on. The summation of all terms $\sum\limits_{n=0,1,2,\ldots} \phi_n$ generates the standard K-matrix factor $[1-iK(b,\xi)]^{-1}$, and we write for the input term corrected by taking into account the initial state interactions:

$$\phi(\mathbf{k}_1^2,\xi_{12};\ \mathbf{k}_3^2,\xi_{23}) = \int \frac{d^2b_1 d^2b_3}{1-iK(b,\xi)} f_0(b_1,\xi_{12}\ ;b_3\ ,\xi_{23})$$
$$\times\ \exp\left(i\mathbf{k}_1\mathbf{b}_1 + i\mathbf{k}_3\mathbf{b}_3\right), \tag{12}$$

Below we use the notation

$$f(b_1,\xi_{12}\ ;b_3\ ,\xi_{23}) = \frac{1}{1-iK(b,\xi)} f_0(b_1,\xi_{12}\ ;b_3\ ,\xi_{23}) \tag{13}$$

The factor $[1-iK(b,\xi)]^{-1}$ is universal for all terms of the amplitude. Hence, it should be introduced into all components of the amplitude. Below, without special emphasizes, we presume that it is done.

2.2.2. *Input term* $\phi(\mathbf{k}_1^2,\xi_{12};\ \mathbf{k}_3^2,\xi_{23})$ *and final state rescatterings*

First, we consider rescatterings in 23- and 12-channels. The input term with the corresponding rescatterings can be written as:

$$\phi^{(23)}(\mathbf{k}_1^2,\xi_{12};\mathbf{k}_3^2,\xi_{23}) = \int d^2b_1 d^2b_3 f^{(23)}(b_1,\xi_{12};b_3\ ,\xi_{23})$$
$$\times\ \exp\left(i\mathbf{k}_1\mathbf{b}_1 + i\mathbf{k}_3\mathbf{b}_3\right),$$
$$\phi^{(12)}(\mathbf{k}_3^2,\xi_{23};\mathbf{k}_1^2,\xi_{12}) = \int d^2b_1 d^2b_3 f^{(12)}(b_3\ ,\xi_{23};b_1,\xi_{12})$$
$$\times\ \exp\left(i\mathbf{k}_1\mathbf{b}_1 + i\mathbf{k}_3\mathbf{b}_3\right), \tag{14}$$

where

$$f^{(23)}(b_1,\xi_{12};b_3\ ,\xi_{23}) = f(b_1,\xi_{12};b_3\ ,\xi_{23})\frac{iK(b_3,\xi_{23})}{1-iK(b_3,\xi_{23})}$$
$$\equiv f(b_1,\xi_{12};b_3\ ,\xi_{23})\ a(b_3,\xi_{23}),$$
$$f^{(12)}(b_3\ ,\xi_{23};b_1,\xi_{12}) = f(b_1,\xi_{12};b_3\ ,\xi_{23})\frac{iK(b_1,\xi_{12})}{1-iK(b_1,\xi_{12})}$$
$$\equiv f(b_1,\xi_{12};b_3\ ,\xi_{23})\ a(b_1,\xi_{12}). \tag{15}$$

Here we introduce a short notation for the two-particle scattering amplitude.

Rescatterings in the 13-channel give us the input term $f^{(13)}$. Since here $\mathbf{k}_1+\mathbf{k}_3=\mathbf{k}_{1'}+\mathbf{k}_{3'}$, it is convenient to use the equation for rescatterings in a somewhat modified form:

$$\phi^{(13)}(\mathbf{k}_1^2,\xi_{12};\ \mathbf{k}_3^2,\xi_{23}) = \int \frac{d^2k_{3'}d^2k_{1'}}{(2\pi)^2}\delta\left(\mathbf{k}_1+\mathbf{k}_3-\mathbf{k}_{1'}-\mathbf{k}_{3'}\right) \times \phi(\mathbf{k}_{1'}^2,\xi_{12};\ \mathbf{k}_{3'}^2,\xi_{23})\, iK\left((\mathbf{k}_{3'_\perp}-\mathbf{k}_{3_\perp})^2,\xi_{13}\right). \quad (16)$$

The transformation of this equality into the impact parameter space leads to

$$f_1^{(13)}(b_1\,,\xi_{12}\,;b_3,\xi_{23}\,;|\mathbf{b}_1+\mathbf{b}_3|\,,\xi_{13}) = f(b_1,\xi_{12}\,;b_3\,,\xi_{23}) \times iK(|\mathbf{b}_1+\mathbf{b}_3|\,,\xi_{13}), \quad (17)$$

where $\xi_{13}\simeq\xi=\xi_{12}+\xi_{23}$. Performing a summation over the complete set of rescattering diagrams in the 13-channel, $\sum\limits_{n=1,2,3\ldots} f_n^{(13)}$, we write:

$$f^{(13)}(b_1\,,\xi_{12};b_3,\xi_{23};|\mathbf{b}_1+\mathbf{b}_3|\,,\xi) = f(b_1,\xi_{12};b_3\,,\xi_{23})\frac{iK(|\mathbf{b}_1+\mathbf{b}_3|\,,\xi)}{1-iK(|\mathbf{b}_1+\mathbf{b}_3|\,,\xi)} \equiv f(b_1,\xi_{12};b_3\,,\xi_{23})\; a\left(|\mathbf{b}_1+\mathbf{b}_3|\,,\xi\right). \quad (18)$$

Formulae (15), (18) give us the pattern for writing other diagrams with final state rescatterings.

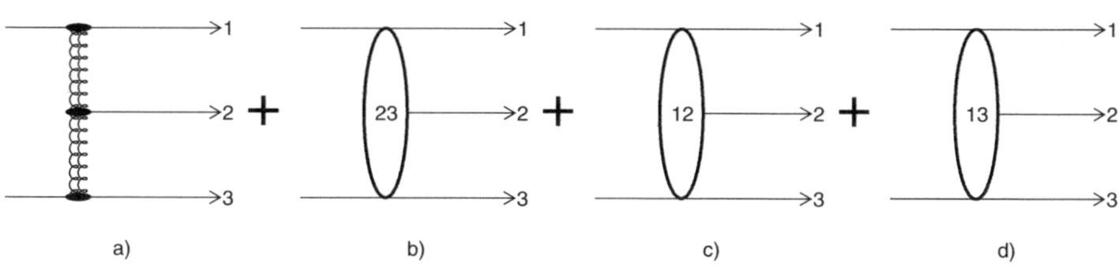

Fig. 2. Four terms for a three particle production amplitude: the input term without final state rescatterings (a), and the terms with rescatterings in the 23-channel (b), in the 12-channel (c), and the 13-channel (d).

3. System of equations for the production amplitude

We present the amplitude as a sum of four terms:
(i) input term $f(b_1,\xi_{12}\,;b_3\,,\xi_{23})$ without final state interactions,

(ii) terms with interactions in final states, that are 12-, 23-states and 13-state with corresponding amplitudes $A^{(12)}(b_1, \xi_{12}\ ; b_3\, , \xi_{23})$, $A^{(23)}(b_1, \xi_{12}\ ; b_3\, , \xi_{23})$, and $A^{(13)}(b_1, \xi_{12}\ ; b_3\, , \xi_{23}; |\mathbf{b}_1 + \mathbf{b}_3|\, , \xi)$.

The index (ij) shows the channel in which the last interaction in the final state takes place. We suppose that initial state interactions in the amplitudes f and $A^{(ij)}$ are taken into account. Then, the total amplitude, see Fig. 2, is written as follows:

$$\begin{aligned} A^{(tot)}_{2\to 3} &= f(b_1, \xi_{12}\ ; b_3\, , \xi_{23}) + A^{(12)}(b_1, \xi_{12}\ ; b_3\, , \xi_{23}) + A^{(23)}(b_1, \xi_{12}\ ; b_3\, , \xi_{23}) \\ &\quad + A^{(13)}(b_1, \xi_{12}\ ; b_3\, , \xi_{23}; |\mathbf{b}_1 + \mathbf{b}_3|,\ \xi). \end{aligned} \tag{19}$$

The equations for $A^{(ij)}$ are shown in Fig. 3, they are written as:

$$\begin{aligned} A^{(23)}(b_1, \xi_{12}\ &;\ b_3\, , \xi_{23}) = f(b_1, \xi_{12}\ ; b_3\, , \xi_{23})\ a_{23}(b_3, \xi_{23}) \\ &+ A^{(12)}(b_1, \xi_{12}\ ; b_3\, , \xi_{23})\ a_{23}(b_{23}, \xi_{23}) \\ &+ A^{(13)}(b_1, \xi_{12}\ ; b_3\, , \xi_{23}; |\mathbf{b}_1 + \mathbf{b}_3|,\ \xi)\ a_{23}(b_{23}, \xi_{23}), \\ A^{(12)}(b_1, \xi_{12}\ &;\ b_3\, , \xi_{23}) = f(b_1, \xi_{12}\ ; b_3\, , \xi_{23})\ a_{12}(b_1, \xi_{12}) \\ &+ A^{(23)}(b_1, \xi_{12}\ ; b_3\, , \xi_{23}) a_{12}(b_1, \xi_{12}) \\ &+ A^{(13)}(b_1, \xi_{12}\ ; b_3\, , \xi_{23}; |\mathbf{b}_1 + \mathbf{b}_3|,\ \xi)\ a_{12}(b_1, \xi_{12}), \\ A^{(13)}(b_1, \xi_{12}\ &;\ b_3\, , \xi_{23}; |\mathbf{b}_1 + \mathbf{b}_3|,\ \xi) = f(b_1, \xi_{12}\ ; b_3\, , \xi_{23})\ a_{13}(|\mathbf{b}_1 + \mathbf{b}_3|,\ \xi) \\ &+ A^{(12)}(b_1, \xi_{12}\ ; b_3\, , \xi_{23})\ a_{13}(|\mathbf{b}_1 + \mathbf{b}_3|,\ \xi) \\ &+ A^{(23)}(b_1, \xi_{12}\ ; b_3\, , \xi_{23})\ a_{13}(|\mathbf{b}_1 + \mathbf{b}_3|,\ \xi). \end{aligned} \tag{20}$$

Let us recall that the two-particle amplitudes, a_{ij}, are introduced in Eq. (15), namely:

$$\begin{aligned} a_{23} \equiv a_{23}(\xi_{23}, b_{23}) &= \frac{iK(b_3, \xi_{23})}{1 - iK(b_3, \xi_{23})}\, , \\ a_{12} \equiv a_{12}(b_1, \xi_{12}) &= \frac{iK(b_1, \xi_{12})}{1 - iK(b_1, \xi_{12})}\, , \\ a_{13} \equiv a_{13}(|\mathbf{b}_1 + \mathbf{b}_3|,\ \xi) &= \frac{iK(|\mathbf{b}_1 + \mathbf{b}_3|,\ \xi)}{1 - iK(|\mathbf{b}_1 + \mathbf{b}_3|,\ \xi)}. \end{aligned} \tag{21}$$

Amplitudes $A^{(23)}$ and $A^{(12)}$ differ only by the permutation of indices $1 \rightleftharpoons 3$. In a short form Eq. (20) reads:

$$\begin{aligned} A^{(23)} &= f a_{23} + A^{(12)} a_{23} + A^{(13)} a_{23}\, , \\ A^{(12)} &= f a_{12} + A^{(23)} a_{12} + A^{(13)} a_{12}\, , \\ A^{(13)} &= f a_{13} + A^{(12)} a_{13} + A^{(23)} a_{13}\, . \end{aligned} \tag{22}$$

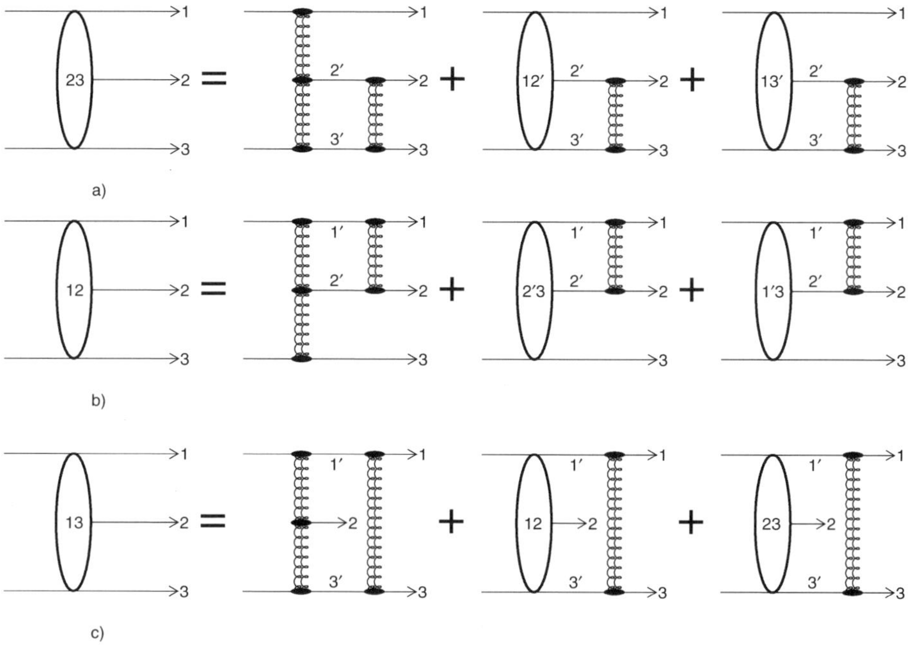

Fig. 3. Diagrams with final state rescatterings in the channels 23, 12 and 13 - figures a, b and c correspondingly. Equalities are graphical realizations of equations for the amplitudes $A^{(ij)}$.

The equations give us

$$A^{(12)} = \frac{f(1 + a_{13} + a_{23} + a_{23}a_{13})a_{12}}{1 - a_{12}a_{23} - a_{12}a_{13} - a_{23}a_{13} - 2a_{12}a_{23}a_{13}},$$
$$A^{(23)} = \frac{f(1 + a_{13} + a_{12} + a_{12}a_{13})a_{23}}{1 - a_{12}a_{23} - a_{12}a_{13} - a_{23}a_{13} - 2a_{12}a_{23}a_{13}},$$
$$A^{(13)} = \frac{f(1 + a_{12} + a_{23} + a_{23}a_{12})a_{13}}{1 - a_{12}a_{23} - a_{12}a_{13} - a_{23}a_{13} - 2a_{12}a_{23}a_{13}}, \qquad (23)$$

with

$$f = \frac{1}{1 - iK(b,\xi)} f_0, \quad \xi = \xi_{12} + \xi_{23}, \quad \mathbf{b} = \mathbf{b}_1 + \mathbf{b}_3\,. \qquad (24)$$

The input term f_0 depends also on $\mathbf{b}_j$ and ξ_{ij}. At moderately high energies it is a two-pomeron term, at ultrahigh energies it can be a two-disk term. So, for the two-pomeron term we write in the momentum and impact parameter spaces, correspondingly:

$$\begin{aligned}\mathbf{k}-\text{space}:\quad f_0 &= g_{2\to 3}\\ &\times ie^{-i\frac{\pi}{2}\Delta}\exp\left[\Delta\xi_{12}-\alpha'\xi_{12}\mathbf{k}_1^2\right]\\ &\times ie^{-i\Delta\frac{\pi}{2}}\exp\left[\Delta\xi_{23}-\alpha'\xi_{23}\mathbf{k}_3^2\right],\end{aligned}$$

$$\begin{aligned}\mathbf{b}-\text{space}:\quad f_0 &= g_{2\to 3}\\ &\times\frac{ie^{-i\frac{\pi}{2}\Delta+\Delta\xi_{12}}}{4\pi\alpha'\xi_{12}}\exp\left[-\frac{\mathbf{b}_1^2}{4\alpha'\xi_{12}}\right]\\ &\times\frac{ie^{-i\frac{\pi}{2}\Delta+\Delta\xi_{23}}}{4\pi\alpha'\xi_{23}}\exp\left[-\frac{\mathbf{b}_3^2}{4\alpha'\xi_{23}}\right].\end{aligned}\tag{25}$$

In the black disk mode the input term reads:

$$\begin{aligned}\mathbf{k}-\text{space}:\quad f_0 &= g_{2\to 3}\cdot iA(\mathbf{k}_1^2,\xi_{12})\,iA(\mathbf{k}_3^2,\xi_{23}),\\ \mathbf{b}-\text{space}:\quad f_0 &= g_{2\to 3}\cdot iT(b_1,\xi_{12})\,iT(b_3,\xi_{23}),\end{aligned}\tag{26}$$

with $A(\mathbf{k}^2,\xi)$, $T(b,\xi)$ determined in Eq. (1).

In the black disk mode the amplitudes a_{ij} behave as $a_{ij}\to -\frac{1}{2}$ in the region $b<R_{black\ disk}\simeq R_0\ln s$ and $a_{ij}\to 0$ at large distances (beyond the black disk area). Therefore the denominator of Eq. (23) is non-zero, and that results in a unique solution.

3.1. *Black disk mode and numerical solution of the three-body equation*

For numerical calculations of f, $A^{(ij)}$ and $A^{(tot)}=f+\sum A^{(ij)}$ we use the Dakhno-Nikonov model[15] with fit results for $(-iK(b,\xi))$ obtained in Refs. 23, 24.

In Fig. 4a we show the profile functions $T(b,\xi)$ found in the fit of data.[1–8] So, we can regard the profile functions at $\sqrt{s}=1,10,100$ TeV as those restored by the data, while $T(b,\xi)$ at $\sqrt{s}=10^n$, $n>2$ are asymptotic values for the black disk regime. The corresponding $-iK(b,\xi)$ are shown in Fig. 4b. The numerical calculation of f, A^{ij}, $A^{(tot)}_{2\to 3}$ are performed using these $(-iK(b,\xi))$, the results are demonstrated in Figs. 5 and 6. Fig. 4 shows typical soliton running with increasing $\ln s$ in the impact parameter space $\mathbf{b}$, similar structure has the amplitude $A_{2\to 3}$, see Figs. 5 and 6.

The comparison of the input amplitude f with $A^{(tot)}_{2\to 3}$ demonstrates that final state scattering corrections do not change principally the shape of $A^{(tot)}_{2\to 3}$ but damp the amplitude by a factor $\sim\frac{1}{4}$. Initial state rescatterings are responsible for additional damping factor $\frac{1}{2}$.

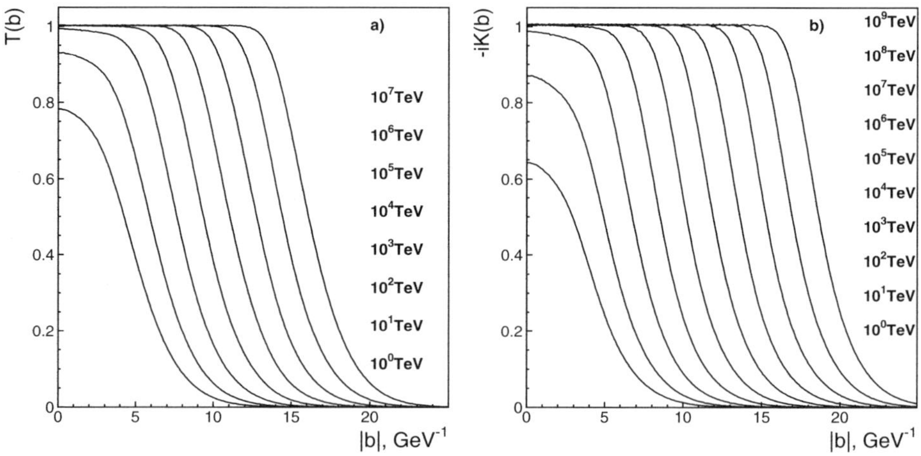

Fig. 4. The profile function, $T(b,\xi)$, and the K-matrix function $(-i)K(b,\xi)$ for the pp scattering amplitude at $\sqrt{s}=1$, 10 TeV obtained by the fit of existing data.[1–8] The curves at $\sqrt{s}=10^n$ TeV at $n\geq 2$ show a continuation of the fit results[23,24] to ultrahigh energies in terms of the black disk picture, here at $b<R_{black\ disk}\simeq R_0\ln s$ one has $(-i)K(b,\xi)\to 1$.

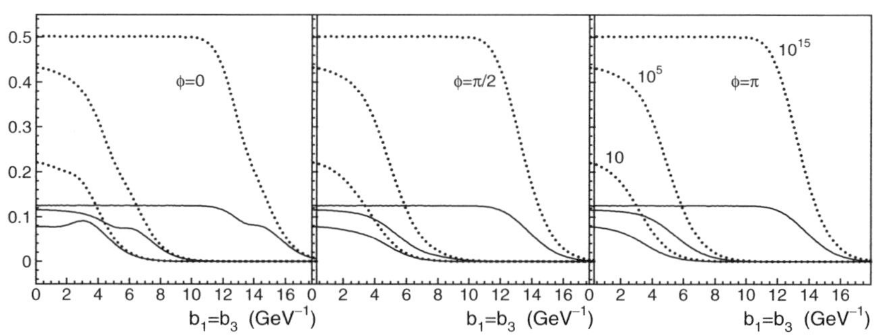

Fig. 5. Black disk mode: Production amplitudes in impact parameter space for colliding energies $\sqrt{s}=10, 10^5, 10^{15}$ TeV at $b_1=b_3$ and $\xi_{12}=\xi_{23}=\xi/2$, ϕ is the angle between $\mathbf{b}_1$ and $\mathbf{b}_3$: f are dotted curves, $A^{(tot)}_{2\to 3}$ are solid curves, see (19), (23), (26).

Amplitude $A^{(tot)}_{2\to 3}$ in the impact parameter space has, similar to $T(b,\xi)$, the soliton structure: impact parameter densities are running toward larger values of $|\mathbf{b}_{ij}|$ with energy increase. Discussion of solitons in physics can be found in Ref. 27.

Investigation of Eq. (22) in general case is complicated by an existence of an infinite set of solutions. Example gives the resonant disk mode[31–33]

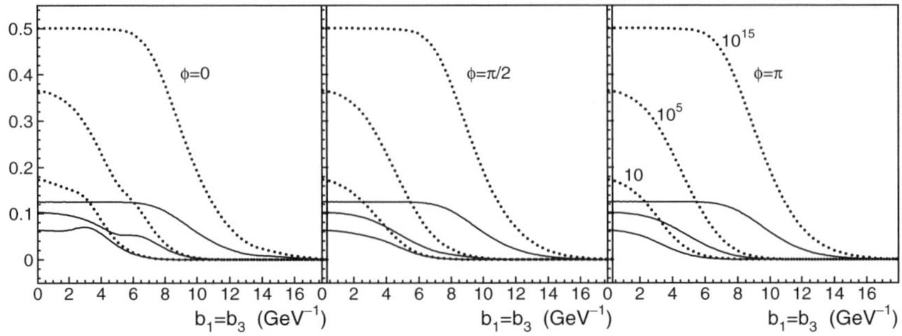

Fig. 6. Black disk mode: Production amplitudes in impact parameter space for colliding energies $\sqrt{s} = 10, 10^5, 10^{15}$ TeV at $b_1 = b_3$ and $\xi_{12} = \xi/3$, $\xi_{23} = 2/3\ \xi$, ϕ is the angle between $\mathbf{b}_1$ and $\mathbf{b}_3$: f are dotted curves, $A^{(tot)}$ are solid curves, see (19), (23), (26).

which corresponds to physical values of the scattering amplitude $a_{ij} \to -1$. However for a verification of this mode the reliable data at energies $\sqrt{s} >$ 100 TeV are needed. Possibly cosmic ray studies can shed a light to this problem but it should be a subject of special investigations.

3.2. *Resonant disk mode*

Within the resonant disk mode, $a_{ij} \to -1$, the determinant of Eq. (22) tends to zero:

$$\det|A_{ij}| = [1 - a_{12}a_{23} - a_{12}a_{13} - a_{23}a_{13} - 2a_{12}a_{23}a_{13}a_{12}]_{a_{ij}\to -1} \to 0\,. \quad (27)$$

However we have unique solution if a_{12}, a_{13}, a_{23} approach to -1 being in the physical region. We write:

$$a_{ij} = \frac{iK_{ij}}{1 - iK_{ij}} = \frac{1}{\kappa_{ij} - 1}, \qquad \kappa_{ij} = \frac{1}{iK_{ij}} \quad (28)$$
$$\text{with} \quad \kappa_{ij} \to 0 \quad \text{at} \quad -iK_{ij} \to \infty$$

The first amplitude of Eq. (23) reads:

$$A_{12} = \frac{\kappa_{13}\kappa_{23}}{\kappa_{12}\kappa_{23}\kappa_{13} - \kappa_{13}\kappa_{12} - \kappa_{13}\kappa_{23} - \kappa_{12}\kappa_{23}}, \quad (29)$$

and analogous formulae are written for A_{13} and A_{23}. That give:

$$A^{(tot)}_{2\to 3} = f_0 \kappa^2_{13}, \quad (30)$$

where, recall, f_0 is input bare amplitude. The factor $\kappa^2_{13} \sim s^{-2\Delta}$ damps strongly the production amplitude, see Eq. (25).

The damping of the production amplitude is in agreement with fact that in the resonant mode $\sigma_{tot} \simeq \sigma_{el}$ and $\sigma_{inel}/\sigma_{tot} \to 0$ with energy increase, see Ref. 33.

4. Conclusion

Shadow effects at collisions of hadrons with light nuclei at high energies were subject of scientific interest of V.N. Gribov, first, we mean his study of the hadron-deuteron scattering[34] and discovery of the reinforcement of shadowing due to inelastic diffractive rescatterings. It turns out that the similar effect exists on hadron level though at ultrahigh energies.

The diffractive scattering amplitudes are growing with energy thus causing the necessity to take into account rescatterings that result in screening effects.

In this paper we calculate these effects in the framework of modified K-matrix technique for eikonal amplitudes. Corresponding calculations of screening effects in diffractive production processes are performed in the impact parameter space.

The key point of the approach is a shrinkage of diffractive cones in hadron reactions at ultrahigh energies. The shrinkage of cones with energy growth demonstrates us the effective suppression of the t-channel singularities that allow us to use quasi-instantaneous interactions. Being more detailed, one can suppose that the long-ranged component of interaction is determined by a cloud of partons which have universal characteristics and properties. The problem of short-range contributions is challenging for the ultra high energies and it is under discussion, see for example Refs. 35–40 and references therein.

The generalization to other ultrahigh energy diffractive production processes is possible within the developed technique.

The authors thank Y.I. Azimov, J. Nyiri and M.G. Ryskin for useful discussions and comments. The work was supported by grants RFBR-13-02-00425 and RSGSS-4801.2012.2.

References

1. G. Latino *et al.* (TOTEM collaboration), *EPJ Web Conf.* **49**, 02005 (2013).
2. P. Abreu *et al.* (Pierre Auger Collaboration), *Phys. Rev. Lett.* **109**, 062002 (2012).
3. S.P. Denisov *et al.*, *Phys. Lett.* **B36**, 415 (1971).

4. UA1 Collaboration, *Phys. Lett.* **B128**, 336 (1982).
5. UA4 Collaboration, *Phys. Lett.* **B147** , 385 (1984).
6. E710 Collaboration, *Phys. Lett.* **B247**, 127 (1990).
7. UA4/2 Collaboration, *Phys. Lett.* **B316**, 448 (1993).
8. CDF Collaboration, *Phys. Rev.* **D50**, 5518 (1994).
9. A. Capella and J. Kaplan, *Phys. Lett.* **B 52**, 448 (1974).
10. M.A. Lapidus, V.I. Lisin, K.A. Ter-Martirosyan, and P.E. Volkovitsky, *Yad. Fiz.* **24**, 1237 (1976) [*Sov. J. Nucl. Phys.* **24**, 648 (1976)].
11. A.B. Kaidalov and K.A. Ter-Martirosyan, *Yad. Fiz.* **39**, 1545 (1984) [*Sov. J. Nucl. Phys.* **39**, 979 (1984)].
12. T.K. Gaisser and T. Stanev, *Phys. Lett.* **B 219**, 375 (1989).
13. M. Block, F. Halzen and B. Margolis, *Phys. Lett.* **B 252**, 481 (1990).
14. R.S. Fletcher, *Phys. Rev.* **D 46**, 187 (1992).
15. L.G. Dakhno and V.A. Nikonov, *Eur. Phys. J.* **A 5**, 209 (1999).
16. M.L. Good, W.D. Walker, *Phys. Rev.* **120**, 1857 (1960).
17. M. Froissart, *Phys. Rev.* **123**, 1053 (1961).
18. Y.I. Azimov, *Phys. Rev.* **D 84**, 056012 (2011).
19. F. Halzen, K. Igi, M. Ishida and C.S. Kim, *Phys. Rev.* **D 85**, 074020 (2012).
20. V. Uzhinsky and A. Galoyan, arXiv:1111.4984v5 (2012) [hep-ph].
21. I.M. Dremin, V.A. Nechitailo, *Phys. Rev.* **D 85**, 074009 (2012).
22. M.M. Block and F. Halzen, *Phys. Rev.* **D 86**, 0501504 (2013).
23. V.V. Anisovich, K.V. Nikonov, and V.A. Nikonov, *Phys. Rev.* **D 88**, 014039 (2013).
24. V.V. Anisovich, V.A. Nikonov, and J. Nyiri, *Phys. Rev.* **D 88**, 094015 (2013).
25. V.V. Anisovich, K.V. Nikonov, V.A. Nikonov, and J. Nyiri, *Int. J. Mod. Phys.* **A 29**, 1450096 (2014).
26. V.N. Gribov, *Nucl. Phys.* **B106**, 189 (1976).
27. L.D. Faddeev, *Physics-Uspekhi*, **56**, 465 (2013).
28. A.V. Anisovich, V.V. Anisovich, M.A. Matveev, V.A. Nikonov, J. Nyiri, A.V. Sarantsev, *Three-particle physics and dispersion relation theory*, World Scientific, Singapore (2013).
29. S.M. Troshin and N.E. Tyurin, *Int. J. Mod. Phys.* **A22**, 4437 (2007).
30. M. Giordano and E. Meggiolaro, *JHEP* **1403**, 002 (2014).
31. S.M. Troshin and N.E. Tyurin, *Int. J. Mod. Phys.* **A29**, 1450151 (2014).
32. I.M. Dremin, arXiv:1404.4142 [hep-ph].
33. V.V. Anisovich, V.A. Nikonov, and J. Nyiri, *Phys. Rev.* **D90**, 074005 (2014).

34. V. N. Gribov, *Sov. Phys. JETP* **29**, 483 (1969) [*Zh. Eksp. Teor. Fiz.* **56**, 892 (1969)].
35. A. Donnachie and P.V. Landshoff, arXiv:1112.2485 (2011) [hep-ph].
36. A.V. Kotikov and L.N. Lipatov, *Nucl. Phys.* **B 661**, 19 (2003).
37. K. Kang and H. Nastase, *Phys. Lett.* **B 624**, 125 (2005).
38. E. Gotsman, E. Levin and U. Maor, *Phys. Rev.* **D85**, 094007 (2012).
39. A.K. Likhoded, A.V. Luchinsky, and A.A. Novoselov, *Phys. Rev.* **D82**, 114006 (2010).
40. V.A. Khoze, A.D. Martin, and M.G. Ryskin, arXiv:1402.2778 [hep-ph] (2014).

DIFFRACTION AS A CRITICAL ELEMENT IN SOFT SCATTERING

URI MAOR
COLLABORATION WITH E. GOTSMAN AND E. LEVIN
Raymond and Beverly Sackler Faculty of Exact Science
Tel Aviv University, Tel Aviv, 69978, Israel

Gribov's partonic Pomeron provides the foundations of updated models which incorporate soft and hard scattering, so as to reproduce the recent LHC p-p cross sections. Explicitly, total, elastic, inelastic and diffrative data. Leading models are: GLM (Gotsman, Levin, Maor), KMR (Khoze, Martin, Ryskin), Kaidalov-Poghosyan and Ostapchenco.

None of these models in their pre-LHC versions reproduced the TOTEM, ALICE, ATLAS and CMS soft LHC data, needing considerable reconstructions, either in the fitting procedures (GLM), or in the details of the theoretical models. In the following, I shall relate mostly to the GLM model, emphasizing the critical role of the diffractive channels.

1. Introduction

The idea of s channel unitarity screenings date back to the ISR, providing a simple way out of paradoxical traps.

(1) Given that non-screened σ_{tot} grows with energy, σ_{el} grows faster (optical theorem). With no screening, σ_{el} will, eventually, be larger than σ_{tot}.
(2) Even though elastic and diffractive scatterings are dynamically similar, the energy dependences of the two channels are different.
(3) The elastic amplitude is central in impact parameter, peaking at b=0. The peripheral diffractive amplitudes peak at large b, which gets larger with energy.
(4) 40 years post ISR, estimates of soft scattering channels at the TeV-scale require a unified analysis of elastic and diffractive scatterings, incorporating s and t unitarity screenings.

2. S-channel unitarity

The simplest s-channel unitarity bound on $a_{el}(s,b)$ is obtained from a diagonal re-scattering matrix, where repeated elastic scatterings secure s-channel unitarity, $2Ima_{el}(s,b) = |a_{el}(s,b)|^2 + G^{in}(s,b)$. i.e. At a given (s,b), $\sigma_{tot} = \sigma_{el} + \sigma_{inel}$. Its general solution is:

$$a_{el}(s,b) = i\left(1 - e^{-\Omega(s,b))/2}\right),$$

$$G^{in}(s,b) = 1 - e^{-\Omega(s,b)}.$$

Ω is model dependent.

The output s-unitarity bound is $| a_{el}(s,b) | \leq 2$, leading to very large total and elastic LHC cross sections, which are not supported by the recent LHC data.

In a Glauber/Gribov eikonal approximation, the input opacity $\Omega(s,b)$ is real. It equals to the imaginary part of the input model Born term, a Pomeron exchange in our context. The output amplitude, $a_{el}(s,b)$, is imaginary.

The consequent black disc bound is $| a_{el}(s,b) | \leq 1$. In a single channel eikonal model, the screened cross sections are:

$$\sigma_{tot} = 2\int d^2b\left(1 - e^{-\Omega(s,b)/2}\right)$$

$$\sigma_{el} = \int d^2b\left(1 - e^{-\Omega(s,b)/2}\right)^2,$$

$$\sigma_{inel} = \int d^2b\left(1 - e^{-\Omega(s,b)}\right).$$

Figure 1 next page shows the s-channel unitarity black bound and the analyticity/crossing bound. Also shown are the non screened and screened elastic b-amplitudes.

The consequent Froissart-Martin bound is:

$$\sigma_{tot} \leq Cln^2(s/s_0), s_0 = 1GeV^2, \quad C \propto 1/2m_\pi^2 \simeq 30mb.$$

C is far too large to be relevant at the TeV-scale.

s-unitarity implies:

$$\sigma_{el} \leq \frac{1}{2}\sigma_{tot} \quad \text{and} \quad \sigma_{inel} \geq \frac{1}{2}\sigma_{tot}.$$

At saturation: $\sigma_{el} = \sigma_{inel} = \frac{1}{2}\sigma_{tot}$.

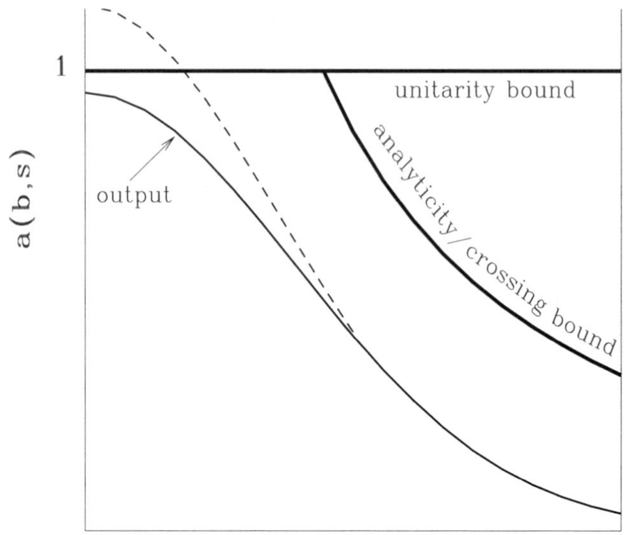

Fig. 1.

Introducing diffraction significantly changes the features of s-unitarity. However, the saturation signatures remain valid.

3. Good-Walker decomposition

Consider a system of two orthonormal states: a hadron Ψ_h and a diffractive state Ψ_D which sums the diffractive Fock states.Ψ_h and Ψ_D do not diagonalize the interaction matrix **T**.

This observation was originally made by Feinberg and Pomeranchuk (1956) and, independantly, by Good and Walker (GW) in a more elegant form (1960).

Let Ψ_1 and Ψ_2 be eigenstates of **T**:

$$\Psi_h = \alpha\Psi_1 + \beta\Psi_2\,,$$
$$\Psi_D = -\beta\Psi_1 + \alpha\Psi_2\,,$$
$$\alpha^2 + \beta^2 = 1\,.$$

The eigenstates initiate 4 $A_{i,k}$ elastic GW amplitudes ($\psi_i + \psi_k \to \psi_i + \psi_k$). i,k=1,2.

For initial $p(\bar{p}) - p$ we have $A_{1,2} = A_{2,1}$.

I shall follow GLM, where the diffractive mass distribution is not defined and requires a specification. The elastic, SD and DD amplitudes in a 2 channel screened GW model are:

$$A_{i,k}(s,b) = \left(1 - e^{\frac{1}{2}\Omega_{i,k}(s,b)}\right) \leq 1\,.$$

$$a_{el}(s,b) = i\{\alpha^4 A_{1,1} + 2\alpha^2\beta^2 A_{1,2} + \beta^4 A_{2,2}\}\,,$$

$$a_{sd}(s,b) = i\alpha\beta\{-\alpha^2 A_{1,1} + (\alpha^2 - \beta^2)A_{1,2} + \beta^2 A_{2,2}\}\,,$$

$$a_{dd}(s,b) = i\alpha^2\beta^2\{A_{1,1} - 2A_{1,2} + A_{2,2}\}\,.$$

Introducing t-channel screening results in a distinction between GW and non GW diffraction.

In the GW sector:

- We obtain the Pumplin bound: $\sigma_{el} + \sigma_{diff}^{GW} \leq \frac{1}{2}\sigma_{tot}$, σ_{diff}^{GW} sums the GW soft diffractive cross sections.
- Below saturation, $\sigma_{el} \leq \frac{1}{2}\sigma_{tot} - \sigma_{diff}^{GW}$, and $\sigma_{inel} \geq \frac{1}{2}\sigma_{tot} + \sigma_{diff}^{GW}$.
- $a_{el}(s,b) = 1$, when and only when: $A_{1,1}(s,b) = A_{1,2}(s,b) = A_{2,2}(s,b) = 1$.
- When $a_{el}(s,b) = 1$, all diffractive amplitudes at the same (s,b) vanish.
- The saturation signature, $\sigma_{el} = \sigma_{inel} = \frac{1}{2}\sigma_{tot}$, in a multi channel calculation is coupled to $\sigma_{diff} = 0$. Consequently, prior to saturation the diffractive cross sections stop growing and start to decrease with energy.
- The above holds only in a multi channel analysis. It does not hold in a single channel model.
- GW saturation signatures are valid also in the non GW sector.

4. Crossed channel unitarity

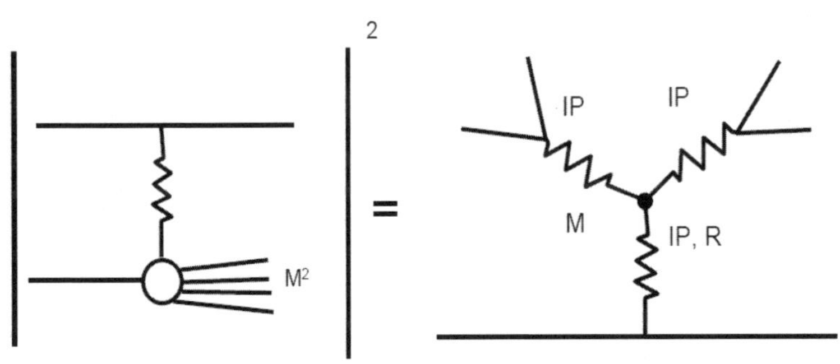

Fig. 2.

Translating the concepts presented into a viable phenomenology requires a specification of $\Omega(s,b)$, for which Regge Pomeron ($I\!P$) theory is a powerful tool. Mueller(1971) applied 3 body unitarity to equate the cross section of: $a + b \to M_{sd}^2 + b$ to the triple Regge diagram $a + b + \bar{b} \to a + b + \bar{b}$, with a leading $3I\!\!P$ vertex term. The $3I\!\!P$ approximation is valid when $\frac{m_p^2}{M_{sd}^2} << 1$ and $\frac{M_{sd}^2}{s} << 1$. The leading energy/mass dependences are:

$$\frac{d\sigma^{3I\!\!P}}{dt\, dM_{sd}^2} \propto s^{2\Delta_{I\!\!P}}, \quad \text{and} \quad (\frac{1}{M_{sd}^2})^{1+\Delta_{I\!\!P}}.$$

Mueller's $3I\!\!P$ approximation for non GW diffraction is the lowest order of t-channel multi $I\!\!P$ interactions, compatible with t-channel unitarity.

Recall that unitarity screening of GW ("low mass") diffraction is carried out explicitly by eikonalization, while the screening of non GW ("high mass") diffraction is carried out by the survival probability.

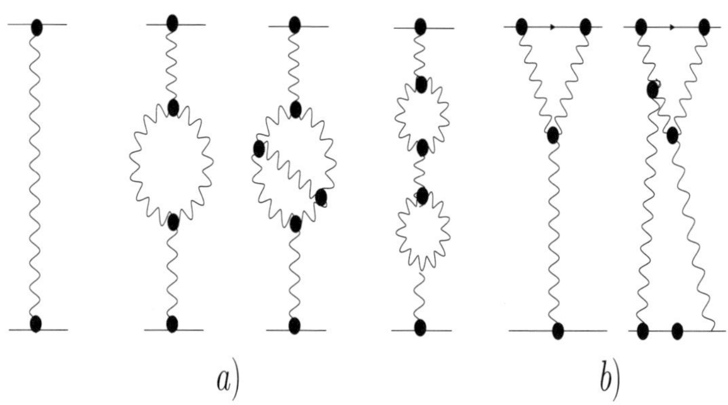

Fig. 3.

Figure 2, shows the $I\!\!P$ Green function. Multi $I\!\!P$ interactions induce large mass diffraction.

Note the analogy with QED:
a) Enhanced diagrams induce the propagator renormalization. b) Semi enhanced diagrams, present the p$I\!\!P$p vertex renormalization.

5. Incorporating Good-Walker and Mueller's diffraction

Both the experimental and theoretical studies of soft diffraction are hindered by conflicting definitions of signatures and bounds.

In our context, I wish to discuss the relationship between GW and non GW diffraction versus Mueller's low and high diffractive mass. Kaidalov, at the time, equated (with no proof) Mueller's low diffractive mass with

GW diffraction, and high diffractive mass with non GW diffraction. The problem is how do we define the bounds of these diffractive mass domains. Following Kaidalov, GW low mass upper bound and Mueller's high mass lower bound, which is 4-5 GeV, coincide. i.e. there is no overlap of low and high mass diffraction. This point of view is shared by KMR, Ostapchenko and Poghosyan. I find this assumption problematic, as it offers no procedure which secures a smooth behaviour of the diffractive mass through this transition.

In GLM model GW diffractive mass is not defined. We presume (also without a proof) that GW and non GW (high mass diffraction) have the same upper bound, commonly taken to be 0.05s. As noted, the main difference between the two diffractive modes is that GW is suppressed by eikonal screenings, while non GW is suppressed by the survival prrobability which has an s-chanel eikonal component initiated by the re-scattering of the initial projectiles, and a t-channel screening induced by the multi $I\!P$ interactions.

In GLM most of the diffraction is GW, while in KMR it is non GW high mass. Originally, GLM did not define a diffractive mass distribution. This has been amended in one of GLM recent papers, where we consider the Pomeron as a partonic probe. In this model:
$I\!P$-q interactions contribute to GW mass
$I\!P$-g interactions contribute to non GW, high mass.

6. The partonic Pomeron

Current $I\!P$ models differ in details, but have in common a relatively large adjusted $\Delta_{I\!P}$ input, and a diminishing $\alpha'_{I\!P}$.

Recall that, traditionally, $\Delta_{I\!P}$ determines the energy dependence of the total, elastic and diffractive cross sections, while $\alpha'_{I\!P}$ determines the forward slopes. This picture is modified in updated $I\!P$ models in which s and t unitarity screenings induce a smaller $I\!P$ intercept at t=0, which gets smaller with energy.

The exceedingly small fitted $\alpha'_{I\!P}$ implies a partonic description of the $I\!P$, which leads to a pQCD interpretation. Gribov's partonic Regge theory provides the microscopic substructure of the $I\!P$ where the slope of the $I\!P$ trajectory is related to the mean transverse momentum of the partonic dipoles constructing the Pomeron.

$$\alpha'_{I\!P} \propto 1/ < p_t >^2,$$

accordingly,

$$\alpha_S \propto \pi / ln \left(< p_t^2 > / \Lambda_{QCD}^2 \right) << 1 \, .$$

We obtain a $I\!P$ with hardness changing continuesly from hard (BFKL like) to soft (Regge like). This is a non trivial relation as the soft $I\!P$ is a moving pole in J-plane, while, the BFKL hard $I\!P$ is a branch cut, approximated, some times, as a simple pole with: $\Delta_{I\!P} = 0.2 - 0.3$, $\alpha'_{I\!P} \simeq 0$.

GLM and KMR models are rooted in Gribov's partonic $I\!P$ theory with a hard pQCD $I\!P$ input. It is softened by unitarity screening (GLM), or the dependence of its partons' transverse momenta on their rapidity (KMR). The two definitions are correlated.

GLM and KMR have a bound of validity, at 60(GLM) and 100(KMR) TeV, implied by their approximations. Consequently, as attractive as updated Pomeron models are, we can not utilize them above 100 TeV at the most. To this end, the only relevant models are single channeled, most of which have a logarithmic parametrization input such as $[Aln(s) + Bln^2(s)]$.

7. DIS: From soft to hard

The single $I\!P$ picture, suggested by GLM and KMR models, implies a smooth transition from the input hard $I\!P$ to a soft $I\!P$.

In a different context, such a transition is supported by HERA dependence of $\lambda = \Delta_{I\!P}$ on Q^2. Note, though, that a smooth transition from a soft to hard $I\!P$ can be reproduced also by a 2 $I\!P$s (soft and hard) model, such as Ostapchenko's.

8. Unitarity saturation

Unitarity saturation is coupled to 4 experimental signatures:

$$\frac{\sigma_{inel}}{\sigma_{tot}} = \frac{\sigma_{el}}{\sigma_{tot}} = 0.5 \, , \quad \frac{\sigma_{tot}}{B_{el}} = 9\pi \, , \quad \text{and } \sigma_{diff} = 0$$

in a multi-channel model.

Following are p-p TeV-scale data, relevant to the assessment of saturation:

CDF(1.8 TeV):
$\sigma_{tot} = 80.03 \pm 2.24 mb$,
$\sigma_{el} = 19.70 \pm 0.85 mb$,
$B_{el} = 16.98 \pm 0.25 GeV^{-2}$.
TOTEM(7 TeV):
$\sigma_{tot} = 98.3 \pm 0.2(stat) \pm 2.8(sys) mb$,

$\sigma_{el} = 24.8 \pm 0.2(stat) \pm 2.8(sys)mb$,
$B_{el} = 20.1 \pm 0.2(stat) \pm 0.3(sys)GeV^{-2}$.
ATLAS(7 TeV):
$\sigma_{tot} = 95.4 \pm 1.4mb$,$\sigma_{el} = 24.0 \pm 0.6mb$.
AUGER(57 TeV):
$\sigma_{tot} = 133 \pm 13(stat) \pm^{17}_{20} sys \pm 16(Glauber/Gribov)mb$,
$\sigma_{inel} = 92 \pm 7(stat) \pm^{9}_{11} (sys) \pm 16(Glauber/Gribov)mb$.
We get: $\frac{\sigma_{inel}}{\sigma_{tot}} =$
0.754(CDF), 0.748(TOTEM),
0.748(ATLAS), 0.692(AUGER).

The numbers above suggest a very slow approach toward saturation, well above the TeV-scale. Consequently, the study of pp saturation depends on information above the TeV-scale.

There are 2 sources from which we may obtain the desired information:

- Cosmic Rays data. Recall that p-p cross sections obtained from p-Air data have relatively large margin of errore. AUGER p-p cross sections are a good example.
- Since updated $I\!\!P$ models are confined to the TeV-scale, p-p cross sections at higher energies can be calculated only in single channeled models, the deficiencies of which have been stated before.

Out of a few single channeled nodels, I shall quote Block and Halzen (BH), which reproduce well the inelastic and total cross sections at the TeV-scale. The BH model can be applied at exceedingly high energies. BH prediction at the Planck-scale ($1.22{\cdot}10^{16}TeV$) is $\sigma_{inel}/\sigma_{tot} = 1131mb/2067mb = 0.547$. It indicates that saturation will be attained, if at all, at non realistic energies.

The predicted multi channel vanishing of the diffractive cross sections at saturation implies that σ_{sd}, which up to the TEVATRON grows slowly with energy, will eventually start to reduce. This may serve as an early signature that saturation is being approached Specifically, the preliminary TOTEM measurement of:

$$\sigma_{sd} = 6.5 \pm 1.3mb$$
$$3.4 < M_{sd} < 1100GeV$$
$$2.4 \cdot 10^{-7} < \xi < 0.025$$

suggests a radical change in the energy dependence of $\sigma_{sd}/\sigma_{inel}$ which is considerably smaller than its value at CDF.

$$\frac{\sigma_{sd}}{\sigma_{inel}} = 0.151(CDF),\ 0.088(TOTEM)\,.$$

This feature, if correct, is, presently, particular to diffraction. It suggests a much faster approach toward unitarity saturation than suggested by $\frac{\sigma_{inel}}{\sigma_{tot}}$.

TOTEM diffractive data is preliminary, as the upper diffractive mass bound of 0.025s is half of the commonly used bound of 0.05s (including GLM).

The GLM SD cross sections (in mb) are

$$\sigma_{sd}(W) = \sigma_{sd}^{GW} + \sigma_{sd}^{nonGW} = 9.2+1.95(1.8),\ 10.7+4.18(7),\ 11.5+5.81(14).$$

ALICE 7 TeV diffractive cross sections:

$$\sigma_{sd} = 14.9 + 3.4 - 5.9mb\,, \quad \text{and} \quad \sigma_{dd} = 9.0 \pm 2.6mb$$

which are significantly different from TOTEM data.

They are compatible with GLM based on 0.05s diffractive mass bound!

Recall that, EL, SD and DD cross section values are obtained from: a b^2 integration of the corresponding amplitude square.

The growth of σ_{sd}, as a function of W, is mainly a consequence of $a_{sd}(s,b)$ moving slowly toward higher b values.

The net result is a continuation of SD moderate increase with energy. Consequently, I do not expect a suppression of σ_{sd} at an energy of 7 TeV.

An explanation of an early reduction of the diffractive channels at relatively low energies, will require, thus, a fundamental change in our understanding of soft scattering at the TeV-scale.

GRIBOV INELASTIC SHADOWING IN THE DIPOLE REPRESENTATION

B.Z. KOPELIOVICH*

Departamento de Física, and
Centro Científico-Tecnológico de Valparaíso,
Universidad Técnica Federico Santa María,
Casilla 110-V, Valparaíso, Chile
E-mail: boris.kopeliovich@usm.cl

The dipole phenomenology, which has been quite successfully applied to various hard reactions, especially on nuclear targets, is applied for calculation of Gribov inelastic shadowing. This approach does not include ad hoc procedures, which are unavoidable in calculations done in hadronic representation. Several examples of Gribov corrections evaluated within the dipole description are presented.

1. Introduction

The Glauber model[1] was the first theoretical approach which calculated the effects of shadowing in hadron-nucleus interactions. The model was, however, essentially non-relativistic. In the pioneering papers by V.N. Gribov[2,3] it was realised that the length scales of interaction rising with energy significantly change the pattern of hadron-nucleus interaction at high energies. In particular, particles created in an inelastic collision with one nucleon, can be subsequently absorbed by other bound nucleons. Such corrections make the nuclear medium more transparent for hadrons, and consequently lead to a reduction of the total hadron-nucleus cross section. These corrections, called Gribov inelastic shadowing, improve the Glauber model, making it well-founded.

2. From the Glauber model to the Gribov-Glauber theory

The amplitude of probability for a hadron to interact with the nucleus is one minus the probability amplitude of no interaction with any of the

*Work partially supported by Fondecyt (Chile) grant 1130543

bound nucleons. So the hA elastic amplitude at impact parameter b has the eikonal form,

$$\Gamma^{hA}(\vec{b};\{\vec{s}_j,z_j\}) = 1 - \prod_{k=1}^{A}\left[1 - \Gamma^{hN}(\vec{b}-\vec{s}_k)\right] , \tag{1}$$

where $\{\vec{s}_j, z_j\}$ denote the coordinates of the target nucleon N_j. $i\Gamma^{hN}$ is the elastic scattering amplitude on a nucleon normalized as,

$$\begin{aligned}\sigma^{hN}_{tot} &= 2\int d^2b\,\mathrm{Re}\,\Gamma^{hN}(b);\\ \sigma^{hN}_{el} &= \int d^2b\,|\Gamma^{hN}(b)|^2 .\end{aligned} \tag{2}$$

In the approximation of single particle nuclear density one can calculate a matrix element between the nuclear ground states,

$$\left\langle 0\left|\Gamma^{hA}(\vec{b};\{\vec{s}_j,z_j\})\right|0\right\rangle = 1 - \left[1 - \frac{1}{A}\int d^2s\,\Gamma^{hN}(s)\int\limits_{-\infty}^{\infty} dz\,\rho_A(\vec{b}-\vec{s},z)\right]^A , \tag{3}$$

where

$$\rho_A(\vec{b}_1,z_1) = \int\prod_{i=2}^{A} d^3r_i\,|\Psi_A(\{\vec{r}_j\})|^2 , \tag{4}$$

is the nuclear single particle density.

Eq. (3) is related via unitarity to the total hA cross section,

$$\begin{aligned}\sigma^{hA}_{tot} &= 2\mathrm{Re}\int d^2b\left\{1 - \left[1 - \frac{1}{A}\int d^2s\,\Gamma^{hN}(s)\,T_A(\vec{b}-\vec{s})\right]^A\right\}\\ &\approx 2\int d^2b\left\{1 - \exp\left[-\frac{1}{2}\sigma^{hN}_{tot}\,(1-i\rho_{pp})\,T^h_A(b)\right]\right\} ,\end{aligned} \tag{5}$$

where ρ_{pp} is the ratio of the real to imaginary parts of the forward pp elastic amplitude,

$$T^h_A(b) = \frac{2}{\sigma^{hN}_{tot}}\int d^2s\,\mathrm{Re}\,\Gamma^{hN}(s)\,T_A(\vec{b}-\vec{s}) , \tag{6}$$

and

$$T_A(b) = \int_{-\infty}^{\infty} dz\,\rho_A(b,z) , \tag{7}$$

is the nuclear thickness function. We use the Gaussian form of $\Gamma^{hN}(s)$ in what follows,

$$\mathrm{Re}\,\Gamma^{hN}(s) = \frac{\sigma_{tot}^{hN}}{4\pi B_{hN}} \exp\left(\frac{-s^2}{2B_{hN}}\right) , \tag{8}$$

where B_{hN} is the slope of the differential hN elastic cross section. Notice that the accuracy of the optical approximation (the second line in Eq. (5)) is quite high for heavy nuclei. For the sake of simplicity, we use the optical form throughout the paper although for numerical evaluations we always rely on the accurate expression (the first line in Eq. (5)). The effective nuclear thickness, Eq. (6) implicitly contains energy dependence, which is extremely weak.

In what follows we also neglect the real part of the elastic amplitude, unless specified, since it gives a vanishing correction $\sim \rho_{pp}^2/A^{2/3}$.

Besides the total cross section Eq. (5) one can calculate within the Glauber model also elastic, quasielastic (break-up of the nucleus without particle production) and inelastic cross sections. One can find details of such calculations, as well as numerical results, in Ref. 4–6, and below in Sect. 3.2.

The Glauber model is intensively used nowadays as a theoretical tool to study heavy ion collisions. However, this model is subject to significant Gribov corrections.[2]

2.1. *Intermediate state diffractive excitations*

The Glauber model is a single-channel approximation, therefore it misses the possibility of diffractive excitation of the projectile in the intermediate state as is illustrated in Fig. 1.

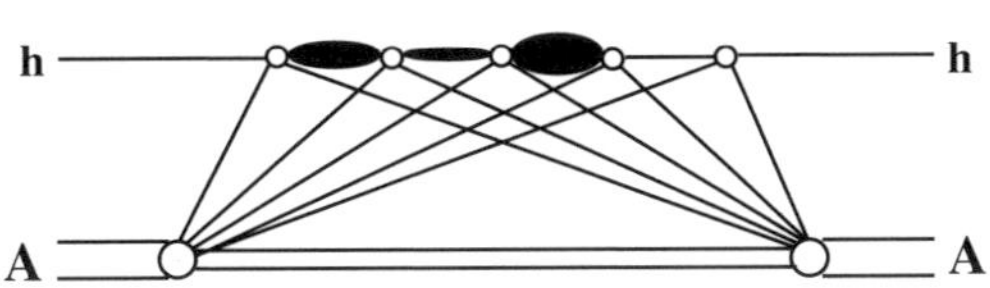

Fig. 1. Diagonal and off-diagonal diffractive multiple excitations of the projectile hadron in intermediate state

The inclusion of multiple diffractive transitions between different excitations, like depicted in Fig. 1, is, obviously, a challenge, because diffractive transitions between different excited states cannot be measured in diffractive processes. There is, however, one case free of these problems, shadowing in hadron-deuteron interactions. In this case no interaction in the intermediate state is possible, and the knowledge of diffractive cross section $hN \to XN$ is sufficient for calculations of the inelastic correction with no further assumptions. In this case the Gribov correction to the Glauber model for the total cross section has the simple form,[2]

$$\Delta\sigma_{tot}^{hd} = -2\int dM^2 \int dp_T^2 \, \frac{d\sigma_{sd}^{hN}}{dM^2 dp_T^2} \, F_d(4t). \tag{9}$$

Here $F(t)$ is the deuteron electromagnetic form factor; σ_{sd}^{hN} is the cross section of single diffractive dissociation $hN \to XN$ with longitudinal momentum transfer

$$q_L = \frac{M^2 - m_h^2}{2E_h}; \tag{10}$$

and $t = -p_T^2 - q_L^2$.

The formula for the lowest order inelastic corrections to the total hadron-nucleus cross section was suggested in Ref. 7,

$$\Delta\sigma_{tot}^{hA} = -8\pi \int d^2b \, e^{-\frac{1}{2}\sigma_{tot}^{hN} T_A(b)} \int\limits_{M_{min}^2} dM^2 \, \frac{d\sigma_{sd}^{hN}}{dM^2\, dp_T^2}\bigg|_{p_T=0}$$
$$\times \int\limits_{-\infty}^{\infty} dz_1 \, \rho_A(b, z_1) \int\limits_{z_1}^{\infty} dz_2 \, \rho_A(b, z_1) \, e^{iq_L(z_2 - z_1)} \,, \tag{11}$$

This correction takes care of the onset of inelastic shadowing via phase shifts controlled by q_L and does a good job describing data at low energies,[8,9] as one can also see in Fig. 2.

In a similar way Gribov corrections can be calculated for neutrino-nucleus cross section at low Q^2, when PCAC hypothesis is at work. Then one can employ the Adler relation, which connects neutrino and pion induced cross sections at $Q^2 = 0$, and write the nucleus-to-proton ration of total neutrino-nucleus cross sections as,[10]

$$R_{A/N}^{\nu} = 1 - \frac{8\pi}{A\sigma_{tot}^{\nu N}} \frac{d\sigma_{diff}^{\nu\to\pi}}{dp_T^2}\bigg|_{p_T=0} \int d^2b \int\limits_{-\infty}^{\infty} dz_1 \, \rho_A(b, z_1)$$
$$\times \int\limits_{-\infty}^{z_1} dz_2 \, \rho_A(b, z_2) e^{-iq_L^{\pi}(z_2 - z_1)} \, e^{-\frac{1}{2}\sigma_{tot}^{\pi N} T_A(b, z_2, z_1)}, \tag{12}$$

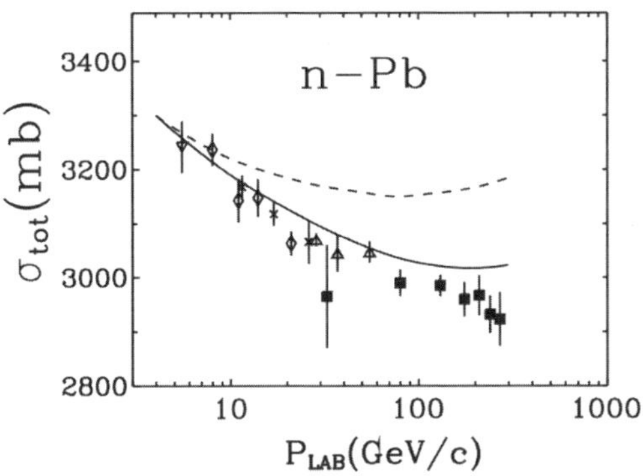

Fig. 2. Data and calculations[8] for the total neutron-lead cross section. The dashed and solid curves correspond to the Glauber model and corrected for Gribov shadowing respectively.

where $q_L^\pi = (Q^2 + m_\pi^2)/2\nu$.

At very low energy where q_L^π is large, the second term in Eq. (12) is suppressed, and the first one, corresponding to the cross section proportional to A, dominates. At high energies $q_L^\pi \ll R_A$ can be neglected and the integrations over $z_{1,2}$ can be performed analytically[10]. Due to the Adler relation the first term in Eq. (12) and the volume part of the second term cancel, and the rest is the "surface" term $\propto A^{2/3}$,

$$\left.\frac{d^2\sigma(\nu A \to l\,X)}{dQ^2\,d\nu}\right|_{q_L \ll 1/R_A} = \frac{G^2}{2\pi^2}\,f_\pi^2\,\frac{E-\nu}{E\nu}\,\sigma(\pi A \to X), \tag{13}$$

as could be anticipated in accordance with the Adler relation. Notice that the high-energy regime actually starts at rather low energies if $Q^2 \lesssim m_\pi^2$.

The results of numerical evaluation of the shadowing effect Eq. (12) for neon are plotted in Fig. 3. The calculations[10] done within the Glauber approximation, Eq. (12), and also including the Gribov's inelastic corrections (important at high energies $\nu > m_{a_1}^2 R_A$) are plotted in Fig. 3 by solid curves as a function of energy, for different Q^2.

The calculated shadowing effects are compared with BEBC data[11] in Fig. 4. As it was anticipated, the shadowing exposes an early onset, and a significant suppression occurs at small Q^2 in the low energy range of hundreds MeV. This is an outstanding feature of the axial current. This seems to be supported by data, although with a rather poor statistics.

Concluding this section, we notice that although the approximation of

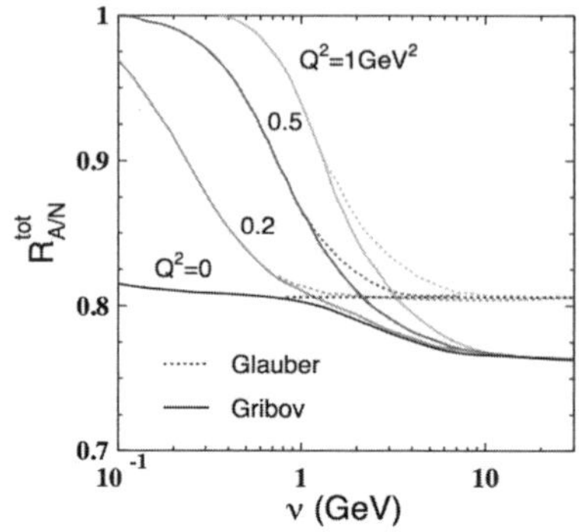

Fig. 3. The neon to nucleon ratio of total neutrino cross sections at different Q^2 (see Ref. 10). Dashed and solid curves correspond to the Glauber and Gribov corrected calculations.

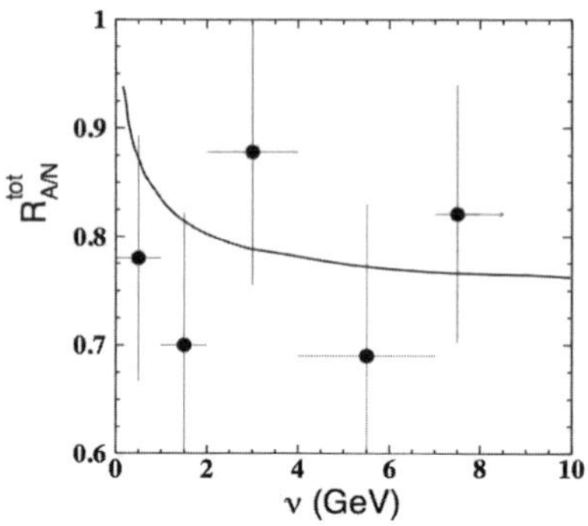

Fig. 4. The neon to proton ratio of the total neutrino cross sections, calculated in Ref. 10 for $x < 0.2$ and $Q^2 < 0.2\,\text{GeV}^2$. The data present the BEBC results.[11]

lowest order Gribov corrections does in some cases a good job, the higher order off-diagonal transitions neglected in Eqs. (11) and (12), might be important, but unknown. Indeed, the intermediate state X has a definite mass M, but no definite cross section. It was fixed in Eq. (11) at σ_{tot}^{hN} with no justification.

Thus, the lowest order Gribov correction Eq. (11) has a nice feature of including phase shifts, which allows to describe the onset of Gribov shadowing, like it is presented in Fig. 2. However, the uncertainties related to the missed higher order corrections and the unknown absorption in Eq. (11) seem to be incurable, while one works in the hadronic (eigenstates of the mass matrix) representation.

2.2. *Interaction eigenstate representation*

If a hadron were an eigenstate of interaction, i.e., could undergo only elastic scattering (as a shadow of inelastic channels) and no diffractive excitation was possible, the Glauber formula would be exact and no inelastic shadowing corrections were needed. This simple observation gives a hint that one should switch from the hadronic basis to a complete set of mutually orthogonal eigenstates of the scattering amplitude operator. This was the driving idea of the description of diffraction in terms of elastic amplitudes,[12,13] and becomes a powerful tool for the calculation of inelastic shadowing corrections in all orders of multiple interactions.[14,15] Notice that this idea was used by Gribov[2] also to explain the role of the coherence length increasing with energy, on the example of an incoming proton, fluctuating to a nucleon-pion pair.

Physical states (including leptons, photons) can be expanded over the complete set of states $|k\rangle$,

$$|h\rangle = \sum_k \Psi^h_k \, |k\rangle \ , \tag{14}$$

which are the eigenstates of the scattering amplitude operator, $\hat{f}\,|k\rangle = f^{kN}_{el}\,|k\rangle$. For the sake of simplicity we neglect the real part of the amplitude, i.e. assume that $f^{kN}_{el} = i\,\sigma^{kN}_{tot}/2$.

Ψ^h_k in Eq. (14) are weight factors (amplitudes) of the Fock state decomposition. They obey the orthogonality conditions,

$$\sum_k \left(\Psi^{h'}_k\right)^\dagger \Psi^h_k = \delta_{h\,h'} \ ;$$
$$\sum_h \left(\Psi^h_l\right)^\dagger \Psi^h_k = \delta_{lk} \ . \tag{15}$$

We also assume that the amplitude is integrated over impact parameter, *i.e.* that the forward elastic amplitude is normalized as $|f^{kN}_{el}|^2 = 4\,\pi\,d\sigma^{kN}_{el}/dt|_{t=0}$. The amplitudes of elastic $f_{el}(hh)$ and off diagonal diffractive $f_{sd}(hh')$ transitions can be expressed as,

$$f^{hN}_{el} = 2i \sum_k \left|\Psi^h_k\right|^2 \sigma^{kN}_{tot} \equiv 2i\,\langle\sigma\rangle \ ; \tag{16}$$

$$f^{hN}_{sd}(h \to h') = 2i \sum_k (\Psi^{h'}_k)^\dagger \, \Psi^h_k \, \sigma^{kN}_{tot} \ . \tag{17}$$

Note that if all the eigenamplitudes were equal, the diffractive amplitude (17) would vanish due to the orthogonality relation, (15). The physical reason is obvious. If all the f^{kN}_{el} are equal, the interaction does not affect the coherence between different eigencomponents $|k\rangle$ of the projectile hadron $|h\rangle$. Therefore, the off-diagonal transitions are possible only due to differences between the eigenamplitudes.

By summing up in the diffractive cross section using the completeness

Eq. (15), and excluding the elastic channels, one gets,[14–16]

$$16\pi \left.\frac{d\sigma^{hN}_{sd}}{dt}\right|_{t=0} = \sum_k \left|\Psi^h_k\right|^2 \left(\sigma^{kN}_{tot}\right)^2 - \left(\sum_i \left|\Psi^h_k\right|^2 \sigma^{kN}_{tot}\right)^2$$
$$\equiv \left\langle \left(\sigma^{kN}_{tot}\right)^2 \right\rangle - \left\langle \sigma^{kN}_{tot} \right\rangle^2 . \qquad (18)$$

Each of the eigenstates propagating through the nucleus can experience only elastic scatterings, so the Glauber eikonal approximation becomes exact for such a state. Then, the cross sections for hadron-nucleus collisions should be averaged over the Fock states, chosen to be interaction eigenstates,[14,15]

$$\sigma^{hA}_{tot} = 2\int d^2b \left\{1 - \left\langle \exp\left[-\frac{1}{2}\sigma^{kN}_{tot}\, T^h_A(b)\right]\right\rangle\right\}. \qquad (19)$$

This is to be compared with the Glauber approximation. The difference is obvious, in Eq. (19) the exponential is averaged, while in the Glauber approximation the exponent is averaged,

$$\left.\sigma^{hA}_{tot}\right|_{Gl} = 2\int d^2b \left\{1 - \exp\left[-\frac{1}{2}\left\langle\sigma^{kN}_{tot}\right\rangle T^h_A(b)\right]\right\}, \qquad (20)$$

where $\langle\sigma^{kN}_{tot}\rangle = \sigma^{hN}_{tot}$. The difference between Eqs. (19) and (20), is the Gribov inelastic correction calculated in all orders of opacity expansion, which was impossible to do within hadronic representation (see above). This result can be compared with the expression (11) for the lowest order correction expanding the exponentials in Eqs. (19) and (20) in number of collisions up to the lowest order. Applying (18) we find,

$$\sigma^{hA}_{tot} - \left.\sigma^{hA}_{tot}\right|_{Gl} = \int d^2b\, \frac{1}{4}\left[\left\langle\sigma^{kN}_{tot}\right\rangle^2 - \left\langle\left(\sigma^{kN}_{tot}\right)^2\right\rangle\right] T^h_A(b)^2$$
$$= -4\pi \int d^2b\, T^h_A(b)^2 \int dM^2 \left.\frac{d\sigma^h_{sd}}{dM^2dt}\right|_{t=0}. \qquad (21)$$

This result is indeed identical to Eq. (11), if neglecting the phase shift vanishing at high energies, and also to expand the exponential.

3. Color dipoles as eigenstates of interactions

The dipole representation in QCD, first proposed in Ref. 15, allows to calculated Gribov corrections in all orders, because a high energy dipole is an eigenstate of interaction. Indeed, the transverse dipole separation, which controls the scattering amplitude, is preserved during propagation and remains intact after multiple soft interactions within a nuclear target.

To perform proper averaging in Eq. (19) one should expend the beam particle (hadrons, vector or axial currents) over Fock states. Each Fock component is a colorless dipole, consisted of two or more partons. After averaging over intrinsic dipole distances, one should sum up different Fock states with proper weights.

3.1. *Deep inelastic scattering*

The parton model interpretation of the space-time development of reactions is not Lorentz invariant. While deep inelastic scattering (DIS) at small $x \ll 1$ is interpreted in the Bjorken reference frame as absorption of the virtual photon by a target parton carrying corresponding fractional momentum x of the target, in the target rest frame it looks differently. The high energy virtual photon fluctuates into a $\bar{q}q$ pair, which interacts with the targets and is produced on mass shell.

The two main contributions to the diffractive cross section in (11) at high energies come from the triple-Regge terms $\mathbb{PPR}$ and $\mathbb{PPP}$, which generate $1/M^3$ and $1/M^2$ mass dependences respectively. The quark-gluon structure of these contributions in diffraction is illustrated in Fig. 5. Apparently, the $\bar{q}q$ and $\bar{q}qg$ states, propagating through the nucleus, attenuate with an absorptive cross section, which cannot be trivially fixed, like it was done in Eq. (11), but should be properly treated within the dipole approach.

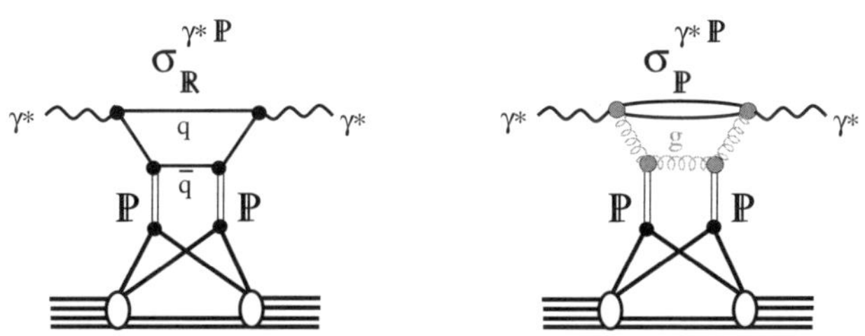

Fig. 5. Gribov corrections with $\bar{q}q$ and $\bar{q}q + g$ intermediate states.

3.1.1. *"Frozen" dipoles*

If the lifetime of partonic fluctuations of a photon significantly exceeds the nuclear size, the dipole approach allows to calculate easily the shadowing effects in DIS. Indeed, in this case one can rely on the "frozen" approximation Eq. (19), which for the interaction of a virtual photon has the form,

$$\sigma_{T,L}^{\gamma^* A}(x,Q^2) = 2\int d^2b \int_0^1 d\alpha \int d^2r_T \left|\Psi_{q\bar{q}}^{T,L}(\alpha,r_T,Q^2)\right|^2 \times \left[1 - e^{-\frac{1}{2}\sigma_{q\bar{q}}(r_T,x)T_A(b)}\right], \tag{22}$$

where r_T is the dipole transverse separation; α is the fractional light-cone momentum of the quark; the distribution functions $\Psi_{q\bar{q}}^{T,L}(\alpha,r_T)$ in the quadratic form read,

$$\left|\Psi_{q\bar{q}}^{T}(\alpha,r_T)\right|^2 = \frac{2N_c\alpha_{em}}{(2\pi)^2}\sum_{f=1}^{N_f} Z_f^2 \left\{\left[1-2\alpha(1-\alpha)\right]\varepsilon^2 \mathrm{K}_1^2(\varepsilon r_T) + m_f^2 \mathrm{K}_0^2(\varepsilon r_T)\right\}; \tag{23}$$

$$\left|\Psi_{q\bar{q}}^{L}(\alpha,r_T)\right|^2 = \frac{8N_c\alpha_{em}}{(2\pi)^2}\sum_{f=1}^{N_f} Z_f^2 Q^2\alpha^2(1-\alpha)^2 \mathrm{K}_0^2(\varepsilon r_T). \tag{24}$$

The advantage of the dipole description is pretty obvious, Eq. (22) includes Gribov inelastic shadowing corrections to all orders of multiple interactions,[15] what is hardly possible within hadronic representation.

On the other hand, the dipoles having a definite size, do not have any definite mass, therefore the phase shifts between amplitudes on different nucleons cannot be calculated as simply as in Eq. (11). A solution for this problem was proposed in Ref. 17, 18.

3.1.2. *Path integral technique*

The lifetime of the "frozen" dipole, or coherence length (or time), is given by,

$$l_c = \frac{2\nu}{Q^2 + M_{\bar{q}q}^2}, \tag{25}$$

where ν is the dipole energy in the target rest frame; $M_{\bar{q}q}^2 = (m_q^2 + k_T^2)/\alpha(1-\alpha)$. At very small Bjorken $x = Q^2/2m_N\nu \ll 1$ the coherence length can significantly exceed the nuclear size, $l_c \gg R_A$, and one can safely rely on the "frozen" approximation. However, if $t_c \lesssim R_A$ such an approximation is not appropriate and one should correct for the dipole size fluctuations during propagation through the nucleus, which corresponds to inclusion of the phase shifts between DIS amplitudes on different bound nucleons in Eq. (11). Within the dipole description this can be done employing the path integral technique,[19] which sums up different propagation paths of the partons.

For a $\bar{q}q$ component of a transversely or longitudinally polarized virtual photon Eq. (22) should be replaced by,

$$\begin{aligned}\left(\sigma_{tot}^{\gamma^* A}\right)^{T,L} &= A\left(\sigma_{tot}^{\gamma^* N}\right)^{T,L} \\ &- \frac{1}{2} Re \int d^2b \int_0^1 d\alpha \int_{-\infty}^{\infty} dz_1 \int_{z_1}^{\infty} dz_2 \int d^2r_1 \int d^2r_2 \\ &\times \Psi_{\bar{q}q}^{T,L^\dagger}(\varepsilon, \vec{r}_2)\, \rho_A(b, z_2)\, \sigma_{q\bar{q}}^N(s, \vec{r}_2)\, G(\vec{r}_2, z_2 \,|\, \vec{r}_1, z_1) \\ &\times \rho_A(b, z_1)\, \sigma_{q\bar{q}}^N(s, \vec{r}_1)\, \Psi_{\bar{q}q}^{T,L}(\varepsilon, \vec{r}_1)\,. \end{aligned} \tag{26}$$

The Green's function $G(\vec{r}_2, z_2 \,|\, \vec{r}_1, z_1)$ describes propagation of a $\bar{q}q$ pair in an absorptive medium, having initial separation $\vec{r}_1$ at the initial position z_1, up to the point z_2, where it gets separation $\vec{r}_2$, as is illustrated in Fig. 6.

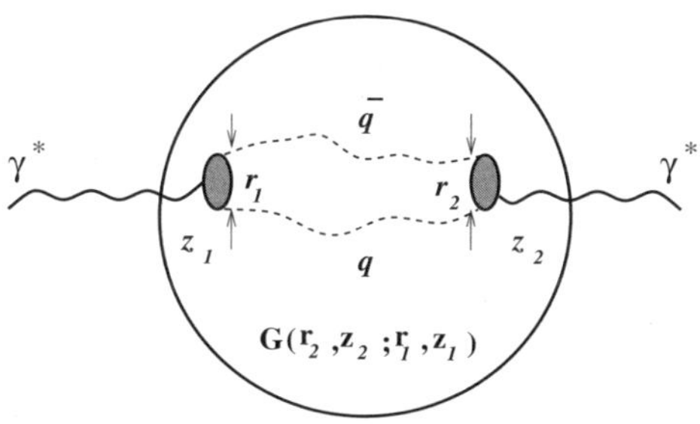

Fig. 6. Propagation of a $q\bar{q}$-pair through a nucleus between points with longitudinal coordinates z_1 and z_2. The evolution of the $\bar{q}q$ separation from the initial, $\vec{r}_1$, up to the final, $\vec{r}_2$, due to the transverse motion of the quarks, is described by the Green's function $G(\vec{\rho}_2, z_2; \vec{\rho}_1, z_1)$, a solution of Eq. (27).

It satisfies the evolution equation,[18,20,21]

$$\left[i\frac{\partial}{\partial z_2} + \frac{\Delta_\perp(r_2) - \varepsilon^2}{2\nu\alpha(1-\alpha)} + U(r_2, z_2)\right] G(\vec{r}_2, z_2 \,|\, \vec{r}_1, z_1) = 0 \tag{27}$$

The light-cone potential in the left-hand side of this equation describes nonperturbative interactions within the dipole, and its absorption in the medium. The real part of the potential responsible for nonperturbative quark interactions was modelled and fitted to data of F_2^p in Ref. 34. Here we fix $\mathrm{Re}\, U(r_2, z_2) = 0$, and treat quarks as free particles for the sake of simplicity. The imaginary part of the potential describes the attenuation of the dipole in the medium,

$$\mathrm{Im}\, U(r,z) = -\frac{1}{2}\,\sigma_{\bar{q}q}(r)\,\rho_A(b,z). \tag{28}$$

The numerical results of the calculations, which are performed either disregarding or including the real part of the potential, modelled in Ref. 34, are plotted in Fig. 7 by dashed and solid curves respectively. One can see that inclusion of the nonperturbative effects does not lead to a significant change of the magnitude of shadowing. Comparison with NMC data[23,24] shows a pretty good agreement. We remind that this is a parameter-free

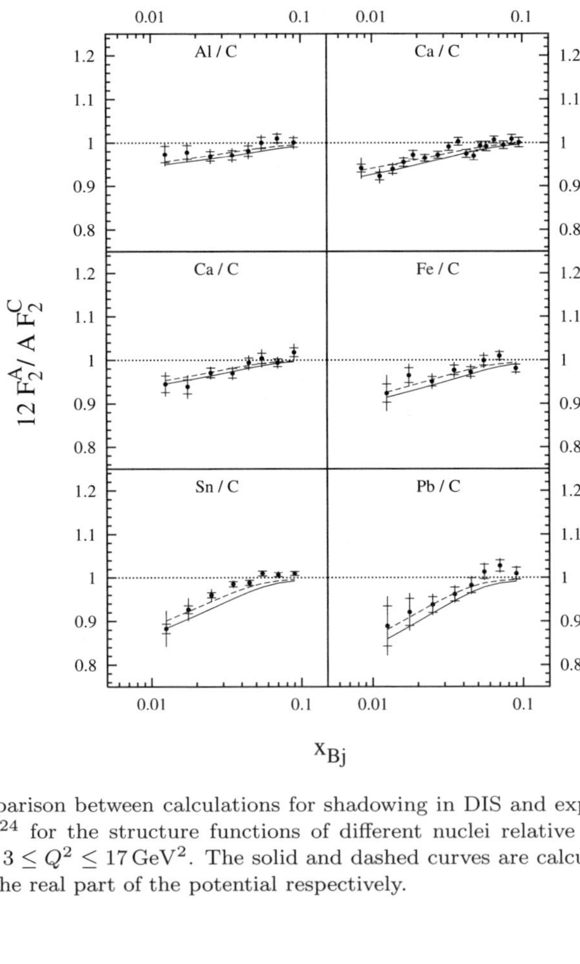

Fig. 7. Comparison between calculations for shadowing in DIS and experimental data from NMC[23,24] for the structure functions of different nuclei relative to carbon. Q^2 ranges within $3 \leq Q^2 \leq 17\,\mathrm{GeV}^2$. The solid and dashed curves are calculated including or excluding the real part of the potential respectively.

description. More example of comparison with data can be found in Ref. 22. Notice also that we did not include any mechanism of nuclear enhancement at $x > 0.1$, the effect called antishadowing, which affect the small-x region as well.

3.1.3. *Gluon shadowing*

As was mentioned above, the contribution of all Fock components in the cross section should be summed up

$$\sigma_{tot}^{\gamma^* A} = A\,\sigma_{tot}^{\gamma^* N} - \Delta\sigma_{tot}(\bar{q}q) - \Delta\sigma_{tot}(\bar{q}qg) - \Delta\sigma_{tot}(\bar{q}q2g) - \ldots \qquad (29)$$

The next after the $\bar{q}q$ Fock component is $\bar{q}q + g$. It has a considerably shorter coherence time, compared with $\bar{q}q$, because of specific nonperturbative effects increasing the mean transverse momentum of gluons.[25,34] Correspondingly, the next component $|\bar{q}q2g\rangle$ has even a much shorter coherence time t_c, which makes the 4th term in Eq. (29) negligibly small within the currently achieved kinematic range. Thus, we keep only the first two low Fock components.

Notice that a successful attempt to sum up all Fock components was done in the form of Balitsky-Kovchegov equation[26,27]

The 3rd term in the total cross section Eq. (29) is illustrated in Fig. 8.

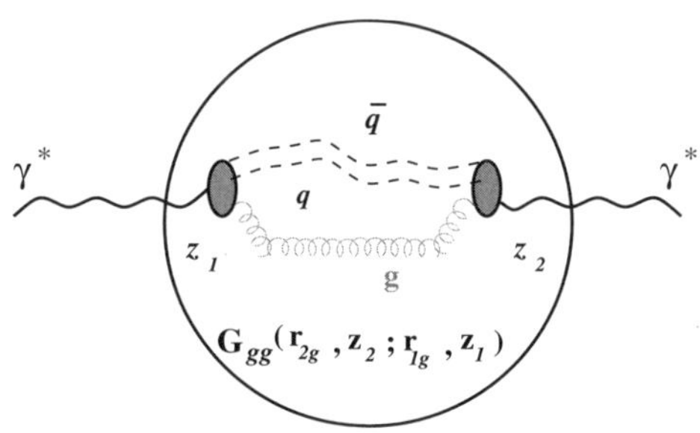

Fig. 8. Propagation through a nucleus of the $q\bar{q} - g$ fluctuation of a longitudinally polarized photon. Neglecting the small, $\sim 1/Q^2$ size of the color-octet $\bar{q}q$ pair, the effective octet-octet dipole propagation is described by the Green's function $G_{gg}(\vec{r}_{2g}, z_2; \vec{r}_{1g}, z_1)$.

Differently from the case of a $|\bar{q}q\rangle$ Fock state, where we found that at high Q^2 perturbative QCD can be safely used for shadowing calculations, the nonperturbative effects remain important for the $|\bar{q}\,q\,g\rangle$ component even for highly virtual photons. High Q^2 squeezes the $\bar{q}q$ pair down to a size $\sim 1/Q$, while the mean quark-gluon separation at $\alpha_g \ll 1$ depends on the strength of nonperturbative gluon interaction which is characterised in this limit by a small separation $r_0 \approx 0.3\,\text{fm}$ (see Ref. 34), which is considerably smaller than the confinement radius $1/\Lambda_{QCD}$. This is confirmed by various experimental observations,[25] in particular by the observed strong suppression of the diffractive gluon radiation.[34] In nonperturbative QCD models this scale is related to the instanton size.[28,29]

The nonperturbative quark-gluon wave function was found in Ref. 34 to have the form,

$$\Psi_{qg}\left(\vec{r}_g, \alpha_g\right)\Big|_{\alpha_g \ll 1} = -\frac{2i}{\pi}\sqrt{\frac{\alpha_s}{3}}\,\frac{\vec{e}\cdot\vec{r}_g}{r_g^2}\,\exp\left(-\frac{r_g^2}{2r_0^2}\right), \tag{30}$$

where $\vec{e}$ is the gluon polarization vector.

For $Q^2 \gg 1/r_0^2$ the $\bar{q}q$ is small, $r_{\bar{q}q}^2 \ll r_g^2$, and one can treat the $\bar{q}qg$ system as a color octet-octet dipole, as is illustrated in Fig. 8. Then the three-body Green's function factorizes,

$$G_{q\bar{q}g}\left(\vec{r}_{2g}, \vec{r}_2, z_2; \vec{r}_{1g}, \vec{r}_1, z_1\right) \Rightarrow G_{q\bar{q}}\left(\vec{r}_2, z_2; \vec{r}_1, z_1\right) G_{gg}\left(\vec{r}_{2g}, z_2; \vec{r}_{1g}, z_1\right). \tag{31}$$

The color octet-octet Green function G_{gg}, describing the propagation of a glue-glue dipole with $\alpha_g \ll 1$ through the medium, satisfies the simplified evolution equation,[34]

$$\left[i\frac{\partial}{\partial z_2} - \frac{Q^2}{2\nu} - V(\vec{r}_{2g}, z_2)\right] G_{gg}\left(\vec{r}_{2g}, z_2; \vec{r}_{1g}, z_1\right) = 0\,. \tag{32}$$

Here

$$\operatorname{Im} V(\vec{r}_{2g}, z) = -\frac{1}{2}\sigma_{gg}(r, x)\rho_A(z), \tag{33}$$

with the color-octet dipole cross section, which reads,

$$\sigma_{gg}\left(r, x\right) = \frac{9}{4}\,\sigma_{q\bar{q}}\left(r, x\right). \tag{34}$$

The real part of the potential must correctly reproduce the wave function Eq. (30).

$$\operatorname{Re} V(\vec{r}_{2g}, z) = \frac{r_{2g}^2}{2\nu\alpha_g r_0^4} \tag{35}$$

Longitudinal photons can be used to disentangle between the effects of higher twist quark shadowing (2d term in Eq. (29)), and leading twist gluon shadowing (3rd term in Eq. (29)). The contribution, which mixes up these two types of Gribov corrections, comes from so called aligned jet configurations[30] of the $\bar{q}q$ pair. Namely the mean $\bar{q}q$ separation $r_{\bar{q}q}^2 \sim$

$Q^2/\alpha_q(1-\alpha_q)$ is small, unless the large value of Q^2 is compensated by the smallness of α_q or $(1-\alpha_q)$. In the wave function of longitudinal photons, Eq. (24) such aligned-jet configurations are suppressed, so shadowing of longitudinal photons should represent the net effect of gluon shadowing.

While the distance between the q and the $\bar{q}$ is small, of the order $1/Q^2$, the gluon can propagate relatively far at a distance $r_g \sim r_0$ from the $q\bar{q}$-pair, which after the emission of the gluon is in a color-octet state. Therefore, the entire $|q\bar{q}g\rangle$-system appears as a octet-octet gg-dipole, and the shadowing correction to the longitudinal cross section directly gives the magnitude of gluon shadowing, which we want to calculate.

Thus, the cross section of longitudinal photons is proportional to the gluon distribution function, therefore,

$$\frac{g_A(x,Q^2)}{g_N(x,Q^2)} \approx \frac{\sigma_A^L(x,Q^2)}{\sigma_N^L(x,Q^2)}\,. \tag{36}$$

The shadowing correction to $\sigma_A^L(x,Q^2)$ has the form (compare with (26)),

$$\begin{aligned}\Delta\sigma_A^L(x,Q^2) &= -\mathrm{Re}\int d^2b \int\limits_{-\infty}^{\infty} dz_1 \int\limits_{z_1}^{\infty} dz_2\, \rho_A(b,z_1)\rho_A(b,z_2)\\ &\times \int d^2r_{2g}\; d^2\, r_{2\bar{q}q}\, d^2r_{1g}\, d^2r_{1\bar{q}q} \int d\alpha_q\, d\ln(\alpha_g) F^{\dagger}_{\gamma^*\to\bar{q}qg}(\vec{r}_{2g},\vec{r}_{2\bar{q}q},\alpha_q,\alpha_g)\\ &\times G_{\bar{q}qg}(\vec{r}_{2g}\;,\;\vec{r}_{2\bar{q}q},z_2;\vec{r}_{1g},\vec{r}_{1\bar{q}q},z_1)F_{\gamma^*\to\bar{q}qg}(\vec{r}_{1g},\vec{r}_{1\bar{q}q},\alpha_q,\alpha_g).\end{aligned} \tag{37}$$

Here

$$F^{\dagger}_{\gamma^*\to\bar{q}qg}(\vec{r}_g,\vec{r}_{\bar{q}q},\alpha_q,\alpha_g) = -\,\Psi^L_{\bar{q}q}\,(\vec{r}_{\bar{q}q},\alpha_q)\;\vec{r}_g\cdot\vec{\nabla}\,\Psi_{qg}\,(\vec{r}_g)\;\sigma^N_{gg}\,(x,r_g)\,. \tag{38}$$

Assuming $Q^2 \gg 1/r_0^2$ we can neglect $r_{\bar{q}q} \ll r_g$. The net diffractive amplitude $F_{\gamma^*\to\bar{q}qg}(\vec{r}_{1g},\vec{r}_{1\bar{q}q},\alpha_q,\alpha_g)$ takes the form of Eq. (38), and we can rely on the factorized relation (31) for the 3-body Green's function, with equation (32) for the evolution of the gluonic dipole.

The results of numerical calculation of (37) for the ratio

$$R_g(x,Q^2) = \frac{g_A(x,Q^2)}{A\,g_N(x,Q^2)}\,, \tag{39}$$

are depicted in Fig. 9 as function of Bjorken x for $Q^2 = 4$ and $40\,\mathrm{GeV}^2$. The predicted small magnitude of gluon shadowing was confirmed by next-to-leading (NLO) global analyses of DIS data.[31,32] It also goes along with the well known smallness of the triple-Pomeron coupling, measured at low virtuality.[33] The latter controls large mass diffraction, which proceeds via

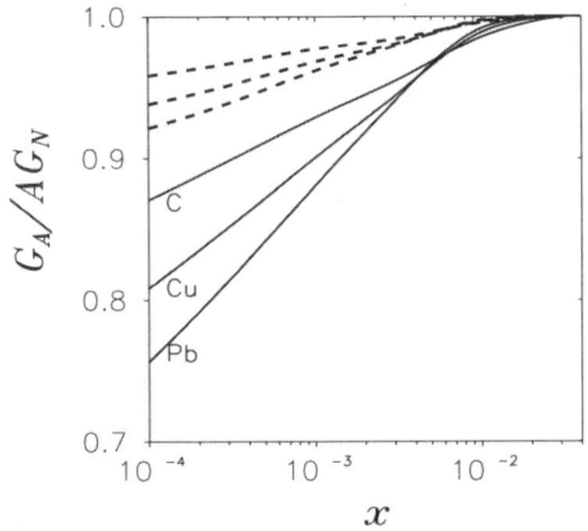

Fig. 9. Ratio (39) for carbon, copper and lead at small Bjorken x and $Q^2 = 4\,\mathrm{GeV}^2$ (solid curves) and $40\,\mathrm{GeV}^2$ (dashed curves).

gluon radiation, and its smallness leads to suppression of gluon radiation and gluon shadowing related to corresponding Gribov corrections. The observed weakness of these effects is interpreted in the dipole approach a smallness of the parameter $r_0 \approx 0.3\,\mathrm{fm}$ in Eq. (30).[25,34]

3.2. *Hadron-nucleus cross sections*

Applications of the dipole approach to calculation of Gribov corrections to the hadron-nucleus total cross sections contains more uncertainties and modelling compared with hard reactions, like DIS. Nevertheless, it allows to make a progress compared with the hadronic representation, which involves ad hoc assumptions about the interaction cross section, of an excited hadronic state, and the unknown higher terms in the opacity expansion.

3.2.1. *Excitation of the valence quark skeleton of the proton*

First of all one should rely on a parametrization of the dipole cross section, which allows an extension to the soft, large separation region. Following Refs. 5, 34 we chose the saturated shape of the cross section, which rises as r_T^2 at small r_T, but levels off at large r_T,

$$\sigma_{\bar{q}q}(r_T, s) = \sigma_0(s)\left[1 - \exp\left(-\frac{r_T^2}{R_0^2(s)}\right)\right], \tag{40}$$

where $R_0(s) = 0.88\,fm\,(s_0/s)^{0.14}$ and $s_0 = 1000\,GeV^2$ (see Ref. 34). The energy dependent factor $\sigma_0(s)$ is defined as,

$$\sigma_0(s) = \sigma_{tot}^{\pi p}(s)\left(1 + \frac{3\,R_0^2(s)}{8\,\langle r_{ch}^2\rangle_\pi}\right), \tag{41}$$

where $\langle r_{ch}^2\rangle_\pi = 0.44 \pm 0.01\,fm^2$ (see Ref. 37) is the mean square of the pion charge radius.

This dipole cross section is normalized to reproduce the pion-proton total cross section, $\langle\sigma_{\bar{q}q}\rangle_\pi = \sigma_{tot}^{\pi p}(s)$. The saturated shape of the dipole cross section is inspired by the popular parametrization given in Refs. 35, 36, which is fitted to the low-x and high Q^2 data for $F_2^p(x, Q^2)$ from HERA. However, that should not be used for our purpose, since it is unable to provide the correct energy dependence of hadronic cross sections. Namely, the pion-proton cross section cannot exceed 23 mb. Besides, Bjorken x is not a proper variable for soft reactions, since at small Q^2 the value of x is large even at low energies. The s-dependent dipole cross section Eq. (40) was fitted[34] to data for hadronic cross sections, real photoproduction and low-Q^2 HERA data for the proton structure function. The cross section (41) averaged with the pion wave function squared (see below) automatically reproduces the pion-proton cross section.

In the case of a proton beam one needs a cross section for a three-quark dipole, $\sigma_{3q}(\vec{r}_1, \vec{r}_2, \vec{r}_3)$, where $\vec{r}_i$ are the transverse quark separation with a condition $\vec{r}_1 + \vec{r}_2 + \vec{r}_3 = 0$. In order to avoid the introduction of a new unknown phenomenological quantity, we express the three-body dipole cross section via the conventional dipole cross section $\sigma_{\bar{q}q}$ (see Ref. 4),

$$\sigma_{3q}(\vec{r}_1, \vec{r}_2, \vec{r}_3) = \frac{1}{2}\Big[\sigma_{\bar{q}q}(r_1) + \sigma_{\bar{q}q}(r_2) + \sigma_{\bar{q}q}(r_3)\Big]\,. \tag{42}$$

This form satisfies the limiting conditions, namely, turns into $\sigma_{\bar{q}q}(r)$ if one of three separations is zero. Since all these cross sections involve nonperturbative effects, this relation hardly can be proven, but should be treated as a plausible assumption.

The 3-quark valence wave function is modelled assuming that the dipole cross section is independent of the sharing of the light-cone momentum among the quarks, so the wave function squared of the valence Fock component of the proton, $|\Phi(\vec{r}_i, \alpha_j|^2$ should be integrated over fractions α_i. The result depends only on transverse separations $\vec{r}_i$. The form of the nonperturbative valence quark distribution is unknown, therefore for the sake of simplicity we assume the Gaussian form,

$$
\begin{aligned}
|\Psi_N(\vec{r}_1,\vec{r}_2,\vec{r}_3)|^2 &= \int\limits_0^1 \prod_{i=1}^{3} d\alpha_i\, |\Phi(\vec{r}_i,\alpha_j)|^2\, \delta\left(1-\sum_{j=1}^{3}\alpha_j\right) \\
&= \frac{2+r_p^2/R_p^2}{(\pi\, r_p\, R_p)^2}\quad \exp\left(-\frac{r_1^2}{r_p^2}-\frac{r_2^2+r_3^2}{R_p^2}\right)\delta(\vec{r}_1+\vec{r}_2+\vec{r}_3)\ , \qquad (43)
\end{aligned}
$$

where $\vec{r}_i$ are the interquark transverse distances. The two scales r_p and R_p characterizing the mean transverse size of a diquark and the mean distances to the third quark.

For the sake of simplicity here we assume that the forces binding the valence quarks are of an iso-scalar nature, therefore the quark distribution is symmetric, i.e. $r_p = R_p$ in Eq. (43). In this case the mean interquark separation squared is $\langle\vec{r}_i^{\,2}\rangle = \frac{2}{3}R_p^2 = 2\langle r_{ch}^2\rangle_p$. See other possibilities of an asymmetric valence structure in Ref. 5.

Apparently, any model for the dipole cross section and valence quark distribution in the proton, must reproduce correctly data for diffractive excitation of the proton, otherwise the Gribov corrections will come out wrong. It was demonstrated that with the above choice of the dipole cross section and proton wave function one reproduces quite well the results of the global analysis[33] of single diffraction data, namely the triple-Reggeon term $\mathbb{PPR}$, as well as the triple-Pomeron one $\mathbb{PPP}$, controlling diffractive gluon radiation.

The total cross sections reads,

$$
\begin{aligned}
\sigma_{tot}^{pA} &= 2\int d^2b\,\left[1-\left\langle e^{-\frac{1}{2}\,\sigma_{3q}(r_i)\,T_A(b)}\right\rangle\right] \\
&\equiv 2\int d^2b\left[1-\int\prod_{i=1}^{3} d^2r_i\,|\Psi_N(r_j)|^2\, e^{-\frac{1}{2}\,\sigma_{3q}(\vec{r}_k)\,T_A(b)}\right]. \qquad (44)
\end{aligned}
$$

Using the wave function Eq. (43) with $r_p = R_p$ and the cross section, Eq. (40), we get the following forward elastic cross section,

$$
\left.\frac{d\sigma_{el}^{pp}}{dp_T^2}\right|_{p_T=0} = \frac{\gamma^2}{(1+\frac{2}{3}\gamma)^2}\,\frac{\sigma_0^2(s)}{16\pi}\,, \qquad (45)
$$

where $\gamma = 3\langle r_{ch}^2\rangle_p/R_0^2(s)$.

4. Gluon shadowing

Eikonalization of the lowest Fock state $|3q\rangle$ of the proton done in (44) corresponds to the Bethe-Heitler regime of gluon radiation. Indeed, gluon bremsstrahlung is responsible for the rising energy dependence of the cross section, and in the eikonal form (44) one assumes that the whole spectrum of gluons is radiated in each of multiple interactions. However, the Landau-Pomeranchuk-Migdal effect[38,39] is known to suppress radiation in multiple interactions. Since a substantial part of the inelastic cross section at high energies is related to gluon radiation, the LPM effect suppresses the cross section. This is a quantum-mechanical interference phenomenon and it is a part of the suppression called Gribov inelastic shadowing. In the QCD dipole picture it come from inclusion of higher Fock states, $|3qg\rangle$, etc. Each of these states represents a colorless dipole and its elastic amplitude on a nucleon is subject to eikonalization.

As we already mentioned, the eikonalization procedure requires the fluctuation lifetime to be much longer than the nuclear size. Otherwise, one has to take into account the "breathing" of the fluctuation during propagation through a nucleus, which can be done by applying the light-cone Green function technique.[20,21,34] In hadronic representation this is equivalent to saying that all the longitudinal momenta transfers must be much smaller than the inverse mean free path of the hadron in the nucleus. Otherwise, one should employ the path-integral technique, described above.

The c.m. energies of HERA-B, RHIC and LHC are sufficiently high to treat the lowest Fock state containing only valence quarks as "frozen" by the Lorentz time dilation during propagation through the nucleus. Indeed, for the excitations with the typical nucleon resonance masses, the coherence length is sufficiently long compared to the nuclear size. This is why we applied eikonalization without hesitation so far. Such an approximation, however, never works for the higher Fock states containing gluons. Indeed, since the gluon is a vector particle, the integration over effective mass of the fluctuation is divergent, dM^2/M^2, which is the standard triple-Pomeron behaviour. Therefore, the energy of collisions can never be sufficiently high to neglect the large-mass tail. For this reason the inelastic shadowing corrections, related to excitation gluonic degrees of freedom never saturates, and keeps rising logarithmically with energy.

There are, however non-linear effects which are expected to stop the rise of inelastic corrections at high energies. This is related to the phenomenon of gluon saturation[40,41] or color glass condensate.[42] The strength of these nonlinear effect is expected to be rather mild due to smallness of the gluonic spots in the nucleus.[25] The reason is simple, in spite of a sufficient longitudinal overlap of gluon clouds originated from different nucleons, there is insufficient overlap in the transverse plane. This fact leads to a delay of the onset of saturation up to very high energies, since the transverse radius squared of the gluonic clouds rise with energy very slowly, logarithmically, with a small coefficient of the order of $0.1\,\mathrm{GeV}^{-2}$.

The details of the calculation of inelastic corrections related to excitation of gluonic degrees of freedom can be found in Ref. 4, 34. The numerical results for nuclear cross sections at the energies of RHIC and LHC can be found in Ref. 5, 6. As an example, the total proton-lead cross section, calculated at $\sqrt{s} = 5\,\mathrm{TeV}$ in the Glauber approximation, and corrected to Gribov shadowing related to excitation of valence quarks and gluons, results in $\sigma_{tot}^{pPb} = 4242.5\,\mathrm{mb}$; $4235.2\,\mathrm{mb}$; and $4207.1\,\mathrm{mb}$ respectively.

5. Summary

The dipole phenomenology in QCD has been intensively developing over the past three decades, due to both theoretical efforts and precise experimental data, in particular from HERA. This theoretical tool allows to calculate the effects of Gribov inelastic shadowing on a more solid ground and in all orders of opacity expansion. In this note we presented several explicit examples.

References

1. R.J. Glauber, in *Lectures in Theoretical Physics*, Eds. W.E. Brittin and L.G. Dunham, Interscience, New York, 1959.
2. V.N. Gribov, *Sov. Phys. JETP* **29**, 483 (1969) [*Zh. Eksp. Teor. Fiz.* **56**, 892 (1969)].
3. V.N. Gribov, *Sov. Phys. JETP* **30**, 709 (1970) [*Zh. Eksp. Teor. Fiz.* **57**, 1306 (1970)].
4. B.Z. Kopeliovich, *Phys. Rev.* **C68**, 044906 (2003).
5. B.Z. Kopeliovich, I.K. Potashnikova and I. Schmidt, *Phys. Rev.* **C73**, 034901 (2006).
6. M. Alvioli, C. Ciofi degli Atti, B.Z. Kopeliovich, I.K. Potashnikova, and I. Schmidt, *Phys. Rev.* **C81**, 025204 (2010).
7. V. Karmanov and L.A. Kondratyuk, *Sov. Phys. JETP Lett.* **18**, 266 (1973).
8. P.V.R. Murthy *et al.*, *Nucl. Phys.* **B 92**, 269 (1975).
9. A. Gsponer *et al.*, *Phys. Rev. Lett.* **42**, 9 (1979).
10. B.Z. Kopeliovich, *Phys. Lett.* **B227**, 461 (1989).
11. P.P. Allport *et al.* [WA59 Collaboration], *Phys. Lett.* **B232**, 417 (1989).
12. E. Feinberg and I.Ya. Pomeranchuk, *Nuovo Cimento Suppl.* **3**, 652 (1956).
13. M.L. Good and W.D. Walker, *Phys. Rev.* **120**, 1857 (1960).

14. B.Z. Kopeliovich and L.I. Lapidus, *Sov. Phys. JETP Lett.***28**, 664 (1978).
15. B.Z. Kopeliovich, L. I. Lapidus, and A.B. Zamolodchikov, *Sov. Phys. JETP Lett.* **33**, 595 (1981).
16. H.I. Miettinen and J. Pumplin, *Phys. Rev.* **D18**, 1696 (1978).
17. B.Z. Kopeliovich, J. Raufeisen, and A.V. Tarasov, *Phys. Lett.* **B440**, 151 (1998).
18. B.G. Zakharov, *Phys. Atom. Nucl.* **61**, 838 (1998).
19. R. Feynman and A.R. Hibbs, *Quantum mechanics and path integrals*, McGraw-Hill, New York, 1965.
20. B.Z. Kopeliovich and B.G. Zakharov, *Phys. Rev.* **D44**, 3466 (1991).
21. B.Z. Kopeliovich, A. Schäfer, and A.V. Tarasov, *Phys. Rev.* **C59**, 1609 (1999). See the extended version in arXiv:hep-ph/9808378.
22. B.Z. Kopeliovich, J. Raufeisen, and A.V. Tarasov, *Phys. Rev.* **C62**, 035204 (2000).
23. The NM Collaboration, M. Arneodo *et al.*, *Nucl. Phys.* **B481**, 23 (1996).
24. The NM Collaboration, P. Amaudruz *et al.*, *Nucl. Phys.* **B441**, 3 (1995).
25. B.Z. Kopeliovich, I.K. Potashnikova, B. Povh, and I. Schmidt, *Phys. Rev.* **D76**, 094020 (2007).
26. I. Balitsky, *Nucl. Phys.* **B463**, 99 (1996).
27. Y.V. Kovchegov, *Phys. Rev.* **D60**, 034008 (1999).
28. T. Schäfer and E.V. Shuryak, *Rev. Mod. Phys.* **70**, 323 (1998).
29. E. Shuryak and I. Zahed, *Phys. Rev.* **D69**, 014011 (2004).
30. J.D. Bjorken and J. Kogut, *Phys. Rev.* **D8**, 1341 (1973).
31. D. de Florian and R. Sassot, *Phys. Rev.* **D69**, 074028 (2004).
32. M. Hirai, S. Kumano, and T.-H. Nagai, *Phys. Rev.* **C70**, 044905 (2004).
33. Y.M. Kazarinov, B.Z. Kopeliovich, L.I. Lapidus, and I.K. Potashnikova, *JETP* **43**, 598 (1976).
34. B.Z. Kopeliovich, A. Schäfer, and A.V. Tarasov, *Phys. Rev.* **D62**, 054022 (2000).
35. K. Golec-Biernat and M. Wüsthoff, *Phys. Rev.* **D59**, 014017 (1999).
36. K. Golec-Biernat and M. Wüsthoff, *Phys. Rev.* **D60**, 114023 (1999).
37. S. Amendolia *et al.*, *Nucl. Phys.* **B277**, 186 (1986).
38. L.D. Landau and I.Ya. Pomeranchuk, JETP**24**, 505 (1953).
39. A.B. Migdal, *Phys. Rev.* **103**, 1811 (1956).
40. L.V. Gribov, E.M. Levin, and M.G. Ryskin, *Phys. Rep.* **100**, 1 (1983).
41. A.H. Mueller, *Eur. Phys. J.* **A1**, 19 (1998).
42. L. McLerran and R. Venugopalan, *Phys. Rev.* **D49**, 2233 (1994).

FINE STRUCTURE OF THE DIFFRACTION CONE: FROM THE ISR TO THE LHC

D.A. FAGUNDES

Universidade Federal de Santa Catarina - Campus Blumenau, Rua Pomerode, 710 Salto do Norte, 89065-300 Blumenau, SC, Brazil, E-mail: daniel.fagundes@ufsc.br

L. JENKOVSZKY

Bogolyubov Institute for Theoretical Physics, National Academy of Sciences of Ukraine, UA-03680 Kiev, Ukraine, E-mail: jenk@bitp.kiev.ua

E.Q. MIRANDA

InSTEC, Quinta de los Marinos, Ave. Salvador Allende y Lucas, La Habana 10400 Cuba, E-mail: equintana@ucf.edu.cu

G. PANCHERI

INFN Frascati National Laboratories, Via E. Fermi 40, 00444, Italy, E-mail: Giulia.Pancheri@lnf.infn.it

P.V.R.G. SILVA

Instituto de Física Gleb Wataghin, Universidade Estadual de Campinas, 13083-859, Campinas, SP, Brazil, E-mail: precchia@ifi.unicamp.br

Following earlier findings, we argue that the low-$|t|$ structure in the elastic diffractive cone, recently reported by the TOTEM Collaboration at 8 TeV, is a consequence of the threshold singularity required by $t-$channel unitarity, such as revealed earlier at the ISR. By using simple Regge-pole models, we analyze the available data on the pp elastic differential cross section in a wide range of c.m. energies, namely those from ISR to LHC8, obtaining good fits of all datasets. This study hints at the fact that the non-exponential behavior observed at LHC8 is a recurrence of the low-$|t|$ "break" phenomenon, observed in the seventies at ISR, being induced by the presence of a two-pion loop singularity in the Pomeron trajectory.

1. Introduction

The recent observation by the TOTEM collaboration[1] of a non-purely exponential behavior of the differential pp elastic cross-section at $\sqrt{s} = 8$ TeV (LHC8), is discussed in this note in the light of two simple Pomeron models[2] inclusive of a threshold singularity in the Pomeron trajectory, as required by t-channel unitarity.[3*] All these models are tested in a wide energy range, namely from ISR to LHC, and in all cases analyzed we find good agreement with the data, in particular in the diffraction cone.

The obstacle to describe pp scattering both in the very small $-t$ region and the subsequent behavior until after the observed dip, was well known, reflecting in the difficulty to obtain the values at the optical point, i.e. the total cross-section, and also a good description of the structure with the dip and the bump. Early attempts in this direction, including the need of a square root singularity near the threshold, can be found in Ref. 6. An empirical model, based on the old Phillips and Barger (PB) model independent description,[7] had recently highlighted this difficulty.[8] Here we revisit the problem of modeling the elastic differential cross section with this model, which in principle has no well-defined s-dependence build in. We shall also test the PB model, augmented by a threshold singularity[8], with the present LHC8 data.[1] While giving our perspective on this problem, we also discuss recent attempts to give a complete description of the available range of $-t$ data, in the framework of Regge Field Theory (RFT) and based on a two-channel eikonal approach, also inclusive of the pion loop singularity.[9]

It is well-known that unitarity constrains the analytic properties of the scattering amplitude. In particular, as shown in Refs. 10, Regge trajectories near the threshold behave as

$$\Im\alpha(t) \sim (t - t_0)^{\Re\alpha(t_0)+1/2}, \tag{1}$$

where t_0 is the lightest threshold, e.g. $t_0 = 4m_\pi^2$ for the f or Pomeron trajectory. A good approximation to the lightest threshold is by a square root[6]:

$$\alpha(t) \sim \alpha_1\sqrt{t_0 - t}, \tag{2}$$

where α_1 is a free parameter, that in Refs. 11 was associated with the pion mass, $\alpha_1 = m_\pi/(1\ \text{GeV}^2)$. While the low-mass $4m_\pi^2$ threshold is responsible for the low-$|t|$ structure (the so-called "break" near $t \approx -0.1$ GeV^2), the otherwise exponential shape of the forward cone is provided by the nearly linear behavior of the Pomeron trajectory beyond the break (in fact, a smooth curvature) and until $t \approx -1$ GeV^2. At large $|t|$ the trajectory tends

*Excellent overviews on the topic can also be found in the Refs. 4, 5

to its logarithmic asymptotics, but this is beyond the scope of the present study (see, e.g., Ref. 2 and earlier references therein).

The non-exponential behavior of the cone at low-t was recently studied in Ref. 8, where two options were scrutinized, one with the above threshold singularity in the trajectory and the other one with the form factor, inspired in early suggestions by Donnachie and Landshoff (see Ref. 12 and earlier references therein). As already mentioned, recent measurements by the TOTEM Collaboration found deviation from the linear exponential behavior of the elastic diffraction cone at 8 TeV (see Ref. 1). We argue that the structure seen at 8 TeV is a recurrence of that seen earlier at the ISR, both resulting from the threshold singularity imposed by t channel unitarity. To this end we use a simple Regge-pole model extrapolating from the ISR energy region to that of the LHC. In doing so, we use a single "effective" Regge trajectory that incorporates leading (Pomeron) and non-leading Regge exchanges. While in the ISR energy region f exchange is important, it is negligible at the LHC (see e.g., Ref. 2).

In the present study we use two simple versions of the Regge-pole model incorporating an "effective" trajectory including the required threshold singularity, with and without a linear term added. We make a number of fits, one based only on the ISR data to see its extrapolation to the LHC, and the other one including also the new LHC data.

2. Simple Pomeron Pole (Model 1)

The first simple model analyzed is the one from Ref. 6, whose parametrization for the elastic differential cross section follows:

$$\frac{d\sigma_{el}}{dt} = r \exp\{bt + 2\ln(s/s_0)[\alpha(t) - 1]\}, \tag{3}$$

where the Pomeron trajectory is written as

$$\alpha(t) = \alpha_0 + \alpha' t - \alpha_1 \sqrt{t_0 - t} \tag{4}$$

and r (mbGeV^{-2}), b (GeV^{-2}), α_0 (dimensionless), α' (GeV^{-2}), α_1 (GeV^{-1}) are free parameters, $s_0 = 1$ GeV^2, $t_0 = 4m_\pi^2$ (GeV^2). On performing data analysis with this model we consider two cases (variants):

i. a linear term is present in the Pomeron ("effective") trajectory (α' free parameter) - denoted as *with Linear term*;
ii. a linear term is absent in $\alpha_P(t)$ ($\alpha' = 0$ fixed) - denoted as *without Linear term*.

In such case, the Pomeron is regarded as a simple 'supercritical' pole in the complex angular momenta plane, with $\Delta = \alpha(0) - 1 > 0$ and $\Delta \sim 0.1$.

3. Double Pomeron Pole (Model 2)

In the framework of the so-called Dipole Pomeron Model (DPM), the Pomeron amplitude is regarded as a two-term amplitude being written as[2]:

$$A_P(s,t) = i\frac{a_P}{b_P}\frac{s}{s_0}[r_1^2(s)e^{r_1^2(s)[\alpha_P-1]} - \epsilon_P r_2^2(s)e^{r_2^2(s)[\alpha_P-1]}], \quad (5)$$

where again

$$\alpha(t) = \alpha_0 + \alpha' t - \alpha_1\sqrt{t_0 - t} \quad (6)$$

represent the Pomeron ("effective") trajectory and

$$r_1^2(s) = b_P + L - i\pi/2, \quad (7)$$

$$r_2^2(s) = L - i\pi/2, \quad (8)$$

$L = \ln(s/s_0)$ and a_P, b_P, α_0 (adim.), α' (GeV^{-2}), α_1 (GeV^{-1}) and ϵ_P are free parameters. In this model, the energy-dependent functions $r_1(s)$ and $r_2(s)$, having logarithm growth in s reflect the unitarization of the Pomeron amplitude at high-energies. Here, $t_0 = 4m_\pi^2$ (GeV^2) and $s_0 = 1$ GeV^2 are fixed. Just like with Model 1, we considered here also two possible cases (variants), namely:

i. a linear term is present in the Pomeron ("effective") trajectory (α' free parameter) - denoted as *with Linear term*;
ii. a linear term is absent in $\alpha_P(t)$ ($\alpha' = 0$ fixed)- denoted as *without Linear term*;

In this framework, from the amplitude in Eq. (5) we calculate the elastic differential cross section through the following expression:

$$\frac{d\sigma_{el}}{dt} = \frac{\pi}{s^2}|A_P(s,t)|^2. \quad (9)$$

While the parameter ϵ_P in Eq. (5) - reflecting absorptive effects - plays a major role in the dip-bump region it can be approximately neglected on dealing with the physics we're interested in, namely the break at very small $-t$.

4. Data Analysis and Fits

In the following we shall briefly discuss our data analysis, presenting in the first place the datasets used in data reductions and then showing the main results achieved.

4.1. *Datasets*

The datasets analyzed here are those of the elastic differential cross section for pp scattering[13] at ISR energies, namely in the interval 23.5−62.5 GeV as well as the recent ones at LHC7, measured by the TOTEM Collaboration.[14] In our first approach to data reductions, the LHC8 data[1] is not taken into account. However, as we explain along the text, it can be used when properly specified. In Table 1 we display the number of points in comprising each dataset used for data reductions. As we are analyzing the effect of deviations of exponential behavior $\sim e^{-b|t|}$ at the diffraction cone, only the data in the interval $|t|$: 0.01 − 0.35 GeV^2 were considered on performing χ^2 goodness of fit analyses.

Table 1. Number of points in each energy in the range of momentum transfer range $|t|$: 0.01 − 0.35 GeV^2 used for data reductions.

$\sqrt{s}$ (GeV)	23.5	30.7	44.7	52.8	62.5	7000	8000
N° points	62	88	139	59	53	81	30

4.2. *Fits - Model 1*

The results obtained with Model 1, shortly described in Section 2, are displayed in Tables 2 and 3, where we consider the two possible cases: with or without linear term. These fits are shown also in Fig. 1. In Fig. 2 we show the behavior of the local slope

$$B(t) = \frac{d}{dt}\left[\ln \frac{d\sigma}{dt}\right] \tag{10}$$

for fits including LHC7-TOTEM data[14].

Table 2. Fit of Model 1 (with $\alpha' = 0$ fixed) with or without the LHC7-TOTEM data added. Statistical information regarding the goodness of fit is also displayed.

	without TOTEM	with TOTEM
r	38.50 ± 0.72	22.76 ± 0.17
b	7.73 ± 0.11	5.594 ± 0.057
α_0	1.1168 ± 0.0026	1.1840 ± 0.0012
α'	0 (fixed)	0 (fixed)
α_1	0.2240 ± 0.0063	0.3399 ± 0.0027
χ^2/DOF	1.77	4.97
DOF	397	478

Table 3. Fit of Model 1 (with α' free) with or without the LHC7-TOTEM data added. Statistical information regarding the goodness of fit is also displayed.

	without TOTEM	with TOTEM
r	36.410 ± 0.94	22.10 ± 0.17
b	6.91 ± 0.28	5.298 ± 0.065
α_0	1.1140 ± 0.0028	1.1277 ± 0.0023
α'	0.084 ± 0.026	0.2597 ± 0.0091
α_1	0.2009 ± 0.0095	0.1371 ± 0.0075
χ^2/DOF	1.76	3.26
DOF	396	477

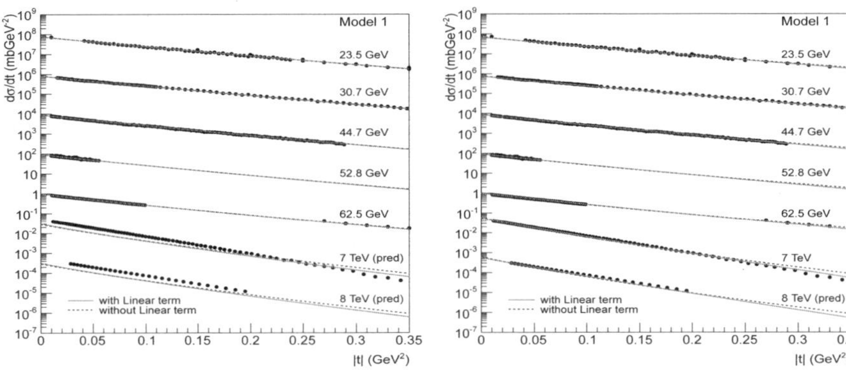

Fig. 1. Fits of Model 1, with or without TOTEM data (*left*) in datasets. In the left panel we show this model prediction for the case where the LHC7 data is not added in the data sets. In the right panel we refit the data, now including the LHC7 data, as given in Ref. 14. In both cases we investigate the effect of a linear term in Pomeron trajectory, finding a slightly better agreement with data for the model with a linear term. Curves and data are multiplied by $10^{\pm 2}$ factors to be shown in the same canvas.

4.3. *Fits - Model 2*

In this section we present the results obtained with Model 2, namely the Dipole Pomeron Model. As explained in Section 3, we consider here only fits with $\epsilon_P = 0$ fixed and, as in Model 1, we investigate the effect of the linear term in the trajectory by considering either $\alpha' = 0$ fixed or α' as a free parameter. For the first variant, the fits without and with LHC7-TOTEM data are shown in Table 4. For the latter, the results are shown in Table 5. For both cases, the comparison among data and curves are in Fig. 3.

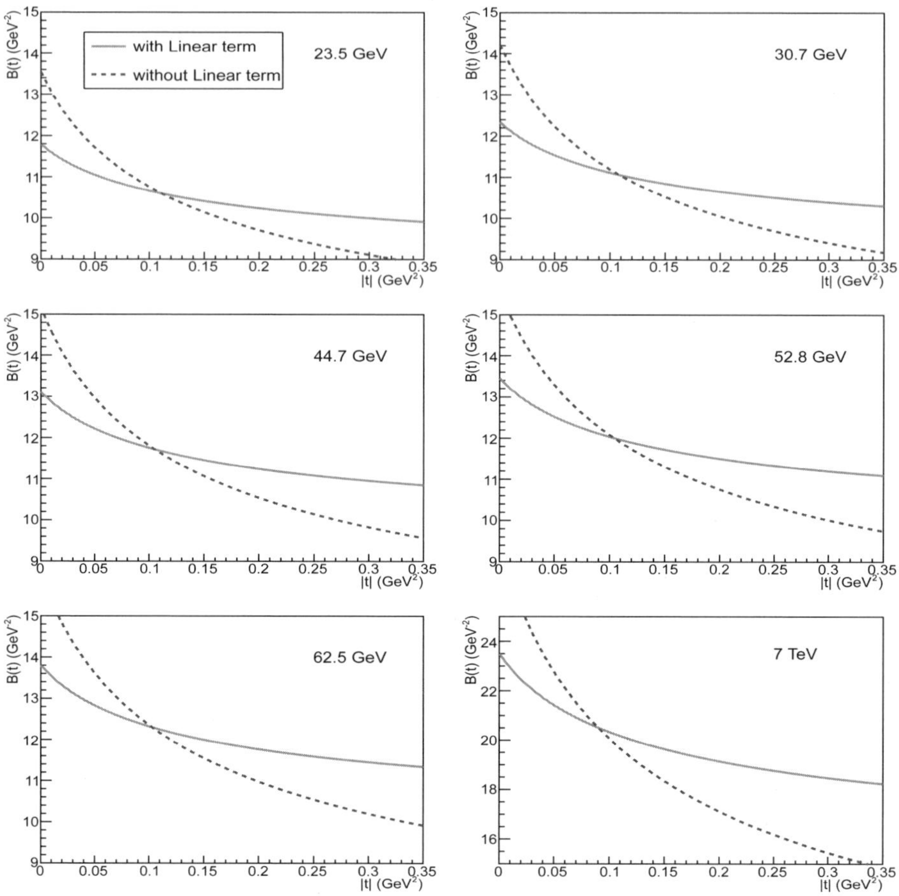

Fig. 2. Local slope $B(t)$ for the fits of Model 1 (fit with TOTEM data) in two variants: with and without linear term in the trajectory.

4.4. *Discussion*

In this section we discuss general aspects of the results obtained with Models 1 and 2 for all variants considered.

Considering the fits with and without the linear term in the trajectory, it is clear from Figs. 1 and 3 that, for both Models, the parametrization with the linear term (α' as a free parameter) describes better the data analyzed, specially when LHC7-TOTEM data is included. It can also be seen in statistical grounds by the values of χ^2 showed in Tabs. 2-5. This effect is more evident for Model 2 (see right panel in Fig. 3 and Tabs. 4

Table 4. Fits using Model 2 with $\alpha' = 0$ fixed and $\epsilon_P = 0$ fixed without and with TOTEM data. Statistical informations are also shown.

	without TOTEM	with TOTEM
a_P	0.660 ± 0.046	1.270 ± 0.012
b_P	1.230 ± 0.097	5.258 ± 0.076
α_0	1.1483 ± 0.0013	1.15325 ± 0.00073
α'	0 (fixed)	0 (fixed)
α_1	0.6026 ± 0.0038	0.4304 ± 0.0023
ϵ_P	0 (fixed)	0 (fixed)
χ^2/DOF	14.43	19.07
DOF	397	478

Table 5. Fits using Model 2 with α' free and $\epsilon_P = 0$ fixed without and with TOTEM data. Statistical informations are also shown.

	without TOTEM	with TOTEM
a_P	2.91 ± 0.10	1.598 ± 0.012
b_P	6.76 ± 0.31	8.03 ± 0.11
α_0	0.9625 ± 0.0028	1.0260 ± 0.0015
α'	0.6014 ± 0.0088	0.4453 ± 0.0058
α_1	-0.0549 ± 0.0088	-0.0096 ± 0.0049
ϵ_P	0 (fixed)	0 (fixed)
χ^2/DOF	1.99	5.93
DOF	396	477

and 5). We conclude that the linear term is necessary to describe the data, even in the low momentum transfer region.

Regarding the extrapolations to 7 and 8 TeV from fits without LHC7-TOTEM, i.e., with only ISR data, from Figs. 1 and 3 it is evident that our predictions underestimate the data, since we have not used data from $\bar{p}p$ scattering to anchor our fits at very high energies.[†] For this reason, in the rest of the text, we will focus only in the results obtained with LHC7-TOTEM data.

5. Fine Structure in the LHC8-TOTEM Data

If one expects to extrapolate the results here presented to higher energies, specially to 13 TeV, it is of great importance to include the data obtained for the differential cross section in 8 TeV by TOTEM Collaboration.[1] In

[†]In our understanding, due to poor statistics, we cannot treat $\bar{p}p$ on the same grounds, at least on treating deviations from the exponential behavior at small $-t$, which was not seeing at the Tevatron 1.8 TeV, for instance.

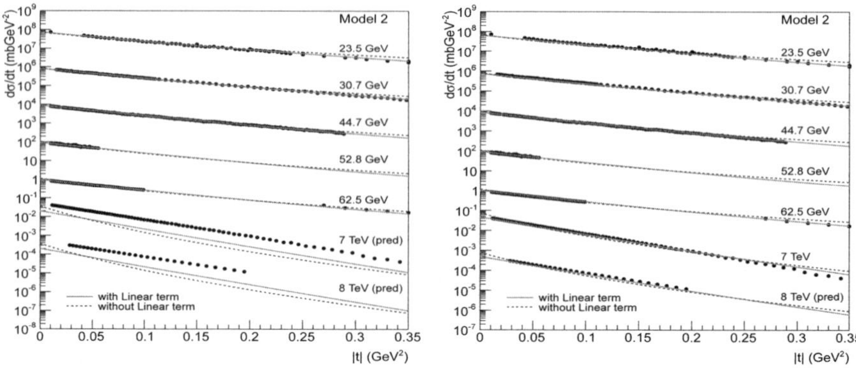

Fig. 3. Fits of Model 2, with or without TOTEM data (*left*) in datasets. In the left panel we show this model prediction for the case where the LHC7 data is not added in the data sets. In the right panel we refit the data, now including the LHC7 data, as given in Ref. 14. In both cases we investigate the effect of a linear term in Pomeron trajectory, finding a slightly better agreement with data for the model with a linear term. Curves and data are multiplied by $10^{\pm 2}$ factors to be shown in the same canvas.

this section, we present results of these updated fits.

The results for Model 1 and 2 are showed in Tables 6 and 7, respectively, and in Fig. 4 for both models. For all cases, we consider two variants: with linear term and without linear term in the trajectory.

Table 6. Fits using Model 1 including LHC8-TOTEM[1] data in two variants: without and with the linear term in the trajectory. Statistical informations are also shown.

	w/o Linear term	w/ Linear term
r	23.008 ± 0.058	22.307 ± 0.058
b	7.193 ± 0.034	6.130 ± 0.034
α_0	1.16548 ± 0.00030	1.09924 ± 0.00078
α'	0 (fixed)	0.3195 ± 0.0035
α_1	0.27321 ± 0.00081	0.0402 ± 0.0027
χ^2/DOF	21.4	4.51
DOF	508	507

In order to be able to see in more details the deviation from a pure exponential form, we compare the data and curves to a reference function by means of the ratio:

$$R = \frac{d\sigma/dt - \text{Ref}}{\text{Ref}}, \tag{11}$$

Table 7. Fits using Model 2 with $\epsilon_P = 0$ fixed including LHC8-TOTEM data[1] in two variants: without and with the linear term in the trajectory. Statistical informations are also shown.

	w/o Linear term	w/ Linear term
a_P	1.3450 ± 0.0024	1.4716 ± 0.0023
b_P	7.613 ± 0.045	8.604 ± 0.055
α_0	1.13539 ± 0.00012	1.02574 ± 0.00058
α'	0 (fixed)	0.4512 ± 0.0025
α_1	0.34239 ± 0.00063	-0.034 ± 0.0020
ϵ_P	0 (fixed)	0 (fixed)
χ^2/DOF	81.6	7.20
DOF	508	507

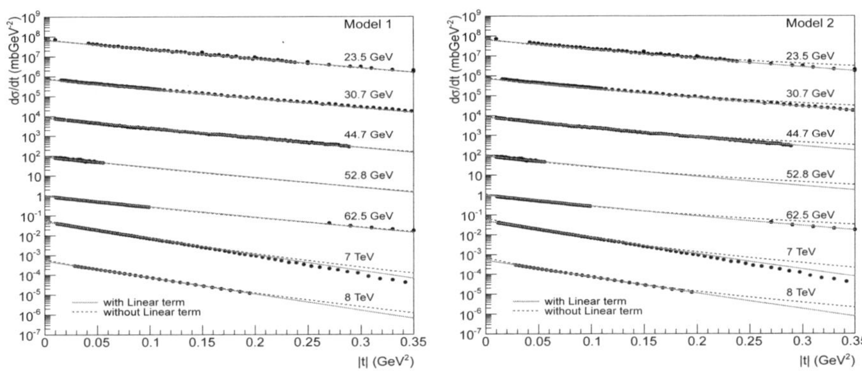

Fig. 4. Fit of Model 1 (left) and Model 2 (right) with and without the linear term in the trajectory with LHC8-TOTEM data[1] in datasets. In both cases we investigate the effect of a linear term in Pomeron trajectory, finding, again, a slightly better agreement with data for the model with a linear term. Curves and data are multiplied by $10^{\pm 2}$ factors to be shown in the same canvas.

where Ref $= Ae^{Bt}$ with A and B determined from a fit to the experimental data. For 8 TeV, we have obtained $A = 518.87 \pm 0.40$ mbGeV^{-2} and $B = 19.3880 \pm 0.0088$ GeV^{-2}.

In Fig. 5 we compare the results obtained with and without the linear term for the two Models considered. Again, it is clear that the linear term is necessary to describe the data analyzed, in special the deviation from a pure exponential behavior presented by 8 TeV data.

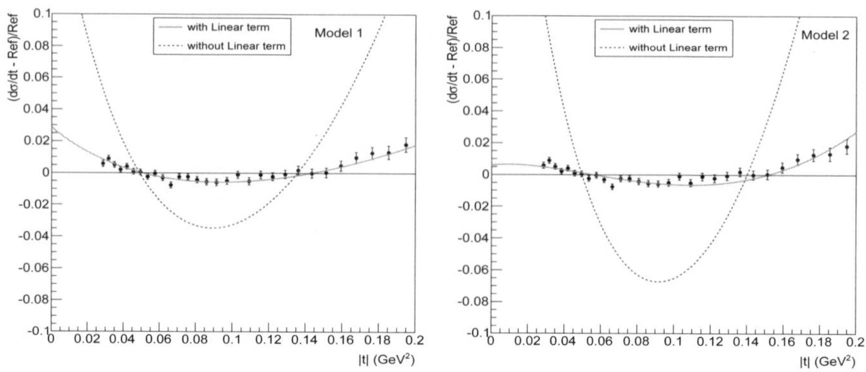

Fig. 5. Ratio R (Eq. (11)) calculated for 8 TeV comparing data and fits with and without linear term for Model 1 (left) and Model 2 (right).

6. The Phillips-Barger Model with the Pion Loop Singularity Added

Here we show the results from a model independent parametrization which can describe both the small and the large $|t|$ region and which had been proposed in 1972 by Phillips and Barger (PB).[7] This model was recently modified as[8]:

$$A_{el} = i\{F_p^2(t)\sqrt{A(s)}e^{B(s)t/2} + \sqrt{C(s)}e^{i\phi(s)}e^{D(s)t/2}\} \tag{12}$$

and three cases can be considered:

(1) $F_p(t) = 1$ which corresponds to the original PB formulation,[7]
(2) $F_p^2(t) = e^{-\gamma(s)\sqrt{4m_\pi^2 - t}}$ which includes a parametrization of the pion loop singularity, labeled as MBP1
(3) $F_p(t) = 1/[1 + |t|/t_0]^2$ which modifies the very small $-t$ behavior with a form factor type behavior, labeled as MBP2

Of interest to the present discussion is the second case, a square root singularity. To argue how this model results can be compared with the TOTEM 8 TeV data, we show a fit of the data using the model MBP1, and the comparison with the REF model (one exponential) in Fig. 6. We find that an independent fit, such as the one indicated by the red curve, can give a very good description of the data, confirming, in this model independent description, the agreement with the presence of the pion loop singularity at very small $-t$-values. The parameter values indicated in the figure correspond to option n.2 in Eq. (12) and to a $\chi^2/dof = 1.03$.

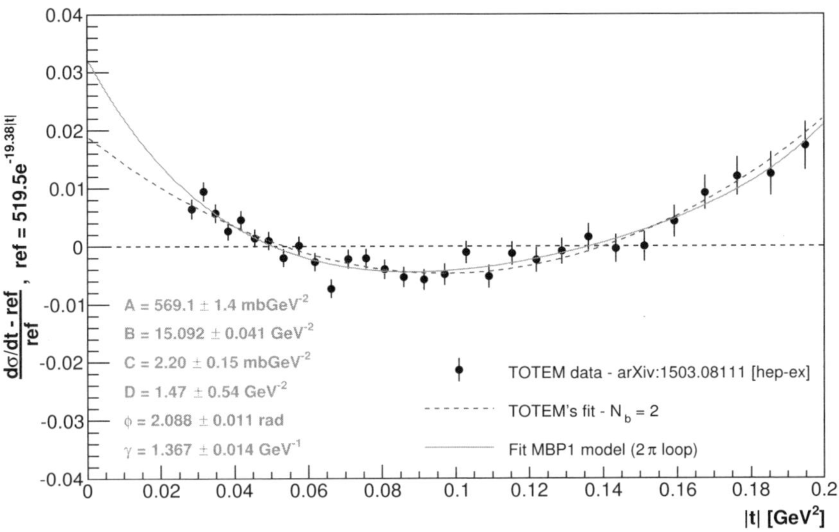

Fig. 6. Fit to TOTEM data with the empirical model of Ref. 8 using the pion loop singularity modification of the PB model, MBP1.

7. Conclusions

Here we study the low-$|t|$ structure of the elastic differential cross section at LHC8, recently reported by TOTEM, through very simple Regge-pole amplitudes with a threshold singularity in the Pomeron trajectory. Our results hints at the possibility of interpreting this phenomenon as a recurrence of a similar effect seen at the ISR, in the seventies. In this first approach we focus on the role of the two-pion loop in determining the fine structure of the diffraction cone, namely in affecting the local slope $B(s,t)$ at the small momentum transfers. Despite the fact that other sources - such as a form factor, eikonal rescatterings and diffractive dissociation[9] - can change the behavior of $B(s,t)$ at small $-t$, our strategy here was to separately investigate the insertion of a two-pion loop in the Pomeron trajectory. Therefore, we have tried to interpolate/extrapolate our fit results for diffraction cone from ISR to LHC. While the interpolation is very good, the extrapolation is only qualitatively satisfactory and shall be investigated in much more detail in a future publication. However, we notice that, the reason of a small mismatch may be attributed mainly to the simplified treatment of multiple Regge-pole exchanges replaced by a single "effective" Regge trajectory. While at the LHC the contribution from secondary trajectories in the nearly forward direction is negligible[2] this is not true at the ISR, where the f trajectory may contribute by half of the total. Two ways how to cure this sufficiency can be envisaged. One is by including all allowed/required Regge exchanges, namely ω, f and the Odderon, apart from the Pomeron. The other one is to account for these by introducing energy-dependent parameters in a single "effective" exchange.

The fine structure of the cone can be better seen on the local slope $B(s,t) = \frac{d}{dt} \ln \frac{d\sigma}{dt}$. It should be, however, remembered that the result depends on the t interval in which the slope is calculated. The relevant bins can be wide or narrow, overlapping or not. This point was recently discussed in Ref. 15. We also notice that simple and elegant formulae extrapolating the forward slope in energy was derived in Ref. 16. They read

$$B(s,t) = k\alpha'(t)\sigma(s), \quad B(s_2,t)/B(s_1,t) = \sigma_t(s_2)/\sigma(s_1), \tag{13}$$

where the coefficient k is determined explicitly in Ref. 16. The virtue of this formula, following[16] from $s-$ channel unitarity is that it is model-independent in the sense that for the total cross section in its r.h.s. on can use either model extrapolations or experimentally measured values of the cross section. The problem with the trajectory is the same as discussed above: in Eq. (13) a single trajectory appears, providing exact predictions when the reaction is dominated by a single Reggeon exchange, as is the case beyond 1 TeV (LHC), dominated by a single Pomeron. Otherwise, an effective trajectory should be used, as discussed above.

We have used two simple and efficient models of the Pomeron amplitude. The first, presented in Sec. 2, is based on a single pole with a supercritical Pomeron intercept, as advocated by Donnachie and Landshoff.[12] The second uses a dipole Pomeron (DP) exchange. Unlike the former, DP promotes logarithmically rising cross sections and consequently the Pomeron is "softer", its "supercriticity" $\alpha(0) - 1 > 0$ being about half compared to the case of a simple pole. Also, it contains absorption corrections, regulated by the parameter producing a diffraction minimum and quantified by the parameter $\epsilon_P > 0$ in Eq. (5). In this work, we fixed $\epsilon_P = 0$ since we are away from the dip region. This choice improves the fits due to the reduction of the number of free parameters.

The parametrization of trajectories is a key issue in the Regge-pole theory. Linear trajectories are popular for their simplicity, however they contradict unitarity both the analytic properties required by the S-matrix theory and the asymptotic constraints, particularly those imposed by the quark model and perturbative quantum chromodynamics (QCD). For practical purposes one chooses a parametrization relevant to the given kinematical region. As shown in the present paper, the economic, parameter-free single square root parametrization, Eq. (2), used e.g. in Ref. 11 produces too strong curvature in the cone, incompatible with the data, as seen in Figs. 1, 3 and 4. The inclusion of a linear term balances this distortion providing good fits to the data.

Finally we note that the reason of non-observation of any fine structure in the cone at the Tevatron ar RHIC may be attributed to poor statistics in relevant experiments. As already mentioned, that is one the reasons we did not perform fits using $\bar{p}p$ datasets. It is also possible that for the same reason it was not observed in diffraction dissociation either. Future experiments may reveal similar effects at the very small $-t$ domain of high-energy diffractive scattering process, either in elastic or inelastic processes. Investigations on this subject are currently in progress.

Acknowledgements

P.V.R.G.S., E.M. and L.J. thank the International Institute of Physics (ITP) at Natal, Brazil, where this work started, for its hospitality and support during the *School and Workshop on New Trends in High-Energy Physics* in October-November, 2014. Research supported by FAPESP, Contract 2013/27060-3 (P.V.R.G.S.) is acknowledged. L.J. was supported also by DOMUS of the Hungarian Academy of Sciences.

References

1. G. Antchev *et al.* (TOTEM Collaboration), *Nucl. Phys.* **B899**, 527-546 (2015), arXiv:1503.08111 [hep-ex].
2. L.L. Jenkovszky, A.I. Lengyel, and D.I. Lontkovskyi, *Int. J. Mod. Phys.* **A26**, 4755-4771 (2011).
3. A.A. Anselm and V.N. Gribov, *Phys. Lett* **B40**, 487 (1972).
4. V.A. Khoze, A.D. Martin, and M.G. Ryskin, *Eur. J. Phys.* **C18**, 167-179 (2000).
5. R. Fiore *et al.*, *Int. J. Mod. Phys.* **A24**, 2551-2599 (2009).
6. G. Cohen-Tannoudji *et al.*, *Lett. Nuov. Cim.* **5**, 957 (1972).
7. R.J.N. Phillips and V.N. Barger, *Phys. Lett* **B46**, 412-414 (1973).
8. D.A. Fagundes *et al.*, *Phys. Rev.* **D88**, 094019 (2013).
9. V.A. Khoze, A.D. Martin, and M.G. Ryskin, *J. Phys.* **G42**, 025003 (2015).
10. A.O. Barut and D.E. Zwanziger, *Phys. Rev.* **127** (1962) 974;
V.N. Gribov and Ya.I. Pomeranchuk, *Nucl. Phys.* **38**, 516 (1962);

R. Oehme, at the *Strong interactions and high energy physics*, Scottish Summer School 1963, ed. R.G. Moorhouse *et al.*, Edinburgh (1964), p. 129.

11. R. Fiore, L.L. Jenkovszky, F. Paccanoni, and A. Prokudin, *Phys. Rev.* **D68**, 013505 (2003); arXiv:hep-ph/0302195.
12. P.V. Landshoff, *Pomeron physics: and updade*, arXiv:hep-ph/0010315; P.V. Landshoff, *Total cross sections at the LHC*, arXiv:hep-ph/0709.0395.
13. http://www.theo.phys.ulg.ac.be/alldata-v2.zip.
14. G. Antchev *et al.* (TOTEM Collaboration), *Europhys. Lett.* **101**, 21002 (2013).
15. L. Jenkovszky and A. Lengyel, *Acta Phys. Pol.* **B46**, 863 (2015), arXiv:1410.4106 [hep-ph].
16. L.L. Jenkovszky and B.V. Struminsky, *Yad. Fiz.* **39**, 1251 (1984) [Engl. translation: *Sov. J. Nucl. Phys.* **39**, 789 (1984)].

CONFINEMENT

COLOR CONFINEMENT FROM FLUCTUATING TOPOLOGY

DMITRI E. KHARZEEV

Department of Physics and Astronomy,
Stony Brook University, New York 11794-3800, USA
E-mail: kharzeev@bnl.gov
and
Department of Physics and RIKEN-BNL Research Center,
Brookhaven National Laboratory, Upton, New York 11973-5000, USA

QCD possesses a compact gauge group, and this implies a non-trivial topological structure of the vacuum. In this contribution to the Gribov-85 Memorial volume, we first discuss the origin of Gribov copies and their interpretation in terms of fluctuating topology in the QCD vacuum. We then describe the recent work with E. Levin that links the confinement of gluons and color screening to the fluctuating topology, and discuss implications for spin physics, high energy scattering, and the physics of quark-gluon plasma.

Keywords: Quantum chromodynamics, confinement, Gribov copies

1. The Gribov copies and knots in the QCD ether

In 1977, Gribov pointed out[1] the fundamental ambiguity in non-Abelian gauge theory due to the existence of multiple solutions of the gauge fixing condition, known today as "Gribov copies". Much of the analysis was carried out in Coulomb gauge, but shortly afterwards Singer[2] showed that the problem persists in all other gauges, and thus no gauge fixing is possible in non-Abelian theories. Gribov has proposed to avoid the problem of copies by restricting the range of functional integration to small amplitude fields for which the solution for the gauge fixing condition is unique. He has demonstrated that such a restriction has a dramatic effect on the Coulomb interaction in QCD – it becomes confining, with Coulomb potential rising linearly with the distance, see Refs. 3, 4 for reviews.

In this contribution (based on a recent work with E. Levin[5]), we will take a different route – namely, we will argue that Gribov copies naturally arise from the compactness of the gauge group, and instead of eliminating them, we have to build a picture of the vacuum that accommodates their existence. The physical quantities in non-Abelian theories should thus be evaluated by summing over Gribov copies. When applied to the gluon propagator in QCD, this recipe leads to the confinement of gluons at large distances.[5]

In fact, shortly after Gribov pointed out the existence of gauge copies, Jackiw, Muzinich and Rebbi[6] showed that the copies, and associated with them discontinuities of gauge potentials are necessary to accommodate the vacuum configurations with arbitrary Chern-Pontryagin index. In an even number of space-time dimensions, such as our (3+1) dimensional Minkowski space, the Chern-Pontryagin index is associated with topology-changing transitions. The discussion of topology of gauge theory is usually restricted to semi-classical topological solutions (instantons, monopoles, dyons, ...) and is thus in the domain of "non-perturbative QCD" traditionally ignored by the practitioners of perturbation theory. However a meaningful perturbartive expansion is based on fixing the gauge, and the existence of Gribov copies thus implies the impossibility to isolate perturbation theory from the topological nature of QCD.

The relevance of topology in gauge theories stems from the compactness of the gauge group. Non-Abelian gauge theories always possess a compact gauge group, but even in the Abelian case we may choose to treat the $U(1)$ group as compact – and in this case the theory acquires a non-trivial topological structure, and the confinement phase, see, e.g., Ref. 7. This fact suggests a link between topology and confinement. Compact gauge groups in general possess a homotopy map to the space-time manifold. For example, the map from the SU(2) $\subset$ SU(N) subgroup of the gauge group to the Euclidean space-time sphere S^3 describes the classical instanton solution.[8] In Minkowski space-time, instanton solutions represent the tunneling events connecting the degenerate vacuum states with different Chern-Simons numbers

$$X(t) = \int d^3x \; K_0(x,t), \tag{1}$$

where K_0 is the temporal component of topological current,

$$K_\mu = \frac{1}{16\pi^2}\epsilon_{\mu\nu\rho\sigma}A^{\nu,a}\left(\partial^\rho A^{\sigma,a} + \frac{1}{3}gC^{abc}A_b^\rho A_c^\sigma\right), \tag{2}$$

and we have absorbed the coupling constant in the definition of the field.

The density of the Chern-Pontryagin number mentioned above is the divergence of topological current, Eq. (2),

$$Q(x) = \partial^\mu K_\mu \,, \tag{3}$$

and is thus given by

$$Q(x) = \frac{1}{32\pi^2} F_{\mu\nu}(x)\tilde{F}^{\mu\nu}(x) \,. \tag{4}$$

The Chern-Pontryagin index is $\int d^4x \; Q(x) \equiv \nu$; for finite action field configurations ν is an integer. As shown in Refs. 2, 6, the presence of Gribov copies is necessary to accommodate configurations with an arbitrary (integer) value of Chern-Pontryagin index.

Prior to proceeding further, let us develop an intuitive picture of the vacuum structure implied by the topology of the compact gauge group. For simplicity, let us begin with the case of the Abelian gauge theory – a familiar electrodynamics. In this case we limit ourselves to the first term in Eq. (2). The corresponding Abelian Chern-Simons number (1) has been known since Gauss's work in the XIX century – it is a so-called "magnetic helicity" introduced in magnetohydrodynamics by Woltjer[9] and Moffatt[10]:

$$\mathcal{H}(t) = \int d^3x \; \boldsymbol{A} \cdot \boldsymbol{B}; \tag{5}$$

to be precise, the Chern-Simons number $X(t)$ (1) is equal to $\mathcal{H}(t)$ divided by $16\pi^2$.

Magnetic helicity measures the "knottedness" of the lines of magnetic flux. To quantify this statement and to characterize the topology of magnetic flux, let us split the magnetic field into N magnetic flux tubes with magnetic fluxes ϕ_i. The magnetic helicity can then be written down[10–13] in terms of the geometric invariant of the links between these flux tubes:

$$\mathcal{H}(t) = \sum_{i=1}^{N} \phi_i^2 \; \mathcal{S}_i + 2 \sum_{i,j} \phi_i \phi_j \; \mathcal{L}_{ij}, \tag{6}$$

where $\mathcal{S}_i$ is the Călugăreanu-White self-linking number, and $\mathcal{L}_{ij}$ is the Gauss linking number.*

In infinite space, the solution of the Maxwell equations characterized by a non-zero value of (6) is very simple – it is a circularly polarized plane wave.

*The linking numbers in (6) do not always detect the topology of the link; the development of the appropriate universal knot invariants is a very active area in modern mathematics.

The magnetic helicity is a parity–odd quantity, and the parity transformation turns a left-handed circularly polarized wave into the right-handed one. In the presence of the boundaries, the solution[14] becomes a Hopfion with linked lines of fields characterized by non-zero value of the second term on the r.h.s. of Eq. (6).

The notion of magnetic helicity is extensively used in magnetohydrodynamics, where it is conserved in the absence of dissipation – in geometrical picture implied by Eq. (6), this means that the lines of magnetic field are "frozen" into the fluid and do not reconnect. Because the time derivative of magnetic helicity is

$$\frac{d}{dt}\mathcal{H}(t) = \frac{d}{dt}\int d^3x\ \boldsymbol{A}\cdot\boldsymbol{B} = -2\int d^3x\ \boldsymbol{E}\cdot\boldsymbol{B} \tag{7}$$

from the definition of electric field $\boldsymbol{E}$, we see that conservation of magnetic helicity implies $\int d^3x\ \boldsymbol{E}\cdot\boldsymbol{B} = 0$. In electrodynamics, this quantity is nothing but the Chern-Pontryagin index ν defined above, since $F_{\mu\nu}\tilde{F}^{\mu\nu} = -4\boldsymbol{E}\cdot\boldsymbol{B}$. The gauge field configurations with non-zero Chern-Pontryagin numbers (such as the ones signaled by the presence of Gribov copies in non-Abelian theory) thus have an important role: they induce *reconnections of magnetic flux*.

In classical magnetohydrodynamics, reconnections of magnetic flux are often dramatic events – for example, in the corona of the Sun they are responsible for solar flares and coronal mass ejections spanning millions of miles (for review, see Ref. 15). In quantum theory, the physics during a magnetic reconnection can be described by adding the Chern-Simons term $\mu_5 K^0$ to the action of electrodynamics,[16] where μ_5 is the chiral chemical potential – since magnetic helicity K^0 is parity-odd, μ_5 is parity-odd as well. The equations of motion of the resulting Maxwell-Chern-Simons theory then exhibit the chiral magnetic effect[17] (see Ref. 18 for review and references) – the electric current $\boldsymbol{J} = e^2/2\pi^2\ \mu_5\boldsymbol{B}$. The spectrum of photons in this theory differs from the one in usual Maxwell electrodynamics and exhibits a "chiral magnetic instability" – the photon energy acquires an imag inary part at soft momenta $k \leq (e^2/2\pi)\mu_5$ reflecting a growth or decay of the photon field.[19–22]

At high momenta $k \gg (e^2/2\pi)\mu_5$ the photons behave as a free quanta of perturbation theory, but in the infrared region the behavior is totally different – the photons either cease to propagate (are confined), or are produced (this is a quantum analog of the solar flare induced by a magnetic reconnection). In the case of the vacuum, which is by definition the state with the lowest energy, the radiation of photons is clearly impossible and the physical choice for the imaginary part of the photon energy corresponds to the photon decay. If the magnetic reconnections take place in the vacuum with a certain rate per unit time per unit volume, then this rate will determine the characteristic length over which the soft photon is allowed to propagate. We will see in the next section that this may be exactly what happens in QCD to soft gluons.

The emerging picture of the vacuum filled by the knots is remarkably close to the one proposed by Lord Kelvin in 1867.[23] Inspired by Helmholtz's work on vortex motion in perfect liquid, Kelvin proposed that the observed atoms were in fact the knots of vortices in the ether. He wrote[23]: *...the endless variety of [knotted or knitted vortex atoms] is infinitely more than sufficient to explain the varieties and allotropies of known simple bodies and their mutual affinities.* While we now know that atoms are definitely not knots in the ether, Kelvin's prophecy may still apply to the structure of the QCD vacuum.

2. Veneziano's ghost and fluctuating topology in the QCD vacuum

The relevance of non-trivial topology in the QCD vacuum is revealed in the physical spectrum of the theory through the chiral anomaly[24,25] and the resulting non-conservation of the axial current:

$$\partial_\mu J_A^\mu = 2N_f Q(x) + \sum_f (2im_f)\bar{q}_f \gamma_5 q_f, \tag{8}$$

where m_f are the masses of quarks, N_f is the number of flavors, and $Q(x)$ is the density of Chern-Pontryagin number given by (3). The vacuum fluctuations in topology and the chiral anomaly thus lead to the explicit breaking of $U_A(1)$ symmetry, and are responsible[26,29] for the anomalously large mass of the would-be Goldstone boson η', as we discuss below.

Veneziano[26] has demonstrated that the fluctuations of topology in the vacuum can be captured by introducing a massless "ghost" in the correlation function of the gauge-dependent topological current (2):

$$K_{\mu\nu}(q) \equiv i \int d^4x \, e^{iqx} \, \langle T\{K_\mu(x) K_\nu(0)\}\rangle \xrightarrow{q^2 \ll \mu^2} -\frac{\mu^4}{q^2} g_{\mu\nu}, \tag{9}$$

where $\mu^4 \equiv \chi_{top}$ is the topological susceptibility of pure Yang-Mills theory. Note that the r.h.s. of Eq. (9) has the "wrong" sign, which means that the ghost does not describe a propagating degree of freedom. Therefore the ghost cannot be produced in a physical process; however the couplings of the ghost (that describe the effect of topological fluctuations) can affect physical amplitudes. A similar "dipole" ghost had been earlier introduced by Kogut and Susskind[27] in the analysis of axial anomaly in the Schwinger model. This procedure has been demonstrated to solve the $U_A(1)$ problem in QCD.[28,29]

The physical meaning of Eq. (9) becomes apparent if one compares it to the correlation function of the electron's coordinate $x(t)$ in a crystal, following Dyakonov and Eides[30]:

$$i \int dt\ e^{i\omega t}\ \langle T\{x(t)x(0)\}\rangle \xrightarrow{\omega\to 0} -\frac{1}{\omega^2\ m^*} = -\frac{1}{\omega^2}\ \frac{\partial^2 E(k)}{\partial k^2}\bigg|_{k=0}, \tag{10}$$

where $E = k^2/2m^*$ is the energy of an electron with an effective mass m^* and quasi-momentum k in a crystal. The emergence of the pole in Eq. (10) signals the possibility of electron's propagation in the periodic potential of the crystal by means of tunneling. It is important to note that the pole emerges not just from a single tunneling event (corresponding to the isolated instanton in QCD), but sums up the effect of many tunnelings throughout the crystalline lattice.

The analogy between Eq. (9) and Eq. (10) becomes even more apparent if we choose the frame with $q^\mu = (\omega, 0)$ and use the analog of coordinate given by Eq. (1). The expression Eq. (9) then takes the form that is completely analogous to Eq. (10):

$$i \int dt\ e^{i\omega t}\ \langle T\{X(t)X(0)\}\rangle \xrightarrow{\omega\to 0} -\frac{\mu^4}{\omega^2}\ V = -\frac{1}{\omega^2}\ \frac{\partial^2 E(\theta)}{\partial \theta^2}\bigg|_{\theta=0}, \tag{11}$$

where V is the volume of the system, and $E(\theta) = \epsilon(\theta)V$ is the energy of the Yang-Mills vacuum. The energy density of the vacuum $\epsilon(\theta)$ is a periodic function of the θ angle that is analogous to the quasi-momentum k in Eq. (10). At small θ, we can expand $\epsilon(\theta)$:

$$\epsilon(\theta) \simeq \mu^4\ \frac{\theta^2}{2}, \tag{12}$$

which exhibits the physical meaning of the quantity μ^4 on the r.h.s. of Eq. (11), which is the topological susceptibility $\chi_{top} = \mu^4$ of the Yang-Mills theory. Note that a term linear in θ is forbidden by the P and CP invariances of QCD.

The expression (11) clearly exhibits the role of the ghost – it describes the effect of topology-changing transitions characterized by non-vanishing Chern-Pontryagin indices in the vacuum. In the Abelian case presented in the previous section, the ghost would describe magnetic reconnections occurring with the rate (per unit volume, unit time) proportional to μ^4.

This interpretation becomes even more apparent if we consider the correlation function of the Chern-Pontryagin number density $Q(x)$, Eq. (3), implied by Eq. (9), as we will now explain. The assumption of the ghost existence, Eq. (9), implies the following Minkowski-space correlation function for $Q(x) = \partial^\mu K_\mu$:

$$i \lim_{q \to 0} \int d^4x \; e^{iqx} \; \langle T\{Q\,(x)\, Q\,(0)\}\rangle \; = -\mu^4 < 0. \tag{13}$$

The quantity $Q(x)$ is gauge-invariant, and excites from the vacuum physical states, the lightest of which is the flavor-singlet η' meson.† By writing down the spectral representation for the correlator (13) it is easy to see that the contribution of these physical states to the r.h.s. should be positive:

$$i \lim_{q \to 0} \int d^4x \; e^{iqx} \; \langle T\{Q\,(x)\, Q\,(0)\}\rangle \; = \int \frac{dM^2}{2\pi} \; \frac{\rho(M^2)}{M^2} > 0. \tag{14}$$

How can we reconcile (13) and (14)? The key to the answer stems from the observation that in the physical world (in which the dispersion integral (14) is written), the theory possesses light quarks of mass m_f, and in the chiral limit of massless quarks the correlation function of topological charge in Eq. (14) must vanish. This is because at $m_f = 0$ the topological charge is a full divergence of the gauge-invariant axial current (see Eq. (8)).

For non-flavor-singlet axial currents $J^\mu_{A,NS}$ that do not mix with gluon states, the vanishing of the correlation function of $\partial_\mu J^\mu_{A,NS}$ at zero momentum is assured by the Goldstone theorem, since the matrix elements of the divergence of the axial current between the vacuum and the Goldstone boson (e.g. a pion)

$$\langle 0|\partial_\mu J^\mu_{A,NS}|\pi\rangle = m_\pi^2 f_\pi \tag{15}$$

vanish in the chiral limit of massless quarks as $m_\pi \to 0$.

In the case of the flavor singlet axial current producing from the vacuum the η' meson with matrix elements $\langle 0|J^\mu_A|\eta'\rangle = \sqrt{2N_f}\; p^\mu f_{\eta'}$ and

†For simplicity we do not discuss here the $\eta' - \eta$ mixing.

$\langle 0|\partial_\mu J^\mu_{A,NS}|\pi\rangle = \sqrt{2N_f}\; m_\eta^2 f_\pi$ (normalization is chosen in accord with Eq. (8)), the situation is different because of the presence of topological fluctuations expressed by the relation (13). To construct the physical correlation function in QCD, we thus have to add the r.h.s. of Eqs. (13) and (14). Saturating the dispersion integral by the η' peak and choosing in accord with Eq. (8), $\langle 0|Q|\eta'\rangle = 1/\sqrt{2N_f}\; m_\eta^2 f_\pi$ and $\rho(M^2) = 4N_f\; m_{\eta'}^4 f_{\eta'}^2\; 2\pi\delta(M^2 - m_{\eta'}^2)$, we get

$$i \lim_{q\to 0} \int d^4x\; e^{iqx}\; \langle T\{Q\,(x)\,Q\,(0)\}\rangle_{\rm QCD} = -\mu^4 + \frac{m_{\eta'}^2 f_{\eta'}^2}{4N_f} = 0, \qquad (16)$$

from which we immediately get the Witten-Veneziano relation[26,29] for the η' mass:

$$m_{\eta'}^2 f_{\eta'}^2 = 4N_f\; \mu^4 = 4N_f\; \frac{\partial^2 \epsilon(\theta)}{\partial\theta^2}\bigg|_{\theta=0}^{YM}. \qquad (17)$$

Here we have indicated explicitly that the topological susceptibility has to be computed for Yang-Mills theory (without light quarks).

Subtracting a constant from the r.h.s. of Eq. (14) may seem strange, however, we have to keep in mind that the dispersion integral in Eq. (14) diverges in the high mass region that we have omitted, since on dimensional grounds $\rho(M^2) \sim M^4$ (as reproduced by the gluon loop) – therefore, it requires subtractions, and this is what the μ^4 term represents.[‡] The Witten-Veneziano prescription described above thus implies that the correlation function (13) should contain a contact negative term, $\langle T\{Q\,(x)\,Q\,(0)\}\rangle \sim -\mu^4\; \delta^4(x)$. There is an ample evidence for this contact term from the lattice simulations, see Fig. 1 from Ref. 31 (the correlator shown in Fig. 1 is Euclidean, hence the sign difference).

The physical interpretation of the contact term is suggested by our discussion in Section 1 – it describes the "white noise" of topological reconnections, with strength given by the topological susceptibility μ^4. Just like the introduction of white noise in kinetic theory allows one to describe the effect of the medium on a test particle, the ghost allows to capture the impact of fluctuating topology. Compared to "microscopic" approaches to QCD vacuum, such as the instanton liquid model and its descendants (for review, see Ref. 32), the approach described here does not identify the dynamics of topological transitions – instead, we parameterize the rate of transitions by

[‡]The T-ordered product in Eq. (14) has to be understood as a Wick, and not as a Dyson one[30].

a single dimensionful parameter, topological susceptibility. On the other hand, it does not rely on a (questionable) semiclassical approximation.

Usually, the discussion of topological effects is limited to the $U_A(1)$ problem and correlation functions of color-singlet correlators, as presented above. However, the fluctuating topology should also affect the propagation of gluons, similarly to the case of photons discussed in the previous section. We will now see that this is indeed the case. The Veneziano ghost defined by Eq. (9) will allow us to address this problem in QCD without resorting to the complicated microscopic dynamics of topological transitions.

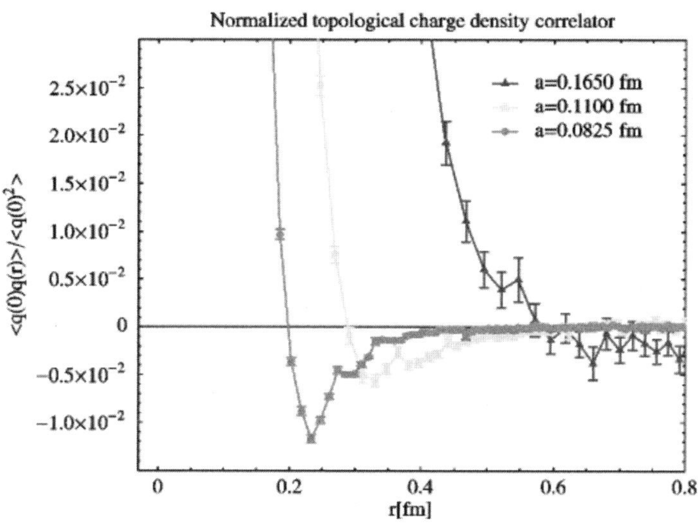

Fig. 1. The lattice QCD data on the correlation function of Chern-Pontryagin charge density at three lattice spacings a; from Ref. 31.

The observation made in our recent paper[5] with E. Levin is that Eqs. (9) and (2) define an effective ghost-gluon-gluon vertex $\Gamma_\mu(q,p)$. Since the topological current (2) is a pseudo-vector, the ghost is a pseudo-vector "particle" – so the interaction of a gluon with, say, a left circular polarization with the ghost will transform it into a right circularly polarized gluon. Because the QCD vacuum is P and CP even, the ghost can only appear in the loops.

Using this vertex, we can re-write the correlator Eq. (9) at small q^2 as follows (see Fig. 2-a):

$$K_{\mu\nu}(q) = \frac{1}{(2\pi)^4 i} \int d^4p\, \Gamma_\mu(q,p) \frac{1}{p^2(q-p)^2} \Gamma_\nu(q,p) = -\frac{\mu^4}{q^2} g_{\mu\nu}, \quad (18)$$

from which we find that

$$\Gamma_\mu(q,p)\, \Gamma_\nu(q,p) \propto -\frac{\mu^4}{p^2} g_{\mu\nu}\ , \text{ for } q \leq p. \quad (19)$$

The vertices $\Gamma_\mu(q,p)$ describe the interactions of soft gluons with topological fluctuations.

Indeed, the gluon propagator is now the solution of the Dyson-Schwinger equation with $\Sigma(p)$ given by

$$\Sigma_{\mu\nu}(p) = \frac{1}{(2\pi)^4 i} \int d^4q\, \Gamma_\mu(q,p) \frac{1}{q^2(q-p)^2} \Gamma_\nu(q,p)\ , \quad (20)$$

where $1/q^2$ is the propagator of the ghost. In evaluating the integral of Eq. (20) we assume that $q \ll p$ since Γ_μ's describe the non-perturbative effects at small momenta. Therefore,

$$\Sigma_{\mu\nu}(p) \simeq \frac{4\pi}{(2\pi)^4} \int_0^p dq^2 q^2\, \Gamma_\mu(q,p) \frac{1}{q^2 p^2} \Gamma_\nu(q,p) = -g_{\mu\nu}\frac{\mu^4}{p^2}, \quad (21)$$

where we used Eq. (19).

We can now write down the Schwinger-Dyson equation for the gluon propagator[§] $G_{\mu\nu}(p) = g_{\mu\nu} G(p)$:

$$G(p) = \frac{1}{p^2} + \frac{1}{p^2}\Sigma(p)\, G(p) \quad (22)$$

with the solution

$$G(p) = \frac{1}{p^2 - \Sigma(p)} = \frac{1}{p^2 + \frac{\mu^4}{p^2}}\,. \quad (23)$$

The propagator (23) has some remarkable properties. First, $G(p)$ has no gluon pole in the physical region. Instead, this propagator has only complex poles at $p^2 = \pm i\mu^2$. As a result, gluons cannot propagate at large distances – in other words, they are confined. Second, the propagator of the type of Eq. (23), as shown by Gribov[1] and Zwanziger,[33] eliminates the gauge copies. Note however that we have obtained this propagator not by restricting the range of functional integration to the region where the copies

[§]This is the expression in Feynman gauge; since the self-energy $\Sigma_{\mu\nu}$ does not change the polarization of the gluon, the Dyson-Schwinger series can be re-summed in the usual way.

do not appear, but by taking all copies into account through the topological fluctuations.

Let us illustrate the meaning of our result by appealing again to the analogy with the crystal, see Eq. (10). The Gribov copies in this example are analogous to the electron states localized around different sites in the crystalline lattice. When we discuss the wave function of the electron around one of the sites, in general we cannot ignore the presence of the tunneling throughout the lattice. The tunneling can be ignored only if we limit ourselves to the tightly bound electron levels. However the tunneling can be consistently taken into account if instead of an individual bound state we consider the electron bands. Electron energies in these bands acquire a finite width, similarly to what we have observed for the gluons.

Our propagator Eq. (23) results from the admixture of the ghost to the perturbative gluon (see Fig. 2-b), with an amplitude defined by the Yang-Mills topological susceptibility μ^4. In paper,[5] we proposed to call this coherent mixture of a **gl**uon and a gh**ost** a "**glost**". Unlike the ghost, the "glost" can be produced in a physical process, but unlike the perturbative gluon, it is confined and cannot propagate at distances larger than $\sim \mu^{-1} \sim$ 1 fm.

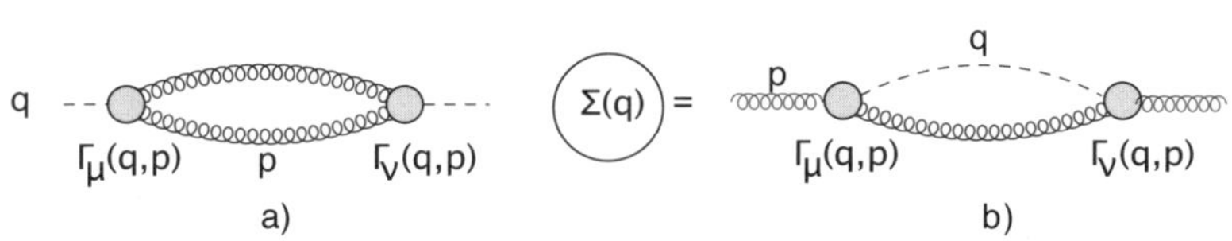

Fig. 2. Fig. 2-a : Eq. (9) in momentum representation; helix lines represent gluons and the dashed line depicts the ghost. Fig. 2-b : the gluon dressed by the interactions with the ghost: a "glost", see Eq. (20) for the corresponding self-energy expression.

At distances $\sim \mu^{-1} \sim 1$ fm, the use of the glost propagator also gives rise to the screening of color charge,[5] leading to the freezing of the effective coupling in the infrared IR limit for pure gauge theory, or to the vanishing of the coupling in the IR in QCD with light quarks, see Fig. 3. It has been shown by Dokshitzer[37] that the experimental data indicate the relative smallness of the QCD coupling in the IR region:

$$\alpha_0 \;=\; \frac{1}{\mu_I}\int^{\mu_I} dk\,\alpha_S\,(k) \;\approx\; 0.5 \quad \text{for } \mu_I \;=\; 2 \text{ GeV}. \tag{24}$$

In our approach we get $\alpha_0 \simeq 0.59$ for renormalization point $k = M_Z$, in reasonable agreement with (24).

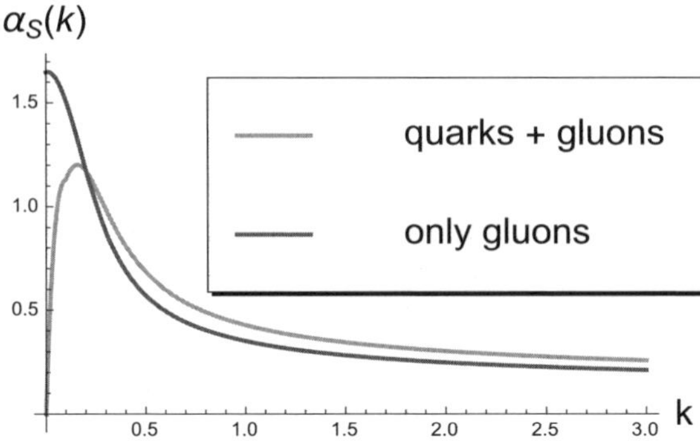

Fig. 3. The running constant α_S as a function of momentum k in our model for QCD with light quarks (red line that goes to zero at $k \to 0$) and gluodynamics (blue line).

Note that close to the deconfinement transition, the topological susceptibility vanishes reflecting the restoration of $U_A(1)$ symmetry.[34,35] Since at $\mu \to 0$ the gluon propagator becomes perturbative, the restoration of $U_A(1)$ symmetry and deconfinement should occur at the same temperature in agreement with the lattice data[34]; however close to T_c the non-perturbative interactions induced by topological fluctuations are important. It would be interesting to explore the implications of this approach to QCD thermodynamics, and to the rates of electromagnetic emission from the quark-gluon plasma – this work has already begun.[38]

It is of fundamental interest to establish the microscopic dynamics responsible for the fluctuating topology captured by the ghost. A recent study within the "deformed QCD" attributes these fluctuations to the topological order in the vacuum.[39] The imaginary part of the gluon self-energy signals an instability w.r.t. decay to the true ground state – what is the structure of this ground state? The clue to the answer may be provided by the Abelian example considered in section 1 – there, the true ground state of the photon field is the (self-)linked knot of magnetic flux. The knots repeatedly emerge in studies of the ground state of non-Abelian gauge theories,[40,41,43] but it is clear that much remains to be done.

In the meantime, the simplistic approach described above captures the link between the confinement and the fluctuating topology in the QCD vacuum, and provides a practical way of computing power-suppressed corrections to QCD amplitudes. In particular, the glost propagator leads to the exponential fall-off of the high-energy hadron scattering amplitude at large impact parameters needed to satisfy the Froissart bound; this can solve the long-standing problem of the perturbative approach in describing high energy scattering.[44] In QCD amplitudes the coupling to the ghost flips the gluon and (through loops) quark polarizations, and can give rise to spin asymmetries[45] that are different from the usual perturbative approach.

3. Acknowledgments

I am indebted to Yuri Dokshitzer and Julia Nyiri for the invitation to contribute to the Gribov-85 Memorial volume. The work presented here was done together with Eugene Levin; I am grateful to him for the enjoyable collaboration. This work was supported in part by the U.S. Department of Energy under Contracts DE-FG-88ER40388 and DE-SC0012704.

References

1. V.N. Gribov, *Nucl. Phys.* **B139**, 1 (1978).
2. I.M. Singer, *Commun. Math. Phys.* **60**, 7 (1978).
3. N. Vandersickel and D. Zwanziger, *Phys. Rept.* **520**, 175 (2012) and references therein.
4. Y.L. Dokshitzer and D.E. Kharzeev, *Ann. Rev. Nucl. Part. Sci.* **54**, 487 (2004) arXiv:hep-ph/0404216.
5. D.E. Kharzeev and E.M. Levin, *Phys. Rev. Lett.* **114**, no. 24, 242001 (2015) arXiv:1501.04622 [hep-ph].
6. R. Jackiw, I. Muzinich, and C. Rebbi, *Phys. Rev.* **D17**, 1576 (1978).
7. M. Panero, *JHEP* **0505**, 066 (2005) arXiv:hep-lat/0503024.
8. A.A. Belavin, A.M. Polyakov, A.S. Schwartz, and Y.S. Tyupkin, *Phys. Lett.* **B59**, 85 (1975).
9. L. Woltjer, *Proc. Nat. Acad. Sci.* **44**, no. 6, 489 (1958).
10. H.K. Moffatt, *J. Fluid Mech.* **35**, 117 (1969).
11. M.A. Berger and G.B. Field, *J. Fluid Mech.* **147**, 133 (1984).
12. H.K. Moffatt and R.L. Ricca, *Proc. R. Soc.* **A439**, 411 (1992).
13. M.A. Berger, *Plasma Phys. Controlled Fusion* **41**, B167 (1999).
14. A.F. Ranada, *Lett. Math. Phys.* **18**, 97 (1989).
15. P.F. Chen, *Living Rev. Solar Phys.* **8**, 1 (2011).
16. D.E. Kharzeev, *Annals Phys.* **325**, 205 (2010) arXiv:0911.3715.
17. K. Fukushima, D.E. Kharzeev, and H.J. Warringa, *Phys. Rev.* **D78**, 074033 (2008) arXiv:0808.3382 [hep-ph].
18. D.E. Kharzeev, *Prog. Part. Nucl. Phys.* **75**, 133 (2014) arXiv:1312.3348; *Ann. Rev. Nucl. Part. Sci.* **65**, 193 (2015) arXiv:1501.01336.

19. A.N. Redlich and L.C.R. Wijewardhana, *Phys. Rev. Lett.* **54**, 970 (1985).
20. K. Tsokos, *Phys. Lett.* **B157**, 413 (1985).
21. M. Joyce and M.E. Shaposhnikov, *Phys. Rev. Lett.* **79**, 1193 (1997).
22. Y. Akamatsu and N. Yamamoto, *Phys. Rev. Lett.* **111**, 052002 (2013) arXiv:1302.2125 [nucl-th].
23. Lord Kelvin (Sir W. Thomson), *Proc. Royal Society of Edinburgh*, **VI**, 94, 1867.
24. S.L. Adler, *Phys. Rev.* **177**, 2426 (1969).
25. J.S. Bell and R. Jackiw, *Nuovo Cim.* **A60**, 47 (1969).
26. G. Veneziano, *Nucl. Phys.* **B159**, (1979) 213.
27. J.B. Kogut and L. Susskind, *Phys. Rev.* **D11**, 3594 (1975).
28. S. Weinberg, *Phys. Rev.* **D11**, 3583 (1975).
29. E. Witten, *Nucl. Phys.* **B156**, 269 (1979).
30. D.I. Dyakonov and M.I. Eides, *Zh. Eksp. Teor. Fiz.* **81**, 434 (1981).
31. I. Horvath *et al.*, *Phys. Lett.* **B617**, 49 (2005) arXiv:hep-lat/0504005.
32. T. Schafer and E.V. Shuryak, *Rev. Mod. Phys.* **70**, 323 (1998) arXiv:hep-ph/9610451.
33. D. Zwanziger, *Nucl. Phys.* **B323**, 513 (1989).
34. C. Bonati, M. D'Elia, H. Panagopoulos, and E. Vicari, *Phys. Rev. Lett.* **110**, no. 25, 252003 (2013) arXiv:1301.7640 [hep-lat].
35. D. Kharzeev, R.D. Pisarski, and M.H.G. Tytgat, *Phys. Rev. Lett.* **81**, 512 (1998) arXiv:hep-ph/9804221.
36. D.E. Kharzeev, arXiv:1501.01336 [hep-ph].
37. Y.L. Dokshitzer, *"Perturbative QCD theory (includes our knowledge of α_s),"* In *Vancouver 1998, High energy physics, vol. 1* 305-324, arXiv:hep-ph/9812252.
38. A. Bandyopadhyay, N. Haque, M.G. Mustafa, and M. Strickland, arXiv:1508.06249 [hep-ph].
39. A.R. Zhitnitsky, *Annals Phys.* **336**, 462 (2013) arXiv:1301.7072.
40. L.D. Faddeev, arXiv:0805.1624 [hep-th].
41. L.D. Faddeev and A.J. Niemi, *Phys. Lett.* **B525**, 195 (2002) arXiv:hep-th/0101078.
42. P. van Baal, A. Wipf, *Phys. Lett.* **B515**, 181 (2001).
43. A. Gorsky, A. Milekhin, and N. Sopenko, arXiv:1506.06695 [hep-th].
44. A. Kovner and U.A. Wiedemann, *Phys. Rev.* **D66**, 051502 (2002); *Phys. Rev.* **D66**, 034031 (2002); *Phys. Lett.* **B551**, 311 (2003).
45. Z.B. Kang and D.E. Kharzeev, *Phys. Rev. Lett.* **106**, 042001 (2011) arXiv:1006.2132 [hep-ph].

INSTABILITIES OF COULOMB PHASES AND THE GRIBOV PICTURE OF CONFINEMENT

MANUEL ASOREY and ALESSANDRO SANTAGATA

Departamento de Física Teórica. Facultad de Ciencias
Universidad de Zaragoza. 50005 Zaragoza. Spain
E-mail: asorey@unizar.es, alessandro.santagata@unizar.es

A new picture of quark confinement based on the instability of Coulomb phase at low energy was introduced by Volodya Gribov in the early nineties. In QCD the effective α coupling constant can reach very large values in the infrared regime what generates Coulomb phase instabilities. In the Gribov picture the instability leads to a vacuum decay into light quarks for coupling constants α larger than a critical value $\alpha_{\rm crit} = 2N\pi(1-\sqrt{2/3})/(N^2-1)$, for SU(N) gauge theories. The instability of Coulomb phase can be derived from first principles in any non-Abelian gauge theory for $\alpha \geq \sqrt{2}$, a value which is larger than the Gribov critical value. In this paper we review the analytic derivation of the Gribov mechanism from first principles and analyze the effects of dynamical quarks in the instability of the Coulomb phase. The instabilities associated to light quarks turn out to appear at larger values of α than the ones induced from pure gluon dynamics, unlike it is expected in the standard Gribov scenario. The analytic results confirm the consistency of the picture where quark confinement is mainly driven by gluonic fluctuations.

Keywords: Quark confinement, Coulomb phases, Gribov mechanism.

1. Introduction

The behavior of QCD is well understood in analytic terms in two extreme regimes: in the weak coupling regime, where perturbation theory shows that quarks interact almost asymptotically free via a weak Coulomb-like potential, and in the strong coupling regime where quarks are confined by an effective a linear potential. However, a complete analytic proof confinement is missing. In other terms it is necessary to show that both UV and IR behaviors arise in the same theory. This means to prove that the coupling constant flows from small values at high energy to large values at low energy without crossing any phase transition, otherwise in QCD we should see just one of the two regimes. Asymptotic freedom indicates that the strong coupling constant α increases as the energy becomes smaller and smaller, but that approach is not valid at scales of energy greater than $\Lambda_{QCD} \simeq$

200Mev, where the perturbative analysis breaks down. On the other hand quark confinement arises in a natural way in the strong coupling limit of lattice gauge theories. The consistency of the lattice theory with the continuum limit requires the absence a phase transition on the RG flow from the strong coupling phase to the weak coupling phase.

The proof that no phase transition occurs for intermediate value of the coupling constant in QCD would be equivalent to a proof of confinement. Although numerical simulations on the lattice do not show any trace of such a transition, the analytic proof of this fact has remained elusive for decades and still is one of the challenging theoretical questions in QCD.

Any step forward in that program, which has been promoted to a Millennium problem,[1] requires at least an heuristic understanding or a clear picture of the confinement mechanism. The first and more popular picture of confinement is based on the dual superconductor scenario. In a normal superconductor the phenomenon of superconductivity is due to the condensation, below the critical temperature, of Cooper electron pairs. A related feature is the Meissner effect: external magnetic fields are expelled out from the bulk superconducting material as the temperature drops below the critical value. In type II superconductors the magnetic flux generated by a monopole-antimonopole pair cannot be completely expelled out. Because of the Meissner effect, in the presence of a monopole-antimonopole pair, the magnetic flux coming out from the monopole into the antimonopole in a type II superconductor can survive confined into a thin flux tube connecting both magnetic sources giving rise to a vortex-like configuration which induces a linear potential between the pair. This implies that the energy required to split the two monopoles is linearly growing with the distance. This phenomenon is reminiscent of the Wilson area law, and the confinement of monopoles induced in this way by Meissner effect suggests a possible scenario for quark confinement. If the QCD vacuum behaves like a dual superconductor generated by the condensation of magnetic monopoles the chromo-electric flux will be expelled from the vacuum by dual Meissner effect. Thus, the immersion of a quark–antiquark pair into such a magnetic superconducting vacuum will generate a vortex of chromo-electric flux along the line connecting the two particles. The flux tube will behave like a string, inducing an effective quark–antiquark potential growing linearly with the distance which would imply quark confinement. The dual superconductor picture inspired many approaches to the confinement problem but none of them achieved a analytic formulation leading to a full proof quark confinement.

An alternative picture came from the AdS/CFT correspondence. The main issue of this correspondence is that strongly coupled gauge theories are dual to perturbative string theories. This connection provided a bridge between two different fields of physics, opening the possibility that some problems that are unaccessible in gauge theory can be, in principle, translated in the language of string theory and reciprocally. This connection inspired the search of a string theory dual to QCD with the hope of getting new insights in non perturbative effects of strong interactions. In particular there is a very interesting duality between type IIB string theory in a background of five-dimensional anti-de Sitter space times a five-sphere and the large N limit of $\mathcal{N}=4$ supersymmetric Yang-Mills theory in four dimensions. However, the resulting dual gauge theory is conformally invariant and it is not confining.

Both pictures can be very useful in understanding how the confinement mechanism should work, but they are not connected with real QCD. There is an alternative picture, due to V. Gribov, that is based on the Coulomb phase instability in real QCD. Gribov's picture was inspired by the instability of bound states in relativistic hydrogenoide atoms with supercritical large charges.

The Gribov approach to confinement[2,3] is based on the idea that a similar instability should appear for large values of the effective coupling constant α in heavy quarks backgrounds. The instability would imply a vacuum decay into light quark–antiquark pairs.[4,5] The analysis of Dyson-Schwinger equations for light quark Green functions in SU(N) gauge theories points to the existence of a critical value of the coupling constant $\alpha_{\rm crit} = 2N\pi(1-\sqrt{2/3})/(N^2-1)$, such that for large values of α ($\alpha > \alpha_{\rm crit}$) the theory becomes unstable.[4–6]

A derivation of the Gribov mechanism from first principles[7,8] pointed out the existence of gluonic instabilities in the heavy quark Coulomb phase for values of the coupling constant α larger than $\alpha_c = \sqrt{2}$ which is beyond the Gribov's critical value $\alpha_{\rm crit}$ (see also Refs. 9–20 for earlier approaches).

However, in the Gribov picture light dynamical quarks seem to play a leading role in the confinement mechanism. In the derivation from first principles this is reflected in the appearance of new instabilities.[21] However, this instabilities start at values of the coupling constant larger than $\alpha_{\rm F} = \sqrt{3}$, that is beyond the critical value of the gluonic fluctuations. This means that quark confinement is essentially generated by gluonic fluctuations. Quark fluctuations are, however, relevant to drive the asymptotic freedom regime at short distances

In this paper, we review the characteristics of the confining instabilities of Coulomb phase and the mechanism of quark confinement mainly driven by gluonic fluctuations.

2. Coulomb phase in a heavy quark–antiquark background

The dynamics of QCD in the presence of external heavy static quark–antiquark is governed by the action

$$S(A,\bar{\psi},\psi) = -\frac{1}{2g^2}\int d^4x \,\mathrm{tr}\, F^{\mu\nu}F_{\mu\nu} + \int d^4x\, \overline{\psi}(i\not{D}_A + m)\psi$$
$$-2i\int dx^0\, \mathrm{tr}\, \mathrm{T}^3 \;\; (qA_0(\vec{x}-L\vec{e}_3, x_0) - qA_0(\vec{x}+L\vec{e}_3, x_0))\,, \qquad (1)$$

where $\pm q$ are the charges of the heavy quarks located at $(0,0,\pm L)$ in the z-axis and $\mathrm{T}^3 = \frac{i}{2}\mathrm{diag}(1,-1,0) \in su(3)$. The Euclidean functional integral

$$Z = \int \delta A \delta\overline{\psi}\delta\psi \mathrm{e}^{S(A,\overline{\psi},\psi)} = \int \delta A \mathrm{e}^{S(A,0,0)} \det(i\not{D}_A + m) \qquad (2)$$

is dominated by the imaginary Coulomb background solution of Euclidean Yang-Mills equations

$$A_0(\vec{x}) = \frac{i\alpha \mathrm{T}^3}{|\vec{x}-L\vec{e}_3|} - \frac{i\alpha\mathrm{T}^3}{|\vec{x}+L\vec{e}_3|}, \qquad \vec{A}=0. \qquad (3)$$

where $\alpha = \frac{g^2}{4\pi}$. Notice that in the Euclidean formalism the Coulomb gauge fields Eq. (3) are imaginary. This solution is a saddle point of the Euclidean action which lives outside the real range of physical gauge fields. Physical instabilities arise from the negative eigenvalues of second order differential operator of gluonic fluctuations

$$-\delta_{\mu\nu}D^2 + D_\mu D_\nu - 2[F_{\mu\nu}\cdot\cdot], \qquad (4)$$

where $D_\mu = (\partial_0 + A_0, \vec{\nabla})$, or from the zero-modes of the Dirac operator

$$i\not{D}_A + m.$$

For instance in Landau gauge $D^\nu\tau_\mu = 0$ one loop corrections are given by

$$Z^{(1)} = \mathrm{e}^{S(A_0,0,0)} \det{}^{-\frac{1}{2}}(-\delta_{\mu\nu}D^2 - 2[F_{\mu\nu}\cdot\cdot]) \det(i\not{D}_A + m), \qquad (5)$$

which is pathological in both cases. However, in the classical vacuum without external fields both determinants are positive. The gluonic operator $-\delta_{\mu\nu}D^2$ is trivially positive; and although the Dirac operator $i\not{D}+m$ is not even selfadjoint its determinant is strictly positive. $i\not{D}+m$ is the sum of an anti-selfadjoint operator $i\not{D}$ and a selfadjoint one $m\mathbb{I}$. Thus, all eigenvalues are of the of the form $i\lambda + m$ with λ real. Now, since $i\not{D}$ anticommutes with γ_5, i.e. $\{i\not{D}, \gamma_5\} = 0$. for any eigenvalue $i\lambda$ of $i\not{D}$ there is another one of the form $-i\lambda$. Therefore, the fermionic determinant

$$\det(i\not{D} + m) = \prod_{\lambda}(i\lambda + m)(-i\lambda + m) = \prod_{\lambda}(\lambda^2 + m^2) > 0 \qquad (6)$$

is strictly positive. However, the Coulomb gauge field background can break the strict positivity of the fermionic determinant because some eingenvalues $i\lambda$ of $i\not{D}$ can become real, implying that the Coulomb phase would become unstable when $i\lambda = -m$. The source of all instabilities either gluonic of fermionic is the singularity of Coulomb potential which require the introduction of consistent boundary conditions. In the case where the boundary conditions introduce a new scale conformal invariance is broken and the stability of the Coulomb phase is lost.

3. Gluon instabilities in Coulomb phases of heavy quark–antiquark backgrounds

Gluon instabilities can arise from negative eigenvalues of second order fluctuations $(\vec{\tau}, \tau_0)$

$$\left(-\delta_{\mu\nu} D^2 + D_\mu D_\nu - 2[F_{\mu\nu}\cdot\cdot]\right) \vec{\tau} = -\lambda^2 \vec{\tau}, \quad \tau_0 = 0. \qquad (7)$$

These eigenvalues arise from non-Abelian static magnetic gauge field perturbations of the form

$$\vec{\tau}(x) = \frac{\vec{x} \times \vec{e}_3}{\rho^{\frac{3}{2}}}\, \phi(\rho, z)\, \mathrm{T}_{12}, \qquad (8)$$

in cylindric coordinates, where T_{12} is any linear combination of T_1 and T_2, and $\phi(\rho, z)$ satisfies the eigenvalue equation

$$\left[\frac{\partial^2}{\partial \rho^2} + \frac{\partial^2}{\partial z^2} - \frac{3}{4\rho^2} + \left(\frac{\alpha}{\sqrt{\rho^2 + (z-L)^2}} - \frac{\alpha}{\sqrt{\rho^2 + (z+L)^2}}\right)^2\right] \phi(\rho, z)$$
$$= \lambda^2 \phi(\rho, z). \qquad (9)$$

The singularities associated to the point-like structure of the heavy quarks can be renormalized by the introduction of appropriate boundary conditions. The differential operator Eq. (7) is essentially selfadjoint for $\alpha^2 < \frac{5}{4}$ and has a unique selfadjoint extension with positive spectrum. In that regime the Coulomb phase is stable because there are not normalizable solutions of Eq. (9). For $\alpha^2 > \frac{5}{4}$ the operator Eq. (7) is not anymore essentially selfadjoint and there is a family of boundary conditions parametrized by an UV scale Λ that are given by

$$\lim_{z\to L}\left(\phi'\,\psi_- - \phi\,\psi'_-\right) = 0, \qquad \lim_{z\to -L}\left(\phi'\,\psi_+ - \phi\,\psi'_+\right) = 0, \tag{10}$$

where ϕ' and ψ' denote the normal derivatives on small spherical balls around the quark singularities and

$$\psi_\pm(\rho, z) = \left(\frac{\rho}{\sqrt{\rho^2 + (z \mp L)^2}}\right)^{\frac{3}{2}} \times \left[\left(\Lambda\sqrt{\rho^2 + (z\mp L)^2}\right)^{\nu} - \left(\Lambda\sqrt{\rho^2 + (z\mp L)^2}\right)^{-\nu}\right] \tag{11}$$

are two asymptotic zero-modes around the heavy quarks.

The appearance of the full-dimensional parameter Λ that breaks conformal invariance provides one of the essential ingredients for the confinement mechanism.

However, there are two different regimes of coupling constants for $\alpha^2 > \frac{5}{4}$: a strong coupling regime $\alpha^2 > \frac{9}{4}$ and a intermediate coupling regime $\frac{5}{4} < \alpha^2 < \frac{9}{4}$. The situation is reminiscent of that with a single quark background[7].

The interesting behavior of the Coulomb phase arises in the intermediate coupling constant regime $\frac{5}{4} < \alpha^2 < \frac{9}{4}$. The spectral problem cannot be solved analytically. However, a numerical simulation shows the existence of two negative modes of Eq. (9) for large values of L: one that is parity symmetric (Fig. 1) and the other is parity odd (Fig. 2).

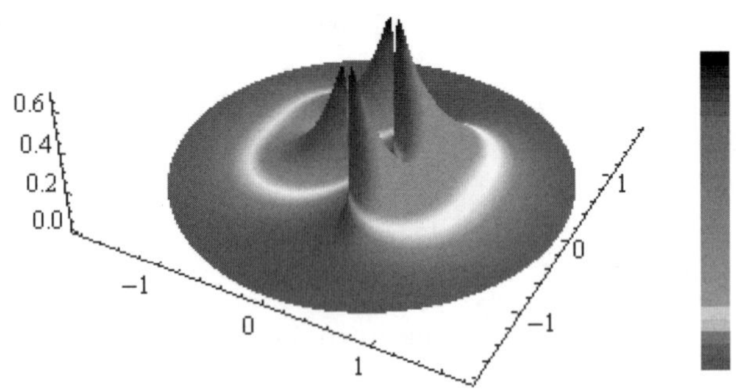

Fig. 1. Parity even unstable gluon mode for $\alpha = 1.495$.

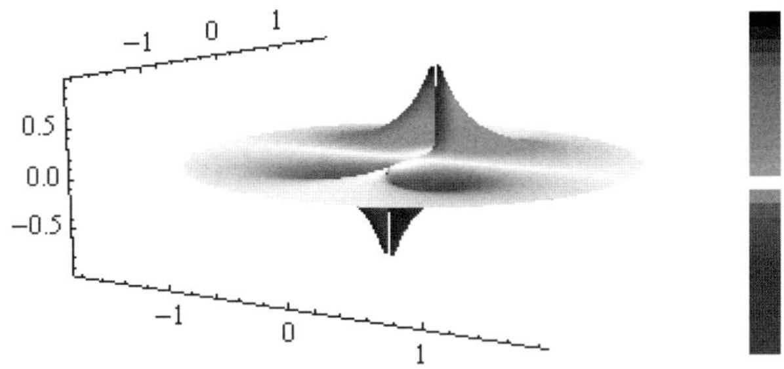

Fig. 2. Parity odd unstable gluon mode for $\alpha = 1.495$.

There is an analytic way of understanding the existence of such unstable modes. In a small neighborhood of each quark $\rho \leq \epsilon$ and $|z \pm L| \leq \epsilon$ with $0 < \epsilon \ll L$, a very good approximation of the interaction potential consists in considering the effect of the other quark as a mere constant shift in the potential. In that case the spectral problem for negative eigenvalues $-\lambda^2$ can be solved analytically. Indeed, the equation

$$\left[\frac{\partial^2}{\partial \rho^2} + \frac{\partial^2}{\partial w^2} - \frac{3}{4\rho^2} + \left(\frac{\alpha}{\sqrt{\rho^2 + w^2}} - \frac{\alpha}{2L}\right)^2\right] \phi(\rho, w) = \lambda^2 \phi(\rho, w) \quad (12)$$

with $w = z \mp L$ has non-trivial solutions. In the intermediate regime of coupling constants $\frac{5}{4} < \alpha^2 < \frac{9}{4}$, for $\alpha < 2L\lambda$ there is only one real solution matching the boundary conditions. It is given in terms of Whittaker function

$$\phi(\rho, w) = \frac{\rho^{3/2}}{\rho^2 + w^2} W\left(\frac{-\alpha^2}{\sqrt{4L^2\lambda^2 - \alpha^2}}, \nu, \frac{\sqrt{4L^2\lambda^2 - \alpha^2}}{L}\sqrt{\rho^2 + w^2}\right). \quad (13)$$

The asymptotic expansion of Whittaker function near $\rho \simeq 0, w \simeq 0$ reduces the boundary conditions matching to a simple spectral equation

$$\frac{\Gamma\left[1 - 2\nu\right]\Gamma\left[\frac{1}{2} + \nu + \frac{\alpha^2}{\sqrt{4L^2\lambda^2 - \alpha^2}}\right]}{\Gamma\left[1 + 2\nu\right]\Gamma\left[\frac{1}{2} - \nu + \frac{\alpha^2}{\sqrt{4L^2\lambda^2 - \alpha^2}}\right]} = \left(\frac{L\Lambda}{\sqrt{4L^2\lambda^2 - \alpha^2}}\right)^{2\nu}. \quad (14)$$

Notice that for a coupling constant $\alpha^2 = 2$ ($\nu = \frac{1}{2}$) the spectral equation Eq. (14) become singular. We shall restrict ourselves to the special intermediate regime $2 < \alpha^2 < \frac{9}{4}$. In that regime the exact solution is very close to the approximate solution Eq. (14) (see Fig. 3). Notice that in this analytic approximation the value of λ^2 increases with the distance $2L$ between the two quarks and possibly vanishes for a finite critical distance $L = L_c$, which is confirmed by the numerical analysis.

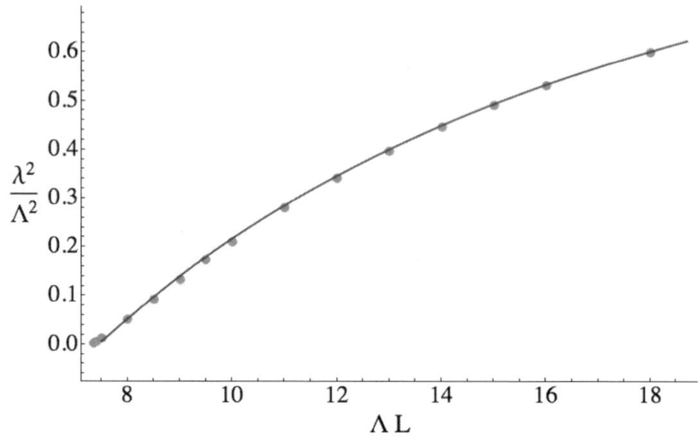

Fig. 3. Behaviour of the leading unstable mode eigenvalue of gluon fluctuations in a heavy quark–antiquark background for $\alpha = 1.495$. The blue curve is the solution of the analytic formula (9) whereas the red dots are exact values. The vanishing of the eigenvalues for critical point critical distance $2L_c$ between the quark means that the Coulomb phase becomes stable at short distances.

The critical distance L_c where the unstable modes vanish can be also obtained by analytic methods. We only have to enter into the short distances domain $\alpha > 2L\lambda$. In such a case the solution Eq. (13) becomes complex and in fact there are two independent real solutions matching the physical boundary conditions: one parity symmetric and another parity antisymmetric. The lowest eigenvalue is associated to the symmetric mode which leads the instability. A similar asymptotic analysis leads to a good approximation to the critical distance L_c. The result is[21]

$$L_c = \frac{\alpha}{\Lambda}\left(\frac{\Gamma[1-2\nu]}{\Gamma[1+2\nu]}\frac{\operatorname{Im}\left[\frac{i^{\frac{1}{2}+\nu}}{\Gamma\left[\frac{1}{2}-i\alpha-\nu\right]}\right]}{\operatorname{Im}\left[\frac{i^{\frac{1}{2}}(-i)^{\nu}}{\Gamma\left[\frac{1}{2}-i\alpha+\nu\right]}\right]}\right)^{\frac{1}{2\nu}}. \tag{15}$$

The exact result is very close to the above approximate value. The critical size L_c can reach arbitrary large values as α approaches the critical

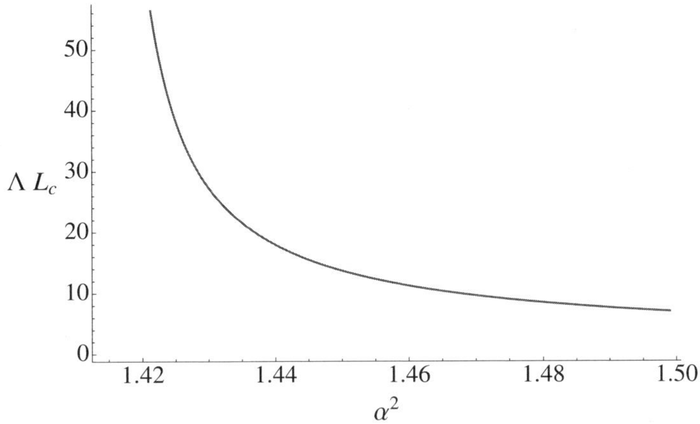

Fig. 4. α-dependence in the intermediate regime of the critical heavy meson size $2L_c$ where negative unstable modes of gluon fluctuations become stable.

value $\alpha_c = \sqrt{2}$ (see Fig. 4), which is consistent with the absence of unstable modes for $\alpha < \sqrt{2}$.

In summary, the behavior of these negative gluonic modes implies that for intermediate values of the coupling constant the Coulomb phase is only stable on a heavy meson background for small sizes $L < L_c$. Whereas, for larger meson sizes ($L > L_c$) the presence of negative modes leads to inconsistency of the background field expansion around Coulomb potentials.

The emerging picture is very appealing. In the intermediate regime of couplings $\sqrt{2} < \alpha < 3/2$ the Coulomb phase is stable for $q - \bar{q}$ pairs with size lower than $2L_c$ and becomes unstable for larger sizes $L > L_c$ of the $q - \bar{q}$ pair.

Thus, there is an explicit compatibility between asymptotic freedom and confinement. The stability properties depend on the distance between the $q - \bar{q}$ pair, pointing out the existence of a smooth interpolation from an asymptotic freedom regime to a confinement regime. These are the very first analytic indications derived from first principles that QCD does not undergo a phase transition at intermediate energy scales.

It is also remarkable that the leading unstable fluctuations of the Coulomb regime (see Fig. 1) exhibit a prominent profile of a thick string connecting the two quarks which points out to a picture of QCD where confinement would be driven by thick strings rather than by fundamental strings.[22,23]

In a stronger coupling regime $\alpha > \frac{3}{2}$ there is an infinity of negative modes[21] independently of meson size. Thus, the Coulomb phase is always unstable in this regime for any meson size, which is compatible with confining behavior of the theory in the strong coupling expansion.

The instability at large distances does not automatically means that the theory is confining. Usually, unstable modes get stabilized by higher order fluctuations. From pure dimensional arguments it follows that in that case the non-perturbative contributions of these modes will induce linear effective potential and quark confinement. However, the peculiar form of the unstable modes Eq. (8) of the gluonic fluctuations is commutative $[\tau_\mu, \tau_\nu] = 0$, which means that the quadratic approximation is exact, and stability cannot be restored by higher order terms in the usual way.

4. Quark instabilities in Coulomb phases

In QCD the critical value of coupling constant, $\alpha_c = \sqrt{2}$ where gluon instabilities arise is larger than Gribov's critical value, $\alpha_{\rm crit} = 3\pi(1-\sqrt{2/3})/4$. However, in the Gribov picture of confinement light quarks play a fundamental role. The Coulomb phase instability leads to a decay of the Coulombian vacuum into light quark–antiquark pairs.[4,5] The derivation of the complete picture from first principles requires, thus, to consider light dynamical quarks. The contribution of those quarks to the vacuum energy is given by the determinant of Dirac operator $i\,\not{D}_A + m$. In absence of heavy quarks, $\not{D}$ is a hermitian operator, but in the presence of a Coulombian background (3) $\not{D}_A$ is not self-adjoint anymore. Thus, it can have some pure imaginary eigenvalues and in such a case $i\not{D}_A$ has some negative real eigenvalues. When one of those eigenvalues is equal to the quark mass m, the determinant of $i\,\not{D}_A + m$ vanishes and a new source of vacuum instability is generated by the effects of dynamical quarks.

The zero modes spectrum of Dirac operator in a Coulombian background

$$(i\,\not{D}_A + m)\psi = 0 \tag{16}$$

can be of the form

$$\psi(x) = e^{-iEx_0} \begin{pmatrix} \psi^+(\mathbf{x}) + \psi^-(\mathbf{x}) \\ \psi^+(\mathbf{x}) - \psi^-(\mathbf{x}) \end{pmatrix}$$

provided that the bi-spinors $\psi^\pm$ verify that

$$\left(-\Delta + E^2 - A_0^2 + 2iEA_0 \mp i\boldsymbol{\sigma}\cdot\boldsymbol{\nabla}A_0 + m^2\right)\psi^\pm = 0 \tag{17}$$

$$m\psi^-(\mathbf{x}) = (-i\boldsymbol{\sigma}\cdot\boldsymbol{\nabla} + iA_0)\psi^+(\mathbf{x}). \tag{18}$$

The equation Eq. (17) can have non-trivial static solutions with $E = 0$

$$\left(-\Delta - A_0^2 \mp i\boldsymbol{\sigma} \cdot \boldsymbol{\nabla} A_0 + m^2\right) \psi^{\pm} = 0, \tag{19}$$

but they cannot be be expressed in a closed analytic form, although numerical simulations show that they do exist. However, there is an analytic argument which confirms the existence of zero-modes. Again, as in the case of gluon fluctuations, a good approximation in a neighborhood of each quark ($\rho \simeq 0$, $z \simeq L$) to the background Coulomb potential is

$$A_0(\boldsymbol{x}) = i\frac{\alpha}{2}\mathrm{T}^3 \left(\frac{1}{|\boldsymbol{x} - L\boldsymbol{e}_3|} - \frac{1}{2L}\right), \tag{20}$$

which reduces the zero-mode Dirac equation to

$$\left(\Delta + \frac{\alpha^2}{4}\left(\frac{1}{r} - \frac{1}{2L}\right)^2 \pm i\frac{\alpha}{2}\frac{\boldsymbol{\sigma} \cdot \boldsymbol{r}}{r^3} - m^2\right)\psi^{\pm}(\boldsymbol{r}) = 0, \tag{21}$$

where we have shifted coordinates $\boldsymbol{x} \to \boldsymbol{x} + L\boldsymbol{e}_3$. Using the spherical symmetry it is possible to search for solutions with the lowest possible total angular momentum ($j = 0$) of the form

$$\begin{aligned}\psi^{\pm}(\boldsymbol{r}) = &\left[\frac{1}{\sqrt{4\pi}}\begin{pmatrix}1\\0\end{pmatrix} + \frac{i}{\sqrt{3}\alpha}\left(\mp 1 - \nu\right)\begin{pmatrix}-Y_1^0(\theta,\phi)\\ \sqrt{2}\;\; Y_1^1(\theta,\phi)\end{pmatrix}\right]\phi^{\pm}(r) + \\ &\left[\frac{1}{\sqrt{4\pi}}\begin{pmatrix}1\\0\end{pmatrix} - \frac{i}{\sqrt{3}\alpha}\left(\pm 1 - \nu\right)\begin{pmatrix}-Y_1^0(\theta,\phi)\\ \sqrt{2}\;\; Y_1^1(\theta,\phi)\end{pmatrix}\right]\phi^{\mp}(r),\end{aligned} \tag{22}$$

where $\nu = \frac{1}{2}\sqrt{4 - \alpha^2}$

$$\left(\frac{d^2}{dr^2} + \frac{2}{r}\frac{d}{dr} - \frac{\alpha^2}{4rL} + \frac{\alpha^2}{16L^2} - m^2 - \frac{\nu(\nu \pm 1)}{r^2}\right)\phi^{\pm} = 0.$$

In the intermediate regime of coupling constants $3 < \alpha^2 < 4$, for $\alpha < 4Lm$ there is only one real solution matching the boundary conditions, which can be given in terms of confluent hypergeometric functions

$$\begin{aligned}\phi^{\pm}(r) = &\, c_{\pm} r^{\pm\nu} e^{-\frac{r}{4L}\sqrt{16L^2m^2-\alpha^2}} \\ &\mathbf{U}\left[\frac{\alpha^2}{2\sqrt{16L^2m^2 - \alpha^2}} + 1 \pm \nu, 2 \pm 2\nu, \frac{r}{2L}\sqrt{16L^2m^2 - \alpha^2}\right]\end{aligned} \tag{23}$$

where

$$c_{\pm} = \pm(2L)^{\mp\nu}(16m^2L^2 - \alpha^2)^{\pm\frac{\nu}{2}}\left[\alpha^2 + 4Lm\alpha + 2(-1 \pm \nu)\sqrt{16m^2L^2 - \alpha^2}\right].$$

From the asymptotic expansion near the origin $\rho \simeq 0, w \simeq 0$ of the confluent hypergeometric function U we get

$$\phi^{\pm}(r) \simeq c_{\pm}(2L)^{\pm 2\nu} r^{\mp\nu - 1} \frac{\left(16L^2m^2 - \alpha^2\right)^{-\frac{1}{2}\mp\nu} \Gamma[1 \pm 2\nu]}{\Gamma\left[1 \pm \nu + \frac{\alpha^2}{2\sqrt{16L^2m^2-\alpha^2}}\right]}. \tag{24}$$

Thus, the only zero mode of the form Eq. (22) compatible with boundary conditions appears only for the critical distance L_c which is solution of the spectral equation

$$\frac{2\sqrt{16m^2L_c^2 - \alpha^2}\,(1-\nu) + \alpha(\alpha - 4mL_c)}{2\sqrt{16m^2L_c^2 - \alpha^2}\,(1+\nu) + \alpha(\alpha - 4mL_c)} \frac{\Gamma(-2\nu)\Gamma\left(\frac{\alpha^2}{2\sqrt{16m^2L_c^2-\alpha^2}} + \nu + 1\right)}{\Gamma(2\nu)\Gamma\left(\frac{\alpha^2}{2\sqrt{16m^2L_c^2-\alpha^2}} - \nu + 1\right)} = -\left(\frac{2\Lambda L_c}{\sqrt{16m^2L_c^2 - \alpha^2}}\right)^{2\nu}. \tag{25}$$

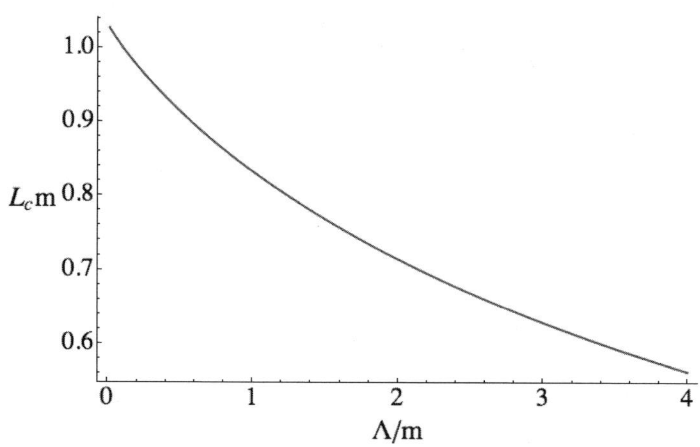

Fig. 5. Dependence of the fermionic critical size $2L_c$ on the boundary condition scale Λ for $\alpha = 1.8$.

The solutions of Eq. (25) point out the existence of a critical distance between the quarks L_c for any value of the boundary condition parameter Λ (see Fig. 5). For that particular separation L_c of the two quarks the fermionic determinant vanishes. The exact result is quite close to the one obtained with the approximation Eq. (25).

The instability can be easily understood in physical terms. The fermionic instability due to zero modes is generated by the spectral flow of eingenvalues of the Dirac operator induced by the change of separation between the two quarks. The fermionic determinant vanishes when one eigenvalue with negative energy crosses the zero level at the critical distance. This phenomenon points out the vacuum instability associated to a quark–antiquark pair creation induced by a zero energy level crossing which generates both a hole in the Dirac sea and a positive energy quark state.

The value of the critical distance L_c where the phenomenon occurs decreases with scale parameter Λ (Fig. 5) of the boundary condition. The dependence of such a critical distance on the strength of the gauge coupling α is also decreasing when coupling increases, becoming negligible at $\alpha = 2$, that is consistent with the fact in that in the strong regime the theory is confining for all distances.

Now, the value of the lowest coupling constant where instabilities arise $\alpha = \sqrt{3}$ is larger than the critical value $\sqrt{2}$ where gluon fluctuations become unstable. Thus, the relevant critical instability is due to gluon fluctuations instead of light quark zero modes unlike it was advocated by Gribov.

Theories with fermions in the adjoint representation of $SU(3)$ can have fermionic instabilities that arises at smaller values of the coupling constant. In that case the instabilities appear even inside the perturbative regime at $\alpha = \frac{\sqrt{3}}{2}$. In particular for supersymmetric gauge theories the leading instabilities of Coulomb phases dues to gaugino fluctuations arise in the perturbative regime.

In QCD the picture is quite different and the leading instability is of gluonic type as in pure gauge theories. Although the instability arises for large values of α, which are beyond the realm of perturbation theory, the couplings are not so large to be considered inside the strong coupling regime which leads to an interesting picture of confinement in the intermediate scale regime providing a further evidence of the absence of phase transition between the two regimes.

Acknowledgments

This work has been partially supported by Spanish DGIID-DGA grant 2015-E24/2, Spanish MINECO grants FPA2012-35453 and CPAN-CSD2007-00042 and European Cooperation in Science and Technology COST Action MP1405 QSPACE.

References

1. J.A. Carlson, A. Jaffe and A. Wiles (Eds.). *The millennium prize problems. Am. Math. Soc.* (2006).
2. V.N. Gribov, Talk at Perturbative QCD Workshop, Lund, Sweden, May 21 024, (1991). Preprint LU-TP-91-7 (unpublished).
3. V.N. Gribov, *Orsay lectures on confinement,* LPTHE-ORSAY 92-60 (1992) arxiv:hep-ph/9403218; LPTHE-ORSAY 94-20 (1994) arxiv:hep-ph/9404332;LPT-ORSAY 99-37 (1999) arxiv:hep-ph/9905285.
4. V.N. Gribov, *Eur. Phys. J.* **C10** (1999)70.
5. V.N. Gribov, *Gauge Theories and Quark Confinement. Collection of works,* Phasis Publishing House, Moscow (2002), 357.
6. Yu.L. Dokshitzer and D.E. Kharzeev, *Ann. Rev. Nucl. Part. Sci.* **54**, (2004) 487.
7. M. Asorey and A. Santagata, *PoS* (ConfinementX) **057** (2012).
8. M. Asorey and A. Santagata, *PoS* (QCD-TNT-III) **004** (2013).
9. J.E. Mandula, *Phys. Rev.* **D 14**, 3497 (1976); *Phys. Lett.* **B67**, 175 (1977); *Phys. Lett.* **B69**, 495 (1977).
10. M. Magg, *Phys. Lett.* **B74**, 246 (1978); *Phys. Lett.***B77**, 199 (1978); *Phys. Lett.* **B78**, 481 (1978); *Nucl. Phys.* **B158**, 154 (1979).
11. J.E. Mandula and L. McLerran, *Phys. Lett.* **B73**, 193 (1978).
12. I.B. Khriplovich, *Sov. Phys. JETP* **47**, 18 (1978) [*Zh. Eksp. Teor. Fiz.* **74**, 37 (1978)].
13. P. Sikivie and N. Weiss, *Phys. Rev. Lett.***40**, 1411 (1978); *Phys. Rev.* **D18**, 3809 (1978); *Phys. Rev.* **D20**, 487 (1979).
14. K.E. Cahill, *Phys. Rev. Lett.* **41**, 599 (1978).
15. A.J.G. Hey and J.E. Mandula, *Phys. Rev.* **D19**, 1856 (1979).
16. P. Sikivie, *Phys. Rev.* **D20**, 877 (1979).
17. R.A. Freedman, L. Wilets, S.D. Ellis, and E.M. Henley, *Phys. Rev.* **D22**, 3128 (1980).
18. C.H. Oh, R. Teh, and W.K. Koo, *Phys. Lett.* **B101**, 337 (1981).
19. L.J. Carson, R. Goldflam, and L. Wilets, *Phys. Rev.* **D28**, 385 (1983).
20. L.J. Carson, *Phys. Rev.* **D29**, 2355 (1984).
21. M. Asorey and A. Santagata, *AIP Conf. Proc.* **1606**, 407-417 (2014).
22. J. Polchinski and L. Susskind, *String theory and the size of hadrons,* arXiv:hep-th/0112204.
23. V. Vyas, *Intrinsic Thickness of QCD Flux-Tubes*, arXiv:1004.2679.

ON THE NONLOCALITY OF THE COULOMB GAUGE EXTERNAL FIELD PROBLEM

PÉTER HRASKÓ

University of Pécs, Hungary, H-7622 Pécs, Vasvári Pál utca 4.
Email: peter@hrasko.com

The apparent nonlocality of the Coulomb gauge external field problem in electrodynamics is illustrated with an example in which nonlocality is especially striking. Explanation of this apparent nonlocal behaviour based on a purely local picture is given. A gauge invariant decomposition of the Lorentz-force into two terms with clear physical meanings is pointed out. Based on this decomposition derivation of the Aharonov–Bohm effect in terms of field strengths alone is given.

1. An introductory example

Consider a rigid body with a given charge distribution $\rho(\mathbf{x})$ on it but of total charge equal to zero. If it is placed into a homogeneous magnetic field $\mathbf{B} = \boldsymbol{\nabla} \times \mathbf{A}$ then in the combined electric field $\mathbf{E}$ of the body and the external magnetic field a certain amount $\mathbf{N}$ of angular momentum will arise. The simplest way to demonstrate this is to slowly remove the magnetic field and to calculate the torque,

$$\mathbf{K}(t) = \int d^3x \, [\mathbf{x}, \rho\mathbf{E}'(t)] \, , \tag{1}$$

acting on the body due to the electric field $\mathbf{E}'(t)$ induced by the changing $\mathbf{B}(\mathbf{t})$. The total amount of angular momentum imparted to the body is equal to the angular momentum $\mathbf{N}$.

In Coulomb gauge the vector potential is chosen so as to make its divergence to vanish and the scalar potential ϕ is equal to the Coulomb field of the charge density $\rho(\mathbf{x})$. For a homogeneous field we have

$$\mathbf{A} = \frac{1}{2}[\mathbf{B}, \mathbf{x}] \, , \qquad (\boldsymbol{\nabla} \cdot \mathbf{A} = 0) \, . \tag{2}$$

The induced electric field is determined by the equation

$$\mathbf{E}'(t) = -\frac{1}{c}\dot{\mathbf{A}}(t), \qquad \mathbf{A}(0) = \mathbf{A}, \qquad \mathbf{A}(\infty) = 0 \, .$$

Then Eq. (1) can be written as

$$\mathbf{K}(t) = -\frac{d}{dt}\int d^3x\,[\mathbf{x},\,\frac{\rho}{c}\mathbf{A}(t)]\,.$$

Since $\dot{\mathbf{N}} = \mathbf{K}$ the total amount of angular momentum received by the body is equal to

$$\mathbf{N} = +\int d^3x\,[\mathbf{x},\,\frac{\rho}{c}\mathbf{A}]\,. \tag{3}$$

The angular momentum stored in the field can, therefore, be written as

$$\mathbf{N} = \frac{1}{2c}\int d^3x\,\rho(\mathbf{x})\Big[\mathbf{x},\,[\mathbf{B},\,\mathbf{x}]\Big]\,.$$

Since

$$\Big[\mathbf{x},\,[\mathbf{B},\,\mathbf{x}]\Big] = -\frac{1}{3}\big(3(\mathbf{x}\cdot\mathbf{B})\mathbf{x} - r^2\mathbf{B}\big) + \frac{2}{3}r^2\mathbf{B}\,,$$

where $r^2 \equiv \mathbf{x}^2$, the Cartesian components of $\mathbf{N}$ are

$$N_\alpha = -\frac{1}{6c}D_{\alpha\beta}B_\beta + \frac{1}{3c}\langle\rho r^2\rangle B_\alpha\,. \tag{4}$$

Here,

$$D_{\alpha\beta} = \int d^3x\,\rho(\mathbf{x})(3x_\alpha x_\beta - r^2\delta_{\alpha\beta})$$

is the quadrupole momentum tensor of the body and

$$\langle\rho r^2\rangle = \int d^3x\,\rho(\mathbf{x})r^2$$

is its scalar part.

Simple as it seems our example reveals its rather peculiar feature when one tries to recalculate $\mathbf{N}$ starting from the equation

$$\mathbf{N} = \frac{1}{4\pi c}\int d^3x\Big[\mathbf{x},\,[\mathbf{E},\,\mathbf{B}]\Big]\,, \tag{5}$$

which defines field angular momentum in terms of field strengths. To this end we replace the fields in Eq. (5) by their respective potentials $\mathbf{B} = \boldsymbol{\nabla}\times\mathbf{A}$ and $\mathbf{E} = -\boldsymbol{\nabla}\phi$ and perform partial integrations within a sphere of arbitrary large radius R. Though, neglecting surface terms, we arrive at the expression (4) again, *the contribution of the surface terms turns out to be of finite magnitude independent of* R (see Appendix A):

$$\text{Surface Terms} = \frac{1}{10c}D_{\alpha\beta}B_\beta. \tag{6}$$

Hence we obtain that, contrary to Eq. (4), within any sphere surrounding the body the amount of angular momentum is equal to

$$N_\alpha^{(R)} = -\frac{1}{15c} D_{\alpha\beta} B_\beta + \frac{1}{3c} \langle \rho r^2 \rangle B_\alpha . \tag{7}$$

Moreover, this momentum is concentrated within the volume of the body since, owing to the independence of Eq. (6) of R, in any spherical shell outside the body the value of the angular momentum is cancelled to zero.

The correct conclusion from this apparent contradiction is that the notion of the homogeneous magnetic field extending to infinity is in general an unacceptable abstraction. In local problems as, e.g., in Zeeman effect, or even in the derivation of Eq. (4), such a concept is a perfectly suitable idealization but when total field momentum or angular momentum are to be calculated the sources of $\mathbf{B}$ should be taken into consideration explicitly

Their simple realization is an infinitely long ideal straight solenoid (an idealization too but this time a harmless one) of very large circular cross section of radius R_s with our body situated on its axis (see Fig. 1). Then an amount of angular momentum given by Eq. (7) is concentrated within the volume of the body while the remaining part of it, the negative of Eq. (6), is found within the solenoid above and below the sphere of radius R_s, arbitrarily far away from the charged body (see Appendix A). The sum total is then given by Eq. (4) as expected.

2. The problem of apparent nonlocality

Assume now that the body we are considering is capable to rotate around a fixed point of it. Then its Lagrangian is

$$L = \frac{1}{2} I_{\alpha\beta} \Omega_\alpha \Omega_\beta + \frac{1}{c} \int d^3x \, \rho(\mathbf{x}) \mathbf{A}(\mathbf{x}) \cdot \mathbf{V}(\mathbf{x}),$$

in which $I_{\alpha\beta}$ and $\mathbf{\Omega}$ are the inertia tensor and the angular velocity and $\mathbf{V} = [\mathbf{\Omega}, \mathbf{x}]$.

In Coulomb gauge, using Eq. (2) for the vector potential and the relation $\mathbf{A}[\mathbf{\Omega}, \mathbf{x}] = \mathbf{\Omega}[\mathbf{x}, \mathbf{A}]$, the interaction term becomes a linear function of the field angular momentum $\mathbf{N}$ of Eq. (3):

$$L = \frac{1}{2} I_{\alpha\beta} \Omega_\alpha \Omega_\beta + \mathbf{N} \cdot \mathbf{\Omega} .$$

In the Hamiltonian $H = \mathbf{J} \cdot \mathbf{\Omega} - L$, the canonical momenta are expressed as

$$J_\alpha = \frac{\partial L}{\partial \Omega_\alpha} = I_{\alpha\beta} \Omega_\beta + N_\alpha . \tag{8}$$

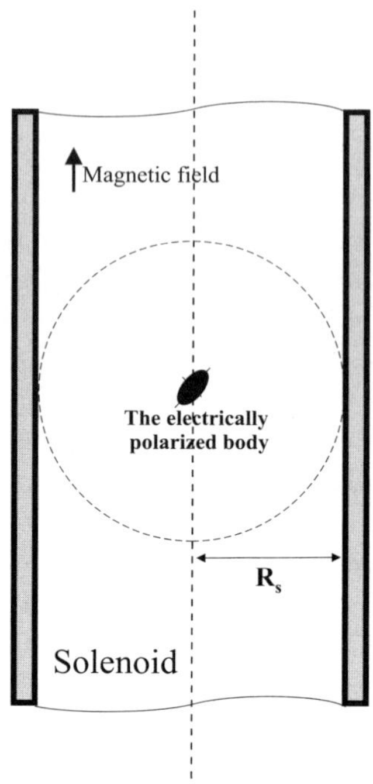

Fig. 1. An electrically polarized body inside a solenoid.

Inverting Eq. (8), we obtain

$$\Omega_\alpha = (I^{-1})_{\alpha\beta}(J_\beta - N_\beta)\,.$$

Substituting this expression into H we obtain, after some rearrangements,

$$H = \frac{1}{2}(I^{-1})_{\alpha\beta}(J_\alpha - N_\alpha).(J_\beta - N_\beta). \tag{9}$$

The canonical variables in this Hamiltonian are the Euler angles which determine the orientation of our rotator and their conjugate momenta. We do not recall the explicit dependence of $\mathbf{J}$ and $\Omega_{\alpha\beta}$ on these variables. According to Eq. (4), $\mathbf{N}$ depends on the Euler angles through the quadrupole momentum tensor $D_{\alpha\beta}$.

The Hamiltonian is the energy of the rotator. The Lorentz force by which $\mathbf{B}$ acts on the rotator does not perform any work on it and so H is equal to its kinetic energy. Hence, the difference $(\mathbf{J} - \mathbf{N})$ must be equal to the angular momentum $\mathbf{R}$ of its rotation. The canonical angular momentum $\mathbf{J}$ can, therefore, be interpreted as the sum of the rotational angular momentum and of the angular momentum stored in the combined external magnetic field and electric field of the body *at rest*. This circumstance expresses the most characteristic feature of the Coulomb gauge: the potential ϕ of a time dependent charge distribution $\rho(\mathbf{x}, t)$ is equal to its *instantaneous* Coulomb potential.[1]

In the previous section we have found that, depending on $D_{\alpha\beta}$ (i.e. on the orientation of the body), a well defined fraction of the field angular momentum is found at a distance arbitrarily far (say light years away) from the body at rest. When our body is allowed to rotate then the potential ϕ determining field angular momentum rotates together with it. Hence, in spite of the huge distance, the amount of this quantity at the far ends of the solenoid *varies synchronously* with the rotation of the body without any delay. Though from a purely computational point of view this apparent nonlocal behaviour presents no difficulty a simple explanation of it in terms of local interactions alone would certainly improve our understanding of this important aspect of electrodynamics.* For this reason in the next section the results arrived at above will be rederived by means of explicitly local considerations.

3. The underlying physical picture

Consider now a point charge moving arbitrarily along the trajectory $\mathbf{x} = \mathbf{r}(t)$ in a stationary magnetic field $\mathbf{B}(\mathbf{x})$ *of finite extension*. The Hamiltonian is

$$H = \frac{1}{2m}\left(\mathbf{p} - \frac{e}{c}\mathbf{A}(\mathbf{r})\right)^2 .$$

Since the Lorentz force acting on the charge does not perform any work on it this H is numerically equal to the kinetic energy $mV^2/2$. Hence $\mathbf{p} = m\mathbf{V} + (e/c)\mathbf{A}(\mathbf{r})$ where $\mathbf{V} = \dot{\mathbf{r}}(t)$ is the instantaneous velocity of the charge.

Any gauge transformation of $\mathbf{A}$ is accompanied by a canonical transformation of $\mathbf{p}$ so as to make the difference $\mathbf{p} - (e/c)\mathbf{A}$ unchanged. As it is well known (see, e.g., Ref. 5) in the special case of the Coulomb-gauge

*Another problem related to the nonlocal behaviour of the Coulomb gauge potential is the finiteness of signal (light) velocity. A substantial amount of literature has been devoted to this subject since W. Heisenberg proposed it to his disciple S. Kikuchi[2] and E. Fermi published his study of quantum electrodynamics.[3] A recent discussion of this problem in a variety of gauges is found in Ref. 4, where references to earlier literature are also found.

the quantity $(e/c)\mathbf{A}$ is equal to the momentum $\mathbf{G}$ stored in the combined electric field of the charge and the external field $\mathbf{B}$ when the charge is *at rest* at the point $\mathbf{r}(t)$ of its trajectory. Hence, this $\mathbf{G}$ seems rigidly attached to the particle as was the case with the field angular momentum $\mathbf{N}$ and the rotator in the example of the previous sections. At any moment of time t and point P in space the electric field contributing to $\mathbf{G}$ is determined by the location of the charge $\mathbf{r}(t)$ at the same moment of time however far away from P, and, therefore, *it obviously cannot be equal to the true momentum* $\mathcal{G}$ *stored in the combined field of the particle and that of the external sources at* t. Since composite charged bodies are built up of pointlike charged constituents the nonlocality found in the case of the rotating body has been inherited from this description of the point charge.

The sole virtue of the example with the rotator in the previous section was that, owing to the splitting of $\mathbf{N}$ into fractions located within disjoint domains of space, its nonlocality appeared more spectacular than that of a point charge. But in order to elucidate the physical picture behind this seemingly nonlocal description it is sufficient to confine ourselves to the case of a single point charge.

The electromagnetic part of the energy-momentum tensor decomposes into the sum of three terms $T^{(i)}_{\alpha\beta}$ according to the power of $\mathbf{B}$ in them. Each term satisfies a balance equation (continuity equation with sources) of its own. This may be verified either by direct computation or by rescaling the external field $\mathbf{B}$ with a scalar factor k. For an arbitrary value of k the electromagnetic energy-momentum tensor becomes equal to the sum $k^0T^{(0)}_{\alpha\beta} + k^1T^{(1)}_{\alpha\beta} + k^2T^{(2)}_{\alpha\beta}$. Since the sum satisfies a balance equation for any value of k the three parts must satisfy it separately.

The term of second order, being constant in time, is obviously of no interest. The term independent of the external field describes radiation and radiation reaction (self-interaction) of the charge and obeys a balance equation with a source equal to the negative of the self-force. This term is of paramount importance of its own right but for the external field problems only the mixed part $T^{(1)}_{\alpha\beta}$ of the tensor is of significance. The sources of the balance equation satisfied by this term in an external magnetic field are the negative of the Lorentz force acting on the charge in the external field and the minus Lorentz force density experienced by the sources of this field from the side of the moving charge. From our point of view a cardinal property of this part of the energy-momentum tensor is the absence of radiation. This follows from the finite extension of the external field in space.

The force experienced by the charge at moment t_0 does not depend on its subsequent motion. Therefore, in calculating this force we can rely on the *truncated* (at $t = t_0$) *trajectory* which is obtained from the true trajectory by bringing the charge to a stop at the position $\mathbf{r}_0 = \mathbf{r}(t_0)$. At this moment the true spatial distribution of the momentum density is certainly different from the momentum density $(1/4\pi c)[\mathbf{E}^{(0)}, \mathbf{B}]$ of a charge in state of rest at $\mathbf{r}_0$ but the total field momentum will tend to the integral

$$\mathbf{G}^{(0)} = \frac{1}{4\pi c} \int d^3x \, [\mathbf{E}^{(0)}, \mathbf{B}] \tag{10}$$

as $t \longrightarrow \infty$.† The reason is that due to the absence of radiation the mixed field momentum emitted by the moving charge before t_0 will be completely absorbed in later times by the sources of the external field.

Let us compare momentum exchange between the particle and the field in the two motions truncated at moments of time t_0 and $t_0 + dt_0$. The field momentum emitted by the particle in the infinitesimal interval $(t_0 < t < t_0 + dt_0)$ consists of two parts: that what remains stored in the form of field momentum forever and an additional part which will be subsequently absorbed by the sources of the external field. Accordingly, the force experienced by the particle consists of two contributions:

$$\mathbf{F}^{(0)} = -\frac{d\mathbf{G}^{(0)}}{dt_0} - \mathbf{F}'^{(0)}, \tag{11}$$

where $\mathbf{F}'^{(0)}$ is the Lorentz force experienced by the sources of $\mathbf{B}$ when some portion of the mixed field momentum emitted at t_0 by the charge is absorbed.

Since $\mathbf{G}^{(0)}$ depends on time through the motion of the charge alone we have

$$\frac{dG_\alpha^{(0)}}{dt_0} = V_\beta(t_0)\partial_\beta G_\alpha^{(0)}, \tag{12}$$

where $\boldsymbol{\partial}$ denotes derivation with respect to $\mathbf{r}$.

The force $\mathbf{F}'^{(0)}$ is related to the momentum difference received by the sources of $\mathbf{B}$ in the motions truncated at the moments t_0+dt_0 and t_0. Hence, in order to evaluate this force we must first calculate the momentum $\mathbf{G}'^{(0)}$ received by the sources via the retarded field of the charge. It is equal to the time integral of the Lorentz force experienced by them:

$$\mathbf{G}'^{(0)} = \frac{1}{c} \int dt \cdot d^3x \, [\mathbf{J}(\mathbf{x}), \mathbf{B}'^{(0)}(\mathbf{x}, t)], \tag{13}$$

†The upper index zero on a quantity indicates that it refers to the truncated motion.

where $\mathbf{J}(\mathbf{x})$ is the current density supporting the external magnetic field. The retarded magnetic field of the charged particle is

$$\mathbf{B}'^{(0)}(\mathbf{x},t) = \boldsymbol{\nabla}\times\mathbf{A}'^{(0)}(\mathbf{x},t).$$

The vector potential here is the Liénard-Wiechert potential of the particle's current density $\mathbf{j}^{(0)}$:

$$\mathbf{A}'^{(0)}(\mathbf{x},t) = \frac{1}{c}\int dt'\cdot d^3x'\,\frac{\delta(t'-t+|\mathbf{x}-\mathbf{x}'|/c)}{|\mathbf{x}-\mathbf{x}'|}\mathbf{j}^{(0)}(\mathbf{x}',t')\,. \tag{14}$$

Since at $t=t_0$ the particle comes to a stop we can write

$$\mathbf{j}^{(0)}(\mathbf{x}',t') = \mathbf{j}(\mathbf{x}',t')\Theta(t_0-t')\,,$$

where $\Theta(t)$ is the step function. Then

$$\int_{-\infty}^{\infty} dt\,\mathbf{A}'^{(0)}(\mathbf{x},t) = \frac{1}{c}\int\frac{d^3x'}{|\mathbf{x}-\mathbf{x}'|}\int_{-\infty}^{t_0} dt'\,\mathbf{j}(\mathbf{x}',t')\,.$$

Now, using the relation

$$\boldsymbol{\nabla}\times(f\mathbf{U}) = [\boldsymbol{\nabla}f,\,\mathbf{U}] + f\boldsymbol{\nabla}\times\mathbf{U}\,, \tag{15}$$

we calculate the curl of this function:

$$\int_{-\infty}^{\infty} dt\,\mathbf{B}'^{(0)}(\mathbf{x},t) = \boldsymbol{\nabla}\times\int_{-\infty}^{\infty} dt\,\mathbf{A}'^{(0)}(\mathbf{x},t) =$$
$$= \frac{1}{c}\int d^3x'\,\left[\boldsymbol{\nabla}|\mathbf{x}-\mathbf{x}'|^{-1},\,\int_{-\infty}^{t_0} dt'\,\mathbf{j}(\mathbf{x}',t')\right]\,.$$

To obtain $\mathbf{F}'^{(0)}$ we have to compute the derivative of $\mathbf{G}'^{(0)}$, as given in Eq. (13), with respect to t_0. At first we write

$$\frac{d}{dt_0}\int_{-\infty}^{\infty} dt\,\mathbf{B}'^{(0)}(\mathbf{x},t) = \frac{1}{c}\int d^3x'\,\left[\boldsymbol{\nabla}|\mathbf{x}-\mathbf{x}'|^{-1},\,\mathbf{j}(\mathbf{x}',t_0)\right]\,.$$

The current density of a pointlike charged particle is

$$\mathbf{j}(\mathbf{x}',t_0) = e\mathbf{V}(t_0)\delta\left(\mathbf{x}'-\mathbf{r}(t_0)\right)\,.$$

Then

$$\frac{d}{dt_0}\int_{-\infty}^{\infty} dt\,\mathbf{B}'^{(0)}(\mathbf{x},t) = \frac{e}{c}\left[\boldsymbol{\nabla}|\mathbf{x}-\mathbf{r}(t_0)|^{-1},\,\mathbf{V}(t_0)\right] = \frac{1}{c}[\boldsymbol{\nabla}\phi^{(0)},\mathbf{V}(t_0)]\,, \tag{16}$$

where

$$\phi^{(0)}(\mathbf{x}) = \frac{e}{|\mathbf{x}-\mathbf{r}(t_0)|}$$

is the potential of a charge at $\mathbf{x} = \mathbf{r}(t_0)$.

Now we can write

$$\mathbf{F}'^{(0)} = \frac{d\mathbf{G}'^{(0)}}{dt_0} = \frac{1}{c^2}\int d^3x \left[\mathbf{J}(\mathbf{x}),\, [\boldsymbol{\nabla}\phi^{(0)},\, \mathbf{V}(t_0)]\right] =$$
$$= \frac{1}{c^2}\int d^3x \left\{(\mathbf{J}\cdot\mathbf{V}(t_0))\,\boldsymbol{\nabla}\phi^{(0)} - (\mathbf{J}\cdot\boldsymbol{\nabla}\phi^{(0)})\mathbf{V}(t_0)\right\}.$$

The second integral on the r.h.s. is equal to zero because the stationary current $\mathbf{J}(\mathbf{x})$ is solenoidal:

$$\int d^3x\,\mathbf{J}\cdot\boldsymbol{\nabla}\phi^{(0)} = -\int d^3x\,\phi^{(0)}\,\boldsymbol{\nabla}\cdot\mathbf{J} = 0\,.$$

In Cartesian coordinates we have

$$\nabla_\alpha\phi^{(0)} \equiv \frac{\partial\phi^{(0)}}{\partial x_\alpha} = -\frac{\partial\phi^{(0)}}{\partial r_\alpha} \equiv -\partial_\alpha\phi^{(0)}\,.$$

Then

$$\mathbf{F}'^{(0)} = -\frac{1}{c^2}\partial_\alpha\int d^3x\,(\mathbf{J}\cdot\mathbf{V}(t_0))\,\phi^{(0)} = -V_\beta(t_0)\partial_\alpha\left[\frac{1}{c^2}\int d^3x\,\phi^{(0)}\,J_\beta\right]. \tag{17}$$

Equation $\boldsymbol{\nabla}\times\mathbf{B} = (4\pi/c)\mathbf{J}$ permits us to replace $\mathbf{J}$ by $\mathbf{B}$:

$$\frac{1}{c^2}\int d^3x\,\phi^{(0)}\,J_\beta = \frac{1}{4\pi c}\int d^3x\,\phi^{(0)}\boldsymbol{\nabla}\times\mathbf{B}\,.$$

Here we can use Eq. (15) again:

$$\frac{1}{c^2}\int d^3x\,\phi^{(0)}\,J_\beta = \frac{1}{4\pi c}\left\{\int d^3x\,[\mathbf{B},\,\boldsymbol{\nabla}\phi^{(0)}] + \int d^3x\,\boldsymbol{\nabla}\times(\phi^{(0)}\mathbf{B})\right\}.$$

The second term is transformed into a surface integral and can be dropped. In the first term $\boldsymbol{\nabla}\phi^{(0)} = -\mathbf{E}^{(0)}$. Then, using Eq. (10), we obtain

$$\frac{1}{c^2}\int d^3x\,\phi^{(0)}\,J_\beta = \frac{1}{4\pi c}\int d^3x\,[\mathbf{E}^{(0)},\,\mathbf{B}] = \mathbf{G}^{(0)}\,.$$

Substitution of this expression into Eq. (17) leads to the concise form

$$F'^{(0)}_\alpha = -V_\beta(t_0)\partial_\alpha G^{(0)}_\beta\,. \tag{18}$$

From now on we may drop upper 0 indices. Inserting Eqs. (12) and (18) into Eq. (11), we obtain the formula for the force acting on the charge,

$$F_\alpha = V_\beta\partial_\alpha G_\beta - V_\beta\partial_\beta G_\alpha\,,$$

where $\mathbf{V}(t) = \dot{\mathbf{r}}(t)$ and $\mathbf{G}$ is equal to the field momentum stored in the combined field of a charge *resting at* $\mathbf{r}(t)$. Since $\mathbf{V}$ is independent of $\mathbf{r}$, the vector form of this equation is

$$\mathbf{F} = [\mathbf{V}, \boldsymbol{\nabla} \times \mathbf{G}] = \big(\boldsymbol{\nabla}(\mathbf{V} \cdot \mathbf{G}) - (\mathbf{V} \cdot \boldsymbol{\nabla})\mathbf{G}\big)\,, \tag{19}$$

where $\boldsymbol{\nabla}$ denotes now differentiation with respect to $\mathbf{r}$ the only coordinate $\mathbf{G}$ depends on.

On the other hand, we know that $\mathbf{F}$ is the Lorentz force which acts on the charge:

$$\mathbf{F} = \frac{e}{c}[\mathbf{V}, \mathbf{B}] = \Big[\mathbf{V}, \boldsymbol{\nabla} \times \frac{e}{c}\mathbf{A}\Big]\,.$$

We see that (up to the factor e/c) $\mathbf{G}$ can be identified with the unique Coulomb gauge vector potential vanishing at infinity together with $\mathbf{G}$. Equation (19) can, therefore, be written also as

$$\mathbf{F} = \frac{e}{c}\big(\boldsymbol{\nabla}(\mathbf{V} \cdot \mathbf{A}) - (\mathbf{V} \cdot \boldsymbol{\nabla})\mathbf{A}\big)\,. \tag{20}$$

At the beginning of the present section we have stressed the apparent nonlocality of this gauge. Now we see explicitly that locality is by no means broken: There is in fact no field momentum attached rigidly to the charge and in the general case the true mixed field momentum $\boldsymbol{\mathcal{G}}$ differs from $\mathbf{G}$ at any moment of time.[‡] The flow of the combined field momentum density obeys continuity equation in space between the moving charge and the sources of the external field and this is what locality means.

4. Splitting of the Lorentz force

A bonus of the preceding investigation is the recognition that in the external field problem the Lorentz force decomposes into two terms each with clear physical meanings. When in Eq. (20) $\mathbf{A}$ is that unique Coulomb gauge vector potential which vanishes at infinity, then the second term describes momentum exchange of the charge with the field, while the first one corresponds to the charge's momentum exchange (via its retarded field) with the currents and magnets supporting the external magnetic field.[§] Naturally, this decomposition is gauge invariant.

[‡]Analogously, the quantity $\mathbf{N}$ in Eq. (9) is in fact different from the true mixed field angular momentum $\boldsymbol{\mathcal{N}}$ at any moment of time.

[§]When the external field is electric, no mixed field momentum arises. It is for this reason that the Coulomb force, $-e\boldsymbol{\nabla}\Phi$, is of purely first type.

An interesting special example is that of a charged particle passing by an infinitely long straight ideal solenoid (a classical 'Aharonov–Bohm situation'). Since the magnetic field outside the solenoid is zero, the two terms in Eq. (20) cancel each other, but neither of them vanishes. That means that *the charge 'catalizes' momentum exchange between the combined field and the coil.* Outside the solenoid we have

$$\mathbf{A} = \frac{S}{2\pi} \cdot \frac{[\mathbf{B}, \mathbf{r}]}{r^2}, \tag{21}$$

where S is the cross section of the solenoid. In this domain the r.h.s. of Eq. (19) is equal to zero and thus Eq. (18) can be written as $F'_\alpha = -V_\beta \partial_\beta G_\alpha$. Since $\mathbf{G} = (e/c)\mathbf{A}$, the total momentum imparted to the coil by the moment t when the charge reaches position $\mathbf{r}$ is equal to

$$\mathbf{p}' = \int_{-\infty}^{t} dt\, \mathbf{F}' = -\frac{eS}{2\pi c} \cdot \frac{[\mathbf{B}, \mathbf{r}]}{r^2}. \tag{22}$$

When the charge leaves freely to infinity no net momentum exchange occurs. But if it is stopped at $\mathbf{r}$, a finite amount of momentum, given by Eq. (22), is transferred from the combined field to the coil. I wonder whether this effect may be experimentally observed.

Inside the solenoid the continuation of the external vector potential, Eq. (21), is given by Eq. (2) and the field there is homogeneous. Inserting this into Eq. (20) we find that, for a point charge revolving in a plane perpendicular to solenoid's symmetry axis on a circle around it, the two terms on the r.h.s. are equal to each other. Therefore, the reaction to the centripetal force acting on the charge is distributed evenly between the field and the coil: the force experienced by the coil is the half of the force experienced by the revolving point charge.

5. Relation to the Aharonov–Bohm effect

When an electron passes by the solenoid no net momentum is transferred to the latter. But the *action* of the coil, considered as the target in the scattering with the electron, changes by a finite amount.

Choose the axis of the solenoid as the z-direction and assume that the trajectory of the electron lies in the xy plane at $y = \pm a$. Then the change of the coil's action is

$$\mathcal{S}'_\pm = \int_{-\infty}^{\infty} p'_x dx = \pm \frac{eSBa}{2\pi c} \int_{-\infty}^{\infty} \frac{dx}{r^2} = \pm \frac{eSB}{2c}. \tag{23}$$

In quantum mechanics this peculiar elastic scattering process can be described in quasiclassical approximation considering both the electron and the coil as quantum systems. The initial wave-function is $\psi\Psi$, where the first factor refers to the electron, the second to the coil. Then the final wave function (the out-state) is given by the superposition

$$\psi_+(\mathbf{r})\Psi_+ + \psi_-(\mathbf{r})\Psi_- \,. \tag{24}$$

The quasiclassical wave functions $\psi_\pm$ for the electron describe its motion in the vicinity of the classical trajectories $y = \pm a$ respectively, while the wave functions $\Psi_\pm$ of the coil are $e^{i\mathcal{S}'_\pm/\hbar}\Psi$. Then Eq. (24) becomes

$$\left(\psi_+(\mathbf{r})e^{i\mathcal{S}'_+/\hbar} + \psi_-(\mathbf{r})e^{i\mathcal{S}'_-/\hbar}\right)\Psi\,. \tag{25}$$

The relative phase between the two terms of Eq. (24) contains obviously the well-known **B**-dependent Aharonov–Bohm phase

$$(\mathcal{S}'_+ - \mathcal{S}'_-)/\hbar = \frac{eSB}{\hbar c}.$$

It is, perhaps, worth pointing out that the reasoning which, beginning with Section 3, led us finally to this formula was based entirely on the fields **E** and **B**. The Coulomb-gauge vector potential **A** appeared merely as a *notation* for the field momentum $\mathbf{G}^{(0)}$ (divided by e/c).

Appendix A. Derivation of Eq. (7)

Let **N** denote the volume integral in Eq. (5) confined to a sphere of radius R. Inserting $\mathbf{E} = -\boldsymbol{\nabla}\phi$ and $\mathbf{B} = \boldsymbol{\nabla} \times \mathbf{A}$ into it, we obtain after some rearrangements

$$N_\alpha = -\frac{1}{4\pi c}\epsilon_{\alpha\beta\gamma}\int_R d^3x\; x_\beta\left\{\nabla_\delta\phi\cdot\nabla_\gamma A_\delta - \nabla_\delta\phi\cdot\nabla_\delta A_\gamma\right\},$$

where $\epsilon_{\alpha\beta\gamma}$ is the Levi-Civita symbol.

In the second term the two ∇ symbols have identical indices. By partial integration we make them to merge into ∇^2:

$$N_\alpha = -\frac{1}{4\pi c}\epsilon_{\alpha\beta\gamma}\int_R d^3x\;\{x_\beta\nabla_\delta\phi\cdot\nabla_\gamma A_\delta - \nabla_\delta(x_\beta\cdot\nabla_\delta\phi\cdot A_\gamma) \\ + \nabla_\beta\phi\cdot A_\gamma + x_\beta\cdot\nabla^2\phi\cdot A_\gamma\}.$$

In the first term the index of the first ∇-symbol and that of the vector potential are identical. By further partial integration we form $\boldsymbol{\nabla}\cdot\mathbf{A}$ out of them:

$$N_\alpha = -\frac{1}{4\pi c}\epsilon_{\alpha\beta\gamma}\int_R d^3x\ \{\nabla_\delta(x_\beta\cdot\phi\cdot\nabla_\gamma A_\delta) - \nabla_\gamma(\phi A_\beta) + \nabla_\gamma\phi\cdot A_\beta -$$
$$-\phi\cdot x_\beta\nabla_\gamma(\boldsymbol{\nabla}\cdot\mathbf{A}) - \nabla_\delta(x_\beta\cdot\nabla_\delta\phi\cdot A_\gamma) + \nabla_\beta\phi\cdot A_\gamma + x_\beta\cdot\nabla^2\phi\cdot A_\gamma\}\ .$$

By Eq. (2) we have $\boldsymbol{\nabla}\cdot\mathbf{A} = 0$. Out of the six terms left we have three pure divergences (surface terms). Two of the three volume terms cancel each other for symmetry reason:

$$\epsilon_{\alpha\beta\gamma}(\nabla_\gamma\phi\cdot A_\beta + \nabla_\beta\phi\cdot A_\gamma) = 0.$$

If we insert $\nabla^2\phi = -4\pi\rho$ into the remaining one we obtain the r.h.s. of Eq. (3) the integration being confined to the sphere of radius R:

$$\mathbf{N}^{(0)} = +\int_R d^3x\left[\mathbf{x},\frac{\rho}{c}\mathbf{A}\right].$$

According to Eq. (4), its value is equal to

$$N_\alpha^{(0)} = -\frac{1}{6c}D_{\alpha\beta}B_\beta + \frac{1}{3c}\langle\rho r^2\rangle B_\alpha$$

for any sphere outside the body.

The three surface integrals are

$$N_\alpha^{(1)} = +\frac{1}{4\pi c}\epsilon_{\alpha\beta\gamma}\int_R d^3x\ \nabla_\gamma(\phi A_\beta)\,,$$
$$N_\alpha^{(2)} = -\frac{1}{4\pi c}\epsilon_{\alpha\beta\gamma}\int_R d^3x\ \nabla_\delta(x_\beta\phi\cdot\nabla_\gamma A_\delta)\,,$$
$$N_\alpha^{(3)} = +\frac{1}{4\pi c}\epsilon_{\alpha\beta\gamma}\int_R d^3x\ \nabla_\delta(x_\beta\cdot\nabla_\delta\phi\cdot A_\gamma)\,.$$

All of them must be of the form $(\eta_i/c)D_{\alpha\beta}B_\beta$ $(i = 1,2,3)$ and our task is to calculate the coefficients η_i.

Using Gauss' theorem,

$$\int_R d^3x\,\nabla_\alpha f = \oint_R dS_\alpha\cdot f = \oint R^2 d\Omega\,\frac{x_\alpha}{R}f\,,$$

where Ω denotes solid angle, we have

$$N_\alpha^{(1)} = +\frac{1}{c}\epsilon_{\alpha\beta\gamma}\frac{R}{4\pi}\oint d\Omega\ x_\gamma\phi A_\beta\,. \qquad \text{(A.1)}$$

The body is uncharged by hypotheses and so no Coulomb potential contributes to ϕ. The contribution of its dipole moment $\mathbf{d}$ to $\mathbf{N}^{(1)}$ is of the form *constant* $\times$ $[\mathbf{d},\,\mathbf{B}]$ but the constant must be zero because it is a polar vector, while $\mathbf{N}^{(1)}$ is an axial one. From moments higher than quadrupole no vector can be formed with $\mathbf{B}$. Hence it is only the quadrupole potential,

$$\phi = \frac{1}{2R^5} x_\alpha x_\beta D_{\alpha\beta} \,, \tag{A.2}$$

which contributes to the surface integrals.

For the vector potential we can substitute

$$A_\beta = -\frac{1}{2} \epsilon_{\beta\rho\sigma} x_\rho B_\sigma \,,$$

and we obtain

$$\begin{aligned} N_\alpha^{(1)} &= -\frac{1}{2c} \epsilon_{\alpha\beta\gamma} \epsilon_{\beta\rho\sigma} B_\sigma \frac{R}{4\pi} \int d\Omega \; x_\gamma x_\rho \phi \\ &= -\frac{R}{8\pi c} \left\{ B_\alpha R^2 \int d\Omega \; \phi + B_\gamma \int d\Omega \; x_\alpha x_\gamma \phi \right\} . \end{aligned}$$

The first integral vanishes because of the zero net charge condition. Substituting (A.2) into the second term and using the relation

$$\int d\Omega \; x_\gamma x_\mu x_\nu x_\alpha = \frac{4\pi R^4}{15} (\delta_{\gamma\mu}\delta_{\nu\alpha} + \delta_{\gamma\nu}\delta_{\mu\alpha} + \delta_{\gamma\alpha}\delta_{\mu\nu})$$

we arrive at the value $\eta_1 = 1/30$.

A completely analogous derivation leads to $\eta_2 = -1/30$. Hence, the first two surface contributions cancel each other.

The remaining surface integral is

$$N_\alpha^{(3)} = +\frac{1}{c} \epsilon_{\alpha\beta\gamma} \frac{R}{4\pi} \int d\Omega \; x_\beta x_\delta \nabla_\delta \phi \cdot A_\gamma \,.$$

Since the potential (A.2) is of degree -3 in r we have

$$x_\delta \nabla_\delta \phi = r \frac{\partial \phi}{\partial r} = -3\phi \,.$$

Hence

$$N_\alpha^{(3)} = -\frac{3}{c} \epsilon_{\alpha\beta\gamma} \frac{R}{4\pi} \int d\Omega \; x_\beta A_\gamma \,.$$

Comparing this with (A.1) we obtain $\eta_3 = +1/10$. Adding together $\mathbf{N}^{(0)}$ and the surface contributions (7) is obtained.

When the body is situated on the axis of an infinitely long straight ideal solenoid of radius R_s, then Eq. (7) remains valid for the field angular momentum within the sphere provided $R \leq R_s$. The total field angular momentum, however, is given by Eq. (4) (i.e. $\mathbf{N}^{(0)}$) because the surface integrals vanish as $R \longrightarrow \infty$. Outside the solenoid the vector potential is given by Eq. (21) rather than Eq. (2) and tends to zero as R increases. Within the solenoid, Eq. (2) remains true at arbitrarily large distances from the body, but the surface of the parts of the sphere remain finite and the integral vanishes on them for this reason.

References

1. J.D. Jackson, *Classical Electrodynamics*, 3rd ed., Wiley, New York, 1999.
2. S. Kikuchi,*Zs. f. Phys.* **66**, 558 (1930).
3. E. Fermi, *Rev. Mod. Phys.* **4**, 87 (1932).
4. J.D. Jackson, *Am. J. Phys.* **70**, 917 (2002), `arXiv:physics/0204034`.
5. N. Kroll, in *Quantum Optics and Electronics, Lectures delivered at Les Houches 1964*, edited by C. Dewitt, Gordon and Breach, New York, 1965.

THE TEMPERATURE-DEPENDENT YANG-MILLS TRACE ANOMALY AS A FUNCTION OF THE MASS GAP

V. GOGOKHIA

MTA Wigner RCP, P.O.B. 49, Budapest, Hungary, 1525
and
A. Razmadze Mathematical Inst. of I. Javakhishvili Tbilisi State University, Depart. Theor. Phys., Tamarashvili str. 6, 0177 Tbilisi, Georgia
E-mail: gogohia.vahtang@wigner.mta.hu, gogokhia@rmi.ge

A. SHURGAIA

A. Razmadze Mathematical Inst. of I. Javakhishvili Tbilisi State University, Depart. Theor. Phys., Tamarashvili str. 6, 0177 Tbilisi, Georgia,
and
European School, IB Word School, 34g A. Kazbegi Av., 0177 Tbilisi, Georgia
E-mail: avsh@rmi.ge

M. VASÚTH

MTA Wigner RCP, P.O.B. 49, Budapest, Hungary, 1525
E-mail: vasuth.matyas@wigner.mta.hu

The trace anomaly or, equivalently, the interaction measure is an important thermodynamic quantity/observable, since it is very sensitive to the non-perturbative effects in the gluon plasma. It has been calculated and its analytic and asymptotic properties have been investigated with the combined force of analytic and lattice approaches to the $SU(3)$ Yang-Mills (YM) quantum gauge theory at finite temperature. The first one is based on the effective potential approach for composite operators properly generalized to finite temperature. This makes it possible to introduce into this formalism a dependence on the mass gap Δ^2, which is responsible for the large-scale dynamical structure of the QCD ground state. The gluon plasma pressure as a function of the mass gap adjusted by this approach to the corresponding lattice data is shown to be a continuously growing function of temperature T in the whole temperature range $[0, \infty)$ with the correct Stefan-Boltzmann limit at very high temperature. The corresponding trace anomaly has a finite jump discontinuity at some characteristic temperature $T_c = 266.5\ MeV$ with latent heat $\epsilon_{LH} = 1.41$. This is a firm evidence of the first-order phase transition in $SU(3)$ pure gluon plasma. It is exponentially suppressed below T_c and has a complicated and rather different dependence on the mass gap and temperature across T_c. In the very high temperature limit its non-perturbative part has a power-type fall off.

1. Introduction

From the very beginning and up to present days, lattice QCD remains the only practical method to investigate QCD at finite temperature and density from first principles. Recently it underwent a rapid progress[1] (and references therein). However, lattice QCD is primarily aimed at obtaining well-defined calculation schemes in order to get realistic numbers for physical quantities. One may therefore get numbers and curves for various thermodynamic quantities/observables, but without understanding what is the physics behind them. Such an understanding can only come from an analytic description of the corresponding lattice data in the whole temperature range and desirably on a general dynamical basis. So the merger between lattice and analytical approaches to QCD at finite temperature and density is unavoidable, i.e., they do not exclude each other: on the contrary, they should be complementary. In other words, numbers and curves come from thermal lattice QCD, while the analytic description of the physics behind them comes from the dynamical theory, which is continuous QCD.

The effective potential approach for composite operators[2,3] turned out to be a very effective analytical and perspective dynamical tool for the generalization of QCD to non-zero temperature[4] (and references therein). In the absence of external sources it is nothing but the vacuum energy density (VED), i.e., the pressure apart from the sign. This approach is non-perturbative (NP) from the very beginning, since it deals with the expansion of the corresponding skeleton vacuum loop diagrams in powers of the Planck constant, and thus allows one to calculate the VED from first principles. The key element in this program is the extension of our paper[3] to non-zero temperature.[4] This makes it possible to introduce the temperature-dependent gluon pressure as a function of the Jaffe-Witten (JW) mass gap.[5,6] It is this which is responsible for the large-scale structure of the QCD ground state (in what follows we will call it as mass gap, for simplicity). The confining dynamics in the gluon matter (GM) is therefore nontrivially taken into account directly through the mass gap and via the temperature-dependent gluon pressure itself, but other NP effects are also present. Being NP, the effective approach for composite operators, nevertheless, makes it possible to incorporate the thermal perturbation theory (PT) expansion in a self-consistent way. In our auxiliary work[7] we have formulated and developed the analytic thermal PT which allows one to calculate the PT contributions in terms of the convergent series in integer powers of a small α_s. In this way, we have explicitly derived and numerically calculated the first PT correction of the α_s-order to the NP part of the gluon pressure investigated and calculated previously in Ref. 4.

2. The gluon pressure $P_g(T)$

Gathering all our results, obtained previously[4,7] and summarized in Ref. 6, the gluon pressure $P_g(T)$ can be written down as follows:

$$P_g(T) = P_{NP}(T) + P_{PT}^s(T) = \Delta^2 T^2 - \frac{6}{\pi^2}\Delta^2 P_1'(T) + \frac{16}{\pi^2} T M(T) + P_{PT}^s(T), \tag{1}$$

where the integrals $P_1'(T)$ and $P_{PT}^s(T)$ are

$$P_1'(T) = \int_0^{\omega_{eff}} d\omega \frac{\omega}{e^{\beta\omega} - 1}, \tag{2}$$

and

$$P_{PT}^s(T) \equiv P_{PT}^s(\Delta^2; T) = \alpha_s \times \frac{9}{2\pi^2}\Delta^2 \int_{\Lambda_{YM}}^{\infty} d\omega\, \omega^2 \, \frac{1}{\bar{\omega}} \frac{1}{e^{\beta\bar{\omega}} - 1}, \tag{3}$$

respectively, while the composition $M(T) = [P_2(T) + P_3(T) - P_4(T)]$ is defined via the following integrals

$$\begin{aligned} P_2(T) &= \int_{\omega_{eff}}^{\infty} d\omega\, \omega^2 \ln\left(1 - e^{-\beta\omega}\right), \\ P_3(T) &= \int_0^{\omega_{eff}} d\omega\, \omega^2 \ln\left(1 - e^{-\beta\omega'}\right), \\ P_4(T) &= \int_0^{\infty} d\omega\, \omega^2 \ln\left(1 - e^{-\beta\bar{\omega}}\right), \end{aligned} \tag{4}$$

where $\omega_{eff} = 1\ GeV$ and the mass gap $\Delta^2 = 0.4564\ GeV^2$ for $SU(3)$ gauge theory have been fixed in Refs. 3, 4, and this choice has been explained as well. Here ω_{eff} is a scale separating the low- and high frequency-momentum regions, while ω' and $\bar{\omega}$ are given by the relations

$$\omega' = \sqrt{\omega^2 + 3\Delta^2} = \sqrt{\omega^2 + m_{eff}'^2}, \quad m_{eff}' = \sqrt{3}\Delta = 1.17\ GeV, \tag{5}$$

and

$$\bar{\omega} = \sqrt{\omega^2 + \frac{3}{4}\Delta^2} = \sqrt{\omega^2 + \bar{m}_{eff}^2}, \quad \bar{m}_{eff} = \frac{\sqrt{3}}{2}\Delta = 0.585\ GeV, \tag{6}$$

respectively. The so-called gluon mean number,[8,9]

$$N_g \equiv N_g(\beta, \omega) = \frac{1}{e^{\beta\omega} - 1}, \quad \beta = T^{-1}, \tag{7}$$

which appears in the integrals (2)-(4), describes the distribution and correlation of massless gluons in the GM. Replacing ω by $\bar{\omega}$ and ω' we can consider the corresponding gluon mean numbers as describing the distribution and correlation of the corresponding massive gluonic excitations in the GM, see integrals $P_3(T)$ and $P_4(T)$ in Eqs. (4). They are of NP dynamical origin, since their masses are due to the mass gap Δ^2. All three different gluon mean numbers range continuously from zero to infinity.[8,9] We have the two different massless excitations, propagating in accordance with the integral (2) and the first of the integrals (4). However, they are not free, since in the PT $\Delta^2 = 0$ limit they vanish (the composition $[P_2(T) + P_3(T) - P_4(T)]$ becomes zero in this case). The gluon mean numbers are closely related to the pressure. Its exponential suppression in the $T \to 0$ limit and the polynomial structure in the $T \to \infty$ limit are determined by the corresponding asymptotics of the gluon mean numbers. The low- and high-temperature expansions for the gluon pressure (1) have been derived in Ref. 6, 7.

It is worth emphasizing that the effective scale ω_{eff} is not an independent scale parameter. From the stationary condition at zero temperature[3] and the scale-setting scheme at non-zero temperature[4] it follows that

$$\omega_{eff}^2 = (0.4564)^{-1}\Delta^2, \tag{8}$$

so it is expressed in terms of the initial fundamental scale parameter - the mass gap. Its introduction is convenient from the technical point of view in order to simplify our expressions which otherwise would be too cumbersome.

Let us note that the term (3) describes the same massive gluonic excitations $\bar{\omega}$, Eq. (6), but their propagation, however, suppressed by the α_s-order. We can consider it as a new massive excitation in the GM, denoted it as $\alpha_s \cdot \bar{\omega}$. In fact, the term $P_{PT}^s(T)$ is NP, depending on the mass gap Δ^2, which is only suppressed by the α_s order. When the interaction is formally switched off, i.e., letting $\alpha_s = \Delta^2 = 0$, the above-defined composition $M(T)$ becomes zero, as it follows from Eqs. (4), and thus the gluon pressure (1) itself. This is due to the normalization condition of the free PT vacuum to zero valid at non-zero T as well.

As mentioned above, the gluon pressure (1) has been calculated and discussed in Ref. 6, 7. It is shown in Fig. 1 and its numerical values in the form of the corresponding Table are present in Ref. 10, where one can find the explicit expansions of its asymptotics as well. It has a maximum at some "characteristic" temperature, $T_c = 266.5\ MeV$. Below T_c the gluon pressure is exponentially suppressed in the $T \to 0$ limit, in accordance with the low-temperature asymptotics of the gluon mean number (7), as noted above. Close to T_c and at moderately high temperatures up to approximately $(4-5)T_c$ the exact functional dependence on the mass gap Δ^2 and the temperature T of the gluon pressure (1) remains rather complicated. This means that the NP effects due to the mass gap are still important up to a rather high temperature. The gluon pressure $P_g(T)$ has a polynomial character in integer powers of T up to T^2 at high temperatures. So it is in agreement with the corresponding asymptotic of the gluon mean number [see Eq. (7)]. At very high temperature its expansion is as follows:

$$P_g(T) \sim B_2 \alpha_s \Delta^2 T^2 + [B_3 \Delta^3 + M^3]T, \quad T \to \infty, \tag{9}$$

up to unimportant parameters here B_2, B_3 and M^3. The term $\sim T^2$ has been first introduced and discussed in the phenomenological equation of state (EoS)[11] (see also Refs. 4, 6, 7, 10, 12–19 and references therein). On the contrary, in our approach both terms $\sim T^2$ and $\sim T$ have not been introduced by hand. They naturally appear on a general ground as a result of the explicit presence of the mass gap from the very beginning in the gluon analytical EoS (1).

Concluding, let us note that the first term $\Delta^2 T^2$ in the gluon pressure (1) plays a dominant role in the region of moderately high temperatures approximately up to $(4-5)T_c$. In the limit of very high temperatures it is exactly cancelled by the term coming from the composition $M(T)$ in Eq. (1), as it has been established in Ref. 7. In other words, the $\sim \Delta^2 T^2$ behavior of $P_g(T)$ is replaced by $\sim \alpha_s \Delta^2 T^2$ behavior at very high temperature, as it should be, in principle. It would be very surprising if a pure NP contribution survived in the limit of a very high temperature, while for its PT counterpart/correction it would be expected/possible. At the same time, the second purely NP term $\sim T$ is suppressed in comparison with the first PT term in the very high temperature limit in Eq. (9), indeed.

3. The full GP EoS

From Fig. 1 it clearly follows that the gluon pressure (1) will never reach the general Stefan-Boltzmann (SB) constant $3P_{SB}(T)/T^4 = 24\pi^2/45$ at very high temperatures. Let us remind that the very high-temperature behavior ($T \to \infty$) of all the thermodynamic quantities is governed just by the SB ideal gas limit, when the matter can be described in terms of non-interacting (i.e., free) massless particles (gluons). That is not a surprise, since the SB term has been cancelled in the gluon pressure from the very beginning due to the normalization condition of the free PT vacuum

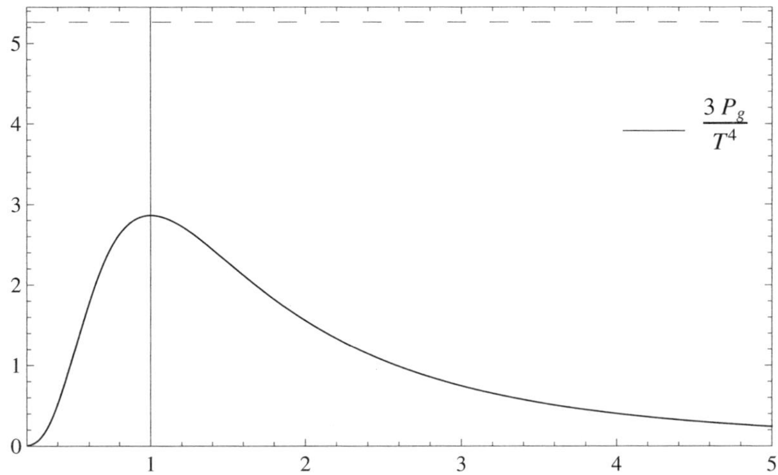

Fig. 1. The gluon pressure (1) scaled (i.e., divided) by $T^4/3$ is shown as a function of T/T_c (solid curve). It has a maximum at $T_c = 266.5\ MeV$ (vertical solid line). The horizontal dashed line is the general SB constant $3P_{SB}(T)/T^4 = 24\pi^2/45$.

to zero.[4,7] Analytically this cancellation/subtraction at high temperatures (above T_c) has been shown in Ref. 7, where it has also been shown that the massless (but not free) gluons may be present at low temperatures (below T_c) in the GM. However, their propagation in this region cannot be described by the SB term itself. All this means that the SB pressure has been already subtracted from the gluon pressure, but in a very specific way, i.e., the above-mentioned normalization condition is not simply the subtraction of SB term. The gluon pressure (1) may change its continuously falling off the regime above T_c only in the near neighborhood of T_c in order for its full counterpart to reach the corresponding SB limit at high temperatures. The SB term is valid only at high temperatures, nevertheless, it cannot be added to Eq. (1) above T_c, even multiplied by the corresponding $\Theta((T/T_c) - 1)$-function. The problem is that in this case the pressure will get a jump at $T = T_c$, which is not acceptable. The full pressure is always a continuous growing function of temperature at any point of its domain. This means that we should add some other terms valid below T_c in order to restore a continuous character of the full pressure across T_c. This can be achieved by imposing a special continuity condition on these terms valid just at T_c. Moreover, the gluon pressure $P_g(T)$ itself should be additionally multiplied by the functions which are always negative below and above T_c.

This guarantees the positivity of the full pressure below T_c, while above T_c it guarantees the approach of the full pressure to the SB limit in the AF way, i.e., slowly and from below. These terms will also contribute to the condition of continuity for the full pressure. All these problems make the inclusion of the SB term into EoS highly non-trivial. The most general way how this can be done is to add to the both sides of Eq. (1) the term $[\Theta((T/T_c)-1)H(T)+\Theta((T_c/T-1))L(T)]$, valid in the whole temperature range, and the auxiliary functions $H(T)$ and $L(T)$ are to be expressed in terms of $P_{SB}(T)$ and $P_g(T)$ (see below).

The previous Eq. (1) then becomes

$$P_{GP}(T) = P_g(T) + \Theta\left(\frac{T_c}{T} - 1\right) L(T) + \Theta\left(\frac{T}{T_c} - 1\right) H(T), \tag{10}$$

and its left-hand side here and below is denoted as $P_{GP}(T)$ (the above-mentioned full counterpart). The gluon plasma (GP) pressure (10) is continuous at T_c if and only if

$$L(T_c) = H(T_c), \tag{11}$$

which can be easily checked. Due to the continuity condition (11), the dependence on the corresponding Θ-functions disappears at T_c, and the GP pressure (10) remains continuous at any point of its domain. The role of the auxiliary function $L(T)$ is to change the behavior of $P_{GP}(T)$ from $P_g(T)$ at low (L) temperatures below T_c, especially in its near neighborhood, as well as to take into account the suppression of the SB-type terms below T_c. The auxiliary function $H(T)$ is aimed to change the behavior of $P_{GP}(T)$ from $P_g(T)$, as well as to introduce the SB term itself and its modification due to AF at high (H) temperatures above and near T_c. These changes are necessary, since in the gluon pressure $P_g(T)$ the SB term is missing (as described above), and it cannot be restored in a trivial way. So the appearance of the corresponding Θ-functions in the GP pressure (10) is inevitable together with the functions $H(T)$ and $L(T)$, playing only an auxiliary role but still useful from the technical point of view.

The gluon pressure $P_g(T)$ (1), which fixes the value of the characteristic temperature $T_c = 266.5\ MeV$, is a necessary analytical and dynamical input information for the GP pressure (10). On the other way around, the lattice pressure[14] is a main numerical input information to use in order to fix the functions $L(T)$ and $H(T)$ in Eq. (10). Our general method[6,10] how to merge analytical and lattice approaches in order to understand the physics of YM quantum gauge theories at non-zero temperature is described in some general terms below. We proposed and developed a method of analytical simulations which allows one to express such introduced auxiliary functions $L(T)$ and $H(T)$ in terms of known basic functions $P_g(T)$ and $P_{SB}(T)$, multiplied by the so-called simulating functions $\phi_l(T)$, $\phi_h(T)$ and $f_l(T)$, $f_h(T)$, respectively. They are to be necessarily represented by the corresponding asymptotics of the gluon mean number (7) in the low and high temperature limits. This makes it possible to reproduce lattice data in any requested temperature interval and to ensure the correct SB limit for all the thermodynamic obsrvables/quantities as well.

Then we have performed the numerical simulation of the GP pressure (1) above T_c in order to fix the function $\phi_h(T)$ in accordance with the lattice pressure[14] in this region by using the Least Mean Square (LMS) method.[20] Our procedure makes it also possible to continue the lattice pressure to the region of very high temperatures. As a next step, we have performed the numerical simulation of the GP pressure (1) below T_c in order to fix the three free fitting parameters, which necessarily appear in the simulating function $f_l(T)$. This has been done in accordance with lattice data[14] in this region, but only very close to T_c. Our procedure makes it also possible to continue the calculation of the GP pressure to very low temperatures, where convincing lattice data does not exists at all. Thus, we can predict the behaviour of the lattice pressure curve up to zero temperature, knowing only its behaviour very close to T_c.

"Sewing" together such obtained two parts with the help of the relation (11), we are coming to the analytical expression reproducing the lattice pressure[14] in the whole temperature range $[0, \infty)$ as a function of the mass gap Δ^2 as follows:

$$\begin{aligned} P_{GP}(T) &= P_g(T) \\ &+ \Theta\left(\frac{T_c}{T} - 1\right)\Big[(0.015732 e^{-\mu_1((T_c/T)-1)} \\ &\quad +0.003884 e^{-\mu_2((T_c/T)-1)})P_{SB}(T) - e^{-\mu(T_c/T)}P_g(T)\Big] \\ &+ \Theta\left(\frac{T}{T_c} - 1\right)\Big[(1-\alpha_s(T))P_{SB}(T) - \phi_h(T)P_g(T)\Big], \end{aligned} \tag{12}$$

where

$$\mu = 0.0001, \quad \mu_1 = 39.1, \; \mu_2 = 3.4, \tag{13}$$

while

$$\alpha_s(T) = \left(0.22037\frac{1}{t} - 0.033\frac{\ln t}{t^2}\right), \tag{14}$$

$$t = 1 + 0.1929\ln(T/T_c), \quad T \geq T_c = 266.5\ MeV, \tag{15}$$

and

$$\phi_h(T) = 1.55 + 0.8482\left(\frac{T_c}{T}\right)^3, \quad \phi_h(T_c) = 2.3982. \tag{16}$$

The GP pressure (12) is completely known now, since the SB pressure is $P_{SB}(T) = (8\pi^2/45)T^4$ and the gluon pressure $P_g(T)$ is also exactly known, Eq. (1). For the numerical evaluation of the GP pressure (12) in detail see Ref. 10. For simplicity, in what follows we will omit the subscript "GP" in the GP pressure (12), i.e., we will put $P_{GP}(T) \equiv P(T)$. The same will be done in the notation for the trace anomaly relation below. According to such obtained analytical expression, the corresponding lattice pressure is exponentially suppressed at low temperatures, it is continuous across T_c and approaches its SB limit at high temperatures, i.e., satisfying thus all thermodynamics limits. In other words, the GP pressure (12) is, in fact, the lattice pressure[14] analytically expressed as a function of the mass gap and temperature and properly continued to the regions of very low and high temperatures, see Fig. 2.

3.1. *Trace anomaly relation*

A thermodynamic quantity of special interest is the thermal expectation value of the trace of the energy momentum tensor. Equivalently, it is known as the interaction measure and defined as follows:

$$I(T) = \epsilon(T) - 3P(T), \tag{17}$$

where $\epsilon(T)$ is the energy density, which in its turn defined as $\epsilon(T) = T(\partial P(T)/\partial T) - P(T)$. So knowing the GP pressure (12), one can calculate any other thermodynamic quantity/observable, see our works.[6,10] The importance of this thermodynamic observable is that it is very sensitive to the NP effects, since the corresponding pure PT contributions are exactly cancelled in the composition (17). This can be explicitly shown by using the GP pressure (12) above T_c, and the relation $3P_{SB}(T) = \epsilon_{SB}(T)$, see below. Properly scaled it is shown in Fig. 3. The rapid rise of the peak (due to the latent heat (LH) in the energy density) is exactly placed at T_c, and it is about 2.5. In all lattice calculations it peaks at about $1.1\,T_c$

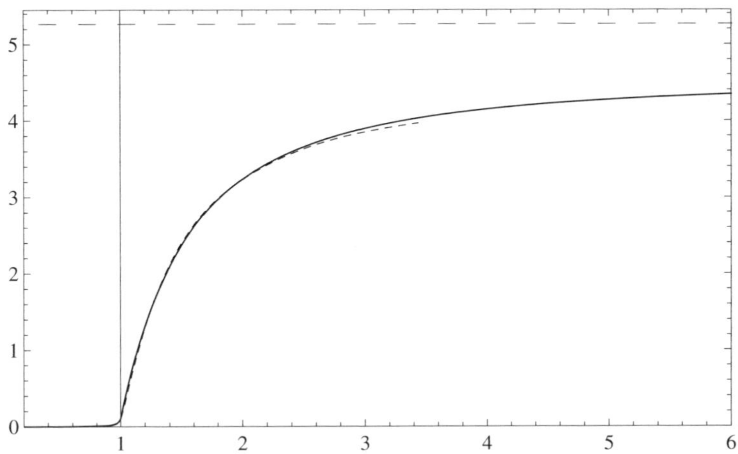

Fig. 2. The GP pressure (12) is shown as a function of $a = T/T_c$ (solid curve). The lattice curve[14] for $SU(3)$ pressure is also shown (dashed curve). The horizontal dashed line is the general SB constant. Both pressures are scaled in the same way, i.e., divided by $T^4/3$.

(see Ref. 14, 15, 17, 18, 21), and it is about 2.6, and almost coincides with our value in Ref. 18. The wrong position of the lattice trace anomaly peak can be due to an ultraviolet cutoff, the finite volume effects, etc. In this connection let us indeed remind that in lattice simulations at any temperature it is necessary finally to go to the continuum (physical) limit, namely lattice spacing goes to zero and then the infinite volume limit should be taken. These are nothing but the removal of the ultraviolet and infrared cutoffs which is the part of the renormalization procedure.[22,23] It seems to us that our analytical method resolves this $SU(3)$ lattice thermodynamics artefact.

Just above T_c and up to rather high temperatures $(4-5)T_c$ the NP effects due to the mass gap are still important in the trace anomaly. Fig. 3 demonstrates rather complicated dependence of the trace anomaly on the mass gap and temperature in this interval. The trace anomaly equation (17), divided by T^4 is

$$\frac{I(T)}{T^4} = -\frac{1}{3}T\alpha_s'(T)(SB) - 0.55\frac{[TP_g'(T) - 4P_g(T)]}{T^4}, \quad T > T_c, \qquad (18)$$

where $(SB) = 24\pi^2/45 \approx 5.2638$ is the above-mentioned SB general constant/limit. So, indeed the main contribution to the trace anomaly comes from the second NP term in Eq. (18), and it is not a simple power-type fall off. It is mainly due to the complicated dependence of the gluon pressure $P_g(T)$ on the mass gap and the temperature in this region, where it cannot be approximated by some simple power-type expression. However, this is possible to do in the limit of very high temperatures, approximately above $(4-5)T_c$. Substituting the asymptotic relation (9) and its derivative into the previous equation and doing some algebra, one obtains

$$\frac{I(T)}{T^4} \sim -\frac{1}{3}T\alpha_s'(T)(SB) + 1.1\bar{B}_2\alpha_s\left(\frac{T_c}{T}\right)^2 + 1.65\bar{B}_3\left(\frac{T_c}{T}\right)^3, \quad T \to \infty, \tag{19}$$

while $\bar{B}_2$ and $\bar{B}_3$ are unimportant here constants.

Concluding, let us briefly discuss the size of the discontinuity in the energy density, the above-mentioned LH. It is

$$\epsilon_{LH} = 1.41 \tag{20}$$

in dimensionless units (for its definition and analytical/numerical evaluation, respectively, see our work[10]). It is worth emphasizing that the same value (20) comes from the independent calculations of the energy density and the trace anomaly, as it should be, since the pressure itself is a continuous function across T_c, i.e., $\epsilon_{LH} = \Delta(\epsilon - 3P)/T_c^4 = \Delta\epsilon/T_c^4$ (here, obviously, Δ is not the mass gap). This means that the first-order phase transition in the GP is analytically confirmed for the first time, in complete agreement with thermal $SU(3)$ YM lattice simulations[14,17,18,24] (and references therein). The reason of such sharp changes at T_c in the derivatives of the GP pressure is that its exponential rise below T_c is changing to the polynomial rise above T_c in order to reach finally the SB limit. The value (20) is in fair agreement with lattice ones in Refs. 14, 18, 21, 25–27 (and references therein). This agreement is not a trivial thing, since we have adjusted our analytical numerical simulations with those of lattice ones in Ref. 14 only for the pressure. First of all, the energy density being derivative of the pressure, nevertheless, is an independent thermodynamic observable, having a discontinuity at T_c, while the pressure is a continuous function across T_c. Secondly, the lattice results heavily depend on how the continuum limit is to be taken and on other details of the above-cited lattice simulations. For example, the lattice data points closest to T_c for the entropy density may still be affected by an upward finite-volume effect.[27]

V.G.'s recollections of, and encouraging discussions with the late Prof. V.N. Gribov will be never forgotten. We especially thank M. Panero for providing us with the lattice data from his paper.[14] The authors are

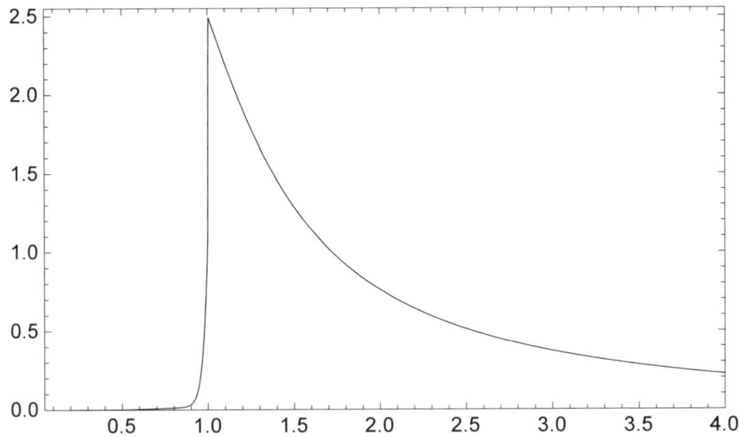

Fig. 3. The trace anomaly (17) divided by T^4 is shown as a function of T/T_c.

also grateful to P. Forgács, J. Nyiri, T. Biró and G.G. Barnaföldi for useful discussions, remarks and help. Support from MTA Wigner Research Centre for Physics and P. Lévai is acknowledged. Partial support comes from "NewCompStar", COST Action MP1304 and OTKA grant No. NK1061019. M.V. was also supported by the János Bolyai Research Scholarship of the HAS.

References

1. *QUARK MATTER 2014*, Proc. of the XXIV Inter. Conf. on Ultra-Relativistic Nucleus-Nucleus Collisions, edited by P. Braun-Munziger, B. Friman, J. Stachel, 19-24, May, 2014, Darmstadt, Germany; *Nucl. Phys.* **A931**, 1-1266 (2014).
2. J.M. Cornwall, R. Jackiw, and E. Tomboulis, *Phys. Rev.* **D10**, 2428 (1974).
3. G.G. Barnaföldi and V. Gogokhia, *J. Phys. G: Nucl. Part. Phys.* **37**, 025003 (2010) arXiv:0708.0163 [hep-th].
4. V. Gogokhia and M. Vasúth, *J. Phys. G: Nucl. Part. Phys.* **37**, 075015 (2010); arXiv:0902.3901 [hep-th].
5. A. Jaffe and E. Witten, *Yang-Mills Existence and Mass Gap*, http://www.claymath.org/prize-problems/, http://www.arthurjaffe.com
6. V. Gogokhia and G.G. Barnafoldi, *The Mass Gap and its Applications*, World Scientific, Singapore, 2013.

7. V. Gogokhia and M. Vasúth, arXiv:1012.4157 [hep-th].
8. J.I. Kapusta and C. Gale, *Finite-Temperature Field Theory*, Cambridge University Press, 2006.
9. J. Letessier and J. Rafelski, *Hadrons and Quark-Gluon Plasma*, Cambridge University Press, 2004.
10. V. Gogokhia, A. Shurgaia, and M. Vasúth, arXiv:1014.3375 [hep-th].
11. P.N. Meisinger, T.R. Miller, and M.C. Ogilvie, *Phys. Rev.* **D65**, 034009 (2002).
12. R.D. Pisarski, *Prog. Theor. Phys. Suppl.* **168**, 276 (2007) arXiv:hep-ph/0612191.
13. E. Shuryak, *Prog. Part. Nucl. Phys.* **62**, 48 (2009) arXiv:0807.3033 [hep-ph].
14. M. Panero, *Phys. Rev. Lett.* **103**, 232001 (2009) arXiv:0907.3719 [hep-lat].
15. E. Megías, E.R. Arriola, and L.L. Salcedo, *Phys. Rev.* **D80**, 056005 (2009).
16. T.S. Biró and J. Cleymans, *Phys. Rev.* **C78**, 034902 (2008) arXiv:hep-ph/9710463.
17. G. Boyd, J. Engels, F. Karsch, E. Laermann, C. Legeland, M. Lutgemeier, and B. Petersson, *Nucl. Phys.* **B469**, 419 (1996) arXiv:hep-lat/9602007.
18. Sz. Borsanyi, G. Endrodi, Z. Fodor, S.D. Katz, and K.K. Szabo, *JHEP* **07**, 056 (2012) arXiv:1204.6184 [hep-lat].
19. O. Andreev, *Phys. Rev.* **D76**, 087702 (2007) arXiv:0706.3120 [hep-ph].
20. I.N. Bronshtein, K.A. Semendyayev, G. Musiol, and H. Muehlig, *Handbook of Mathematics*, Springer, 2004.
21. S. Datta and S. Gupta, *Phys. Rev.* **D82**, 114505 (2010); arXiv:1006.0938 [hep-lat].
22. A.S. Kronfeld, arXiv:hep-ph/0209231, Jour. ref.: eConfCO20620:FRBT05, 2002.
23. A.S. Kronfeld, *Ann. Rev. Nucl. Part. Sci.* **62**, 265 (2012), arXiv:1203.1204 [hep-lat].
24. M. Fukugita, M. Okawa, and A. Ukawa, *Phys. Rev. Lett.* **63**, 1768 (1989).
25. B. Lucini, M. Teper, and U. Wenger, *J. High Enery Phys.* **02**, 003 (2005).
26. B. Beinlich, F. Karsch, and A. Peikert, *Phys. Lett.* **B390**, 268 (1997).
27. H.B. Meyer, *Phys. Rev.* **D80**, 051502(R) (2009); arXiv:0905.4229 [hep-lat].

GENERAL EXACT SOLUTIONS FOR THE FULL GLUON PROPAGATOR IN QCD WITH THE MASS GAP

V. GOGOKHIA

MTA Wigner RCP, P.O.B. 49, Budapest, Hungary, 1525
and
A. Razmadze Mathematical Inst. of I. Javakhishvili Tbilisi State University, Tamarashvili str. 6, 0177 Tbilisi, Georgia
E-mail: gogohia.vahtang@wigner.mta.hu, gogokhia@rmi.ge

G.G. BARNAFÖLDI

MTA Wigner RCP, P.O.B. 49, Budapest, Hungary, 1525
E-mail: barnafoldi.gergely@wigner.mta.hu

We have explicitly shown that Quantum Chromodynamics is a color gauge invariant theory with non-zero mass gap, which has been defined as the value of the regularized full gluon self-energy at a finite scale point. The mass gap itself is mainly generated by the nonlinear interaction of massless gluon modes. All this allows one to establish the structure of the full gluon propagator in the explicit presence of the mass gap. In this case, the two independent general types of formal solutions for the full gluon propagator as a function of the regularized mass gap have been found: (i) The nonlinear iteration solution at which the gluons remain massless is explicitly present. (ii) Existence of the solution with an effective gluon mass is also demonstrated.

Keywords: Quantum Chromodynamics (QCD), gauge theory, mass gap

1. Introduction

The Quantum Chromodynamics (QCD)[1,2] is widely accepted as a realistic quantum field gauge theory of the strong interactions not only at the fundamental (microscopic) quark-gluon level but at the hadronic (macroscopic) level as well. However, Lagrangian of QCD does not contain explicitly any of the mass scale parameters which could have a physical meaning even after the corresponding renormalization program is performed. This means that the only place where it can appear is the QCD ground state. So it is important to know the true dynamical structure of the QCD ground state especially at large distances, which may be source of the above-mentioned mass scale parameter – the so called mass gap Δ as in Refs. 3, 4. The propagation of gluons is one of the main dynamical effects in the QCD vacuum and it is described by the corresponding Schwinger – Dyson (SD) equation of motion for the full gluon propagator.[1] It should be complemented by the corresponding Slavnov – Taylor (ST) identity. The importance of this equation is due to the fact that its solutions reflect the quantum-dynamical structure of the QCD ground state. It is a strongly nonlinear (NL) equation, and therefore the number of independent solutions, which should be considered on equal footing, is not fixed *a priori.* It is well known that any deviation of the full gluon propagator from the free one requires the existence of the mass gap in order to describe both short- and large-scale structures of QCD ground state, the *true vacuum.* If indeed in the QCD vacuum such a mass gap exists then any massive excitation of it has to satisfy the condition $q^2 \geq \Delta^2$, where q^2 is the squared gluon momentum. This means that, in general, the full gluon propagator should contain the two types of compositions (Δ^2/q^2) at different powers and $\sim (q^2 - m^2_{eff})^{-1}$, where the gluon effective mass somehow should depend on the mass gap. So two general types of solutions are to be expected: massless singular and massive regular.

The color gauge structure of this equation has been investigated in detail in Ref. 5. We have explicitly shown that the color gauge invariance of QCD is consistent with the mass gap, generated in the gluon sector of the QCD. Our primary goal in this investigation is to find formal exact solutions for the full gluon propagator as a function of the mass gap, in addition, present a recipe for this method. However, for the sake of completeness and further clarity, it is instructive to describe briefly the main results in the subsequent section based on Ref. 5.

2. The color gauge invariance of QCD at non-zero mass gap

Quantum Chromodynamics is a $SU(3)$ color gauge invariant theory. As underlined above, its dynamical context is determined by the corresponding equations of motion, among which the Schwinger – Dyson equation for the full gluon propagator plays an important role. This can be written as:

$$D_{\mu\nu}(q) = D^0_{\mu\nu}(q) + D^0_{\mu\rho}(q) i\Pi_{\rho\sigma}(q; D) D_{\sigma\nu}(q) \quad , \tag{1}$$

where

$$D^0_{\mu\nu}(q) = i \left\{ T_{\mu\nu}(q) + \xi L_{\mu\nu}(q) \right\} \frac{1}{q^2} \tag{2}$$

is the free gluon propagator, and ξ is the gauge-fixing parameter. Also, here and everywhere below $T_{\mu\nu}(q) := \delta_{\mu\nu} - (q_\mu q_\nu/q^2) = \delta_{\mu\nu} - L_{\mu\nu}(q)$, as usual. The $\Pi_{\rho\sigma}(q; D)$ is the full gluon self-energy which depends on the full gluon propagator due to the non-Abelian character of QCD. Thus the gluon Schwinger – Dyson equation is strongly non-linear. Evidently, we omit the color group indices, since for the gluon propagator and hence for its self-energy they factorize, such as here: $D^{ab}_{\mu\nu}(q) = D_{\mu\nu}(q)\delta^{ab}$. Diagrammatic representation of the gluon Schwinger – Dyson equation (1) is shown in Ref. 5, as well as the detailed description of the full gluon self-energy $\Pi_{\rho\sigma}(q; D)$. It is the sum of a few terms which are tensors, having the dimensions of mass squared.

All these skeleton loop integrals, which represent the above-mentioned tensor terms, are therefore quadratically divergent in perturbation theory (PT), and so they are assumed to be regularized, as discussed below. Let us note in advance that here and below the signature is Euclidean, which implies $q_i \to 0$ when $q^2 \to 0$ and *vice-versa*.

2.1. *The mass gap*

Let us introduce the general mass scale parameter $\Delta^2(D)$, having the dimensions of mass squared, by the subtraction from the full gluon self-energy its value at $q = 0$. Thus, one obtains

$$\Pi^s_{\rho\sigma}(q; D) = \Pi_{\rho\sigma}(q; D) - \Pi_{\rho\sigma}(0; D) = \Pi_{\rho\sigma}(q; D) - \delta_{\rho\sigma}\Delta^2(D), \qquad (3)$$

which is the definition of the subtracted full gluon self-energy $\Pi^s_{\rho\sigma}(q; D)$. Contrary to Quantum Electrodynamics (QED), QCD being a non-Abelian gauge theory can suffer from infrared (IR) singularities in the $q^2 \to 0$ limit due to the self-interaction of massless gluon modes. Thus the initial subtraction at zero in the definition (3) should be made carefully.[1] That is why in all the quantities below the dependence on the finite (slightly different from zero) dimensionless subtraction point α is to be understood. From a technical point of view, however, it is convenient to set formally $\alpha = 0$ in all the derivations below, and to restore the explicit dependence on non-zero α in all the quantities only at the final stage. In parallel, in all the quantities where the dependence on the dimensionless ultraviolet (UV) regulating parameter, λ and α is not shown explicitly, nevertheless, it should be assumed. For example, $\Delta^2(D) \equiv \Delta^2(\lambda, \alpha; D)$ and similarly for all other quantities. So all the expressions are regularized. For our purpose, in principle, it is not important how λ and α have been introduced. They can be removed at the final stage only as a result of the renormalization program.

By the mass gap we understand some fixed mass squared which is related to $\Delta^2(D)$ as follows:

$$\Delta^2(D) = \Delta^2 \times c(D), \qquad (4)$$

where the dimensionless constant $c(D)$ depends on D, while the fixed mass squared, Δ^2 does not depend on D. This will be called the *mass gap* and constant $c(D)$ parameters may depend on all other dimensionless parameters of the theory, namely $c(D) \equiv c(\lambda, \alpha, \xi, g^2)$, where g^2 is the coupling constant squared.[a] From the subtraction (3) it follows that the mass gap Δ^2, having the dimensions of mass squared, is dynamically generated within the QCD gluon sector. It is defined as the value of the full gluon self-energy at some finite point as we explained in the discussion above. It is mainly due to the nonlinear interaction of massless gluon modes. Let us remind that no truncations/approximations/assumptions/, as well as no special gauge choice are made for the regularized skeleton loop integrals, contributing to the full gluon self-energy.[5] Furthermore, one can assign a physical meaning to $\Delta^2(D)$ only after the separation from the unphysical parameter $c(D)$.

2.2. *The transversality of the full gluon self-energy*

Let us continue with the general decompositions of the full gluon self-energy and its subtracted counterpart, which enter the subtraction (3), as follows:

$$\begin{aligned} \Pi_{\rho\sigma}(q; D) &= T_{\rho\sigma}(q)\, q^2 \Pi(q^2; D) + q_\rho q_\sigma \tilde{\Pi}(q^2; D), \\ \Pi^s_{\rho\sigma}(q; D) &= T_{\rho\sigma}(q)\, q^2 \Pi^s(q^2; D) + q_\rho q_\sigma \tilde{\Pi}^s(q^2; D), \end{aligned} \tag{5}$$

where all the invariant functions of q^2 are dimensionless ones, while in addition the invariant functions $\Pi^s(q^2; D)$ and $\tilde{\Pi}^s(q^2; D)$ cannot have pole-type singularities in the $q^2 \to 0$ limit, since $\Pi^s_{\rho\sigma}(0; D) = 0$ by definition; otherwise they remain arbitrary.

Contracting them with q_ρ along with the subtraction (3), one obtains

$$\tilde{\Pi}(q^2; D) = \tilde{\Pi}^s(q^2; D) + \frac{\Delta^2(D)}{q^2}, \tag{6}$$

and

$$\Pi(q^2; D) = \Pi^s(q^2; D) + \frac{\Delta^2(D)}{q^2}. \tag{7}$$

It is worth emphasizing that the full gluon self-energy has a massless single particle singularity due to the non-zero mass gap $\Delta^2(D)$, which is of non-perturbative origin. At the same time, its subtracted counterpart cannot have such a singularity, as mentioned above. In other words, this

[a] Note, for simplicity, in this Section we will not distinguish between $\Delta^2(D)$ and Δ^2, calling both the mass gap.

means that in the explicit presence of the mass gap both invariant functions of the full gluon self-energy gain additional contributions due to it (of course, not only at some finite subtraction point $q^2 = \mu^2 \neq 0$). If the mass gap is welcome in the transversal invariant function $\Pi(q^2; D)$, it is not welcome in its longitudinal counterpart $\tilde{\Pi}(q^2; D)$, since just it violates the Slavnov – Taylor identity in Section 2.3. Let us note that it is the initial subtraction (3) which leaves the subtracted gluon-self energy logarithmical divergent only for any D and hence the invariant function $\Pi^s(q^2; D)$ is free of the quadratic divergences.

2.3. *Slavnov – Taylor identity for the full gluon propagator*

In order to calculate the physical observables in QCD from first principles, we need the full gluon propagator rather than the full gluon self-energy. The basic relation to which the full gluon propagator should satisfy is the corresponding Slavnov – Taylor identity

$$q_\mu q_\nu D_{\mu\nu}(q) = i\xi. \tag{8}$$

It is a consequence of the color gauge invariance/symmetry of QCD, and therefore *"is an exact constraint on any solution to QCD"*,[1] which is valid for any other Slavnov – Taylor identities as well. Being a result of this exact symmetry, it is the general one and it is important for the renormalization of QCD. If some equation, relation or the regularization scheme, etc. do not satisfy it automatically, i.e., without any additional conditions, then they should be modified and not this identity. In other words, all the relations, equations, regularization schemes, etc. should be adjusted to it and not *vice versa*. It implies that the general tensor decomposition of the full gluon propagator is

$$D_{\mu\nu}(q) = i\left\{T_{\mu\nu}(q)\, d\left(q^2\right) + \xi L_{\mu\nu}(q)\right\}\frac{1}{q^2}, \tag{9}$$

where the invariant function, $d(q^2)$ is the corresponding Lorentz structure of the full gluon propagator. This is usually called simply as the *full effective running charge* as well. Let us emphasize once more that these basic relations are to be satisfied in any case, for example, whether the mass gap or any other mass scale parameter is set formally zero or not.

On account of the exact relations (5), (6) and (7), the initial gluon Schwinger – Dyson equation (1) can be equivalently re-written as follows:

$$\begin{aligned} D_{\mu\nu}(q) = D^0_{\mu\nu}(q) + D^0_{\mu\rho}(q)\, iT_{\rho\sigma}(q)\left[q^2\Pi^s\left(q^2; D\right) + \Delta^2(D)\right] D_{\sigma\nu}(q) + \\ + D^0_{\mu\rho}(q)\, iL_{\rho\sigma}(q)\, q^2\tilde{\Pi}\left(q^2; D\right) D_{\sigma\nu}(q)\,. \end{aligned} \tag{10}$$

Contracting this equation with $q_\mu q_\nu$, one arrives at $q_\mu q_\nu D_{\mu\nu}(q) = i\xi - i\xi^2 \tilde{\Pi}\left(q^2; D\right)$, so the Slavnov – Taylor identity (8) is not automatically satisfied. In order to get from this relation the Slavnov – Taylor identity, one needs to put $\tilde{\Pi}\left(q^2; D\right) = 0$, which is equivalent to $\tilde{\Pi}^s\left(q^2; D\right) = -\left(\Delta^2(D)/q^2\right)$, as it follows from the relation (6). This, however, is impossible since $\tilde{\Pi}^s(q^2; D)$ cannot have the power-type singularities at small q^2, as underlined above. The only solution to the previous relation is to disregard $\Delta^2(D)$ from the very beginning, i.e., put formally zero $\Delta^2(D) = 0$ everywhere. In this case from all the relations it follows that the gluon full self-energy coincides with its subtracted counterpart, and both quantities become purely transversal, i.e., $\Pi\left(q^2; D\right) = \Pi^s\left(q^2; D\right)$ and $\tilde{\Pi}\left(q^2; D\right) = \tilde{\Pi}^s\left(q^2; D\right) = 0$. See Eqs. (5)-(7).

One way to satisfy the Slavnov – Taylor identity and thus to maintain the color gauge structure of QCD is to discard the mass gap $\Delta^2(D)$ from the very beginning, i.e., put it formally zero $\Delta^2(D) = 0$ in all the equations, relations, etc. In this limit the initial gluon Schwinger – Dyson equation (10) is modified to

$$D^{PT}_{\mu\nu}(q) = D^0_{\mu\nu}(q) + D^0_{\mu\rho}(q)\, iT_{\rho\sigma}(q)\, q^2 \Pi^s\left(q^2; D^{PT}\right) D^{PT}_{\sigma\nu}(q), \tag{11}$$

and the corresponding Lorentz structure which appears in Eq. (9) becomes

$$d^{PT}\left(q^2\right) = \frac{1}{1 + \Pi^s\left(q^2; D^{PT}\right)}. \tag{12}$$

It is easy to see that the gluon SD equation (11) automatically satisfies the ST identity (8) now. Evidently, in the formal $\Delta^2(D) = 0$ limit we denote $D_{\mu\nu}(q)$ and $d(q^2)$ as $D^{PT}_{\mu\nu}(q)$ and $d^{PT}(q^2)$, respectively. As it has been pointed out in Ref. 5, in this case there will be no problems for ghosts to accomplish their role, namely, to cancel the longitudinal component in the perturbative gluon propagator (11).

2.4. *The general structure of the full gluon propagator*

The formal $\Delta^2(D) = 0$ limit is a real way how to preserve the color gauge invariance in QCD. Then a natural question arises why does the mass gap $\Delta^2(D)$ exist in this theory at all? There is no doubt that the color gauge invariance of QCD should be maintained at non-zero mass gap as well, since it is explicitly present in the full gluon self-energy, and hence in the full gluon propagator – the so-called tadpole term Ref. 5. However, by keeping it 'alive', a problem arises: how to replace the original gluon Schwinger – Dyson equation (10), since it is not consistent with the Slavnov – Taylor identity unless the mass gap is discarded from the very beginning.

By introducing the spurious technics we were able to show that the Slavnov – Taylor identity (8) can be automatically satisfied at non-zero mass gap $\Delta^2(D)$ as well. In other words, our aim is to save the mass gap in the transversal invariant function (7), while removing it from the longitudinal invariant function (6), but without going formally to the PT $\Delta^2 = 0$ limit. In order to keep the mass gap in the framework and – at the same time – to satisfy the Slavnov – Taylor identity (8), we introduced a temporary dependence on $\Delta^2(D)$ in the free gluon propagator, thus making it an auxiliary (spurious) free gluon propagator. Substituting it into the initial gluon Schwinger – Dyson equation (10) and restoring again the dependence on the free gluon propagator, such obtained gluon Schwinger – Dyson equation should satisfy the Slavnov – Taylor identity (8), from which:

$$D_{\mu\nu}(q) = D^0_{\mu\nu}(q) + D^0_{\mu\rho}(q)\, iT_{\rho\sigma}(q)\left[q^2\Pi^s\left(q^2; D\right) + \Delta^2(D)\right] D_{\sigma\nu}(q)\,. \tag{13}$$

The modified gluon Schwinger – Dyson equation (13) is satisfied by the same term of the Lorentz structure, $d(q^2)$ in Eq. (9) as the original gluon SD equation (10),

$$d(q^2) = \frac{1}{1 + \Pi^s\left(q^2; D\right) + \left(\Delta^2(D)/q^2\right)}, \tag{14}$$

which is not surprising, since the original gluon SD equation (10) and its modified version (13) differ from each other only by the longitudinal (unphysical) part. In other words, the spurious method makes it possible to exploit the color gauge invariance at the level of the dynamical equation of motion for the gauge propagator. Because of this invariance the gluon SD equation (10) is defined up to its longitudinal part, while its transversal ("physical") part cannot be affected.

Thus, *the important observation is that now it is not required to put the mass gap $\Delta^2(D)$ formally zero everywhere.* The spurious mechanism does not affect the dynamical context of the original gluon Schwinger – Dyson equation. It makes it possible to retain the mass gap in the transversal part of the gluon Schwinger – Dyson equation, and at the same time, to cancel the term in its longitudinal part, which violates the Slavnov – Taylor identity. In this way, the modified gluon SD equation (13) satisfies automatically the Slavnov – Taylor identity (8).

Due to asymptotic freedom (AF) in QCD the perturbative regime is realized at $q^2 \to \infty$. In this limit all the Green's functions are possible to approximate by their free perturbative counterparts (up to the corresponding perturbative logarithms). However, from the relation (14) it follows that in this limit the mass gap term contribution $\Delta^2(D)/q^2$ is only next-to-next-to-leading order one. The leading order contribution is the subtracted gluon self-energy $\Pi^s(q^2;D)$, which behaves like $\ln q^2$ in this limit, as mentioned above. The constant 1 is the next-to-leading order term in the $q^2 \to \infty$ limit. Such a special structure of the relation (14), namely that the mass gap enters it through the combination $\Delta^2(D)/q^2$ in its denominator only, explains immediately why the mass gap $\Delta^2(D)$ is not important in perturbative. From this structure it follows that the perturbative regime at $q^2 \to \infty$ is effectively equivalent to the formal $\Delta^2(D) = 0$ limit and *vice versa*. That is the reason why this limit can be called the perturbative limit. And that is why we denote $D_{\mu\nu}\left(q;\Delta^2=0\right) = D_{\mu\nu}\left(q;0\right) \equiv D^{PT}_{\mu\nu}(q)$, and hence $d\left(q^2;\Delta^2=0\right) = d\left(q^2;0\right) \equiv d^{PT}\left(q^2\right)$, etc., in accordance with the previous notations. Let us note, however, that sometimes it is useful to distinguish between the asymptotic suppression of the mass gap contribution Δ^2/q^2 in the $q^2 \to \infty$ limit and the formal perturbative $\Delta^2 = 0$ limit.

Thus the formal perturbative $\Delta^2(D) = 0$ limit exists, and it is a regular one. As it follows from above, in this limit one recovers the perturbative QCD system of equations (11)-(12) from the non-perturbative QCD one (13)-(14). So, we distinguish between the perturbative and non-perturbative phases in QCD by the explicit presence of the mass gap. Its aim is to be responsible for the NP QCD dynamics, since it dominates at $q^2 \to 0$ in the 'solution' (14). When it is put formally zero, then the perturbative phase survives only. Evidently, when such a scale is explicitly present then the QCD coupling constant plays no role in the non-perturbative QCD dynamics. How to ensure the transversality of the relevant full gluon propagator in the explicit presence of the mass gap is described in detail in Ref. 5.

3. The non-linear iteration procedure

In order to find a formal solution for the full gluon propagator (9), on account of its effective charge (14), let us rewrite the latter one in the form of the corresponding transcendental (i.e., not algebraic) equation, namely

$$d\left(q^2\right) = 1 - \left[\Pi^s\left(q^2;d\right) + \frac{\Delta^2}{q^2}c\left(d\right)\right] d\left(q^2\right) = 1 - P\left(q^2;d\right) d\left(q^2\right), \quad (15)$$

where Eq. (4) has been already used, and instead of D an equivalent dependence on d is introduced. This is necessary to do since Eq. (14) is not a solution for $d(q^2)$, but it is the non-linear (NL) relation between the two invariant functions $d(q^2)$ and $\Pi^s(q^2;d)$. Eq. (15) is suitable for the formal NL iteration procedure. For future purposes, it is convenient to introduce short-hand notations as follows:

$$\Pi^s\left(q^2; d = d^{(0)} + d^{(1)} + d^{(2)} + ... + d^{(m)} + ...\right) = \Pi^s_m\left(q^2\right), \quad \text{and} \qquad (16)$$
$$c\left(d = d^{(0)} + d^{(1)} + d^{(2)} + ... + d^{(m)} + ...\right) = c_m \equiv c_m\left(\lambda, \alpha, \xi, g^2\right),$$

which leads to the expanded series up to the mth order:

$$P_m\left(q^2\right) = \left[\Pi^s_m\left(q^2\right) + zc_m\right], \ z = \frac{\Delta^2}{q^2}, \ m = 0, 1, 2, 3, ... \ . \qquad (17)$$

Via the corresponding subscript m it is explicitly seen which iteration for the gluon form factor d is actually done in $c(d)$, $\Pi^s(q^2;d)$ and $P(q^2;d)$. Let us also point out that all the invariant functions $\Pi^s_m(q^2)$ can be expand in a formal Taylor series near the finite subtraction point α, since they are regular functions of q^2. This means that they depend on the variable $z^{-1} = q^2/\Delta^2$. If it were possible to express the full gluon form factor $d(q^2)$ in terms of these quantities then it would be the formal solution for the full gluon propagator. In fact, this is nothing but the skeleton loops expansion, since the regularized skeleton loop integrals, contributing to the gluon self-energy as mentioned above, have to be iterated. This is the so-called general non-linear iteration solution. This formal expansion is not a perturbative series. The magnitude of the coupling constant squared and the dependence of the regularized skeleton loop integrals on it is not known. It is the so-called 'cluster' expansion in Ref. 3. In fact, it is the expansion in integer powers of z (positive and negative, see below) and it makes sense for $0 < z < 1$.

It is instructive to describe the general iteration procedure in some details. Evidently, $d^{(0)} = 1$, and this corresponds to the approximation of the full gluon propagator by its free counterpart. Doing the first, second and third iterations respectively in Eq. (15), one thus obtains

$$\begin{aligned}(1^{\text{st}})\ d\left(q^2\right) &= 1 - P_0\left(q^2\right) + \cdots = 1 + d^{(1)}\left(q^2\right) + \ldots \\ &= 1 - P_0\left(q^2\right).\end{aligned} \qquad (18)$$

$$\begin{aligned}(2^{\text{nd}})\ d\left(q^2\right) &= 1 - P_1\left(q^2\right)\left[1 + d^{(1)}\left(q^2\right)\right] + ... \\ &= 1 + d^{(1)}\left(q^2\right) + d^{(2)}\left(q^2\right) + ... \\ &= 1 + d^{(1)}\left(q^2\right) - d^{(1)}\left(q^2\right) - P_1\left(q^2\right)\left[1 - P_0\left(q^2\right)\right] + \ldots .\end{aligned} \qquad (19)$$

$$(3^{\rm rd})\quad d\left(q^2\right) = 1 - P_2\left(q^2\right)\left[1 + d^{(1)}\left(q^2\right) + d^{(2)}\left(q^2\right)\right] + \dots \tag{20}$$
$$= 1 + d^{(1)}\left(q^2\right) + d^{(2)}\left(q^2\right) + d^{(3)}\left(q^2\right) + \dots$$
$$= 1 + d^{(1)}\left(q^2\right) + d^{(2)}\left(q^2\right) - d^{(1)}\left(q^2\right) - d^{(2)}\left(q^2\right)$$
$$-P_2\left(q^2\right)\left[1 - P_1\left(q^2\right)\left(1 - P_0\left(q^2\right)\right)\right]\dots,$$

and so on for the next iterations. Up to the third iteration, we arrive at

$$d\left(q^2\right) = \sum_{m=0}^{\infty} d^{(m)}\left(q^2\right)$$
$$= 1 - \left[\Pi_2^s\left(q^2\right) + zc_2\right]\left[1 - \left[\Pi_1^s\left(q^2\right) + zc_1\right]\left[1 - \Pi_0^s\left(q^2\right) - zc_0\right]\right] + \dots \tag{21}$$

We restrict ourselves by the iterated gluon form factor up to the third term, since this already allows to show explicitly some general features of the NL iteration solution.
Expression (21) can be rewritten as:

$$d\left(q^2\right) = \left[1 - \Pi_2^s\left(q^2\right) + \Pi_1^s\left(q^2\right)\Pi_2^s\left(q^2\right) - \Pi_0^s\left(q^2\right)\Pi_1^s\left(q^2\right)\Pi_2^s\left(q^2\right) + \dots\right]$$
$$+z\left[\Pi_2^s\left(q^2\right)c_1 + \Pi_1^s\left(q^2\right)c_2 - \Pi_0^s\left(q^2\right)\Pi_1^s\left(q^2\right)c_2\right.$$
$$\left.-\,\Pi_0^s\left(q^2\right)\Pi_2^s\left(q^2\right)c_1 + \dots\right]$$
$$-z^2\left[\Pi_0^s\left(q^2\right)c_1c_2 + \Pi_1^s\left(q^2\right)c_0c_2 + \Pi_2^s\left(q^2\right)c_0c_1 + \dots\right]$$
$$-z\left[c_2 - zc_1c_2 + z^2c_0c_1c_2 + \dots\right]. \tag{22}$$

This formal expansion contains two different types of terms. The first type are terms which contain only different combinations of $\Pi_m^s(q^2)$, and they do not depend on z; the second type of terms depends on z in different integer powers. Such structure of terms will be present in each iteration term for the full gluon form factor. However, we already know that all terms which do not depend on z are to be summarized into the $d^{PT}(q^2)$, so the previous expansion can be present as the sum of the two terms

$$d\left(q^2\right) = d^{INP}\left(q^2\right) + d^{PT}\left(q^2\right), \tag{23}$$

where superscript indicates intrinsically NP (INP)

$$d^{INP}(z) \equiv d^{INP}\left(q^2\right)$$
$$= z\left[\Pi_2^s\left(q^2\right)c_1 + \Pi_1^s\left(q^2\right)c_2\right.$$
$$\left.-\,\Pi_0^s\left(q^2\right)\Pi_1^s\left(q^2\right)c_2 - \Pi_0^s\left(q^2\right)\Pi_2^s\left(q^2\right)c_1 - c_2 + \dots\right]$$
$$-z^2\left[\Pi_0^s\left(q^2\right)c_1c_2 + \Pi_1^s\left(q^2\right)c_0c_2 + \Pi_2^s\left(q^2\right)c_0c_1 - c_1c_2 + \dots\right]$$
$$-z^3c_0c_1c_2 \tag{24}$$

In order to proceed, the formal Taylor expansions for $\Pi^s_m(q^2)$ around the finite subtraction point α should be used. Let us remind that $\Pi^s_m(q^2)$ are regular functions of q^2, so they depend on $x = z^{-1} = q^2/\Delta^2$, $\alpha = \mu^2/\Delta^2$, where $\Delta^2 \neq 0$ and μ^2 is the fixed point close to $q^2 = 0$. So one can write,

$$\begin{aligned}
\Pi^s_m(x) &= \sum_{n=0}^{\infty} (x-\alpha)^n \, \Pi^{(n)}_m(\alpha) = \sum_{n=0}^{\infty} \left[\sum_{k=0}^{n} p_{nk} x^k \alpha^{n-k} \right] \Pi^{(n)}_m(\alpha) \\
&= \left[p_{00}\Pi^{(0)}_m(\alpha) + p_{10}\alpha\Pi^{(1)}_m(\alpha) + p_{20}\alpha^2\Pi^{(2)}_m(\alpha) + \ldots \right] \\
&\quad + x\left[p_{11}\Pi^{(1)}_m(\alpha] + p_{21}\alpha\Pi^{(2)}_m[\alpha) + ... \right] \\
&\quad + x^2\left[p_{22}\Pi^{(2)}_m(\alpha) + \ldots \right] + O\left(x^3\right), \qquad (25)
\end{aligned}$$

where and everywhere below $O\left(x^3\right)$ denotes an infinite number of terms of order x^3 and higher. Similarly to the previous expansion, one can write

$$\begin{aligned}
\Pi^s_m(x)\Pi^s_{m'}(x) &= \Pi^s_{mm'}(x) = \sum_{n=0}^{\infty} (x-\alpha)^n \, \Pi^{(n)}_{mm'}(\alpha) \\
&= \sum_{n=0}^{\infty} \left[\sum_{k=0}^{n} b_{nk} x^k \alpha^{n-k} \right] \Pi^{(n)}_{mm'}(\alpha) \\
&= \left[b_{00}\Pi^{(0)}_{mm'}(\alpha) + b_{10}\alpha\Pi^{(1)}_{mm'}(\alpha) + b_{20}\alpha^2\Pi^{(2)}_{mm'}(\alpha) + \ldots \right] \\
&\quad + x\left[b_{11}\Pi^{(1)}_{mm'}(\alpha) + b_{21}\alpha\Pi^{(2)}_{mm'}(\alpha) + \ldots \right] \\
&\quad + x^2\left[b_{22}\Pi^{(2)}_{mm'}(\alpha) + \ldots \right] + O\left(x^3\right). \qquad (26)
\end{aligned}$$

Evidently, such kind of expansions can be written down for any other combinations of initial $\Pi^s_m(x)$, which may appear for the higher orders in the initial expansion (25). Substituting these into Eq. (24), for the $0 < z < 1$ case,

$$d^{INP}(z) = z\sum_{k=0}^{\infty} z^k \Phi_k(\alpha) + z\sum_{k=1}^{\infty} z^{-k}\Phi_{-k}(\alpha) = z \sum_{k=-\infty}^{\infty} z^k \Phi_k(\alpha), \qquad (27)$$

where each coefficient $\Phi_k(\alpha)$ itself is an infinite sum of terms, namely $\Phi_k(\alpha) = \sum_{m=0}^{\infty} \Phi_{km}(\alpha)$. The sum over m indicates that an infinite number of iterations of the above-mentioned corresponding regularized skeleton loop integrals, contributing into the full gluon self-energy, invokes each term labeled by k. From the relations (2), (9) and (26) and again going back to the momentum variable q^2, it follows that the full gluon propagator becomes the exact sum of the two terms, namely

$$D_{\mu\nu}(q) = D^{INP}_{\mu\nu}(q) + D^{PT}_{\mu\nu}(q), \tag{28}$$

where $D^{PT}_{\mu\nu}(q)$ is determined by the relations (11) and (12). Let us emphasize that the two terms in the right-hand-side in this relation are not asymptotics, since they were constructed to be valid in the whole momentum interval. Here $D_{\mu\nu}(q)$ can be considered as the solution for the full gluon propagator because the functional dependence of $D^{INP}_{\mu\nu}(q)$ is fixed up to unknown coefficients (see below how to deal with them), while $D^{PT}_{\mu\nu}(q)$ is assumed to be known because it can be solved by the PT method. Thus solution (31) through $D^{INP}_{\mu\nu}(q)$ provides the summation over all the IR singularities at $q^2 \to 0$ possible in QCD. This means that all the QCD vertices become regular at zero gluon momenta involved. So there are no other IR singularities in QCD apart from those summarized into the full gluon propagator (28).

3.1. *The intrinsically non-perturbative gluon propagator*

The first term in the above relation (28) is

$$D^{INP}_{\mu\nu}(q) = iT_{\mu\nu}(q)d^{INP}\left(q^2\right)\frac{1}{q^2} = iT_{\mu\nu}(q)\frac{\Delta^2}{(q^2)^2}L\left(q^2\right) \tag{29}$$

where we have introduced the following function,

$$L(q^2) = \sum_{k=-\infty}^{\infty}\left(\frac{\Delta^2}{q^2}\right)^k \Phi_k(\lambda, \alpha, \xi, g^2), \ \text{for } 0 < \frac{\Delta^2}{q^2} < 1, \tag{30}$$

which is the Laurent expansion in integer powers of variable Δ^2/q^2 with residues $\Phi_k\left(\lambda, \alpha, \xi, g^2\right)$. The characteristic features of this solution are:

(1) it is transversal by construction and depend on the mass gap Δ^2,
(2) no perturbative IR singularity at $\sim (q^2)^{-1}$,
(3) there are severe NP IR singularities at $\sim (q^2)^{-2-k}$, $k = 0, 1, 2, \dots$,
(4) essential singularities both at $q^2 \to 0$ and $q^2 \to \infty$.

Thus, the INP gluon propagator (29)-(30) makes it possible to sum up all non-perturbative (severe) IR singularities with the help of the mass gap into the full gluon propagator. Its functional dependence is fixed up to unknown residues, which depend on unphysical parameters. Secondly, it seems not to be tractable due to an infinite summation of the non-perturbative IR singularities. However, as it has been shown for the first time in Ref. 5, the Laurent expansion is convergent in every point of the ring $0 < (\Delta^2/q^2) < 1$. At the same time, its behavior close to its essential singularities at $q^2 \to 0$ and $q^2 \to \infty$ is governed by the following theorem from the theory of functions of complex variable,[6,7] namely Picard or Weierstrass – Sokhatsky – Casorati (WSC) theorem, which state that the function $f(z)$ in the close neighborhood of its essential singularity can be replaced by the constant Z, which value depends only on how precisely $z \to z_0$. Based on this, one obtains equations,

$$D^{INP}_{\mu\nu}(q;\Delta^2) = iT_{\mu\nu}(q)\frac{\Delta^2}{(q^2)^2} \times Z, \quad q^2 \to 0\ , \tag{31}$$

$$D^{INP}_{\mu\nu}(q;\Delta^2) = iT_{\mu\nu}(q)\frac{\Delta^2}{(q^2)^2} \times Z', \quad q^2 \to \infty\ . \tag{32}$$

The two different renormalized constants $\Delta^2 \times Z = \Delta^2_R$ and $\Delta^2 \times Z' = \Delta'^2_R \equiv \Lambda^2_{YM}$ appear in the two different limits at $q^2 \to 0$ and $q^2 \to \infty$, respectively, but the mass gap Δ^2 itself comes from the IR terms. Thus, QCD with the mass gap undergoes a phase-transition like behavior between INP QCD and PT QCD in the strong and weak coupling regimes respectively, as presented on Fig. 1. *An important observation is that in QCD survives only the simplest NP (severe) IR singularity* $\sim (q^2)^{-2}$, *while all others*

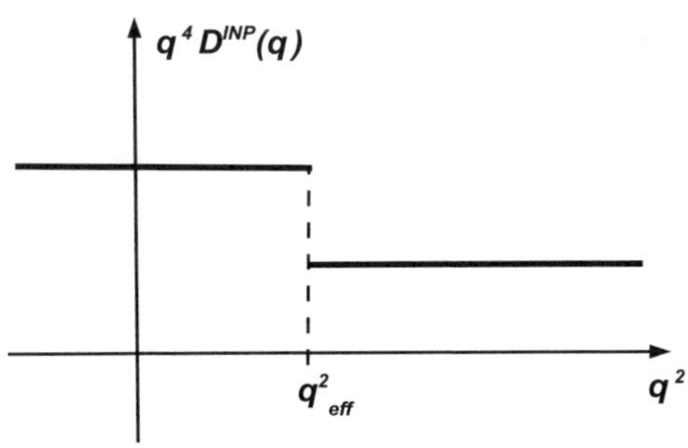

Fig. 1. The phase-transition like behavior of QCD between the strong and weak coupling regimes.

$\sim (q^2)^{-2-k}$ *for* $k = 1, 2, ...$ *are suppressed by the Picard or Weierstrass – Sokhatsky – Casorati theorem.*

In the region important for confinement $q^2 \to 0$ the full gluon propagator $D_{\mu\nu}(q)$ (28) becomes $D^{INP}_{\mu\nu}(q)$ and at $q^2 \to \infty$ it becomes $D^{PT}_{\mu\nu}(q)$ which can be solved by the PT method and should possess asymptotic freedom behavior, thus

$$D^{INP}_{\mu\nu}(q; \Delta^2) = iT_{\mu\nu}(q)\frac{\Delta^2_R}{(q^2)^2}, \quad q^2 \in [0, \infty[\tag{33}$$

where, in principle, we can assign a physical meaning to the mass gap, since it is now separated from the un-physical parameters, Φ_k. That is why it has been labeled by subscript 'R'. In addition the dimensional regularization method (DRM)[8] the non-perturbative IR singularity shown in Eq. (33), is to be regularized in terms of the IR regularization parameter ϵ, which distinguishes dimension from 4. In terms of ϵ any non-perturbative IR singularity $\sim (q^2)^{-2-k}$, $k = 0, 1, 2, ...$, in particular the simplest one of Eq. (33) is always pole-type, $1/\epsilon$, so not depending on k. Precisely this important fact of the theory of generalized functions,[9] which for the first time has been emphasized in our Book,[5] makes the theory IR renormalizable. By the corresponding re-scaling of the mass gap the pole of the first order $1/\epsilon$ can be removed, making thus the theory IR finite, i.e., not depending on ϵ in the final $\epsilon \to 0$ limit. Details of the multiplicative renormalization procedure are to be found in Ref. 5, where the full gluon propagator is replaced by its $D^{INP}_{\mu\nu}(q)$ and $D^{PT}_{\mu\nu}(q)$ parts in order to describe low- and high energies structure, respectively, of strong interactions. In summary, Eq. (33) is an exact solution for the full gluon propagator in the strong coupling regime; it describes singular, massless but, nevertheless, depending on the mass gap excitations in the QCD ground state.

4. The massive solution

One of the direct consequences of the explicit presence of the mass gap in the full gluon propagator is that the gluon may indeed acquire an effective mass.[10] From Eq. (14) it follows that

$$\frac{1}{q^2}d\left(q^2\right) = \frac{1}{q^2 + q^2\Pi^s\left(q^2;\xi\right) + \Delta^2 c(\xi)}, \tag{34}$$

where in D only the dependence on ξ is explicitly shown. The full gluon propagator (9) may have a pole-type solution at the finite point if and only if the denominator in Eq. (34) has a zero at this point $q^2 = -m_g^2$ (Euclidean signature), i.e.,

$$-m_g^2 - m_g^2 \Pi^s\left(-m_g^2;\xi\right) + \Delta^2 c(\xi) = 0, \tag{35}$$

where $m_g^2 \equiv m_g^2(\xi)$ is an effective gluon mass. The previous equation is a transcendental equation for its determination. Evidently, the number of its solutions is not fixed, *a priori*. Excluding the mass gap, one obtains that the denominator in the full gluon propagator becomes

$$q^2 + q^2\Pi^s\left(q^2;\xi\right) + \Delta^2 c(\xi) = q^2 + m_g^2 + q^2\Pi^s\left(q^2;\xi\right) + m_g^2\Pi^s\left(-m_g^2;\xi\right). \tag{36}$$

Let us now expand $\Pi^s(q^2;\xi)$ in a Taylor series near m_g^2:

$$\Pi^s\left(q^2;\xi\right) = \Pi^s\left(-m_g^2;\xi\right) + \left(q^2+m_g^2\right)\Pi'^s\left(-m_g^2;\xi\right) + O\left(\left(q^2+m_g^2\right)^2\right). \tag{37}$$

Substituting this expansion into the previous relation, one obtains

$$\begin{aligned} q^2 + m_g^2 + q^2\Pi^s\left(q^2;\xi\right) + m_g^2\Pi^s\left(-m_g^2;\xi\right) = \\ = \left(q^2+m_g^2\right) Z_3^{-1}\left(m_g^2\right)\left[1+\Pi^{s,R}\left(q^2;\xi\right)\right], \end{aligned} \tag{38}$$

where $\Pi^{s,R}(q^2;\xi) = 0$ at $q^2 = -m_g^2$ and it is regular at small q^2; otherwise it remains arbitrary, while $Z_3^{-1}(m_g^2) = \left[1+\Pi^s\left(-m_g^2;\xi\right) - m_g^2\Pi'^s\left(-m_g^2;\xi\right)\right]$.

The full gluon propagator (9) thus now looks

$$D_{\mu\nu}\left(q;m_g^2\right) = iT_{\mu\nu}(q)\frac{Z_3\left(m_g^2\right)}{\left(q^2+m_g^2\right)\left[1+\Pi^{s,R}\left(q^2;m_g^2\right)\right]} + i\xi L_{\mu\nu}(q)\frac{1}{q^2}, \tag{39}$$

where, for future purpose, in the invariant function $\Pi^{s,R}\left(q^2;m_g^2\right)$ instead of the gauge-fixing parameter ξ we introduced the dependence on the gluon effective mass squared m_g^2, which depends on ξ itself. The gluon propagator's renormalization constant is

$$Z_3\left(m_g^2\right) = \frac{1}{1+\Pi^s\left(-m_g^2;\xi\right) - m_g^2\Pi'^s\left(-m_g^2;\xi\right)}. \tag{40}$$

In the formal perturbative limit $\Delta^2 = 0$, an effective gluon mass is also zero, $m_g^2(\xi) = 0$, as it follows from Eq. (35). So an effective gluon mass is the non-perturbative effect. At the same time, it cannot be interpreted as the 'physical' gluon mass, since it remains explicitly gauge-dependent quantity. In other words, we were unable to renormalize it along with the gluon propagator (39). In the formal perturbative $\Delta^2 = m_g^2(\xi) = 0$ limit the gluon propagator's renormalization constant (40) becomes the standard one,[11,12] namely

$$Z_3(0) = \frac{1}{1+\Pi^s(0;\xi)}. \tag{41}$$

It is interesting to note that Eq. (35) has a second solution in the formal perturbative $\Delta^2 = 0$ limit. In this case an effective gluon mass remains finite, but $1 + \Pi^s\left(-m_g^2; \xi\right) = 0$. So a scale responsible for the NP dynamics is not determined by an effective gluon mass itself, but by this condition. Its interpretation from the physical point of view is not clear. The massive solution (39) is difficult to use for the solution of the color confinement problem, since it is smooth in the $q^2 \to 0$ limit. However, its existence shows the general possibility for a vector particles to acquire masses dynamically, i.e., without so-called Higgs mechanism. The above-mentioned possibility is due only to the internal dynamics and symmetries of the corresponding gauge theory.

The general procedure described above can be directly applied to the massive solution (39). So it becomes

$$D_{\mu\nu}\left(q; m_g^2\right) = D_{\mu\nu}^{TNP}\left(q; m_g^2\right) + D_{\mu\nu}^{PT}(q), \tag{42}$$

where the two terms are,

$$D_{\mu\nu}^{TNP}(q; m_g^2) = \tag{43}$$

$$= iT_{\mu\nu}(q)\left[\frac{Z_3\left(m_g^2\right)}{\left(q^2+m_g^2\right)\left[1+\Pi^{s,R}\left(q^2; m_g^2\right)\right]} - \frac{Z_3(0)}{q^2\left[1+\Pi^{s,R}\left(q^2; 0\right)\right]}\right]$$

$$D_{\mu\nu}^{PT}(q) = i\left[T_{\mu\nu}(q)\frac{Z_3(0)}{\left[1+\Pi^{s,R}\left(q^2; 0\right)\right]} + \xi L_{\mu\nu}(q)\right]\frac{1}{q^2}. \tag{44}$$

Let us remind that in the massive solution the role of the mass gap is played by an effective gluon mass, so the formal perturbative limit is $m_g^2 = 0$. Instead of the superscript 'INP' we introduce superscript truly non-perturbative ('TNP') since it has only perturbative IR singularity $\sim (q^2)^{-1}$, while in the above-mentioned formal perturbative limit it vanishes.

5. Conclusions

The structure of the full gluon propagator in the presence of the regularized mass gap has been firmly established. We have shown explicitly that in its presence at least two independent and different types of formal solutions for the regularized/renormalized full gluon propagator exist. No truncations/approximations/assumptions are made in order to show the existence of these general types of solutions. Also, our approach, in general, and the above-mentioned solutions, in particular, is gauge-invariant, since no special gauge has been chosen. Let us emphasize that before the renormalization program is performed the gauge invariance should be understood in this sense only.

In the presence of the mass gap the gluons may acquire an effective gluon masses, depending on the gauge choice – the so-called massive solution (39) –, but a gauge-fixing parameter remains arbitrary, thus a gauge is not fixed by hand. Its relation to the solution of the color confinement problem is not clear, even after the renormalization program is performed.

The general non-linear iteration solution (29)-(30) for the full gluon propagator depends explicitly on the mass gap. It is always severely singular in the $q^2 \to 0$ limit, so the gluons remain massless, and this does not depend on the gauge choice.

It is worth emphasizing that there exists only one general restriction on the behavior of $\Pi^s(q^2; D)$, which enters the corresponding gluon SD equation (13), in the explicit presence of the mass gap within our approach, namely

$$q^2\Pi^s\left(q^2; D\right) \to 0, \quad q^2 \to 0, \tag{45}$$

at any D. It stems from the exact decomposition (3), since the subtracted gluon self-energy in this limit (or more precisely at $q^2 \to \mu^2$) should go to zero. Otherwise the invariant function $\Pi^s(q^2; D)$ remains arbitrary (but it maybe logarithmic divergent at infinity). Both general types of formal solutions the massive solution and the non-linear iteration one satisfy it. The existence of some other solution(s) for the full gluon propagator, satisfying the general condition (45), should not be excluded *a priori*. Let us remind that the gluon Schwinger – Dyson equation (13) is highly non-linear, so the number of independent solutions is not fixed. Any concrete solution obtained by lattice QCD or by the analytical approach based on the Schwinger – Dyson system of equations is a particular case of these general types (finite or singular at zero gluon momentum) of the formal solutions established here. They are subject to the different truncations/approximations/assumptions and the concrete gauge choice imposed on the invariant function $\Pi^s\left(q^2; D\right)$. However, it is worth emphasizing that the solution (33) describes confinement of gluons as it was explicitly shown in detail in Ref. 5 as it was mentioned at the end of Section 3.

Acknowledgement

V.G.'s recollections and encouraging discussions with the late Prof. V.N. Gribov will be never forgotten. The authors are also grateful to G. 't Hooft, P. Forgács, J. Nyíri, T.S. Biró and M. Vasúth for useful discussions, remarks and help. This work was supported by Hungarian OTKA grants, NK106119, K104260, and TéT_12_CN-1-2012-0016. Author G.G. Barnaföldi also thanks the Bolyai Research Scholarship of the Hungarian Academy of Sciences. Support from MTA Wigner Research Centre for Physics and P. Lévai is acknowledged.

References

1. W. Marciano and H. Pagels, *Phys. Rep.* **C36**, 137 (1978).
2. M.E. Peskin and D.V. Schroeder, *An Introduction to Quantum Field Theory*,
   Addison-Wesley Advanced Book Program, Reading MA, 1995.
3. A. Jaffe and E. Witten, Yang-Mills Existence and Mass Gap,
   http://www.claymath.org/prize-problems/
   http://www.arthurjaffe.com
4. F. Strocchi, *Phys. Lett.* **B62**, 60 (1976).
5. V. Gogokhia and G.G. Barnaföldi, *The Mass Gap and Its Applications*, Word Scientific, Singapore, 2013.
6. G.A. Korn and T.M. Korn, *Mathematical Handbook*, McGraw-Hill Book Company, New York, 1968.
7. M.A. Lavrentiev and B.V. Shabat, *Methods of the Theory of Functions of Complex Variable*, Nauka, Moscow, 1987.
8. G. 't Hooft and M. Veltman, *Nucl. Phys.* **B44**, 189 (1972).
9. I.M. Gel'fand and G.E. Shilov, *Generalized Functions*, Vol. I, Academic Press, New York, 1968.
10. J.M. Cornwall, *Phys. Rev.* **D26**, 1453 (1982).
11. C. Itzykson and J.-B. Zuber, *Quantum Field Theory*, McGraw-Hill Book Company, New York, 1984.
12. T. Muta, *Foundations of QCD*, Word Scientific, Singapore, 1987.

GRAVITATION AND COSMOLOGY

UNCERTAINTY RELATION FOR QUANTUM GRAVITY

YA. I. AZIMOV

Petersburg Nuclear Physics Institute,
National Research Center "Kurchatov Institute",
Gatchina, 188300, Russia
E-mail: azimov@thd.pnpi.spb.ru

Discussion of the physical realization of coordinates demonstrates that the quantum theory of gravity (still absent) should be non-local and, probably, non-commutative as well.

Various fields became familiar objects for contemporary physics. But their description may be different. Classical fields are usually described as numerical functions $A(x, y, z, t)$ of coordinates and time (and, m.b., of some other parameters). It is implicitly assumed that both the coordinates, time, and the field values can be fixed with arbitrary good precision. The above approach is applicable, in particular, to classical gravitation field.

Quantum fields are generally described as operator functions $A(x, y, z, t)$, also depending on coordinates and time. Usually, the field operator A may be diagonalized. And it is assumed again that both the coordinates and time, and the diagonal field values can be fixed with arbitrary good precision.

Quantum theory of gravitation is still absent, though the literature contains many models and attempts to construct it. Their authors act usually in the familiar way and try to formulate the quantum gravity as a local field theory. Here we discuss, whether this is possible.

Having a particular coordinate frame, one may mark a particular physical point by means of a point-like test-body being at rest in the frame. By definition, the test-body should be chosen so to minimize its influence on the external fields. This can certainly be realized in the classical case or for the non-relativistic quantum case. In both cases, space coordinates can be fixed with arbitrary precision.

The situation looks different for the relativistic quantum case accounting for the finite velocity of light c. Position of a test-body of mass M, even if it is point-like, can not be fixed with better precision than its Compton length

$$\lambda_C = \frac{\hbar}{M\,c} \ . \tag{1}$$

For most fields, this fact does not restrict precision of coordinates: for a better precision one should use a heavier test-body. However, if the gravitation field is switched on, one encounters the inconsistency: a heavier test-body provides a better precision for its position, but stronger distorts the external gravitation field.

The same conclusion may be obtained in other way(s). If consideration begins with gravitation field in non-relativistic case and/or without quantum effects, then one could take a test-body to have very small mass, to not perturb the external field. However, accounting for both relativistic and quantum effects in this situation generates large uncertainty in the position of such test-body.

Let us consider the situation in some more detail. To fix the space point (x, y, z) up to the sphere of the radius δ_r one should use a test-body with mass

$$M_\delta \approx \frac{\hbar}{\delta_r\,c} \ . \tag{2}$$

Then, to the gravitation field φ near the point (x, y, z), in the Newtonian approximation, the test-body appends the uncertainty

$$\delta_\varphi \approx \frac{G\,\hbar}{\delta_r^2\,c} \ , \tag{3}$$

where G is the Newton constant. This uncertainty relation between coordinates and gravitation field may be rewritten as

$$\left(\frac{\delta_\varphi}{c^2}\right)\left(\frac{\delta_r}{\lambda_P}\right)^2 \approx 1 \ , \tag{4}$$

with $\lambda_P = \sqrt{G\hbar/c^3}$ being the Planck length. Note that the uncertainty relation (3) disappears if at least one of the following three conditions is satisfied: non-quantum case ($\hbar \to 0$); non-relativistic case ($c \to \infty$); absence of gravitation ($G \to 0$).

Let us discuss the meaning of these relations. When gravity is present, the test-body of mass M generates, in addition to the Compton length (1), one more length scale, the Schwarzschild radius,

$$\lambda_S = \frac{2\,GM}{c^2} \ . \tag{5}$$

At small M, $\lambda_S < \lambda_C$. But at $M > M_P/\sqrt{2}$, where

$$M_P = \sqrt{\frac{\hbar\, c}{G}} \tag{6}$$

is the Planck mass, the inequality changes its sign. If λ_S exceeds the proper dimension of the body, then it corresponds to the radius of the Schwarzschild horizon. The domain inside this surface, where the body is at rest, is invisible for the frame. Therefore, it is λ_S that determines the uncertainty of the position for a too heavy test-body. The minimal physical uncertainty is achieved when

$$\lambda_S = \lambda_C = \sqrt{2}\,\lambda_P \, . \tag{7}$$

Thus, if quantum theory describes gravitation by a field, the field should be non-local, since its coordinates may be physically fixed not better than with finite precision $\sim \lambda_P$.

Furthermore, the combination φ/c^2 is just that appearing in the metric tensor. Therefore, Eq. (4) means that the uncertainty of the coordinates is related to the uncertainties in the metric tensor and, maybe, in other coordinates (and/or time). This may hint at the non-commutativity of operators for different space-time coordinates in the future quantum theory of gravity.

Such properties are not necessary for quantum theory of other known interactions in the absence of gravity (though might work as well in some particular cases). This is related to a unique feature of gravitation: in contrast to any other interactions, there cannot exist a test-body which would be neutral (sterile) in respect to gravity.

Acknowledgements

I thank I.T. Dyatlov for useful discussion. This work is supported by the Russian Science Foundation (Grant No. 14-22-00281).

Ya. Azimov and B. Ioffe, Chernogolovka, 2015.

EARLY FORMED ASTROPHYSICAL OBJECTS AND COSMOLOGICAL ANTIMATTER

ALEXANDER D. DOLGOV

Polo Scientifico e Technologico - Edifico C INFN, Dipartimento di Fisica
Universita di Ferrara 1 via Saragat, 44100 Ferrara, Italy
and
NSU, Novosibirsk, 630090, Russia, E-mai:dolgov@fe.infn.it

Astronomical observations of recent years show that the universe at high red-shifts (about ten) is densely populated by early formed objects: bright galaxies, quasars, gamma-bursters, and it contains a lot of metals and dust. Such a rich variety of early formed objects have not been expected in the standard model of formation of astrophysical objects. There is serious tension between the standard theory and the observations. We describe the model which relaxes this tension and nicely fits the data. The model naturally leads to the creation of cosmologically significant antimatter which may be abundant even in the Galaxy. Phenomenological consequences of our scenario and the possibility of distant registration of antimatter are discussed.

1. Introduction

Probably the most impressive prediction of quantum field theory is the prediction of antimatter. Antiparticles are predicted and observed in experiments, but it is unknown if antimatter, i.e., antistars, antiplanets, anti-galaxies exist anywhere in the universe. Presently cosmological antimatter (the real one not just antiparticles) is actively searched for by several groups and more sensitive detectors are proposed for future observations.

In his famous work,[1] P.A.M. Dirac, predicted "with the tip of his pen" a whole world of antimatter (not just a small planet). He assumed initially that the positively charged "electron" was the proton!? Critique by Oppenheimer, who noticed that hydrogen would be unstable if the proton was a hole in a negative continuum, forced Dirac to conclude that "anti-electron" is a *new* particle with the same mass as the electron (1931). At that time a hypothesis about a new particle was not as blameless as today.

Very soon after that, in 1933, Carl Anderson announced the discovery of the positron, which was awarded by the Nobel prize in 1936. According to the Anderson's words: it was not difficult, simply nobody looked for that.

Paul A.M. Dirac justly got his Nobel prize in 1933 immediately after Anderson's discovery. In his Nobel lecture "Theory of electrons and positrons" (December 12, 1933) Dirac, in particular, said: "If we accept the view of complete symmetry between positive and negative electric charge so far as concerns the fundamental laws of Nature, we must regard it rather as an accident that the Earth (and presumably the whole Solar System) contains a preponderance of negative electrons and positive protons. It is quite possible that for some of the stars it is the other way about, these stars being built up mainly of positrons and negative protons. In fact, there may be half of stars of each kind. The two kinds of stars would both show exactly the same spectra, and there would be no way of distinguishing them by present astronomical methods." However, as we see in what follows, now we know ways to distinguish a star from an antistar by observations from the Earth. The spectra are not exactly the same, even if CPT is unbroken and polarization of the electromagnetic radiation or the type of emitted neutrinos/antineutrinos from supernovae can be good indicators to a star made of antimatter.

It is striking that Dirac was the second person to talk about antimatter. In 1898, 30 years before Dirac and one year after the discovery of electron (J.J. Thomson, 1897) Arthur Schuster (another British physicist) conjectured that there might be other sign electricity, *antimatter*, and supposed that there might be entire solar systems, made of antimatter, *indistinguishable* from ours.

In his paper "Potential Matter. Holiday Dream" [2] A. Schuster wrote "When the year's work is over and all sense of responsibility has left us, who has not occasionally set his fancy free to dream about the unknown, perhaps the unknowable?" and later "Astronomy, the oldest and yet most juvenile of the sciences, may still have some surprises in store. May antimatter be commended to its case".

Schuster made one more fantastic and correct guess. He envisaged that matter and antimatter are capable to annihilate and to produce *vast* energy. Remember that this was done before Einstein established his famous relation between mass and rest energy, $E = mc^2$.

To eliminate antimatter in the close vicinity of the Earth Schuster conjectured that matter and antimatter are gravitationally repulsive because the antimatter has a negative mass. Two such objects in close contact should have a vanishing mass. As we know now, all that happened to be incorrect: matter and antimatter are gravitationally attractive and masses of particles and antiparticles have the same sign, and, due to the CPT-theorem, their masses are equal by magnitude. Presently we have other explanations why antimatter is not seen in our neighbourhood. Moreover, as we argue in what follows, abundant stellar-like antimatter objects may be hidden even in the Galaxy not far from the Earth. Other galaxies may be also abundant with antimatter.

It would be great if these two prophets were right and anti-worlds did exist. However, the standard, simplest approach to baryogenesis (Sakharov) leaves no room for anti-worlds. Still, a simple generalization of baryogenesis scenarios could lead to abundant anti-worlds. Very early formed and/or too old astronomical objects, observed presently, especially in the last few years, may signify the existence of anti-worlds.

An observation of cosmic antimatter will give a clue to physics of baryogenesis, to the mechanism of cosmological C and CP breaking, and present an extra argument in favor of inflation. Since generalized scenarios predict not just a single number of baryon to photon ratio, $\beta = (N_B - N_{\bar{B}})/N_\gamma$, but a whole function of space point, $\beta(x)$, the models are falsifiable.

2. Traditional ways of antimatter search and observational bounds

There are essentially two main methods for the search of cosmic antimatter used up to now. The first, older one, is a registration of the products of e^+e^- and $\bar{p}p$ annihilations into electromagnetic radiation. In the first case especially spectacular is the 0.511 MeV line from e^+e^--annihilation at rest, which is in fact observed in the Galactic center, but its origin is still unclear. Proton-antiproton annihilation would create ~ 100 MeV photons with a rather wide spectrum resulting from the decays of pions originally created by the annihilation.

A registration of excessive cosmic antiprotons and positrons at low energies could also be an indication for antimatter but at the present time the flux of low energy antiprotons and positrons can be explained by the conventional mechanism.

These are of course indirect methods and even if excessive antiprotons, positrons, or gamma-rays are observed, a further analysis is necessary to exclude traditional mechanisms of their production.

Even less direct methods include an analysis of spectrum distortion of CMB (both frequency and angular fluctuations) of the impact of antimatter on big bang nucleosynthesis (BBN), and large scale structure formation.

During the last decade there is a burst of experimental activity for direct search of cosmic antimatter, namely for the search of antinuclei in cosmic rays, which have a negligible probability to be secondary-produced in the universe. There are presently the following missions for the direct search of cosmic antinuclei:

(1) BESS: Japanese Balloon Borne Experiment with Superconducting Solenoidal Spectrometer.
(2) PAMELA (Italian-Russian space mission): Payload for Antimatter Matter Exploration and Light-nuclei Astrophysics.
(3) AMS: AntiMatter Spectrometer (Alpha Magnetic Spectrometer), CERN-MIT-NASA.

The following two, more sensitive, detectors are under discussion now:
PEBS (Positron Electron Balloon Spectrometer,) search for cosmic positrons and antiprotons.
GAPS (Gaseous Antiparticle Spectrometer), search for X-rays from de-excitation of exotic atoms, it may reach a two orders of magnitude better sensitivity than AMS for the ratio $\overline{\mathrm{He}}/\mathrm{He}$.

At the present time the following bound on the flux of the anti-helium-4 in cosmic rays is found : BESS: $\overline{\mathrm{He}}/\mathrm{He} < 3 \times 10^{-7}$. More restrictive limits are expected: PAMELA: $\overline{\mathrm{He}}/\mathrm{He} < 3 \times 10^{-8}$; AMS-2: $\overline{\mathrm{He}}/\mathrm{He} < 10^{-9}$ but not yet reported. The fluxes above are normalized to the observed flux of the cosmic ray helium which at $E < 10$ GeV/nuclei is equal to:

$$\mathrm{d}N_{He}/\mathrm{d}E = 10^2/\mathrm{m}^2/\mathrm{str}/\mathrm{sec}/\mathrm{GeV}. \tag{1}$$

According to the calculations of Ref. 3 the expected fluxes of the secondary produced anti-nuclei are the following. Anti-deuterium is produced in $\bar{\mathrm{p}}\,\mathrm{p}$ or $\bar{\mathrm{p}}\,\mathrm{He}$ collisions The predicted flux of anti-deuterium produced by this mechanism is $\sim 10^{-7}/m^2/s^{-1}/sr/(\mathrm{GeV}/n)$, i.e., 5 orders of magnitude lower than the observed flux of antiprotons. The expected fluxes of secondary produced $^3\overline{\mathrm{He}}$ and $^4\overline{\mathrm{He}}$ are respectively four and eight orders of magnitude smaller than the flux of anti-D.

Observations and bounds, summary:

- $\bar{\mathrm{p}}/\mathrm{p} \sim 10^{-5} - 10^{-4}$, is observed and can be explained by secondary production;
- observed $\mathrm{He}/\mathrm{p} \sim 0.1$; the upper limit: $\overline{\mathrm{He}}/\mathrm{He} < 3 \times 10^{-7}$.
- Theoretical predictions: $\overline{\mathrm{D}} \sim 10^{-5}\bar{\mathrm{p}}$, $^3\overline{\mathrm{He}} \sim 10^{-9}\bar{\mathrm{p}}$, $^4\overline{\mathrm{He}} \sim 10^{-13}\bar{\mathrm{p}}$.
- From the upper limit on $\overline{\mathrm{He}}$: the nearest single antigalaxy should be at the distance larger than 10 Mpc (very crudely).

The production of antinuclei was studied at LHC at ALICE detector.[4] According to its data the production of an antinucleons with an additional antinucleon is suppressed only by a factor about 1/300 which is by far milder than the suppression factors presented above. Probably the difference is related to much higher energies at which data of ALICE are taken. The events with such energies are quite rare in cosmic rays and do not have a significant impact on the total rate of the antinuclei production.

Traditional bounds of the amount of the cosmic antimatter derived from the fluxes of the cosmic gamma rays did not change during the last few decades because they are essentially valid for our galactic supercluster. According to Ref. 5 the nearest anti-galaxy could not be closer than at $\sim$10 Mpc, as follows from the analysis of $\bar{p}p$-annihilation in common intergalactic cloud. The fraction of antimatter in colliding galaxies in Bullet Cluster is bounded from above by $< 3 \times 10^{-6}$ as written in Ref. 6.

Note that CMB excludes *large* isocurvature fluctuations at distances $d > 10$ Mpc and BBN excludes large "chemistry" fluctuations at $d > 1$ Mpc. It means in particular that in a simple model the antimatter domains or objects should be sufficiently small.

There is relatively recent bound on the amount of antistars in the galaxy.[7] The Bondi accretion of the interstellar gas to the surface of an antistar leads to the annihilation luminosity:

$$L_\gamma \sim 3 \cdot 10^{35} (M/M_\odot)^2 v_6^{-3} \tag{2}$$

It allow to put the limit $N_{\bar{*}}/N_* < 4 \cdot 10^{-5}$ inside 150 pc from the Sun, where M is the mass of the star, $M_\odot$ is the solar mass, and v_6 is the star velocity in 10^6 cm/sec.

Note in conclusion that all the presented bounds are true if antimatter makes the same type objects as the *observed* matter ones. For example, as we show in what follows, compact fast antistars may be abundant in the Galaxy but still escape observations. A recent analysis of bounds on different type of antimatter objects was performed in Ref. 8. Earlier works on cosmology with domains of matter and antimatter was discussed in the lectures,[9] where a list of relevant references can be found.

3. Observations of antistars from the Earth

According to Dirac and Schuster cited in the Introduction, antistars are indistinguishable from stars by astronomical observations. The basis for that is the CPT theorem, though it was unknown to the both quoted persons. This theorem demands equality of the level positions in atoms and antiatoms, so spectral measurements at first sight do nor allow to distinguish if the light is emitted by an atom or an antiatom. However, we will see below that this is not exactly true.

Of course if CPT is broken, the masses of particles and antiparticles may be different, though it is not necessarily so. If the particle-antiparticle mass difference is non-zero, $\Delta m \neq 0$, the spectra of anti-atoms might be different in an irregular way, not described by the redshift. However, even if $\Delta m \neq 0$, it must be extremely tiny. As it is argued in Ref. 10, non-zero mass difference between, say, electron and positron or between proton and antiproton breaks gauge invariance of electrodynamics, which in turn leads to non-zero photon mass. The upper bounds on the latter are extremely restrictive and they demand very strong upper bounds on the mass difference $\Delta m_{ee} \equiv |m_e - m_{\bar{e}}| < 2 \cdot 10^{-17}$ eV or even $\Delta m_{ee} < 4 \cdot 10^{-29}$ eV based on Ref. 10. So even if $\Delta m_{ee} \neq 0$, its effect on the spectrum should be tiny.

Below we assume that CPT is respected and show that there are still chances for distant observation of antiworlds.

Since C and CP are broken, the widths of spectral lines of atoms and anti-atoms must be different as it is explained in Ref. 11. Unfortunately the magnitude of the effect for hydrogen is extremely small, maybe because of an accidental cancellation:

$$\Delta\Gamma/\Gamma \sim 10^{-28}. \tag{3}$$

An amplification of the effect, in particular, in heavier atoms and in external magnetic or electric fields might be possible.

Possibilities discussed in Ref. 12 look much more realistic, One evident way is to establish radio communication with inhabitants of the stellar system under scrutiny. It may help to determine if they are or are not antimatter creatures. It is possible but looks more proper for science fiction.

A perspective way, maybe even in not too distant future, is a study of stellar neutrinos versus antineutrinos, especially from energetic SN explosions. At the first stage of SN explosion neutrinos from the neutronization reaction $pe^- \to n\nu$ are emitted, while an anti-SN would emit antineutrinos.

An interesting possibility is a search for antistars through the detection of polarized photons produced in weak interaction processes. These photons are longitudinally polarized and their energy can be well defined if they are created in two body decays. Mono-energetic photons produced e.g. in $B \to K^\star\gamma$ should be left-handed because of dominance of $b \to s\gamma$ penguin transition with left-handed s-quark. However, one can hardly expect noticeable abundance of B-mesons in stars, and most probably there is an equal amount of B and $\bar{B}$ mesons.

Stars with abundant strange quarks look more promising. The outer shell of such stars is populated by Σ-hyperons and the polarization of photons emitted in $\Sigma^+ \to p\gamma$ decay could indicate if the photons are emitted by hyperon or anti-hyperon. The polarization is large, $\alpha = -0.76 \pm 0.08$ and the branching ratio is non-negligible $(1.23 \pm 0.05) \times 10^{-3}$.

Circular polarization of photons in the γ-transitions of nuclei was observed in the terrestrial experiments: $P_\gamma = (4 \pm 1) \cdot 10^{-5}$ in ^{175}Lu transition with the emission of 395 keV photon, $P_\gamma = -(6 \pm 1) \cdot 10^{-6}$ for 482 keV photon emitted in ^{181}Ta transition, and $P_\gamma = (1.9 \pm 0.3) \cdot 10^{-5}$ for 1290 keV photon emitted in transition of ^{41}K. Measurement of circular polarization of such photon lines is ideally suited for search of antistars.

4. Baryogenesis

The prevailing point of view at the present time is that all the universe is made only of matter, while primeval antimatter is absent. Still the fact that antimatter exists creates fundamental cosmological puzzle: why the observed universe is 100% dominated by matter? Antimatter exists but not antiworlds, why? The problem deepened because of approximate symmetry between particles and antiparticles. In fact before 1956, the common faith in exact C, P, and T symmetries looked unbreakable, but with passing years more and more out of these discrete symmetries were discovered to be broken.

The puzzle of the baryon asymmetry, i.e., of the observed predominance of matter over antimatter is resolved by three Sakharov's principles[13]:

(1) Nonconservation of baryons.
(2) Violation of symmetry between particles and antiparticles, i.e., C and CP.
(3) Breaking of thermal equilibrium.

There is a plethora of baryogenesis scenarios which can explain the single number:

$$\beta_{observed} = \frac{N_B - N_{\bar{B}}}{N_\gamma} \approx 6 \times 10^{-10} . \tag{4}$$

The usual outcome: $\beta = const$, which makes it impossible to distinguish between models and does not leave space for cosmological antimatter. But the situation is not so pessimistic, simple and natural generalizations of the simplest models of baryogenesis allow for a lot of antimatter almost at hand.

Of all the models, the particularly interesting one is the Affleck–Dime (AD), or supersymmetric baryogenesis.[14] It is the only known scenario of baryogenesis which may create large baryon asymmetry, $\beta \sim 1$, while all other models are able to create only a very small β, so theoretical efforts are usually done to enhance the asymmetry as much as possible to make it compatible with the observed value.

Supersymmetry (SUSY) predicts the existence of scalar bosons with non-zero baryonic number, $B \neq 0$. Such bosons may condense along flat directions of the quartic potential, which is a generic property of the potential in such models:

$$U_\lambda(\chi) = \lambda|\chi|^4\left(1 - \cos 4\theta\right) . \tag{5}$$

There can be also the mass term, $m^2\chi^2 + m^{*\,2}\chi^{*\,2}$, which has another set of flat directions:

$$U_m(\chi) = m^2|\chi|^2[1 - \cos(2\theta + 2\alpha)] . \tag{6}$$

where $\chi = |\chi|\exp(i\theta)$ and $m = |m|\exp(\alpha)$. If $\alpha \neq 0$, C and CP are broken.

In grand unified (GUT) SUSY models the baryonic number of χ is naturally non-conserved. It is expressed by the non-invariance of $U(\chi)$, Eqs. (5) and (6), with respect to the phase rotation, $\chi \to \exp(i\theta)\chi$.

Initially (at inflation) χ is naturally away from the origin and when inflation is over, it starts to evolve down to the equilibrium point, $\chi = 0$, according to the equation of Newtonian mechanics for point-like particle:

$$\ddot{\chi} + 3H\dot{\chi} + U'(\chi) = 0. \tag{7}$$

The Baryonic charge of χ,

$$B_\chi = \dot{\theta}|\chi|^2 , \tag{8}$$

is analogous to the mechanical angular momentum in the two-dimensional complex plane [$\mathrm{Re}\,\chi$, $\mathrm{Im}\,\chi$]. The decays of χ transferred the baryonic charge accumulated in the rotational motion of χ into that of quarks in B-conserving processes.

If $m \neq 0$, the angular momentum, or what is the same the baryonic number, B, is generated by possibly different directions of the quartic and quadratic valleys at low χ. If CP-odd phase α is small but non-vanishing, both baryonic and antibaryonic regions are possible with dominance of one of them. Matter and antimatter domain may exist but globally $B \neq 0$. This is briefly the canonical version of the AD-scenario which results in normal value of the baryon asymmetry with possible domains of antimatter of arbitrary sizes and values of β. Such a model of antimatter creation is strongly restricted by the observational data.

A minor modification of AD-scenario can lead to a very early formation of compact stellar-type objects and naturally to a comparable amount of anti-objects, such that the bulk of baryons and (equal) antibaryons are in the form of compact stellar-like objects, e.g., QSOs or primordial black holes (PBH), formed in the regions with a huge baryon asymmetry, plus sub-dominant observed homogeneous baryonic background. The amount of antimatter may be comparable or even larger than of *known* baryons, but such "compact" (anti)baryonic objects would not contradict any existing observations.

Note that the astronomical data accumulated during the last few years demonstrate a rich population of superheavy black holes in the very early universe, which indeed could be created in the framework of the modified AD-scenario, but remain mysterious otherwise.

The necessary generalization of the AD-baryogenesis was suggested in Ref. 15 and further developed in Ref. 16. To this end a general renormalizable coupling of the Affleck – Dine field χ to inflaton, Φ, has been introduced (the first term in the equation below), so the total potential of χ included the interaction with the inflaton, Φ, gets the form:

$$U = g|\chi|^2(\Phi - \Phi_1)^2 + \lambda|\chi|^4 \ln\left[\frac{|\chi|^2}{\sigma^2}\right] + \lambda_1\left(\chi^4 + h.c.\right) + (m^2\chi^2 + h.c.). \tag{9}$$

The second term in this expression is the so called Coleman – Weinberg[17] potential obtained by summation of one-loop radiative corrections.

When Φ is far from Φ_1, the effective mass squared of χ is positive and large, so the gate to the flat directions is closed. It is open only during a relatively short period, when $\Phi \approx \Phi_1$, but still at the inflationary epoch. Because of that cosmologically small but possibly astronomically large bubbles with very high β could be created, occupying a small fraction of the universe volume, while the rest of the universe has the normal baryon asymmetry $\beta \approx 6 \cdot 10^{-10}$, created in the regions with small initial χ, occupying the bulk of the universe volume. Such process looks as phase transition of 3/2 order. In the simplest version of the model the high-β objects would have positive and negative baryon asymmetry with practically equal probability. Interesting anti-objects should be astronomically large, so the inflation is necessary and not too large to avoid problems with the existing observations. So some, rather mild fine-tuning of the model parameter values is necessary.

The distribution of high baryon density bubbles over mass (and size) has very simple log-normal form:

$$\frac{\mathrm{dN}}{\mathrm{d}M} = C_M \exp\left[-\gamma \ln^2(M/M_0)\right], \tag{10}$$

where C_M, γ, and M_0 are constant model dependent parameters. The spectrum is practically model independent, because it is essentially determined by the exponential expansion (inflation).

There are several (essentially two) periods when the density contrast between the regions with high baryonic density and those with the normal low one arose:

(1) Firstly, after the formation of the domains (bubbles) with high value of χ and correspondingly with huge baryon (anti-baryon) asymmetry the initial density contrast between the high-B bubbles and the rest of the universe would be zero or very small, because quarks, which carry baryonic number at this stage were massless. These are the so called isocurvature perturbations. Somewhat later due to different equations of state inside and outside of the domains (or what is the same of high-B bubbles) some density contrast would evolve because the matter inside the bubbles were more non-relativistic due to higher amplitude of non-relativistic field χ inside.

(2) The second period of perturbation rise is much more efficient. It takes place after the QCD phase transition at $T \sim 100$ MeV when quarks made non-relativistic protons. During this stage compact bubbles with high baryonic mass density would be formed. With properly chosen parameters such bubbles could be primordial black holes (PBHs), or dense stellar like objects. The PBH masses could range from a few $M_\odot$ up to $10^{6-7} M_\odot$.

The model successfully explains quite a few mysterious facts observed by astronomers, with the data especially accumulated during the recent few years. It presents a mechanism of superheavy black hole formation, and what is especially important, at very early time. Correspondingly it explains existence of quasars observed at high redshifts, $z \sim 10$, which is mysterious in the standard model. Moreover, the near-quasar space is enriched with metals. An explanation of this fact demands an unusual BBN (see below) or an early supernova formation. High redshift gamma bursters also demand early created supernovae — this can be simply arranged in the considered model. Early formed superheavy BHs can be seeds of galaxy formation. In the accepted nowadays picture the process is supposed to be the opposite: first galaxies were formed and superheavy BHs in the center of practically all observed galaxies were created later by the matter accretion to the galactic center. However, in the standard approach the rate of superheavy BH formation is too low to explain their formation during the cosmological time. Moreover, several small galaxies are known with the central superheavy BH having masses similar by the magnitude to the total mass of the galaxy. This surely does not fit the standard accretion picture.

BBN in high-B environment. If $\beta \equiv \eta \gg 10^{-9}$, light (anti)element abundances would be anomalous: the clouds with high β would contain much less (anti)deuterium and more (anti)helium, than the normal one. Evolved chemistry in the so early formed QSOs can be explained, at least to some extend, by more efficient production of metals during BBN due to much larger ratio $\beta = N_B/N_\gamma$. The standard BBN essentially stops at ^{4}He because of the very small β. However, in the model considered here β is much larger than the canonical value, even being close or exceeding unity. BBN with anomalously high $\beta \gg \beta_{obs} = 6 \cdot 10^{-10}$ was considered in Ref. 18 but still only up to $\beta \ll 1$. An extension to $\beta \sim 1$ would be very interesting.

To summarize the model explains in the unique way the formation of high-redshift supernovae, gamma-bursters, stars older than the universe observed in the Milky Way, cut-off BH distribution at $M < 6M_\odot$ in galaxies, and more. It is discussed in detail in the next section.

According to our estimates the compact anti-objects mostly survived in the early universe, (especially if they are PBHs). The model predicts an existence of a kind of early dense stars formed when the initial pressure outside was larger than internal pressure. Astrophysics of such objects may be quite unusual. Such "stars" may evolve quickly and, in particular, make early SNs, enrich the universe with heavy (anti)nuclei and reionize the universe.

Phenomenology and bounds on compact antimatter objects and disperse anti-clouds was studied in the papers.[8,19] The early formed stars and black holes are discussed in Ref. 20.

5. Mysteries in the sky

On the background of a quite successful description of the gross features of the universe the standard ΛCDM cosmology encounters plenty of disturbing and interesting problems. These troubling data became especially prominent during the last several years. In short, there is an avalanche of astronomical discoveries of different kind of astrophysics objects which could not be created and evolved in the available cosmological time. They are observed in the very early universe at redshifts $z \sim 10$ and in the contemporary universe, e.g. in our Galaxy there are stars, which look older than the universe. Rephrasing the famous words by Marcellus from "Hamlet" we can say: "Something is rotten in the state of ~~Denmark~~ the Universe". It is tempting to explain all these data by our model described in the previous section. It is especially interesting that the suggested scenario predicts, as a byproduct abundant cosmological antimatter in close vicinity to the Earth.

To see how old are the objects observed at different redshifts let us present the expression of the universe age t_U, as a function of the cosmological redshift z:

$$t(z) = \frac{1}{H} \int_0^{1/(z+1)} \frac{dx}{\sqrt{1 - \Omega_{tot} + \Omega_m/x + x^2\,\Omega_v}}, \tag{11}$$

where Ω_m is the fractional (in terms of the critical energy density) energy density of matter (i.e., baryonic plus dark matter), and Ω_v is the fraction of the density of dark energy. The total fraction of the cosmological energy density, Ω_{tot}, is supposed to be unity (spatially flat universe), since the energy density of the CMB photons and neutrino can be neglected during essential time interval in the integral.

According to the Planck data, the present day values of theses parameters are: $\Omega_{tot} = 1$, $\Omega_m = 0.317$, and $\Omega_v = 0.683$. There is some tension between the values of the Hubble parameter measured by Planck, $H_{pl} = 67.3$ km/sec/Mpc and that measured by the traditional astronomical methods, which can lead to H as large as $H_{astr} = 74$ km/sec/Mpc, see Ref. 21 for discussion. Possibly both results might be correct and the difference in the measured values of H emerges because the Planck and the astronomical measurements are sensitive to H at different moments of the cosmological history. This e.g. can occur in a model where a fraction of dark matter is unstable and decays at some moment between the recombination and the present day.[22]

We present a few examples of the universe age in Giga-years for different z. The first number corresponds to the Planck measured value of H, and the other, shorter one, in brackets to the larger astronomical value: $t_U \equiv t(0) = 13.8\,(12.5.)$; $t(3) = 2.14\,(1.94)$; $t(6) = 0.93; (0.82)$; $t(10) = 0.47\,(0.43)$; and $t(12) = 0.37;\ (0.33)$.

5.1. *Strange phenomena in the Milky Way*

Old stars in the Milky Way. With an increased accuracy of the nuclear chronology the ages of several stars have been recently determined to be much older than expected. Below we quote a few recent results.

Employing thorium and uranium abundances in comparison with each other and with several stable elements the age of metal-poor, halo star BD+17° 3248 was estimated as 13.8 ± 4 Gyr, see in Ref. 23. For comparison the age of inner halo of the Galaxy is 11.4 ± 0.7 Gyr.

The age of a star in the galactic halo, HE 1523-0901, was estimated to be about 13.2 Gyr based on Ref. 25. First time many different chronometers, such as the U/Th, U/Ir, Th/Eu and Th/Os ratios to measure the star age have been employed.

The metal deficient *high velocity* subgiant in the solar neighbourhood HD 140283 has the age 14.46 ± 0.31 Gyr (see Ref. 26). The central value exceeds the universe age by two standard deviations, if $H = 67.3$ and $t_U = 13.8$, while for $H = 74$, and thus $t_U = 12.5$ the excess is by nine standard deviations. So this could really looks as a star older than the universe.

A possible explanation of this discrepancy could be an unusual initial chemical content of such stars. Normally a pre-stellar cloud consists of 25% of He^4 and 75% of hydrogen. However, in our case the primordial nucleosynthesis is able to create much heavier elements, as is mentioned in the previous sections or/and the early supernovae could enrich the interstellar gas with heavy elements, so the initial chemical content of some stars would be much different from the traditional one. Such stars could evolve to their present state considerably faster than the usual ones. Of course this conjecture should be verified by the stellar nucleosynthesis calculations.

The mass distribution of BHs observed in the Milky way demonstrates some peculiar features not understood by the conventional theory. It is found that their masses are concentrated in the narrow range $(7.8 \pm 1.2)M_\odot$ as in Ref. 27. This result agrees with another paper where a peak around $8M_\odot$, a paucity of sources with masses below $5M_\odot$, and a sharp drop-off above $10M_\odot$ are observed.[28] Such facts are indeed very strange, if these BHs were formed by the stellar collapse, as it is usually assumed. However the log-normal mass distribution (10) of the primordial black holes naturally explains such a form of the mass spectrum of the stellar mass black holes in the Galaxy. Astronomical data also indicate a two-peak mass distribution of the PBHs and compact stars, which is probably observed, but not explained up to now.[29] Quoting this work: "sample of black hole masses provides strong evidence of a gap between the maximum neutron star mass and the lower bound on black hole masses". These results also fit the mass distribution of our model, assuming that the neutron stars were mostly created by the conventional mechanism.

5.2. *Supermassive BHs and quasars in (near) contemporary universe*

It seems that every large galaxy and some smaller ones contain a central supermassive BH with masses larger than $10^9 M_\odot$ in giant elliptical and compact lenticular galaxies and $\sim 10^6 M_\odot$ in spiral galaxies like Milky Way.

The mass of BH is typically 0.1% of the mass of the stellar bulge of galaxy[30] but some galaxies may have huge black holes: e.g., NGC 1277 has the central BH of $1.7 \times 10^{10} M_\odot$, or 60% of its bulge mass.[31] This fact creates serious problems for the standard scenario of formation of central supermassive BHs by accretion of matter in the central part of a galaxy. An inverted picture looks more plausible, when first a supermassive black hole was formed and attracted matter serving as seed for subsequent galaxy formation.

More similar examples can be found in Ref. 32. As the authors say, although supermassive black holes correlate well with their host galaxies, there is an emerging view that outliers exist. Henize 2-10, NGC 4889, and NGC1277 are examples of super massive black holes at least an order of magnitude more massive than their host galaxy suggests. The dynamical effects of such ultramassive central black holes is unclear.

A discovery of an ultra-compact dwarf galaxy older than 10 Gyr, enriched with metals, and probably with a massive black in its center also seems to be at odds with the standard model.[33] The dynamical mass of the galaxy is $2 \times 10^8 M_\odot$ and the radius is $R \sim 24$ pc, so its density is very high. The Chandra data reveal a variable central X-ray source with $L_X \sim 10^{38}$ erg/s that could be an active galactic nucleus associated with a massive black hole or a low-mass X-ray binary. The analysis of optical spectroscopy shows the object to be old, $\gtrsim 10$ Gyr and of solar metallicity, with elevated [Mg/Fe] and strongly enhanced [N/Fe] that indicates light element self-enrichment; such self-enrichment may be generically present in dense stellar systems.

Recently an observation of a quasar quartet embedded in giant nebula was reported[34] in a survey for Lyman-α emission at redshift $z \approx 2$. According to the authors, it reveals rare massive structure in distant universe.

All galaxies once passed through a hyperluminous quasar phase powered by accretion onto a supermassive black hole. But because these episodes are brief, quasars are rare objects separated by cosmological distances, so the chance of finding a quadruple quasar is $\sim 10^{-7}$. It implies that the most massive structures in the distant universe have a tremendous supply ($\sim 10^{11} M_{\odot}$) of cool dense ($n \approx 1/\mathrm{cm}^3$) gas, in conflict with current cosmological simulations.

5.3. *Young objects at high redshifts*

There is a large "zoo" of evolved astronomical objects formed in the universe in surprisingly short times. Several galaxies have been observed at high redshifts, with natural gravitational lens "telescopes". There is a galaxy at $z \approx 9.6$ which was formed when the universe was approximately 0.5 Gyr old.[35]

Moreover a galaxy at $z \approx 11$ has been observed[36] which was formed earlier than the universe age was 0.41 Gyr (or even shorter with larger H).

An observation of a not so young but extremely luminous galaxy was recently reported by WISE[40]: $L = 3 \cdot 10^{14} L_{\odot}$; age: ~ 1.3 Gyr. For the creation of such galaxy the galactic seeds, or embryonic black holes, might be bigger than thought possible. As one of the authors of the paper, P. Eisenhardt, said : "How do you get an elephant? One way is start with a baby elephant." The BH was already billions of $M_{\odot}$, when our universe was only a tenth of its present age of 13.8 billion years.

Thus we smoothly come to another and even more striking example of early formed objects, i.e., to quasars observed at high z. A quasar with maximum $z = 7.085$ has been observed.[37] It means that the quasar was formed at $t < 0.75$ Gyr. Its luminosity and mass are respectively $L = 6.3 \cdot 10^{13} L_{\odot}$ and $M = 2 \cdot 10^9 M_{\odot}$. The quasars are supposed to be supermassive black holes (BH) and their formation in such a short time looks problematic by a conventional mechanisms.

As it is summarized in Ref. 38 "rapid emergence of high-z galaxies so soon after the big bang may actually be in conflict with current understanding of how they came to be. This problem is very reminiscent of the better known (and probably related) premature appearance of supermassive black holes at $z \sim 6$. It is difficult to understand how $10^9 M_{\odot}$ black holes appeared so quickly after the big bang without invoking non-standard accretion physics and the formation of massive seeds, both of which are not seen in the local Universe."

Very recently another monster was discovered[39]: "An ultraluminous quasar with a twelve billion solar mass black hole at redshift 6.30". There is already a serious problem with formation mechanism of lighter and less luminous quasars which is now multifold deepened with this new "creature". As it is stated in the paper, about 40 quasars with $z > 6$ are known, each quasar containing BH with $M \sim 10^9 M_\odot$. Such black holes, when the Universe was less than one billion years old, present substantial challenges to theories of the formation and growth of black holes and the coevolution of black holes and galaxies. Now we have an order of magnitude heavier one, $M \approx 10^{10} M_\odot$. It has an optical and nearinfrared luminosity a few times greater than those of previously known $z > 6$ quasars.

As we have already mentioned, the universe at $z \sim 10$ is quite dusty, for a recent analysis and references see the paper.[41] The medium around the observed early quasars contains considerable amount of "metals" (elements heavier than He). According to the standard picture, only elements up to ^{4}He and traces of Li, Be, B were formed in the early universe by BBN, while heavier elements were created by stellar nucleosynthesis and dispersed in the interstellar space by supernova explosions. If so, prior to or simultaneously with the QSO formation a rapid star formation should take place. These stars could produce plenty of supernovae which enriched interstellar space by metals.

Moreover, observations of high redshift gamma ray bursters (GBR) also indicate a high abundance of supernova at large redshifts. The highest redshift of the observed GBR is $z = 9.4$ and there are a few more GBRs with smaller but still high redshifts. The necessary star formation rate for explanation of these early GBRs is at odds with the canonical star formation theory.

On the other hand, our model may naturally lead to high density of compact stars, which in particular can be early supernova created much earlier than $z = 10$.

6. Conclusion

The observational data at high redshifts, $z \sim 10$, which became available in the recent years show that the universe is populated by the early formed bright galaxies, quasars, gamma-bursters, and contains a lot of metals and dust. Such rich early formed varieties have not been expected in the standard model of formation of astrophysical objects. There is serious tension between the standard theory and observations.

Probably the most striking is the existence of supermassive black holes both in the early and in the present day universe. The standard scenario is that such black holes have been formed through accretion of matter to an initial solar mass black hole. However, it demands much more time than that available during all the cosmological history. This is especially troubling for the supermassive black holes (quasars) created at the dawn of the universe at high z. The inverted picture when primordial BHs were first created and then served as seeds for the galaxy formation looks much more probable. Black holes found in small galaxies, having masses about half of the total galaxy mass surely cannot be created by the conventional mechanism.

The model,[15] suggested almost a quarter of century ago, very nicely fits all the new observations, at least qualitatively. Of course more detailed quantitative work is necessary. A natural, though not obligatory, outcome of this model is a prediction of cosmic antimatter, mostly in the form of compact high velocity stars in the Galaxy. The observational bounds on the abundance of such stars are quite vague, so we may have a lot of antimatter practically at hand. Such early formed stars would behave as cold dark matter and so should have much larger velocity than the normal stars and hence they would populate the galactic halo. The mysterious MACHOs[42] can be just such very early and almost non-luminous stars.

An observation of antistars in the Galaxy would present a very strong support of the suggested scenario. Since antimatter was created in clouds with the unusually high baryon-to-photon ratio β, the BBN in these clouds would produce anomalous abundances of light elements. So it may be helpful for the discovery of cosmic antimatter to look firstly for the clouds with anomalous chemistry in the galaxy. Though with 50% probability it may be the normal matter with anomalous n_B/n_γ, there is a good chance that this region is made of antimatter and so it should be thoroughly investigated, especially by searching for gamma radiation produced by the matter-antimatter annihilation.

Another interesting feature of this model of antimatter formation is that heavy antinuclei, not only anti-He^4 can also be found in the cosmic rays. Their abundances could be similar to those observed in SN explosions.

Acknowledgement

This work was supported by the grant of the Russian Federation government 11.G34.31.0047.

References

1. P.A.M. Dirac, *Proc. Royal Soc. London*, **A117**, 610 (1928).
2. A. Schuster, *Nature*, **58**, 367 (1898).
3. R. Duperray, B. Baret, and D. Maurin *et al.*, *Phys. Rev.* **D71**, 083013 (2005);
   B. Baret, A. Barrau, and G. Boudoul, *et al. AIP Conf. Proc.* **842**, 1004 (2006).
4. N. Martin [ALICE Collaboration], *Journal of Physics: Conference Series* **455**, 012007 (2013).
5. G. Steigman, *Ann. Rev. Astron. Astrophys.* **14**, 339 (1976).
6. G. Steigman, *JCAP* **0810**, 001 (2008).
7. P. von Ballmoos, *Hyperfine Interact.* **228**, 3, 91 (2014).
8. S.I. Blinnikov, A.D. Dolgov, and K.A. Postnov, *Phys. Rev.* **D92**, 2, 023516 (2015)
9. F.W. Stecker, *The matter-antimatter asymmetry of the universe, keynote lecture at XIVth Rencontres de Blois 2002, Matter-Antimatter Asymmetry*, arXiv:hep-ph/0207323;
   A.D. Dolgov, *Cosmological matter antimatter asymmetry and antimatter in the universe, keynote lecture at XIVth Rencontres de Blois 2002, Matter-Antimatter Asymmetry*, arXiv:hep-ph/0211260, and references therein.
10. A.D. Dolgov and V.A. Novikov, *Phys. Lett.* **B732**, 244 (2014).
11. A.D. Dolgov, I.B. Khriplovich, and A.S. Rudenko *JETP Lett.* **96**, 421 (2012).
12. A.D. Dolgov, V.A. Novikov, and M.I. Vysotsky, *JETP Lett.* **98**, 519 (2013).
13. A.D. Sakharov, *Pis'ma Zh. Eksp. Teor. Fiz.* **5**, 32 (1967) [*JETP Lett.* **5**, 24 (1967)].
14. I. Affleck and M. Dine, *Nuclear Physics* **B249**, 361 (1985).
15. A. Dolgov and J. Silk, *Phys. Rev.* **D47**, 4244 (1993).
16. A.D. Dolgov, M. Kawasaki, and N. Kevlishvili, *Nuclear Physics* **B807**, 229 (2009).
17. S.R. Coleman and E.J. Weinberg, *Phys. Rev.* **D7**, 1888 (1973).
18. R. Nakamura, M.-A. Hashimoto, S.-I. Fujimoto, N. Nishimura, and K. Sato, *arXiv:1007.0466*;
    S. Matsuura, S.-I. Fujimoto, S. Nishimura, M.-A. Hashimoto, and K. Sato, *Phys. Rev.* **D 72**, 123505 (2005), arXiv:astroph/0507439;
    S. Matsuura, S.-I. Fujimoto, M.-A. Hashimoto, and K. Sato, *Phys. Rev.*

D 75, 068302 (2007), arXiv:0704.0635 [astro-ph];
S. Matsuura, A.D. Dolgov, S. Nagataki, and K. Sato, *Prog. Theor. Phys.* **112**, 971 (2004), arXiv:astro-ph/0405459.
19. C. Bambi and A.D. Dolgov, *Nuclear Physics* **B784**, 132 (2007), arXiv:astro-ph/0702350.
20. A.D. Dolgov and S.I. Blinnikov, *Phys. Rev.* **D89**, 2, 021301 (2014).
21. Planck Collaboration, *A. & A.* **571**, A16 (2014), arXiv:1303.5076 [astro-ph.CO].
22. Z. Berezhiani, A.D. Dolgov, and I.I. Tkachev, arXiv:1505.03644 [astro-ph.CO].
23. J.J. Cowan, C. Sneden, and S. Burles, *et al. Astrophys. J.* **572**, 861 (2002), arXiv:astro-ph/0202429.
24. J. Kalirai, *Nature*, **486**, 90 (2012), arXiv:1205.6802 [astro-ph].
25. A. Frebe, N. Christlieb, J.E. Norris, and C. Thom, *Astrophys. J.* **660**, L117 (2007), arXiv:astro-ph/0703414.
26. H.E. Bond, E.P. Nelan, D.A. VandenBerg, G.H. Schaefer, and D. Harmer, *Astrophys. J. Lett.* **765**, L12 (2013), arXiv:1302.3180 [astro-ph].
27. F. Ozel, D. Psaltis, R. Narayan, and J.E. McClintock, *Astrophys. J.* **725**, 1918 (2010), arXiv:1006.2834 [astro-ph.GA].
28. L. Kreidberg, C.D. Bailyn, W.M. Farr, and V. Kalogera, *astrophys. J.* **757**, 36 (2012), arXiv:1205.1805 [astro-ph.HE].
29. W.M. Farr, N. Sravan, A. Cantrell, L. Kreidberg, C.D. Bailyn, I. Mandel, and V. Kalogera, *Astrophys. J.* **741**, 103 (2011), *arXiv:1011.1459.*
30. N. Häring and H.-W. Rix, *Astrophys. J. Lett.* **604**, L89 (2004), arXiv:astro-ph/0402376;
E. Sani, A. Marconi, L.K. Hunt, and G. Risaliti, *Mon. Not. R. Astron. Soc.* **413**, 1479 (2011), arXiv:1012.3073 [astro-ph.CO].
31. R.C.E. van den Bosch, K. Gebhardt, K. Gültekin, G. van de Ven, A. van der Wel, and J.L. Walsh, *Nature*, **491**, 729 (2012), arXiv:1211.6429 [astro-ph.CO].
32. F. Khan, K. Holley-Bockelmann, and P. Berczik, *Astrophys. J.* **798**, 103 (2015), arXiv:1405.6425 [astro-ph.GA].
33. J. Strader, *et al*, *Astrophys. J. Lett.* **775**, L6 (2013), arXiv:1307.7707 [astro-ph.CO].
34. J.F. Hennawi *et al.*, *Science*, **348**, 779 (2015).
35. W. Zheng, M. Postman, A. Zitrin, *et al.*, *Nature* **489**, 406 (2012), arXiv:1204.2305 [astro-ph.CO].
36. D. Coe, A. Zitrin, M. Carrasco, *et al. Astrophys. J.* **762**, 32 (2013),

arXiv:1211.3663 [astro-ph.CO].

37. D.J. Mortlock, *et al.*, *Nature* **474**, 616 (2011), arXiv:1106.6088 [astro-ph.CO].
38. F. Melia, *Astrophys. J* **147**, 120 (2014), arXiv:1403.0908 [astro-ph.GA].
39. X.-B. Wu *et al.*, *Nature* **518**, 512 (2015).
40. C.-W. Tsai, P.R.M. Eisenhardt *et al.*, *Astrophys, J.* **805**, 90 (2015), arXiv:1410.1751 [astro-ph.GA].
41. L. Mattsson, arXiv:1505.04758 [astro-ph.GA].
42. C. Alcock, R.A. Allsman, D.R. Alves, *et al.*, *Astrophys. J.* **542**, 281 (2000), arXiv:astro-ph/0001272;
    P. Tisserand, L. Le Guillou, C. Afonso, *et al.*, *A&A* **469**, 387 (2007), arXiv:astro-ph/0607207;
    A. Riffeser, S. Seitz, and R. Bender, *Astrophys. J.* **684**, 1093 (2008), arXiv:0805.0137 [astro-ph].

MANIFESTATIONS OF THE ROTATION AND GRAVITY OF THE EARTH IN SPIN PHYSICS EXPERIMENTS

YURI N. OBUKHOV

Theoretical Physics Laboratory, Nuclear Safety Institute, RAS,
52 B. Tulskaya, 115191 Moscow, Russia, E-mail: obukhov@ibrae.ac.ru

ALEXANDER J. SILENKO

Research Institute for Nuclear Problems, Belarusian State University,
Minsk 220030, Belarus
and
Bogoliubov Laboratory of Theoretical Physics,
Joint Institute for Nuclear Research, Dubna 141980, Russia
E-mail: alsilenko@mail.ru

OLEG V. TERYAEV

Bogoliubov Laboratory of Theoretical Physics,
Joint Institute for Nuclear Research, Dubna 141980, Russia
E-mail: teryaev@theor.jinr.ru

An influence of the rotation and gravity of the Earth on the particle motion and the spin evolution is not negligible and it should be taken into account in spin physics experiments. The Earth rotation brings the Coriolis and centrifugal forces in the lab frame and also manifests in the additional rotation of the spin and in the change of the Maxwell electrodynamics. The change of the Maxwell electrodynamics due to the Earth gravity is much smaller and can be neglected. One of manifestations of the Earth rotation is the Sagnac effect. The electric and magnetic fields acting on the spin in the Earth's rotating frame coincide with the corresponding fields determined in the inertial frame instantly accompanying a lab. The effective electric field governing the particle motion differs from the electric field in the instantly accompanying frame. Nevertheless, the difference between the conventional Lorentz force and the actual force in the Earth's rotating frame vanishes on average in accelerators and storage rings due to the beam rotation. The Earth gravity manifests in additional forces acting on particles/nuclei and in additional torques acting on the spin. The additional forces are the Newton-like force and the reaction force provided by a focusing system. The additional torques are caused by the corresponding focusing field and by the geodetic effect. As a result, the Earth gravity leads to the additional spin rotation about the radial axis which may not be negligible in EDM experiments.

1. Introduction

A high precision of actual experiments gives rise to a necessity of taking into account effects caused by the Earth's rotation and gravity. The experiments are performed in a non-Minkowskian spacetime. The non-diagonal components of the metric are equal to $g_{0i} = -(\boldsymbol{\omega} \times \boldsymbol{R})^i/c$, where $\boldsymbol{\omega}$ is the angular velocity of the Earth rotation ($\omega = 7.27 \times 10^{-5}$ rad/s) and $\boldsymbol{R}$ is the radius vector of the given point on the Earth surface ($R \approx 6370$ km). These components are of the order of 10^{-6} and the effects conditioned by the Earth rotation cannot be *a priori* neglected. We need to take into account the spin precession in the rotating frame and a change of the Maxwell electrodynamics due to the non-Minkowskian spacetime metric. The corresponding change of electrodynamics due to the Earth gravity is much less. However, a spin motion in the gravitational field of the Earth and an influence of the Newton-like force on particle dynamics should be considered.

We use the system of units $\hbar = 1,\ c = 1$. Nevertheless, we include these constants into some formulas. The world indices are labeled by Greek letters, whereas we reserve Latin letters from the beginning of the alphabet for tetrad indices. The Latin letters $i, j, k \ldots$ denote spatial world indices.

2. Electromagnetic interactions of a Dirac particle in a rotating frame

Since we consider spin-1/2 particle, the quantum-mechanical description is the most adequate. The perfect agreement between the quantum-mechanical and classical descriptions of a spinning particle in arbitrary gravitational and inertial fields has been stated in Ref. 1.

The initial covariant Dirac equation has the form (see Ref. 2 and references therein)

$$(i\hbar\gamma^a D_a - m)\Psi = 0, \qquad a = 0, 1, 2, 3, \tag{1}$$

where D_a is the spinor covariant derivative. The Dirac matrices γ^a defined in local Lorentz frames have the usual form. Equation (1) describes gravitational (inertial) and electromagnetic interactions.

In Ref. 2, an exact Hermitian Hamiltonian in the Dirac representation corresponding to Eq. (1) has been determined. To give a more detailed description of electromagnetic interactions of a Dirac particle, an anomalous magnetic moment (AMM) of the particle should be taken into account. For this purpose, we use the generalization[3] of the initial equation, introducing the AMM μ' and the electric dipole moment (EDM) d:

$$\left(i\hbar\gamma^a D_a - m + \frac{\mu'}{2}\sigma^{ab}F_{ab} + \frac{d}{2}\sigma^{ab}G_{ab}\right)\Psi = 0. \tag{2}$$

The use of tetrad indices for Dirac matrices is caused by the definition of the three-component spin (pseudo)vector in the particle rest frame. This frame belongs to local Lorentz frames. After passing to a Minkowski spacetime, Eq. (2) coincides with the corresponding equation for a Dirac particle with the AMM and EDM (see Ref. 4). Quantities F_{ab} and G_{ab} are defined by $F_{ab} = e^\mu_a e^\nu_b F_{\mu\nu}$ and $G_{ab} = e^\mu_a e^\nu_b G_{\mu\nu}$, where $F_{\mu\nu} = (\boldsymbol{E}, \boldsymbol{B})$ is the electromagnetic field tensor and the tensor $G_{\mu\nu} = (-\boldsymbol{B}, \boldsymbol{E})$ is obtained with the dual transformation ($G^{\mu\nu} = \frac{1}{2}\eta^{\mu\nu\lambda\rho}F_{\lambda\rho}$). We can mention that $F_{ab} = (\boldsymbol{\mathfrak{E}}, \boldsymbol{\mathfrak{B}})$, $G_{ab} = (-\boldsymbol{\mathfrak{B}}, \boldsymbol{\mathfrak{E}})$, where

$$\boldsymbol{\mathfrak{E}} = \boldsymbol{E} + \boldsymbol{G}, \quad \boldsymbol{\mathfrak{B}} = \boldsymbol{B}, \tag{3}$$

$$\boldsymbol{G} = \boldsymbol{B} \times (\boldsymbol{\omega} \times \boldsymbol{r}), \quad \boldsymbol{E} = -\nabla\Phi - \frac{\partial \boldsymbol{A}}{\partial t}, \quad \boldsymbol{B} = \nabla \times \boldsymbol{A}. \tag{4}$$

The quantities $\boldsymbol{\mathfrak{E}}$ and $\boldsymbol{\mathfrak{B}}$ are the effective fields in the rotating frame.

The general form of the Dirac Hamiltonian found in Ref. 2 does not contain additional terms characterizing the AMM and EDM. In the considered case, an inclusion of these terms lead to the following Dirac Hamiltonian[3]:

$$\mathcal{H} = \beta m + e\Phi + \boldsymbol{\alpha}\cdot\boldsymbol{\pi} - \boldsymbol{\omega}\cdot(\boldsymbol{r}\times\boldsymbol{\pi}) - \frac{\hbar}{2}\boldsymbol{\omega}\cdot\boldsymbol{\Sigma} - \boldsymbol{\Pi}\cdot\boldsymbol{\mathcal{M}} + i\boldsymbol{\gamma}\cdot\boldsymbol{\mathcal{P}}, \tag{5}$$

where

$$\boldsymbol{\mathcal{M}} = \mu'\boldsymbol{\mathfrak{B}} + d\boldsymbol{\mathfrak{E}}, \quad \boldsymbol{\mathcal{P}} = \mu'\boldsymbol{\mathfrak{E}} - d\boldsymbol{\mathfrak{B}}, \quad \boldsymbol{\pi} = -i\hbar\nabla - e\boldsymbol{A}, \tag{6}$$

and $\boldsymbol{\pi}$ is the kinetic momentum operator.

Now we fulfill the transformation to the FW representation[5] with the method developed in Refs. 6, 7. This representation holds a special place in quantum mechanics due to some unique properties. In this representation, quantum mechanical operators for relativistic particles in an external field have the same form as in the nonrelativistic quantum theory. In particular, the position operator[8] and momentum operator are equal to $\boldsymbol{r}$ and $\boldsymbol{p} = -i\hbar\nabla$, and the polarization operator for spin-1/2 particles is expressed by the Dirac matrix $\boldsymbol{\Pi}$. In other representations, these operators are expressed by much more cumbersome formulas (see Refs. 5, 9). The relations between the operators in the FW representation are analogous to the relations between the corresponding classical quantities. The simple form of operators corresponding to classical observables is a great advantage of this representation. The above-mentioned properties of the FW representation allow one its successful use for passing to the semiclassical approximation and to the classical limit of relativistic quantum mechanics.[5,10] We note that the Hamiltonian and all other operators are diagonal in two spinors (block-diagonal) in this representation.

When the FW representation is used, the passage to the classical limit is usually accomplished by replacing the operators in expressions for the Hamiltonian and in operator equations of motion with the corresponding classical quantities. The possibility of such a replacement, explicitly or implicitly used in practically all works devoted to the relativistic FW transformation, was recently rigorously proved in Ref. 11. This possibility radically simplifies an interpretation of basic quantum-mechanical equations, especially in the relativistic case.

Initial Hamiltonian (5) can be presented in the following general form:

$$\mathcal{H}=\beta m+\mathcal{E}+\mathcal{O}, \quad \beta\mathcal{E}=\mathcal{E}\beta, \quad \beta\mathcal{O}=-\mathcal{O}\beta, \tag{7}$$

where $\mathcal{E}$ and $\mathcal{O}$ are even and odd (diagonal and off-diagonal in two spinors) operators, respectively.

The Hamiltonian transformed to the FW representation is exact for terms proportional to the zeroth and first powers of the Planck constant and also for terms describing contact interactions and proportional to $\hbar^2$. It is given by Ref. 6, 12

$$\begin{aligned}\mathcal{H}_{FW}&=\beta\epsilon+\mathcal{E}-\frac{1}{8}\left\{\frac{1}{\epsilon(\epsilon+m)},[\mathcal{O},[\mathcal{O},\mathcal{F}]]\right\},\\ \mathcal{F}&=\mathcal{E}-i\frac{\partial}{\partial t}, \qquad \epsilon=\sqrt{m^2+\mathcal{O}^2}.\end{aligned} \tag{8}$$

In the considered case, it is convenient to present this Hamiltonian in the form[3]

$$\mathcal{H}_{FW}=\mathcal{H}_{FW}^{(D)}+\mathcal{H}_{FW}^{(add)}, \tag{9}$$

$$\begin{aligned}\mathcal{H}_{FW}^{(D)}&=\beta\epsilon'+e\Phi-\boldsymbol{\omega}\cdot(\boldsymbol{r}\times\boldsymbol{\pi})-\frac{\hbar}{2}\boldsymbol{\omega}\cdot\boldsymbol{\Sigma}-\frac{e\hbar}{4}\left\{\frac{1}{\epsilon'},\boldsymbol{\Pi}\cdot\boldsymbol{B}\right\}\\ &+\frac{e\hbar}{8}\left\{\frac{1}{\epsilon'(\epsilon'+m)},\left[\boldsymbol{\Sigma}\cdot(\boldsymbol{\pi}\times\boldsymbol{\mathfrak{E}}-\boldsymbol{\mathfrak{E}}\times\boldsymbol{\pi})-\hbar\nabla\cdot\boldsymbol{\mathfrak{E}}\right]\right\},\end{aligned} \tag{10}$$

$$
\begin{aligned}
\mathcal{H}_{FW}^{(add)} &= \frac{1}{4}\left\{\frac{1}{\epsilon'}, \left[\boldsymbol{\Sigma}\cdot(\boldsymbol{\pi}\times\boldsymbol{\mathcal{P}}-\boldsymbol{\mathcal{P}}\times\boldsymbol{\pi})-\hbar\nabla\cdot\boldsymbol{\mathcal{P}}\right]\right\}-\boldsymbol{\Pi}\cdot\boldsymbol{\mathcal{M}} \\
&+\frac{1}{4}\left\{\frac{1}{\epsilon'(\epsilon'+m)}, \left[(\boldsymbol{\Pi}\cdot\boldsymbol{\pi})(\boldsymbol{\pi}\cdot\boldsymbol{\mathcal{M}})+(\boldsymbol{\mathcal{M}}\cdot\boldsymbol{\pi})(\boldsymbol{\Pi}\cdot\boldsymbol{\pi})\right.\right. \\
&\left.\left.+2\pi\beta\hbar\,(\boldsymbol{\pi}\cdot\boldsymbol{\mathcal{J}}+\boldsymbol{\mathcal{J}}\cdot\boldsymbol{\pi})+\beta\frac{\hbar}{2}\left\{\left([\boldsymbol{\omega}\times\boldsymbol{r}]\cdot\nabla\right),(\boldsymbol{\pi}\cdot\boldsymbol{\mathcal{P}})\right\}\right]\right\},
\end{aligned}
\tag{11}
$$

where $\epsilon' = \sqrt{m^2+\boldsymbol{\pi}^2}$ and $\boldsymbol{\mathcal{J}} = [\boldsymbol{\nabla}\times\boldsymbol{\mathcal{M}} - \partial\boldsymbol{\mathcal{P}}/(\partial t)]/(4\pi)$. The operator $\mathcal{H}_{FW}^{(D)}$ results from a transformation of the Hamiltonian, corresponding Eq. (1), and the operator $\mathcal{H}_{FW}^{(add)}$ contains terms, proportional to the AMM and EDM. In the last term of Eq. (11), the nabla operator acts on $\boldsymbol{\mathcal{P}}$ and defines derivatives of this quantity.

It is important that Eqs. (9)–(11) are exact in relation to the above-mentioned terms proportional to the zeroth, first, and second powers of $\hbar$. We do not calculate other terms of the second power of $\hbar$ and neglect terms of higher powers of $\hbar$. We are to underline that μ', d, $\boldsymbol{\mathcal{P}}$, and $\boldsymbol{\mathcal{M}}$ are proportional to $\hbar$.

3. Effective fields acting on particle and spin

The results obtained in the previous section allow us to determine effective fields acting on a particle and a spin. Let us use the simplest way of this determination.

Let $\boldsymbol{r}$ be the radius-vector of a spinning particle possessing an EDM. We can compare the spin dynamics in the frame rotating with the angular velocity $\boldsymbol{\omega}$ and in the instantly accompanying frame moving with the velocity $\boldsymbol{V} = \boldsymbol{\omega}\times\boldsymbol{r}$ relative to the lab frame.[13] It is convenient to consider the simplest case when the particle velocity is equal to zero in the both frames in the considered moment of time. To describe the spin motion in the instantly accompanying frame, we suppose $\boldsymbol{\omega} = 0$ in Eqs. (9)–(11). For a particle at rest, the spin-dependent terms in the two Hamiltonians are given by

$$
\begin{aligned}
\mathcal{H}_{FW}^{(acc)} &= -\mu\boldsymbol{\Pi}\cdot\boldsymbol{B}' - d\boldsymbol{\Pi}\cdot\boldsymbol{E}', \\
\mathcal{H}_{FW}^{(rot)} &= -\frac{\hbar}{2}\boldsymbol{\omega}\cdot\boldsymbol{\Sigma} - \mu\boldsymbol{\Pi}\cdot\mathfrak{B} - d\boldsymbol{\Pi}\cdot\mathfrak{E},
\end{aligned}
\tag{12}
$$

where $\boldsymbol{E}'$ and $\boldsymbol{B}'$ are the field acting on the particle in the instantly accompanying frame.

In the general case, the spin-dependent terms in the FW Hamiltonian have the form $\boldsymbol{\Omega}\cdot\boldsymbol{s}$, where $\boldsymbol{\Omega}$ is the operator of the angular velocity of spin precession and $\boldsymbol{s}$ is the spin operator equal to $\hbar\boldsymbol{\Pi}/2$ or $\hbar\boldsymbol{\Sigma}/2$. A difference between the angular velocities of spin precession in the two frames can be caused by (i) the frame rotation with the angular velocity $\boldsymbol{\omega}$ and (ii) the Thomas precession with the angular velocity $\boldsymbol{\omega}_T = -[\gamma^2/(\gamma+1)]\boldsymbol{\beta}\times\dot{\boldsymbol{\beta}}$, $\boldsymbol{\beta} = \boldsymbol{v}/c$. However, the Thomas precession is zero for a particle at rest. Therefore, Eq. (12) results in

$$\boldsymbol{\mathfrak{E}} = \boldsymbol{E}', \quad \boldsymbol{\mathfrak{B}} = \boldsymbol{B}', \quad \boldsymbol{E} = \boldsymbol{E}' - \boldsymbol{B}' \times (\boldsymbol{\omega}\times\boldsymbol{r}). \tag{13}$$

These relations have a simple physical sense. Let us consider the Earth's rotating frame. In this case, the effective fields acting on the particle spin (but not on the particle momentum!) are equal to fields defined with formulas of special relativity for a given distribution of charges and currents.

To determine the particle motion, we are to find the classical limit of the Hamiltonian (9) and to calculate the force. This force is given by[14] $\boldsymbol{F} = d\boldsymbol{\mathfrak{P}}/(dt)$, where $\mathfrak{P}^i = -\delta^{ij}p_j + g^{0i}p_0$ is a spatial part of the *covariant* four-momentum $\mathfrak{P}^\mu = g^{\mu\nu}p_\nu$ as written in Ref. 15. A sufficient precision is ensured by taking into account terms linear in $\boldsymbol{E}$ and $\boldsymbol{B}$ and leading inertial terms. After replacing of the operators in Eqs. (9)–(11) with corresponding classical quantities, we determine this force:

$$\boldsymbol{F} = \boldsymbol{E} + \boldsymbol{\beta}\times\boldsymbol{B} + 2\boldsymbol{\pi}\times\boldsymbol{\omega} + \epsilon'\boldsymbol{\omega}\times(\boldsymbol{\omega}\times\boldsymbol{r}). \tag{14}$$

Here $\boldsymbol{F}$ is the sum of the Lorentz, Coriolis and centrifugal forces. Since $\boldsymbol{\pi} = \epsilon'(\boldsymbol{v} + \boldsymbol{\omega}\times\boldsymbol{r})$ (see Ref. 29), the inertial accelerations are given by

$$\boldsymbol{w}_{Cor} = 2\boldsymbol{v}\times\boldsymbol{\omega}, \quad \boldsymbol{w}_{cf} = -\boldsymbol{\omega}\times(\boldsymbol{\omega}\times\boldsymbol{r}). \tag{15}$$

Therefore, the effective electric field acting on the particle momentum is equal to $\boldsymbol{E}$. It differs from the effective electric field acting on the particle spin. The corresponding effective magnetic fields coincide.

We need also to mention the Sagnac effect.[16] In storage rings, it manifests in a nonequivalence of different parts of a particle orbit. The time spent by the particle to pass a certain part of the orbit depends on a direction of its velocity relative to the velocity of the Earth rotation. Thus, the particle passes two halves of the orbit during different intervals of time. While the *coordinate* velocity of the particle depends on the Earth rotation, the *measurable* velocity of the particle cannot exceed c (see explanations given in Refs. 17, 18).

4. Manifestations of the Earth rotation

The Earth rotation influences the particle motion and the spin precession. The angular velocity of the spin precession acquires the additional term

$-\boldsymbol{\omega}$. However, the change of the angular velocity of the particle revolution is equal to $2\boldsymbol{\omega}$. Therefore, the resulting angular velocity of the spin precession relative to immobile detectors is given by

$$\boldsymbol{\Omega}_{res} = \boldsymbol{\Omega} + \boldsymbol{\omega} + \boldsymbol{\Omega}'_g, \tag{16}$$

where $\boldsymbol{\Omega}$ is the angular velocity of spin rotation in the instantly accompanying frame *without regard to gravity* and $\boldsymbol{\Omega}'_g$ defines the total contribution of the Earth gravity. This contribution will be considered in the next section. The quantity $\boldsymbol{\Omega}$ depends on the magnetic and electric dipole moments and is given by the generalized Thomas-Bargmann-Michel-Telegdi equation[19] extended on the electric dipole moment:

$$\begin{aligned}\boldsymbol{\Omega} = -\frac{e}{mc}\Bigg[&\left(G+\frac{1}{\gamma}\right)\boldsymbol{B} - \frac{\gamma G}{\gamma+1}(\boldsymbol{\beta}\cdot\boldsymbol{B})\boldsymbol{\beta} - \left(G+\frac{1}{\gamma+1}\right)\boldsymbol{\beta}\times\boldsymbol{E} \\ &+ \frac{\eta}{2}\left(\boldsymbol{E} - \frac{\gamma}{\gamma+1}(\boldsymbol{\beta}\cdot\boldsymbol{E})\boldsymbol{\beta} + \boldsymbol{\beta}\times\boldsymbol{B}\right)\Bigg].\end{aligned} \tag{17}$$

Here $G = (g-2)/2$, $g = 2mc\mu/(e\hbar s)$, $\eta = 2mcd/(e\hbar s)$, $\boldsymbol{\beta} = \mathbf{v}/c$, γ is the Lorentz factor, μ and d are the magnetic and electric dipole moments, and s is the spin quantum number. An extension of the T-BMT equation due to the EDM has already been discussed in the original paper of Bargmann, Michel and Telegdi.[19] Then, the equation of spin motion of the particle with the anomalous magnetic moment μ' and the EDM has been obtained in Refs. 20, 21 by the dual transformation $\mu' \to d$, $\boldsymbol{B} \to \boldsymbol{E}$, $\boldsymbol{E} \to -\boldsymbol{B}$. The rigorous derivation of this equation has been presented in Ref. 22. The resulting equation of spin motion coincides with that presented in Refs. 20, 21. However, the derivation fulfilled in Ref. 22 has not used the supplementary assumption of dual symmetry.

The term $\boldsymbol{\omega}$ in Eq. (16) is rather small and can usually be neglected in storage ring experiments. In particular, it is negligible in all $g-2$ experiments. This term can be of minor importance in storage-ring EDM experiments based on the frozen spin method. However, the contribution of the Earth rotation is very important in EDM experiments with atoms at rest[23] and in some other experiments with atoms.[24] In these experiments, it is equal to $-\boldsymbol{\omega}$. The contribution of the Earth rotation was taken into account to obtain some restrictions on hypothetical interactions. In particular, a reconsideration of the experimental data[24] has allowed us to maintain new bounds on the spacetime torsion.[25] It is important that the Cartan spacetime torsion couples only to the particle spin and is not attached to the orbital angular momentum of a test particle.[26] If the spacetime torsion is nonzero, the angular velocity of atom spin rotation in the Earth's rotating frame differs from $-\boldsymbol{\omega}$.

As a rule, the vectors $\boldsymbol{\Omega}$ and $\boldsymbol{\omega}$ are not collinear. When the vector $\boldsymbol{\Omega}$ is vertically directed (this always takes place in storage rings), only the vertical component of the angular velocity of the Earth rotation and the radial component of the Coriolis acceleration are important. The contribution of the Earth rotation to the angular frequency of the spin precession is equal to $\omega \sin\varphi$, where φ is the geographical latitude. For atoms at rest, the horizontal component of $\boldsymbol{\omega}$ can imitate the presence of the EDM. For particles and nuclei in storage rings, this component may be disregarded because the radial and longitudinal projections of the angular velocity of the Earth rotation, ω_r and ω_ϕ, are zeroth on average.

The appearance of the effective electric field originated from the magnetic field in the instantly accompanying frame may be important. This field is created by magnets moving together with the rotating Earth. Any charged particle or ion moves in this field with an acceleration. As a result, the particle or ion drifts in the plane orthogonal to the magnetic field. For particles and nuclei in storage rings, this effect is, however, eliminated by a magnetic or electric focusing field. Since the quasi-electric field has a constant direction in the particle orbit plane, an average radial projection of this field is zero. Therefore, the quasi-electric field $-\boldsymbol{B} \times (\boldsymbol{\omega} \times \boldsymbol{r})$ does not bring any systematical corrections to the cyclotron frequency and to the angular velocity of spin precession in storage rings.

The Sagnac effect may lead to observable effects for particles and nuclei in storage rings. While particles and nuclei pass two halves of the orbit during different intervals of time, the main correction proportional to the radius of the Earth vanishes. However, the smaller correction proportional to the ring radius is nonzero and can be important.[27,28]

5. Manifestations of the Earth gravity

Main manifestations of the Earth gravity in storage rings and accelerators have been first determined in Ref. 29. The Earth attraction manifests in additional forces acting on particles/nuclei and in additional torques acting on the spin. The additional forces are the Newton-like force (a relativistic generalization of the Newton force) and the reaction force provided by a focusing system. The additional torques are caused by the corresponding focusing field and by the geodetic effect (the spin precession in a gravitational field). In storage ring EDM experiments, the latter torque leads to the spin rotation about the radial axis with the angular velocity[30]

$$\boldsymbol{\Omega}_g = \frac{2\gamma+1}{c(\gamma+1)}\boldsymbol{\beta}\times\boldsymbol{g}, \quad \boldsymbol{\beta} = \frac{\boldsymbol{v}}{c}, \tag{18}$$

where $\boldsymbol{g}$ is the Newtonian acceleration and γ is the Lorentz factor. The gravitational force acting on the particle is given by[29,30]

$$\boldsymbol{F}_g = \frac{d\boldsymbol{p}}{dt} = \frac{m^2c^2+2\boldsymbol{p}^2}{c\sqrt{m^2c^2+\boldsymbol{p}^2}}\boldsymbol{g} = \frac{2\gamma^2-1}{\gamma}m\boldsymbol{g}. \tag{19}$$

When magnetic focusing is used, this gravitational force defines the nonzero radial magnetic field

$$B_r = \frac{(2\gamma^2-1)mg}{\sqrt{\gamma^2-1}|e|}. \tag{20}$$

This field causes the spin rotation with the average angular velocity[29]

$$\boldsymbol{\Omega}_m = -\frac{(1+G\gamma)(2\gamma^2-1)}{c(\gamma^2-1)}\boldsymbol{\beta}\times\boldsymbol{g}. \tag{21}$$

When electric focusing is used, the gravitational force (19) gives rise to the vertical electric field

$$E_z = \frac{(2\gamma^2-1)mg}{\gamma e}. \tag{22}$$

The corresponding average angular velocity of the spin rotation is given by

$$\boldsymbol{\Omega}_e = -\frac{2\gamma^2-1}{c\gamma}\left(G+\frac{1}{\gamma+1}\right)\boldsymbol{\beta}\times\boldsymbol{g}. \tag{23}$$

The total effect of the Earth gravity on the spin motion reads

$$\begin{aligned}\boldsymbol{\Omega}_g+\boldsymbol{\Omega}_m &= -\frac{\gamma\left[1+G(2\gamma^2-1)\right]}{c(\gamma^2-1)}\boldsymbol{\beta}\times\boldsymbol{g},\\ \boldsymbol{\Omega}_g+\boldsymbol{\Omega}_e &= \frac{1-G(2\gamma^2-1)}{c\gamma}\boldsymbol{\beta}\times\boldsymbol{g}.\end{aligned} \tag{24}$$

Thus, the total contribution of the Earth gravity is equal to

$$\boldsymbol{\Omega}'_g = \boldsymbol{\Omega}_g + \boldsymbol{\Omega}_m$$

and

$$\boldsymbol{\Omega}'_g = \boldsymbol{\Omega}_g + \boldsymbol{\Omega}_e$$

when one uses magnetic and electric focusing, respectively.

This effect may not be negligible. Calculation shows that the Earth gravity can bring the same effect as the deuteron EDM of $d = 1.5\times10^{-29}$

e·cm in the planned deuteron EDM experiment with magnetic focusing.[31] The effect of the Earth gravity can be important, because the expected sensitivity of this experiment[31] is of the same order.

In $g-2$ experiments, the influence of the Earth gravity on spin dynamics can be disregarded because the spin motion about the vertical axis is measured.

It is also instructive to estimate a change of electrodynamics due to the Earth gravity. It is sufficient to compare it with the corresponding change due to the Earth rotation. The change of diagonal components of the metric tensor caused by the Earth gravity is of the order of $r_g/R = 2gR/c^2$, where r_g is the gravitational radius of the Earth. The nondiagonal components of the metric tensor caused by the Earth rotation are of the order of $\omega R/c$. Thus, the change of electrodynamics due to the Earth rotation is $\omega c/g \approx 2200$ times greater and the influence of the Earth gravity on electromagnetic interactions can be neglected.

6. Summary

We considered the influence of the rotation and gravity of the Earth on the particle motion and on the spin evolution. This influence is not negligible and should be taken into account in high-energy-physics experiments. The Earth rotation brings the Coriolis and centrifugal forces in the lab frame and also manifests in the additional rotation of the spin and in the change of the Maxwell electrodynamics. The change of the Maxwell electrodynamics due to the Earth gravity is much smaller and can be neglected. The electric and magnetic fields acting on the spin in the Earth's rotating frame coincide with the corresponding fields determined in the inertial frame instantly accompanying a lab. The effective electric field governing the particle motion differs from the electric field in the instantly accompanying frame. Nevertheless, the difference between the conventional Lorentz force and the actual force in the Earth's rotating frame vanishes on average in accelerators and storage rings due to the beam rotation.

The Earth gravity manifests in additional forces acting on particles/nuclei and in additional torques acting on the spin. The additional forces are the Newton-like force and the reaction force provided by a focusing system. The additional torques are caused by the focusing field and by the geodetic effect. In storage ring EDM experiments, the additional forces and torques lead to the additional spin rotation about the radial axis. This effect may not be negligible in the EDM experiments.

Acknowledgments

This work was supported in part by the Belarusian Republican Foundation for Fundamental Research (Grant No. Φ14D-007) and by the Heisenberg-Landau program of the German Ministry for Science and Technology (Bundesministerium für Bildung und Forschung). A.S. acknowledges the hospitality and support of the Center for Axion and Precision Physics Research at the Korea Advanced Institute of Science and Technology.

References

1. Yu.N. Obukhov, A.J. Silenko, and O.V. Teryaev, *Phys. Rev.* **D88**, 084014 (2013); arXiv:1308.4552 [gr-qc].
2. Yu.N. Obukhov, A.J. Silenko, and O.V. Teryaev, *Phys. Rev.* **D84**, 024025 (2011).
3. Yu.N. Obukhov, A.J. Silenko, and O.V. Teryaev, to be published.
4. A.J. Silenko, *Russ. Phys. J.* **48**, 788 (2005).
5. L.L. Foldy and S.A. Wouthuysen, *Phys. Rev.* **78**, 29 (1950).
6. A.J. Silenko, *Phys. Rev.* **A77**, 012116 (2008).
7. A.J. Silenko, *Phys. Rev.* **A91**, 022103 (2015).
8. T.D. Newton and E.P. Wigner, *Rev. Mod. Phys.* **21**, 400 (1949).
9. A.J. Silenko, *J. Math. Phys.* **44**, 2952 (2003).
10. J.P. Costella and B.H.J. McKellar, *Am. J. Phys.* **63**, 1119 (1995).
11. A.J. Silenko, *Phys. Part. Nucl. Lett.* **10**, 91 (2013).
12. A.J. Silenko, *Theor. Math. Phys.* **176**, 987 (2013).
13. The results obtained in the present work can be applied for any inertial frame, in particular, for the center-of-mass one.
14. L.D. Landau and E.M. Lifshitz, *The Classical Theory of Fields*, 3rd ed., Pergamon Press, Oxford 1973, p. 251.
15. Alternatively, the force can be defined as $-d\boldsymbol{p}/(dt)$, where $\boldsymbol{p}$ is a spatial part of p_ν.
16. G. Sagnac, *CR Acad. Sci.* **157**, 708 (1913).
17. A.A. Logunov and Yu.V. Chugreev, *Sov. Phys. Usp.* **31**, 861 (1988).
18. G.B. Malykin, *Phys.-Usp.* **43**, 1229 (2000).
19. L.H. Thomas, *Nature (London)* **117**, 514 (1926); *Phil. Mag.* **3**, 1 (1927); V. Bargmann, L. Michel, and V.L. Telegdi, *Phys. Rev. Lett.* **2**, 435 (1959). This equation has also been obtained by J. Frenkel, *Z. Phys.* **37**, 243 (1926).
20. D.F. Nelson, A.A. Schupp, R.W. Pidd, and H.R. Crane, *Phys. Rev. Lett.* **2**, 492 (1959).

21. I.B. Khriplovich, *Phys. Lett.* **B444**, 98 (1998).
22. T. Fukuyama and A.J. Silenko, *Int. J. Mod. Phys.* **A28**, 1350147 (2013).
23. B.J. Venema, P.K. Majumder, S.K. Lamoreaux, B.R. Heckel, and E.N. Fortson, *Phys. Rev. Lett.* **68**, 135 (1992).
24. C. Gemmel *et al.*, *Eur. Phys. J.* **D57**, 303 (2010); C. Gemmel *et al.*, *Phys. Rev.* **D82**, 111901(R) (2010); M. Burghoff *et al.*, *J. Phys.: Conf. Ser.* **295**, 012017 (2011); W. Heil *et al.*, *Ann. Phys.* (Berlin) **525**, 539 (2013); F. Allmendinger, W. Heil, S. Karpuk, W. Kilian, A. Scharth, U. Schmidt, A. Schnabel, Yu. Sobolev, and K. Tullney, *Phys. Rev. Lett.* **112**, 110801 (2014).
25. Yu.N. Obukhov, A.J. Silenko, and O.V. Teryaev, *Phys. Rev.* **D90**, 124068 (2014).
26. F.W. Hehl, Y.N. Obukhov, and D. Puetzfeld, *Phys. Lett.* **A377**, 1775 (2013).
27. A.C. Melissinos, *"The Sagnac effect and the Tevatron"*, arXiv:1101.4008 [physics.acc-ph].
28. Y.K. Semertzidis, *"The Sagnac effect in the proton EDM experiment"*, CAPP/IBS, KAIST, 5 June 2014.
29. A.J. Silenko and O.V. Teryaev, *Phys. Rev.* **D76**, 061101(R) (2007).
30. A.J. Silenko and O.V. Teryaev, *Phys. Rev.* **D71**, 064016 (2005).
31. D. Anastassopoulos *et al* (EDM Collaboration), *"AGS Proposal: Search for a Permanent Electric Dipole Moment of the Deuteron Nucleus at the* 10^{-29} *e· cm Level,"* http://www.bnl.gov/edm/deuteron_proposal_080423_final.pdf

O.V. Teryaev (middle), Chernogolovka, 2015.

TESTING A POSSIBLE WAY OF GEOMETRIZATION OF THE STRONG INTERACTION BY A KALUZA – KLEIN STAR

SZILVIA KARSAI and PÉTER PÓSFAY

MTA Wigner RCP, P.O.B. 49, Budapest, Hungary, 1525,
and
Eötvös Loránd University,
Pázmány Péter sétány 1/A, Budapest, Hungary, 1117,
E-mail: karsai.szilvia@wigner.mta.hu, posfay.peter@wigner.mta.hu

GERGELY GÁBOR BARNAFÖLDI and BÉLA LUKÁCS

MTA Wigner RCP, P.O.B. 49, Budapest, Hungary, 1525,
E-mail: barnafoldi.gergely@wigner.mta.hu, lukacs.bela@wigner.mta.hu

Geometrization of the fundamental interactions has been extensively studied during the century. The idea of introducing compactified spatial dimensions originated by Kaluza and Klein. Following their approach, several model were built representing quantum numbers (e.g. charges) as compactified space-time dimensions. Such geometrized theoretical descriptions of the fundamental interactions might lead us to get closer to the unification of the principle theories. Here, we apply a $3+1_C+1$ dimensional theory, which contains one extra compactified spatial dimension 1_C in connection with the flavor quantum number in Quantum Chromodynamics. Within our model the size of the 1_C dimension is proportional to the inverse mass-difference of the first low-mass baryon states. We used this phenomena to apply in a compact star model – a natural laboratory for testing the theory of strong interaction and the gravitational theory in parallel. Our aim is to test the modification of the measurable macroscopical parameters of a compact Kaluza – Klein star by varying the size of the compactified extra dimension. Since larger the R_C the smaller the mass difference between the first spokes of the Kaluza – Klein ladder resulting smaller-mass stars. Using the Tolman – Oppenheimer – Volkov equation, we investigate the M-R diagram and the dependence of the maximum mass of compact stars. Besides testing the validity of our model we compare our results to the existing observational data of pulsar properties for constraints.

Keywords: compact star, neutron star, compactified extra dimensions, general relativity, Kaluza – Klein model

1. Introduction

Unification of the two successful theories: gravitational and quantum fields can happen at the Planck's scale in $3+1$ dimensional spacetime, however, these energy-regions are still far from the scope of the recent highest-energy experimental tests. By introducing extra compactified spatial dimension(s) in Kaluza – Klein theories, related observations might occur even at few-TeV energies.[1,2] Therefore geometrization of quantum fields, where charges are associated with compactified extra dimensions, are under investigations.

Since, compactified extra dimension(s) or such topological issues have no direct observables in the classical physics cases, thus search for these can start at the known highest energy systems: (i) collisions of nuclei in high-energy particle accelerators or (ii) inside extreme-energetic celestial objects such as compact stars or black holes.[3]

So far, we have already presented a simplified alternative Kaluza – Klein-like description for interactions via including a compactified extra dimension (1_C), which we could compare to the observables of strangeness-content compact objects. This idea is extended here in comparison to compact objects involving high-mass baryon states with charm and bottomness flavor content.

In this paper we would like to investigate how the maximum mass of a compact object depends on the size of the compactified extra dimension, R_C. We test the modification of the measurable macroscopical parameters of a compact Kaluza – Klein star by varying the size of the compactified extra dimension. We connect the R_C value to charm- and bottomness-content baryonic matter. Using the Tolman – Oppenheimer – Volkov equation, we study the M-R diagram and the dependence of the calculated maximum mass of compact stars. Beside testing the validity of our model we compare our results to the existing observational data of known pulsar properties for constraints and suggest new directions for the future theoretical developments.

2. The Kaluza – Klein Star in $3 + 1_C + 1$ Dimensional Spacetime

Observed symmetries of the standard 'low-energy' physical world suggest a space-time with $3+1$ macroscopical dimensions, however this can not exclude the existence of compactified extra dimensions at microscopical scales at extreme energies. Due to this fact constructing a generalized theory of compact stars living in a $3+1_C+1$ dimensional Kaluza – Klein spacetime must behave as a standard $3+1$ dimensional object as well. One can consider a $3+1_C+1$ dimensional space-time, where the particles have enough energy to move into the extra compactified space dimension indicated by 1_C.[4–7]

In the $3+1_C+1$ dimensional space-time, particles can move freely along the extra x^5 direction, however we require a periodic boundary condition, which results a Bohr-type quantization condition for k_5 momentum component. Generalization of the Heisenberg uncertainty relation induces an uncertainty in the position with the size of $2\pi R_C$, where R_C is the compactification radius as introduced above. Moreover, motion into the 5th dimension generates an extra mass term appears as 'excited mass', $\widehat{m}$ in the pure 4-dimensional description:

$$k_5 = \frac{n\,\hbar c}{R_C} \qquad \longrightarrow \qquad \widehat{m} = \sqrt{m^2 + \left(\frac{n\,\hbar c}{R_C}\right)^2} . \tag{1}$$

Considering a compactified radius $R_C \sim 10^{-12} - 10^{-13}$ cm this extra 'mass' is $\widehat{m} \sim 100$ MeV, which is a familiar quantity in hadron spectroscopy.[8]

Based on Eq. (1) one can introduce a 'pseudo charge' generated by k_5:

$$\widehat{q} = \pm\, n \cdot \frac{2\hbar\sqrt{G}}{c\, R_C} \ , \tag{2}$$

which acts as vector-scalar interaction. We can directly see that $\widehat{q}$ is *not* the electric charge, since it is connected to the gravitational constant and the compactified radius. Indeed $\widehat{q}^2 < 16\pi G m^2$, where G is the gravitational constant and m is the rest mass of the elementary particle. This idea is supported by Lukács and Pacher,[9] which paper reviewed the Brans – Dicke theory[10] and pointed out: the generation of extra mass by movement in the 5th direction is limited due to the charge-to-mass ratio of the observed particles such as e.g. proton. This model helped to rule out the electro-magnetic being of the charge. On the other hand, applying the model for the weak charge, opened a new direction showing the possibility for the unification and geometrization of the weak and gravitational theories. In this picture the extra compactified 5th dimension represents the e.g. hypercharge or other flavor quantum number (strangeness, charm or bottomness).

3. Theoretical Background of the Model

Describing a compact object one can assume symmetries, which help us to simplify both equations and their solution in the $1+3+1_C$ dimensional space-time. The first and simplest approximation is the *spherical symmetry* of the object in the $1+3+1_C$ dimensional space-time. Such an object is typically *static*, since stays in equilibrium with temperature close to *zero*. Assuming perfect fluid for the inner structure is a standard description including the *isotropy* in the standard 3 spatial dimensions. We generalized the *Killing-symmetries* in the 5D space-time, but an *anisotropy* needs to be chosen for the fluid in the microscopical 5th dimension. These symmetries yield to the metric *

$$ds^2 = \mathrm{e}^{2\nu}\,\mathrm{d}t^2 - \mathrm{e}^{2\lambda}\,\mathrm{d}r^2 - r^2\mathrm{d}\Omega^2 - \mathrm{e}^{2\Phi}(\mathrm{d}x^5)^2\ , \tag{3}$$

where functions ν, λ, and Φ depend on the radius r only. Here, we denote the usual spherical elementary surface by $\mathrm{d}\Omega^2$. From this metric, the 5D Einstein equation remains in the usual standard general form:

$$-\gamma\ T_{ik} = R_{ik} - \frac{1}{2}R^l{}_l\ g_{ik}\ . \tag{4}$$

This includes the applied 5-dimensional energy-momentum tensor containing energy density ε and pressure components P and P_5 in an anisotropic way $P \neq P_5$:

$$T^{ik} = \mathrm{diag}\ (\varepsilon\,\mathrm{e}^{2\nu},\ P\,\mathrm{e}^{2\lambda},\ P\,r^2,\ P\,r^2\sin^2\theta,\ P_5\,\mathrm{e}^{2\Phi})\ . \tag{5}$$

Substituting Eqs. (3) and (5) into the 5D Einstein equation in Eq. (4) one obtains the following differential-equation system:

$$-\gamma\ \varepsilon = -\frac{1}{r^2} + \mathrm{e}^{-2\lambda}\left[\Phi'' + \Phi'^2 - \lambda'\Phi' + \frac{2\Phi'}{r} - \frac{2\lambda'}{r} + \frac{1}{r^2}\right] \tag{6}$$

$$-\gamma P = \frac{1}{r^2} + \mathrm{e}^{-2\lambda}\left[-\nu'\Phi' - \frac{2\Phi'}{r} - \frac{2\nu'}{r} - \frac{1}{r^2}\right] \tag{7}$$

$$-\gamma P = \mathrm{e}^{-2\lambda}\left[\nu'\lambda' - \nu'' - \nu'^2 - \Phi'' - \Phi'^2 - \Phi'(\nu' - \lambda') - \frac{2\Phi'}{r} - \frac{\nu'}{r} + \frac{2\lambda'}{r}\right] \tag{8}$$

$$-\gamma P_5 = \frac{1}{r^2} + \mathrm{e}^{-2\lambda}\left[-\nu'' - \nu'^2 + \nu'\lambda' - \frac{2\nu'}{r} + \frac{2\lambda'}{r} - \frac{1}{r^2}\right] \tag{9}$$

where $\gamma = 8\pi G/c^4$ and G is the gravitational constant.

A typical neutron star has a surface temperature of $T = 10^6 - 10^8$ K, which is much more below the binding energy of nuclei, $T \approx 10^{10}$ K. Thus one can consider these objects as cold nuclear matter, especially inside the fermion star we can use the $T = 0$ approximation. The equation of state

---

*Coordinates x^i with Latin indices ($i = 0, 1, 2, 3$ and 5) follow the Einstein convention in the $1+3+1_C$ dimensional space-time.

(EoS) connects the thermodynamical quantities of the fluid, which can be e.g. the particle density, ρ. Thus, for the matter in a local co-moving frame we can write

$$\varepsilon = \varepsilon(\rho); \qquad P = P(\rho); \qquad P_5 = P_5(\rho) \ . \tag{10}$$

From the appropriate Bianchi identity we obtain

$$T^{ir}{}_{;r} = 0 \qquad \longrightarrow \qquad P' = -\nu'(\varepsilon + P) + (P_5 - P)\Phi' \ . \tag{11}$$

The influence of the extra dimensional behavior is seen in the latter Eq. (11), where in parallel to the normal isotropic pressure, $P = P_1 = P_2 = P_3$ and the pressure in the extra compactified dimension, P_5 appears too.

The 5-dimensional Einstein equations (6)–(9) include two extra variables, compared to the 4-dimensional formalism: P_5 and Φ. However, $P_5(\rho)$ is a known function of density and specified by the actual interaction in the matter. Thus $\Phi(r)$ is the *only new degree of freedom* determined by Eq. (9) as detailed in Ref. 4.

We generalize the equation of state of the non-interacting 5-dimensional fermionic matter using the thermodynamic potential formally,

$$\widehat{\Omega} = -\widehat{V}\frac{\widehat{g}}{\beta}\int_0^\infty \frac{\mathrm{d}^4\widehat{k}}{(2\pi)^4}\left[\ln\left(1 + \mathrm{e}^{-\beta(\sqrt{\mathbf{k}^2+\widehat{m}}-\widehat{\mu})}\right) + \ln\left(1 + \mathrm{e}^{-\beta(\sqrt{\mathbf{k}^2+\widehat{m}}+\widehat{\mu})}\right)\right], \tag{12}$$

where the total volume in the 5-dimensional spacetime is $\widehat{V} = 2\pi R_C\ V$ including the compactification radius, R_C and volume of the 3-dimensional subspace, V. The $\widehat{g} = 2$ stands for the multiplicity of the n^0 neutron. Further physical quantities denoted by '$\ \widehat{}\ $' are the same as in the usual thermodynamical potential in the $1+3$ dimensional spacetime.

Due to the quantization of the 5-momenta, $\widehat{k}$, the integral over its 5th component, k_5 in Eq. (12) can be rewritten as sum over the excited states, $n = 0, 1, ...$:

$$\int_0^\infty \mathrm{d}^4\widehat{k} = \int_0^\infty \mathrm{d}^3\mathbf{k}\,\mathrm{d}k_5 \qquad \longrightarrow \qquad \frac{1}{R_C}\sum_{i=\min(n)}^{\max(n)}\int_0^\infty \mathrm{d}^3\mathbf{k}. \tag{13}$$

Substituting this latter Eq. (13) into Eq. (12), we can get back the usual thermodynamical potential:

$$\widehat{\Omega} = -\widehat{V}\frac{\widehat{g}}{\beta}\sum_{i=\min(n)}^{\max(n)}\int_0^\infty \frac{\mathrm{d}^3\mathbf{k}}{(2\pi)^3}\times$$
$$\times\left[\ln\left(1 + \mathrm{e}^{-\beta(\sqrt{\mathbf{k}^2+\widehat{m}_i}-\widehat{\mu}_i)}\right) + \ln\left(1 + \mathrm{e}^{-\beta(\sqrt{\mathbf{k}^2+\widehat{m}_i}+\widehat{\mu}_i)}\right)\right]. \tag{14}$$

We note, the standard, usual thermodynamical potential, $\Omega = \Omega(m_i, \mu_i, \beta)$ and the $\widehat{\Omega} = \widehat{\Omega}(\widehat{m}_i, \widehat{\mu}_i, \widehat{\beta})$ given by Eq. (14) looks formally the same, but their variables and sums over the excited states are different.

Thermodynamical variables can be obtained using the usual definitions in the zero temperature approximation, $T = 0$, in a two-fermion component model:

$$\widehat{\epsilon} = \frac{\widehat{g}}{(2\pi)^4} \int_0^{\widehat{k_F}} \widehat{\varepsilon} \mathrm{d}^4 \widehat{\mathbf{k}} \bigg|_{T=0} = \tag{15}$$

$$= \frac{\widehat{g}}{16\pi^3 R_C} \sum_{n=0}^{1} \left[\widehat{\mu} \sqrt{\widehat{\mu}^2 - \widehat{m}^2} \left(\widehat{\mu}^2 - \frac{1}{2} \widehat{m}^2 \right) + \frac{\widehat{m}^4}{2} \ln \left| \frac{\widehat{m}}{\widehat{\mu} + \sqrt{\widehat{\mu}^2 - \widehat{m}^2}} \right| \right],$$

$$\widehat{P} = -\frac{1}{2\pi R_C} \frac{\partial \widehat{\Omega}}{\partial V} \bigg|_{T=0} = \tag{16}$$

$$= \frac{\widehat{g}}{48\pi^3 R_C} \sum_{n=0}^{1} \left[\widehat{\mu} \sqrt{\widehat{\mu}^2 - \widehat{m}^2} \left(\widehat{\mu}^2 - \frac{5}{2} \widehat{m}^2 \right) + \frac{3}{2} \widehat{m}^4 \ln \left| \frac{\widehat{m}}{\widehat{\mu} + \sqrt{\widehat{\mu}^2 - \widehat{m}^2}} \right| \right].$$

$$\widehat{P_5} = -\frac{1}{2\pi V} \frac{\partial \widehat{\Omega}}{\partial R_C} \bigg|_{T=0} = \tag{17}$$

$$= \frac{\widehat{g}}{48\pi^3 R_C^3} \left[\widehat{\mu} \sqrt{\widehat{\mu}^2 - \bar{m}^2} \left(5 \frac{\widehat{\mu}^2}{\bar{m}^2} - 11 \right) + 6 \bar{m}^2 \ln \left| \frac{\bar{m}}{\widehat{\mu} + \sqrt{\widehat{\mu}^2 - \bar{m}^2}} \right| \right].$$

It is easy to see, that based on Eqs. (16) and (17) can be connected to the usual description via these expressions: $\epsilon = 2\pi R_C \widehat{\epsilon}$ and $P = 2\pi R_C \widehat{P}$.

4. A Special 5 Dimensional Solution

The solution of Einstein equations (6)–(9) are quite similar in the 4- and 5-dimensional spacetimes, however there is no analytic solution only numerical. In case of a specially-chosen pressure component P_5, there is a unique solution of the Eqs. (6)–(9), especially with the $\mathrm{d}\Phi/\mathrm{d}r = 0$ constraint. This means the 5th metric component is constant, which choice in Eqs. (6)–(8) lead to the Tolman – Oppenheimer – Volkov equation[11] and can be solved separately with the $\Phi =$ const. condition. In addition, Eq. (9) gives P_5 as well.

We note, the extra dimension has its influence on $\varepsilon(\rho)$ and $P(\rho)$,[12] however solution in the 5-dimensional spacetime does not differ formally from the 4 dimensional neutron star solution (except for P_5). Furthermore, applying condition $\Phi \neq$ const., one must turn to solve Einstein equations directly numerically.

Neglecting the effects of electromagnetic charge, let us start with a neutral, single massive fermion (e.g. 'neutron', with mass m) as elementary building block of an extra dimensional compact star. Since the minimal nonzero fifth momentum component is given by Bohr-type quantization (1), then the extra direction of the phase space is not populated until the Fermi-momentum $k_F < \hbar c/R_C$. However, at the threshold both $k_5 = \pm\hbar c/R_C$ states appear. One can represent this as another ('excited') particle with excited mass, $\widehat{m}$ and with non-electric 'charge', $\widehat{q}$. We assume this second particle appears in complete chemical equilibrium with the ground-state 'neutron': $\mu_{\widehat{m}} = \mu_m$.

Increasing the non-electric charge $\widehat{q}$ i.e. the excited mass, $\widehat{m}$, this recipe can be repeated for every integer n, when k exceeds a threshold $n\hbar c/R_C$, and these higher excitations are introduced into the equation of state.

5. Results and Discussion

Applying our model, first we calculated the mass-radius relation, M-R, which is the key macroscopical observable of compact objects, especially, to get constraints for their inner structure. Due to the lack of precise $M(R)$ data,[13–15] the maximal mass of the measurable object can give restrictions to the initial conditions and therefore the inner structure as well. Besides the dependence of the M-R relation on the R_C we carry out the M_{max}, which were found to be also determined by the size of the extra compactified dimension, represented by the compactified radius.

In Fig. 1, the mass vs. radius relation, M/M_{sol}-R of compact stars calculated for 1-component non-interacting fermionic matter in $1+3+1_C$ dimensional space-time for various radii of the extra compactified dimension, R_C (*lines with dots*). Here M_{sol} is for the Solar-mass unit. We compared this to a reference calculation including a non-interacting, two-component fermion star (n^0 and Λ_s) (*thick solid green line*). To obtain this curve we set the $R_C \approx 0.33$ fm using Eq. (1) with $m_{\Lambda_s} = 1115.638$ MeV.

By varying the R_C from 0.1 fm to 10 fm, compact star masses obtained from the extra dimensional model are relatively small: M is between 0.2 and 0.7 M_{sol}, with radius and R is in the $2-25$ km region. As Fig. 1 presents, increasing the size of the compactified extra dimension results smaller-mass objects, which is in agreement with the fact, that mass difference between ground (m) and excited ($\widehat{m}$) states gets smaller inducing several excited

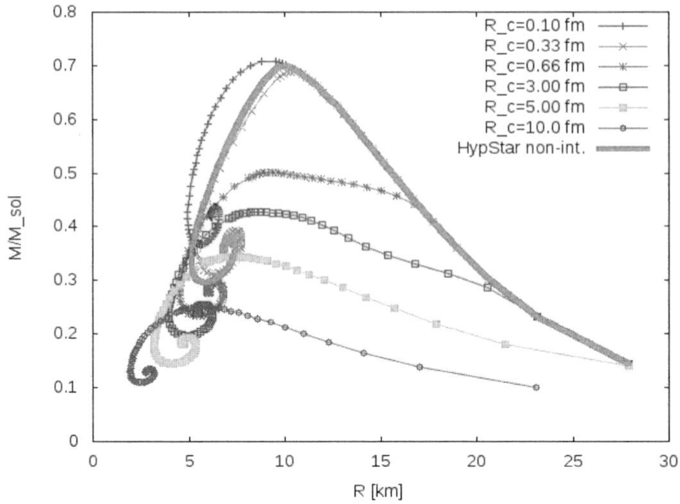

Fig. 1. The mass vs. radius relation M/M_{sol}-R of compact stars calculated for 1-component non-interacting fermionic matter in $1{+}3{+}1_C$ dimensional space-time for various radii of the extra compactified dimension, R_C (*lines with dots*). A non-interacting, two-component fermion star (n^0 and Λ_s), is also plotted for comparison (*thick solid green line*).

degree of freedom by the excitation number, n. On the other hand, in case of smaller compactification radii, high-mass excited states appear resulting massive compact objects. We note, this simple model does not contain interaction between the fermion states, so relatively light compact stars are obtained. These results can be changed by introducing interaction, as in realistic equation of state parametrizations like e.g. CompOSE.[16]

Using the obtained M/M_{sol}-R curves, we extracted the maximum mass, M_{max}/M_{sol} of compact stars for various radii of the extra compactified dimension, R_C in Fig. 2. We highlighted some R_C values associated with baryon masses of Λ_s, Σ_c, and Λ_b.

As Fig. 2 presents, the maximum mass is saturating at $M_{max} \approx 0.7\,M_{sol}$ as R_C gets smaller than ~ 0.5 fm. On the other hand the larger the R_C the maximum mass decreasing, which results less-and-less massive stars.

Increasing the R_C decreases the mass difference between the first spokes of the Kaluza – Klein ladder. In this case the masses of the excited states are getting closer to each other with small mass gaps between them. This induces several low-mass degrees of freedom which results small-mass stars. For small R_C values mass difference between the excited states are getting

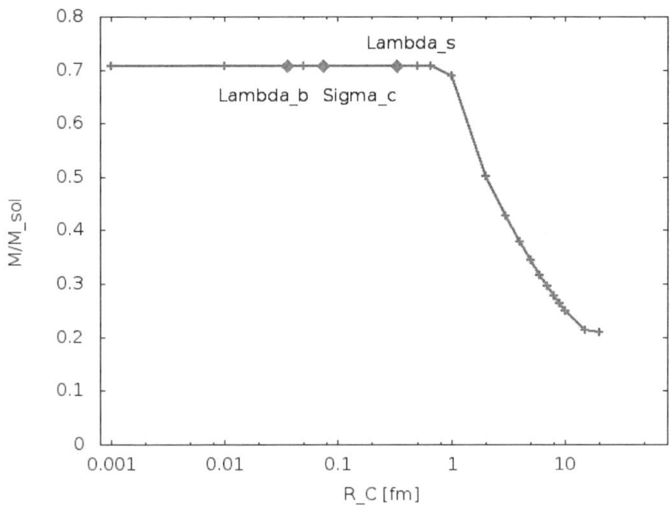

Fig. 2. Maximum mass, M_{max}/M_{sol} of compact stars calculated for 1-component non-interacting fermionic matter in $1+3+1_C$ dimensional space-time for various radii of the extra compactified dimension, R_C. We highlighted R_C values associated with baryons Λ_s, Σ_c, and Λ_b.

larger. In this case the less, but massive excited particles result the saturation at the highest $M_{max} \approx 0.7\, M_{sol}$.

Besides testing the validity of our model we compare our results to the existing observational data of pulsar properties from Refs. 13–15, which exhibits that our numerical result on M_{max}/M_{sol} is much more below the measured maximal-mass values. This originates from the lack of interaction in our model, what we would like to include into our model in the future.

6. Summary

Here we presented a simple model for a possible geometrization of the strongly interacting matter in a Kaluza – Klein-like spacetime. We applied our approach for modeling the inner structure of a compact star including a compactified extra dimension, with size R_C. This was compared to a two-component fermion model, where mass difference between ground- and excited-state particle are connected via the compactified radius. We have found that within non-interacting fermionic matter, the $M_{max} \approx 0.7\, M_{sol}$, which is much more below the predictions of either the interacting fermion gas model or the measured maximum mass values $\sim 2.0 M_{sol}$. In order to make our model more realistic, in the future we plan to extend our model with the generalized interaction between fermionic matters, and thus understand more the geometrized excited states in both theoretical framework.

Acknowledgements

This work was supported by: Hungarian National Research Found (OTKA), NK106119, K104260, K104292 and theNewCompStar COST Action MP1304. Author GGB was partially supported by the János Bolyai Research Scholarship of the Hungarian Academy of Sciences.

References

1. Th. Kaluza, *Sitzungsber. Preuss. Akad. Wiss. Phys. Math. Kl.* **LIV** 966 (1821).
2. O. Klein, *Z. Phys.* **37**, 875 (1926).
3. G.G. Barnaföldi, B. Lukács, and P. Lévai, *Astron. Nachr.* **328**, 809 (2007).
4. G.G. Barnaföldi, B. Lukács, and P. Lévai, arXiv:astro-ph/0312330v1 (2003).
5. G.G. Barnaföldi, B. Lukács, and P. Lévai, arXiv:astro-ph/0312332v1 (2003).
6. G.G. Barnaföldi, B. Lukács, and P. Lévai, *J. Phys. Conf. Ser.* **218**, 012010 (2010).
7. B. Lukács, *The Nature of Time*, eds. R. Buccheri *et al.*, Kluver Academic, p. 277 (2003).
8. A. Arkhipov, arXiv:astro-ph/0309327v5 (2004).
9. B. Lukács and T. Pacher *Phys. Lett.* **A113**, 200 (1985).
10. C. Brans and R.H. Dicke *Phys. Rev.* **124**, 925 (1961).
11. N.K. Glendenning, *Compact Stars*, Springer, New York, 2000.
12. N. Kan and K. Shiraishi, *Phys. Rev.* **D66**, 105014 (2002).
13. P.B. Demorest, *Nature* **467**, 1081 (2010).
14. M.H. van Kerkwijk, R. Breton, S.R. Kulkarni, *Astrophys. J.* **728**, 95 (2011).
15. S. Guillot *et al.*, *Astrophys. J.* **772**, 7 (2013).
16. `https://compstar.uni-frankfurt.de/compose/`

SOME QCD/GRAVITY INTERSECTIONS*

O.V. TERYAEV

Joint Institute for Nuclear Research, Dubna, 141980 Russia

Gravitational form factors are the matrix elements of the Belinfante energy momentum tensor (EMT) which naturally incorporate the hadron structure and the equivalence principle. The relocalization property allowing to transform EMT to the Belinfante form provides the "kinematical" counterpart of the famous $U_A(1)$ problem. The equivalence principle may be approximately valid for quarks and gluons separately in non-perturbative (NP)QCD, and this conjecture is supported by the experimental and lattice data. The extra-dimensional gravity leading to holographic AdS/QCD is supporting the relation of quark transverse momentum to the Regge slope, discovered by V.N. Gribov.

1. Introduction

QCD and gravity correspond to the strongest and weakest of fundamental interactions and hopefully might be unified in the future Theory of Everything. They manifest deep similarities in their mathematical structure, in particular, in the context of Gribov Reggeon calculus. [1]

Here I will discuss the more simplistic similarities, emerging from the fact that QCD EMT appears both in the context of strong interactions themselves and in that of coupling to gravity. Already at the stage of relocalization (required to get the Belinfante tensor which plays the role of current describing the coupling to gravity) the constraint emerges being the counterpart of QCD $U_A(1)$ problem. If the hadrons are considered, this similarity allows one to relate their structure to the famous equivalence principle (EP) (in particular to its generalization for spin discovered [2] by I.Yu. Kobzarev and L.B. Okun). It can be generalized for particles of any spin (in particular for vector mesons and deuterons) and may be approximately valid separately for quarks and gluons. Finally, for effective extra-dimensional gravity the relations between the transverse structure of different parton distributions support the discoveries of V.N. Gribov.

*This work is partially supported by grant RFBR 14-01-00647.

2. Nucleon in the external gravitational field.

It is instructive to examine the gravitational form factors (being the straightforward analogs of electromagnetic ones) by considering[3] (which was first discussed at the **Gribov-70** meeting in Paris) the action of electromagnetic and gravity fields in a uniform manner and comparing them.

Let us start with the more common case of the interaction with the electromagnetic field, described by the matrix element of electromagnetic current,

$$M = \langle P'|J_q^\mu|P\rangle A_\mu. \tag{1}$$

This matrix element at zero momentum transfer is fixed by the fact, that the interaction is due to the *local* $U(1)$ symmetry, whose *global* counterpart is producing the conserved charge (and of course is depending on the normalization of eigenvectors $\langle P|P'\rangle = (2\pi)^3 2E\delta(\vec{P} - \vec{P}')$),

$$\langle P|J_q^\mu|P\rangle = 2e_q P^\mu\,, \tag{2}$$

so that in the rest frame the interaction is completely defined by the scalar potential

$$M_0 = \langle P|J_q^\mu|P\rangle A_\mu = 2e_q M\phi\,. \tag{3}$$

At the same time, the interaction with the weak classical gravitational field is

$$M = \frac{1}{2}\sum_{q,G}\langle P'|T_{q,G}^{\mu}|P\rangle h_{\mu\nu}\,, \tag{4}$$

where h is a deviation of the metric tensor from its Minkowski value.

One can already observe at this point the first QCD/gravity intersection. The transition from the canonical to the Belinfante form of EMT describing the coupling to gravity requires to drop the surface terms in coordinate space. For the matrix elements in the momentum space this requires the smooth behavior in the small momentum transfer limit. Since the quark spin density is proportional to the (singlet) axial current, one has[4,5]

$$q^2\frac{\partial}{\partial q^\alpha}\langle P|J_{5S}^\alpha|P+q\rangle = (q^\beta\frac{\partial}{\partial q^\beta} - 1)q_\gamma\langle P|J_{5S}^\gamma|P+q\rangle\,. \tag{5}$$

Performing the standard decomposition

$$\langle P,S|J_\mu^5(0)|P+q,S\rangle = 2MS_\mu G_1 + q_\mu(Sq)G_2\,, \tag{6}$$

one may conclude that

$$q^2 G_2|_0 = 0. \tag{7}$$

This property is well-known as the particular manifestation[6] of the famous $U_A(1)$−problem of QCD (I had a chance to discuss some of that in a 10 minutes talk in Gatchina in 1988, when V.N. Gribov was in the audience) when the $\eta^{'}$−meson responsible for the pole in the singlet channel remains massive even in the chiral limit.

It is interesting that this property is not valid for gluon spin contribution described by the topological current and the most simplistic solution of this problem[7] is the impossibility to separate the spin and orbital parts of the gluon angular momentum.

Let us come back to (4). The relative factor 1/2, which will play a crucial role, comes from the fact that the variation of the action with respect to the metric is producing an energy-momentum tensor with the coefficient 1/2, while the variation with respect to the classical source A^μ, is producing the current without such a coefficient. It is this coefficient that guarantees the correct value for the Newtonian limit, fixed by the *global* translational invariance

$$\sum_{q,G} \langle P|T_i^{\mu\nu}|P\rangle = 2P^\mu P^\nu \,, \tag{8}$$

which, together with the approximation for h (with factor of 2 having the geometrical origin),[8]

$$h_{00} = 2\phi(x) \,, \tag{9}$$

results in the rest frame expression,

$$M_0 = \sum_{q,G} \langle P|T_i^{\mu\nu}|P\rangle h_{\mu\nu} = 2M \cdot M\phi \,, \tag{10}$$

where we used the same notation for gravitational and scalar electromagnetic potentials, and identified the normalization factor $2M$ in order to make the similarity between Eqs. (3) and (10) obvious. One can see that the interaction with the gravitational field is described by the charge, equal to the particle mass, which is just the equivalence principle. It is appearing here as a low energy theorem, rather than a postulate. The similarity with the electromagnetic case allows one to clarify the origin of such a theorem.

The situation with the terms linear in Δ is different for electromagnetism and gravity. While such a term is defined by the specific dynamics in the electromagnetic case, producing the anomalous magnetic moment, the similar contribution in the gravitational case is entirely fixed by the angular momentum conservation, which was known in the context of gravity for many years since the ingenious observation of Kobzarev and Zakharov.[9,†] It means that in terms of the gravitational interaction the **Anomalous Gravitomagnetic Moment (AGM) is identically equal to zero**.

In the context of hadron structure, when EMT matrix elements are considered as a probe for parton (angular) momenta and are expressed as the moments of Generalized Parton Distributions (GPDs) this property follows from Ji's Sum Rules.[10] The comparison to gravity provides another interpretation for them.[3,11]

3. Extension of Equivalence Principle

Another application of the discussed QCD/gravity intersection is the possibility to extract the separate couplings od quarks and gluons to gravity.[3] One may pass then to the Extended EP(ExEP)[12] that separates contributions of quarks and gluons to the Anomalous Gravitomagnetic Moment of nucleon are zero in NPQC.

This property is violated in perturbation theory.[13–15] It is not surprising from the qualitative side, as the quarks and gluons correspond to matrix elements of non-conserved operators which cannot be separated due to operators mixing, while beyond PT one may consider the dominant non-perturbative (up to PT corrections) separation.

The reason for PT violation of ExEP value was discussed[13] as emerging due to use of the unsubtracted dispersion relations for the relevant form factors, while performing the subtraction, resulting in the zero AGM, leads to the absence of the smooth transition of electron mass $m_e \to 0$. At the same time, such limit is unphysical in the case of (almost) massless quarks because of phenomenon of Spontaneous Chiral Symmetry Breaking and ExEP may be restored.[12]

This picture is supported already by the well-known fact of approximate cancellation of proton and neutron anomalous magnetic moments. Their singlet combination differs from AGM because they correspond to first and second moments of GPDs E, respectively. Also, this picture is supported by the lattice data, including the most recent ones.[16] Still, the interesting support is coming from the consideration of spin-1 hadrons.

---

†The reason is that the structure of the Poincaré group is richer than that of the $U(1)$ group.

4. Equivalence Principle for Deuterons and Vector Mesons

The notion of AGM and (Ex)EP may be easily generalized[3] for a particle of any spin. Like for spin 1/2 particles the smallness of AGM should correspond to the smallness of isoscalar magnetic moment, for $\rho-$mesons it should lead[5] to $g \approx 2$. Recently, the gyromagnetic ratio of charged ρ mesons was calculated in lattice QCD.[17] The obtained value is fairly close to 2, the one supported by ExEP.

There are more spin degrees of freedom for vector mesons (and, say, deuterons). The tensor polarization is $P-$even, contrary to the vector ones. It does not require the Levi-Civita tensor so that EP is manifested[18] also for forward matrix elements.

It leads to a zero sum rule for the sum of second moments of quark and gluon tensor spin structure functions derived many years ago,[19,20] and the suggested manifestation[18] of EP is related to them like that of Kobzarev and Okun is related to Ji's Sum Rules:

$$\sum_q \langle P,S|T_i^{\mu\nu}|P,S\rangle_{\mu^2} = 2P^\mu P^\nu(1-\delta(\mu^2)) + 2M^2 S^{\mu\nu}\delta_1(\mu^2)$$
$$\langle P,S|T_g^{\mu\nu}|P,S\rangle_{\mu^2} = 2P^\mu P^\nu \delta(\mu^2) - 2M^2 S^{\mu\nu}\delta_1(\mu^2)\,. \tag{11}$$

ExEP, in turn, means that it is valid separately for quarks and for gluons so that $\delta(\mu^2) \approx 0$. The available data on deuteron tensor spin structure function[21] provide just the required singlet combination for quarks and indicate that the second moment is compatible to zero better than the first one. This means that collective gluons in deuteron are suppressed in comparison to the collective sea and ExEP provides the explanation of this fact.

5. AdS/QCD and GPD/TMD Relation

The well-known QCD-gravity intersection is AdS/QCD (see Ref. 22 and references therein). Let me pass to some observation in its framework supporting the discoveries of V.N. Gribov. Recently,[23] in the framework of light-front AdS/QCD the (approximate) relation between GPDs and Transverse Momentum Dependent Distributions (TMDs) was obtained:

$$\frac{\partial}{\partial|t|}[\ln(\mathrm{GPD})] = \frac{(1-x)^2}{4}\frac{\partial}{\partial p_\perp^2}[\ln(\mathrm{TMD})]\,. \tag{12}$$

The universality in the t and $p_\perp^2$ dependence of the GPDs and TMDs comes from the AdS/QCD wave function. The GPDs obtained from the model satisfy the Regge behavior with the Regge slope $\alpha' = 1/(4\kappa^2)$. The average value of Gaussian transverse momentum distribution is then obtained as

$$\langle p_\perp^2 \rangle = \frac{(1-x)^2}{4\alpha' \ln 1/x} = \frac{\kappa^2(1-x)^2}{\ln 1/x}\,, \tag{13}$$

which takes the maximum value of $\langle p_\perp^2 \rangle_{max} \sim (250\text{MeV})^2$ for $x \sim 1/3$.

I learned at this meetings from the discussions with L.N. Lipatov and U. Maor that the relation between the quark transverse momentum and the Regge slope was first derived by V.N. Gribov in the framework of (scalar) ladder summation. That gives a hope that the observed similarity is not a mere coincidence but expresses an interesting particular feature of AdS/QCD duality.

6. QCD Factorization and Black Hole Production

The extra-dimensional black hole (BH) production is usually described by convoluting the classical cross section (defined by Schwarzschild radius) with parton distributions.[24] This, however, is a bit strange from the point of view of QCD factorization. One may use as a reference point the production of a heavy (say, quarkonia, or Brout-Englert-Higgs) meson. In that case its mass serves as a factorization parameter,[25] that is, a "microscope" being the necessary ingredient for parton distributions appearance. If the meson is produced in two gluons fusion and its hadronic decay proceeds via two gluons, its coupling to gluons is obviously related to the hadronic width,

$$A_M^2 = C\Gamma_M / M_M^3\,. \tag{14}$$

Consider the black hole as a quantum state $|T>$ with a Hawking thermal decay. It seemingly has no overlap with a two-gluon state. However, due to gluon-hadron duality (and its interplay with gravity is crucial here) one should instead consider the overlap with jet state which should be non-zero so that

$$A_{BH}^2 = C\Gamma_{BH}/M_{BH}^3 | < T|2j > |^2\,. \tag{15}$$

Note that such an expression would mean the reversibility and absence of the famous information paradox. Probabilistically, the jet-thermal overlap factor should be exponentially suppressed by jet multiplicity, and such an exponential suppression is known to appear in a completely different quasiclassical approach.[26] If one still likes to keep the classical mechanism of BH production, one may consider the scattering with the production of gluon pair with a large invariant mass, serving as a factorization parameter, with subsequent classical formation of BH. The extra suppression in gluon scattering $\sim \alpha_S^2(M_{BH}^2)$ may affect the obtained exclusion limits.[27]

7. Conclusions.

There are some connections of gravity with QCD on the "phenomenological" level. They include:

- The properties of axial current form factors implied by relocalization invariance.
- The relation of Ji's Sum rules for Generalized Parton Distributions s with Equivalence Principle.
- The Extension of Equivalence Principle for quarks and gluons separately manifested also for spin 1 hadrons.
- Relation between Generalized Parton Distributions and Transverse Momentum Dependent Distributions in light-front AdS/QCD.
- QCD factorization influence on existing approaches to Black Holes production.

Finally, I would like to thank Kolya Nikolaev and Julia Nyiri for the invitation to participate in this important Conference.

References

1. L.N. Lipatov, *Nucl. Phys. Proc. Suppl.* **219-220**, 78 (2011).
2. I.Y. Kobzarev and L.B. Okun, *Zh. Eksp. Teor. Fiz.* **43**, 1904 (1962) [*Sov. Phys. JETP* **16**, 1343 (1963)].
3. O.V. Teryaev, arXiv:hep-ph/9904376.
4. O.V. Teryaev, arXiv:hep-ph/9303228.
5. O.V. Teryaev, *AIP Conf. Proc.* **915**, 260 (2007), arXiv:hep-ph/0612205.
6. A.V. Efremov, J.Soffer, and O.V. Teryaev, *Nucl. Phys.* **B346**, 97 (1990).
7. O.V. Teryaev, *Phys. Part. Nucl.* **45**, 57 (2014).
8. L.D. Landau and E.M. Lifshitz, *Course of Theoretical Physics, v.2., The Classical Theory of Fields*, Butterworth-Heinemann, Oxford, 1980.
9. I.Yu. Kobzarev and V.I. Zakharov, *Ann. Phys.* **37**, 1 (1966).
10. X. Ji, J. Tang, and P. Hoodbhoy, *Phys. Rev. Lett.* **76**, 740 (1996); X. Ji, *Phys. Rev. Lett.* **78**, 610 (1997).

11. S.J. Brodsky, M.Diehl, and D.S. Hwang, *Nucl. Phys.* **B596**, 99 (2001), arXiv:hep-ph/0009254.
12. O.V. Teryaev, *"Spin structure of nucleon in QCD: Inclusive and exclusive processes,"* Prepared for the *International Conference on New Trends in High-Energy Physics: Experiment, Phenomenology, Theory,* Yalta, Crimea, 22-29 Sep 2001, *Czech. J. Phys.* **53**, 47 (2003), arXiv:hep-ph/0306301.
13. K. Milton, *Phys. Rev.* **D4**, 3579 (1971); *ibid.* **D15**, 538 (1977).
14. F.A. Berends and R. Gastmans, *Ann. Phys.* **98**, 225 (1976).
15. S.J. Brodsky, D.S. Hwang, B.Q. Ma, and I. Schmidt, *Nucl. Phys.* **B593**, 311 (2001).
16. M. Deka *et al.*, *Phys. Rev.* **D91**, no. 1, 014505 (2015), arXiv:1312.4816 [hep-lat].
17. E.V. Luschevskaya, O.A. Kochetkov, O.V. Teryaev, and O.E. Solovjeva, *JETP Lett.* **101**, 10, 674 (2015).
18. O.V. Teryaev, *Mod. Phys. Lett.* **A24**, 2831 (2009).
19. A.V. Efremov and O.V. Teryaev, *Sov. J. Nucl. Phys.* **36**, 557 (1982) [*Yad. Fiz.* **36**, 950 (1982)].
20. A.V. Efremov and O.V. Teryaev, arXiv:hep-ph/9910555.
21. A. Airapetian *et al.* [HERMES Collaboration], *Phys. Rev. Lett.* **95**, 242001 (2005) arXiv:hep-ex/0506018.
22. S.J. Brodsky, G.F. de Teramond, H.G. Dosch, and J. Erlich, *Phys. Rept.* **584**, 1 (2015), arXiv:1407.8131 [hep-ph].
23. T. Maji, C. Mondal, D. Chakrabarti, and O.V. Teryaev, arXiv:1506.04560 [hep-ph].
24. M.V. Savina, *Phys. Atom. Nucl.* **78**, no. 4, 532 (2015) [*Yad. Fiz.* **78**, no. 6, 571 (2015)].
25. A.V. Efremov and A.V. Radyushkin, *Theor. Math. Phys.* **44**, 664 (1981) [*Teor. Mat. Fiz.* **44**, 157 (1980)].
26. M.B. Voloshin, *Phys. Lett.* **B524**, 376 (2002) [Erratum: *Phys. Lett.* **B605**, 426 (2005)], arXiv:hep-ph/0111099.
27. S. Chatrchyan *et al.* [CMS Collaboration], *JHEP* **1307**, 178 (2013), arXiv:1303.5338 [hep-ex].

STANDARD MODEL PHYSICS

NATURALNESS AND RENORMALIZATION GROUP IN THE STANDARD MODEL

GRIGORII B. PIVOVAROV

Theory Division, Institute for Nuclear Research,
Moscow, 117312, Russia
E-mail: gbpivo@ms2.inr.a.ru

I define a naturalness criterion formalizing the intuitive notion of naturalness discussed in the literature. After that, using ϕ^4 as an example, I demonstrate that a theory may be natural in the MS-scheme and, at the same time, unnatural in the Gell-Mann–Low scheme. Finally, I discuss the prospects of using a version of the Gell-Mann–Low scheme in the Standard Model.

Keywords: Naturalness; Renormalization Group; Standard Model.

1. Introduction

By now the renormalization group functions of the Standard Model have been computed up to three loops.[1] This computation is performed within the 't Hooft–Weinberg approach, which is the present-day standard.

Below I will argue that the 't Hooft–Weinberg approach may be misleading for models with scalar fields. My argument is based on my interpretation of the naturalness criterion.

There is some disagreement in the literature on the naturalness of the Standard Model. On the one hand, it is claimed[2] that Higgs may be unnatural, and "the multiverse offers the most plausible answer at our disposal." On the other hand, it seems to be claimed[3] that the Standard Model may be self-consistent up to the Planck scale.

This disagreement may be caused by a difference in definitions of the notions used. So, in the next section, I give my definition of the naturalness criterion.

Next, using the example of ϕ^4 theory, I demonstrate how different renormalization schemes may lead to opposite results on the naturalness of the Standard Model.

In conclusion, I comment the plausibility of developing an analogue of the Gell-Mann–Low scheme for the Standard Model.

2. Naturalness

Naturalness is related to the presence of large quantum corrections to scalar masses squared.[4] Formulae for these corrections may involve the ultraviolet cut-off, or some large masses of the model replacing the cut-off. In any case, the large corrections are proportional to a square of a large parameter having the dimension of mass.

It was stressed[2] that in the case when the cut-off is involved in the large corrections, it plays the role of a physical parameter. The presence of the large corrections to the scalar mass should not and does not depend on the regularization used in the computation. The presence of the large corrections to the scalar masses is considered troublesome, because, if we want to keep the Higgs mass to the value measured by now, it implies the need for some fine-tuning between the parameters involved in the corrections.

My opinion is that this trouble requests an analysis of its meaning. Indeed, any computation of the scalar masses involves ultraviolet divergences. Because of this, the masses can be taken as input parameters of the theory. In this case they are not computed, but just taken from an experiment. If so, how come that a computation of masses is considered troublesome? Just take the needed values from experiment, and no computation of the quantum corrections to the masses is needed regardless of their size.

Here is my answer. Indeed, it is possible to take the measurable masses as input parameters. But frequently it is computationally inconvenient. Instead of physical masses, the theoretical predictions are parametrised with a set of input parameters related to a particular renormalization scheme (e.g. MS-scheme, or momentum subtraction scheme). So, the necessity appears to find relationships between different sets of input parameters used in different renormalization schemes. In these relationships large quantum corrections may appear.

There is the next question. How come that a particular change of input parameters may be considered as unnatural? Mathematically, any one-to-one change of input parameters is admissible. But this is not the case in physics. A physicist supposes that the input parameters she/he are using are measurable. If a one-to-one relationship between two sets of input parameters is established within a theory, the relative accuracies with which the parameters of a set #1 are known are translated unambiguously to relative accuracies for the parameters of the set #2. Let us imagine that we measured the parameters of a set #1, and it appears that the parameters of a set #2 are known because of this with relative accuracies exceeding any expectation of experimentalists. This situation seems to be unnatural. Or rather, it means that we somehow overestimated the number of input parameters required in the theory, because the parameters of the set #2 should be fine-tuned to provide the parameters of the set #1 with the measured values.

In high energy physics, the input parameters are determined in conjunction with a particular renormalization scheme. The latter always involves an energy scale. For example, momentum subtractions involve a momentum squared normalization point, the MS-scheme involves the dimensional regularization mass parameter, etc. The physical meaning of this parameter is that it gives the energy scale at which the input parameters are convenient to measure.

Therefore, each set of input parameters of a quantum field theory comes with the energy scale at which it should be measured. The theory predicts the dependence of the relative accuracies of the input parameters on the energy scale. This prediction is provided by the renormalization group evolution of the input parameters.

Let us consider the simplest possible model example: assume that our theory has the only input parameter — the mass squared of a scalar particle measured at the energy scale Q^2, $m^2(Q^2, m_{ph}^2)$, where m_{ph}^2 is the physical mass squared. (A formal definition will be given in the next Section.) That is, assume that the exact values of all other parameters (masses and couplings) are somehow sent to us by the Gods and can be excluded from the list of the input parameters subject to measurements. Take also that $m^2(m_{ph}^2, m_{ph}^2) = m_{ph}^2$, which is the definition of the physical mass squared, m_{ph}^2 (the mass squared measured at physical mass squared equals the physical mass squared). Under these assumptions,

$$\frac{\delta m^2(Q^2, m_{ph}^2)}{m^2(Q^2, m_{ph}^2)} = r\Big(\frac{m_{ph}^2}{Q^2}\Big)\frac{\delta m_{ph}^2}{m_{ph}^2}, \tag{1}$$

where

$$r\Big(\frac{m_{ph}^2}{Q^2}\Big) = \frac{m_{ph}^2}{Q^2}\frac{\partial \log(m^2(Q^2, m_{ph}^2)/Q^2)}{\partial(m_{ph}^2/Q^2)}. \tag{2}$$

The dimensionless function $r(x)$ introduced above seems to be a useful characterization of the theory. Its value at the physical mass squared taken in the units of the energy scale Q^2 gives the factor transforming the relative error in the physical mass squared to the relative error in the mass squared measured at the energy scale Q^2.

Here is my definition of the naturalness criterion: *A theory is natural if it satisfies the condition*

$$\lim_{x\to 0} r(x) \neq 0.$$

Otherwise, if

$$\lim_{x\to 0} r(x) = 0,$$

a theory is unnatural, because in this case one is able to improve infinitely the relative accuracy of the mass squared simply by increasing the energy scale Q^2 of the measurement.

It is also possible to refine this definition: I call a theory strongly unnatural if its $r(x) \sim x$ at $x \to 0$, and weakly unnatural if $r(x) \sim 1/\log(1/x^p)$, $p > 0$ at $x \to 0$.

This refinement may be useful in the following way: the strongly unnatural theories become inapplicable when the reaction energies become moderately higher than the scalar masses of the theory, because the relative accuracy of the knowledge of the mass squared grows in this case as Q^2/m_{ph}^2. This is in contrast to the weakly unnatural theories, where the growth of the accuracy is much slower, as $\log(Q^2/m_{ph}^2)^p$.

In the next Section I will try to determine the r-function for the ϕ^4 theory. I will show that different answers will be obtained for different renormalization group evolution of the mass squared present in the literature. I also try to determine the source of these differences.

Concluding this section I point out that my naturalness criterion does not require the presence of a second mass in the theory whose value would be much larger when the one of the above m_{ph}^2. This may seem to be in contradiction with the claim[2] that such a mass is needed to make the notion of naturalness meaningful. But in fact, there is no contradiction, because the second parameter of the right dimension is provided in my consideration by the energy scale of the mass measurement, Q^2.

3. The r-function for ϕ^4

It follows from the definition (2) that to determine the r-function one has to determine the running mass squared of the scalar particle, $m^2(Q^2, m_{ph})$. Remarkably, three different answers for this running are available in the literature. Correspondingly, one has three options for the r-function depending on which answer for the running mass one uses. I will first present all three options, and point out my preference after that.

But first I define the running mass squared. Its definition can be extracted from the behaviour of the renormalized (UV-finite) propagator near a Euclidean momentum Q_0:

$$D(Q^2) = \frac{Z(Q_0^2)}{Q^2 + m^2(Q_0^2, m_{ph}^2)} + \mathcal{O}\big((Q^2 - Q_0^2)^2\big). \tag{3}$$

This defines unambiguously the field normalization constant $Z(Q_0^2)^{-\frac{1}{2}}$ and the running mass squared $m^2(Q_0^2, m_{ph}^2)$, if the propagator $D(Q^2)$ is known. The physical mass squared is defined from the equation

$$m_{ph}^2 = m^2(m_{ph}^2, m_{ph}^2). \tag{4}$$

So, to get the running mass squared, we need to determine the renormalized propagator. I start with the propagator renormalized within the MS-scheme.[5,6]

The propagator renormalized in the MS-scheme can be read in the above two papers:

$$D(Q^2) = \frac{1}{(Q^{2(1-\gamma_\phi)} + m_{ph}^{2(1-\gamma_\phi)})\mu^{2\gamma_\phi}}. \tag{5}$$

Here μ is the mass unit of dimensional regularisation, and γ_ϕ is the so-called anomalous dimension of the scalar field. It was calculated in the leading order in the quartic self-coupling g[6]:

$$\gamma_\Phi = \frac{g^2}{12(16\pi^2)^2}. \tag{6}$$

The definition (3) and the formula for the propagator (5) imply that MS-scheme leads to the following running of the mass squared:

$$m_{ms}^2(Q^2, m_{ph}^2) = Q^2\left(\frac{m_{ph}^2}{Q^2}\right)^{1-\gamma_\phi}. \tag{7}$$

The subscript is to remind that this evolution is determined within the MS-scheme.

Now, using the definition (2), one obtains

$$r_{ms}(x) = 1 - \gamma_\phi, \tag{8}$$

which means that the r-function calculated within MS-scheme is constant.

I conclude that there is no naturalness problem in ϕ^4 renormalized within the MS-scheme. The naturalness criterion is fulfilled.

Let us compare this with the treatment in a paper not using the MS-scheme, but discussing the naturalness problem.[4] (This may be the first paper considering closely the naturalness problem.)

After an adaptation to the notions of the present treatment, the reasoning of this paper looks as follows. First, the propagator is presented in the form

$$D(Q^2) = \frac{1}{Q^2 + m_{ph}^2 - \Sigma(Q^2, m_{ph}^2)}, \tag{9}$$

where $\Sigma(Q^2, m_{ph}^2)$ is the self-energy of the scalar field.

Next, I need some expression for the self-energy. In the original paper[4] the author uses an UV cut-off and assumes that the the self-energy is proportional to this cut-off squared. Within the present reasoning, I emulate this by assuming that the leading behaviour of the self-energy at large Q^2 is

$$\Sigma(Q^2, m_{ph}^2) = -kQ^2 + o(Q^2). \tag{10}$$

I note here that if the propagator is renormalized with subtractions at momentum Q, this assumption can be justified in the leading order in the coupling, and the constant from (10) turns out to be[7]

$$k = \gamma_\phi. \tag{11}$$

Now I take Eqs. (10) and (11), neglect the term sub-leading at large Q^2, and substitute it in the definitions from the previous Section. This yields for the running mass

$$m_S^2(Q^2, m_{ph}^2) = m_{ph}^2 + \gamma_\phi Q^2, \tag{12}$$

and for the r-function

$$r_S(x) = \frac{x}{1+\gamma_\phi}. \tag{13}$$

Here the subscripts on m_S^2 and r_S mean Susskind's running mass squared and r-function.

I conclude that, within this treatment, ϕ^4 is strongly unnatural, because $r_S(x) \sim x$.

There is a third treatment in the literature of the mass squared evolution in ϕ^4 (see Ref. 7). It is obtained by applying a version of the Gell-Mann–Low scheme to the ϕ^4 theory. As a result, a nonlinear equation for the renormalized propagator is obtained. It implies a running of the scalar mass squared at large Q^2:

$$m_{GML}^2(Q^2, m_{ph}^2) = \frac{\gamma_\phi Q^2}{1 + 4\gamma_\phi \log \frac{m_{ph}^2}{Q^2}} + o\Big(\frac{Q^2}{1 + 4\gamma_\phi \log \frac{m_{ph}^2}{Q^2}}\Big). \tag{14}$$

Here the subscript means "Gell-Mann–Low running."

Two remarks are in order about this result. First, it is in agreement with Eq. (12): if the first term in its right-hand-side is expanded in powers of γ_ϕ, the leading term will be $\gamma_\phi Q^2$, as in the right-hand side of Eq. (12). Second, the resummation of the leading logs of Q^2 has resulted in a Landau pole in the running mass squared taken in the units of Q^2, m^2_{GML}/Q^2.

Now I can determine the r-function corresponding to the Gell-Mann–Low running of the mass squared:

$$r_{GML}(x) = \frac{-1}{1 + 4\gamma_\phi \log x}. \tag{15}$$

Interpretation of this result is complicated by the presence of the Landau pole. If Q^2 is smaller than Q^2_{LP}, the momentum squared corresponding to the location of the Landau pole, the absolute value or the negative r_{GML} increases with Q^2 to infinity. If Q^2 is larger than Q^2_{LP}, the positive r_{GML} decreases to zero.

I conclude that, if I use ϕ^4 as an effective theory at momenta below the Landau pole, there is no naturalness problem in this theory. This change with respect to the naive treatment ignoring the logarithms of Q^2 is due to the resummation of the leading logs of Q^2.

Is it sensible to consider ϕ^4 at momenta above the Landau pole in the mass squared, where the naturalness problem reappears? I don't know the answer.

I suppose that the presence of the Landau pole in the mass squared is a feature of ϕ^4, and that in a more realistic model it may be replaced by the asymptotic freedom behaviour: the scalar mass in the units of the energy tends to zero as the inverse logarithm of Q^2.

For the r-function that would mean a change of the sign in front of the logarithm in Eq. (15). Such an r-function would have the asymptotic $1/\log(1/x^p)$ at zero x, where p is the coefficient by the logarithm replacing within the more realistic model the factor $-4\gamma_\phi$ of the ϕ^4.

I conclude that the resummation of the leading logarithms of Q^2 in the mass squared may result in a weakly unnatural behaviour of mass squared.

In the further discussion I assume that the Landau pole in the mass squared is replaced by the asymptotic freedom behaviour

The first consequence I draw from the above consideration is that the logarithms of the energy scale in the mass corrections should be taken into account. Neglecting them may result in a qualitative error in a consideration of the naturalness problem.

Next, I need to compare the Gell-Mann–Low scheme result (15) to the MS-scheme result (8). The results differ qualitatively. The MS-scheme result is that there is no naturalness problem, and the Gell-Mann–Low result is that a theory with scalar fields may be weakly unnatural.

Thus, I am forced to accept that the fulfilment of the naturalness criterion depends on the renormalization scheme.

Looking for a reason of this qualitative difference between the schemes I recall the claim[8] that dimensionally regularized theory with quadratic divergences has a series of poles in the plane of complex dimension at $4-2/N$, where N is the number of loops in a contributing Feynman diagram. This is how the quadratic divergences appear within dimensional regularization.

The MS-scheme does not subtract these poles, because they are away from the physical dimension. (Notice, however, that these poles approach the physical dimension infinitely close as the number of loops grows to infinity.)

I conclude that MS-scheme does not subtract the poles related to quadratic divergences, while momentum subtractions do subtract them. This is a possible reason for the qualitative difference between the MS-scheme and the Gell-Mann–Low scheme.

If I have to chose between the two schemes, I would prefer the Gell-Mann–Low scheme, because its set of input parameters has a clear physical meaning.

The only problem with the the Gell-Mann–Low scheme is that it is not developed for non-Abelian gauge theories.

So, the second lesson I draw from the above consideration is that the MS-scheme may be unreliable in theories with scalar particles.

What should be done to develop an alternative to MS-scheme for non-Abelian gauge theories? This is discussed in the next section.

4. Prospects of Gell-Mann–Low schemes for the Standard Model

In view of the above, it seems to be an imperative to develop a version of the Gell-Mann–Low scheme for the Standard model.

Let me remind what such a scheme should be. The Green's functions of the Standard Model should be expressed in therms of input parameters extracted with explicitly given rules from the Green's functions themselves. The parameters will be dependent on the energy scale(s) parametrizing the way of the extraction. After that, evolution equations for the input parameters describing their dependence on the energy scale(s) should be derived from the requirement of independence of the Green's functions from the arbitrary choice of the energy scale(s).

This program has been outlined in an early paper[9] on renormalization of spontaneously broken gauge theories. To my knowledge it has never been realized, because the MS-scheme approach has always seemed after the appearance of this paper much more promising. There are papers[10,11] proving the existence of momentum subtraction schemes, but no papers trying to apply really the Gell-Mann–Low schemes to study the evolution of the input parameters of the Standard Model.

I think the Gell-Mann–Low scheme should be fully developed for the Standard Model after more than forty years of negligence. At the very least, one would obtain in this way an alternative to the MS-scheme. My hope is, however, that I would prove in this way that the Standard Model is weakly unnatural, and obtain a new estimate for the scale up to which it is self-consistent. Hopefully, this new estimate will not lead to expectations of new physics at 1 TeV, as the present naive naturalness treatments do.

References

1. A.V. Bednyakov, A.F. Pikelner, and V.N. Velizhanin, *Phys. Lett.* **B737**, 129 (2014), arXiv:1406.7171 [hep-ph].
2. G.F. Giudice, *PoS* ***EPS-HEP2013***, 163 (2013), arXiv:1307.7879 [hep-ph].
3. A. Boyarsky, O. Ruchayskiy, and M. Shaposhnikov, *Ann. Rev. Nucl. Part. Sci.* **59**, 191 (2009), arXiv:0901.0011 [hep-ph].
4. L. Susskind, *Phys. Rev.* **D20**, 2619 (1979).
5. J.C. Collins and A.J. Macfarlane, *Phys. Rev.* **D10**, 1201 (1974).
6. J.C. Collins, *Phys. Rev.* **D10**, 1213 (1974).
7. G. Pivovarov, *Phys. Rev.* **D81**, 076007 (2010), arXiv:0912.1521 [hep-ph].
8. M.J.G. Veltman, *Acta Phys. Polon.* **B12**, 437 (1981).
9. B.W. Lee and J. Zinn-Justin, *Phys. Rev.* **D5**, 3121 (1972).
10. E. Kraus, *Acta Phys. Polon.* **B29**, 2647 (1998), arXiv:hep-th/9807102.
11. E. Kraus, *Annals Phys.* **262**, 155 (1998), arXiv:hep-th/9709154.

HYPERFINE SPLITTING IN MUONIUM AND POSITRONIUM

MICHAEL I. EIDES

Department of Physics and Astronomy, University of Kentucky, Lexington, KY 40506, USA
and
Petersburg Nuclear Physics Institute, Gatchina, St. Petersburg 188300, Russia,
E-mail: eides@pa.uky.edu, eides@thd.pnpi.spb.ru

VALERY A. SHELYUTO

D.I. Mendeleyev Institute for Metrology, St. Petersburg 190005, Russia,
E-mail:shelyuto@vniim.ru

Calculation of hard three-loop corrections of order $m\alpha^7$ to hyperfine splitting in muonium and positronium is reviewed. All these contributions are generated by the graphs with photon, electron and/or muon loop radiative insertions in the two-photon exchange diagrams. We calculate contributions of six gauge invariant sets of diagrams.

1. Introduction

Experimental and theoretical research on hyperfine splitting (HFS) in muonium and positronium was for many years and remains nowadays an active field of theoretical and experimental research. The hyperfine splitting in muonium is measured with an accuracy of 1.1 parts in 10^8 (see Ref. 1)

$$\Delta E_{HFS}(Mu) = 4\ 463\ 302.776\ (51)\ \text{kHz}. \tag{1}$$

A new higher accuracy measurement of muonium HFS is now planned at J-PARC, Japan.[2]

The current theoretical prediction[3] for the muonium hyperfine splitting is (see also Ref. 4)

$$\Delta E_{HFS}(Mu) = 4\ 463\ 302.904\ (518)\ (3)\ (70)\ \text{kHz}, \tag{2}$$

where the first error comes from the experimental error of the electron-muon mass ratio m/M, the second comes from the error in the value of the fine structure constant α (see Ref. 5), and the third is an estimate of the yet unknown theoretical contributions.

The theoretical status of the expression for hyperfine splitting in Eq. (2) is comprehensively discussed in Ref. 3 (see also Ref. 4). The agreement between theory and experiment is excellent. The error of the theoretical prediction in Eq. (2) is dominated by the experimental error of the direct measurement of the electron-muon mass ratio.[1,6] As a result hyperfine splitting in muonium serves nowadays as the source of the most accurate value of the electron-muon mass ratio[4] that is six time more accurate than the result of direct mass ratio measurement.[3]

The purely theoretical error in Eq. (2) is larger than the experimental uncertainty in Eq. (1), and further improvement of the theory would allow to further reduce the error of the electron-muon ratio. The current objective of the theory is to reduce the contribution of the theoretical uncertainty in the error bars in Eq. (2) to the level determined by the uncertainty of the fine structure constant. Another perspective on the theory of hyperfine splitting in muonium arises if we recall that the weak interaction contribution to hyperfine splitting is equal -0.065 kHz (see Ref. 7–10). We see that after reduction of the theoretical error below 10 Hz, muonium could become the first atom where a shift of atomic energy levels due to weak interaction will be observed experimentally.[11] More experimental and theoretical work is needed to achieve this goal.

The theoretical error about 70 Hz in Eq. (2) is determined by the estimate of the still uncalculated single-logarithmic and nonlogarithmic radiative-recoil corrections of order $\alpha^2(Z\alpha)(m/M)\widetilde{E}_F$, as well as of the nonlogarithmic contributions of order $(Z\alpha)^3(m/M)E_F$, $\alpha(Z\alpha)^2(m/M)E_F$, and of some other smaller corrections (see a more detailed analysis in Refs. 3, 4). Calculation of these contributions and reduction of the theoretical uncertainty of the hyperfine splitting in muonium below 10 Hz is the current task of the theory. To improve the theory of hyperfine splitting in muonium we had started a program of calculating all corrections to hyperfine splitting of the order of $1-10$ Hz (see Refs. 12–26).

The experimental research on HFS in positronium started with the discovery of positronium[27] and the first HFS measurement,[28] both in 1951. Comparable in accuracy results were obtained in the eighties[29–31]

$$\Delta E_{exp} = 203\ 388.5\ (1.0)\ \text{MHz}, \tag{3}$$

and much later in 2013[32]

$$\Delta E_{exp} = 203\ 394.2\ (1.6)_{stat}\ (1.3)_{sys}\ \text{MHz}. \tag{4}$$

An even later but less accurate result in Ref. 33 is compatible with this last number. We see that the recent result in Ref. 32 is about three standard deviations higher than the earlier results. In this situation new high precision measurements of the positronium HFS are warranted.

Theoretical work started with calculation of the leading contribution of order $m\alpha^4$ to positronium HFS splitting in the end of forties and beginning of fifties.[34–36] The full quantum electrodynamic theoretical expression for the positronium HFS splitting is a series in the fine structure constant α with the coefficients that are polynomials in $\ln\alpha$. During the years many corrections to the leading contribution were calculated. Nowadays all terms up to and including $m\alpha^7\ln\alpha$ are already calculated, see the state of the art theoretical expression in.[37] Calculation of the nonlogarithmic corrections of order $m\alpha^7$ is the next theoretical goal. First results for these corrections were published recently.[37–43]

The current theoretical uncertainty of the positronium HFS theory can be estimated by comparison with the known results for HFS in muonium. There are two major differences between HFS in muonium and positronium. First, additional annihilation channel arises in the case of positronium, and second, the masses of constituents coincide in the case of positronium. After calculation of the one-photon annihilation contribution of order $m\alpha^7$ in Ref. 37, the dominant contribution to the theoretical uncertainty of HFS in positronium is generated by the unknown nonlogarithmic terms of order $m\alpha^7$ that are similar to the terms of order $\alpha(Z\alpha)^2E_F$ in muonium. These corrections in muonium are generated by the one-loop radiative insertions in the electron line. Structurally they are similar to the classic Lamb contributions and are represented by the series in $\ln(Z\alpha)$, $[c_1\ln^2(Z\alpha)+c_2\ln(Z\alpha)+c_3]\alpha(Z\alpha)^2E_F$, see, e.g., review in Refs. 3, 44. The nonlogarithmic term with the coefficient c_3 arises from the ultrasoft momenta of order $m(Z\alpha)^2$ and is about $2\times10^{-6}E_F$ in muonium.[45,46] Similar ultrasoft terms in positronium arise due to one-loop radiative insertions in each of the fermion lines and also due to an exchange of an ultrasoft photon between the fermion lines. In this situation we accept $2\times10^{-6}E_F^{Ps}\sim 250$ kHz as a fair estimate of the ultrasoft nonlogarithmic contribution in positronium. This term is still uncalculated and after the recent calculation of the one-photon annihilation contribution in Ref. 37 it can be used as an estimate of the theoretical uncertainty of the current positronium HFS theory.

2. Hard Nonlogarithmic Three-Loop Corrections to Hyperfine Splitting

Hard nonlogarithmic contributions to HFS in positronium of order $m\alpha^7$ are similar to the radiative and radiative-recoil corrections to HFS in muonium of orders $\alpha^2(Z\alpha)E_F$ and $\alpha^2(Z\alpha)(m/M)E_F$, respectively. Both in muonium and positronium these corrections are generated by seven gauge invariant sets of nonannihilation diagrams, six of them are presented in Figs. 2–4 and in Figs. 6–8.* One more set of diagrams with two-photon exchanges and two radiative photon insertions in one and the same fermion line also generates hard nonlogarithmic corrections of order $m\alpha^7$. We have calculated these corrections in muonium some time ago.[19–21,23,24] The corrections in muonium are power series in the electron-muon mass ratio with the coefficients enhanced by large logarithms of this mass ratio. The goal of the old work on muonium was to calculate the coefficients in this expansion, at least the factors before the logarithms, analytically. In the case of positronium the masses are equal and the hard corrections of order $m\alpha^7$ are pure numbers. We apply the approach developed for muonium to positronium. We consider an electromagnetically bound system of two particles with arbitrary masses M and m, and obtain general expressions for the hard corrections to HFS of order $m\alpha^7$ without expansion in the mass ratio of the constituents. We check numerically that in the case of a small mass ratio these formulae reproduce with high accuracy the sum of all already known terms in the expansion in the small mass ratio for muonium. We use the general expressions for the case of equal masses and calculate all hard three-loop contributions to HFS in positronium of order $m\alpha^7$ that are due to the diagrams with closed electron loops.

We start with the infrared divergent contribution to HFS by the two-photon exchange diagrams in Fig. 1 calculated in the scattering approximation,

$$\Delta E = \frac{Z\alpha}{\pi} E_F \frac{mM}{M^2 - m^2} \int_0^\infty dq^2 f_\mu(q). \tag{5}$$

The dimensionless weight function $f_\mu(q)$ in terms of an auxiliary function,

$$f(q) = -\frac{1}{4} + \frac{\sqrt{q^2+4}}{4q} - \frac{2\sqrt{q^2+4}}{q^3} \tag{6}$$

*We systematically omit diagrams with crossed exchanged photons in all figures.

has the form

$$f_\mu(q) = f(q) - 4\mu^2 f(2\mu q)\,. \tag{7}$$

Fig. 1. Diagrams with two-photon exchanges

The integral in Eq. (5) is a sum of an infrared linearly divergent integral and a finite one. In a more accurate approximation (with the off mass shell external fermion lines) the linear divergence is cutoff at the characteristic atomic scale $\sim m\alpha$ and turns into a contribution of a lower order in α. The remaining finite part of the integral originates at hard integration momenta $\sim m$ (or in the interval from m to M in the case of unequal masses) and generates a contribution of order αE_F. In the case of unequal masses, for muonium, the linearly infrared divergent contribution turns into the leading nonrecoil Fermi contribution E_F to HFS, while the finite part generates the leading recoil correction of order $\alpha(m/M)E_F$, see, e.g., Refs. 3, 44. Let us emphasize that due to the linear (as opposed to logarithmic) nature of the apparent divergence it leaves no finite remnant of order αE_F and should be simply thrown away. No need in matching of high and low integration momenta arises.

Radiative insertions in Figs. 2–4 and in Figs. 6–8 make these diagrams infrared convergent and justify validity of the scattering approximation for their calculation. All corrections calculated below are obtained by some modifications of the basic integral in Eq. (5).

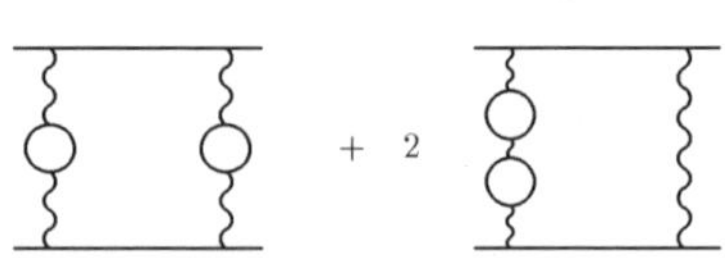

Fig. 2. Diagrams with two one-loop polarization insertions

2.1. *Analytic Results for One- and Two-Loop Polarization Insertions in the Exchanged Photons*

Consider first the diagrams in Fig. 2 with two one-loop polarization loops. Insertion of a polarization operator in a photon line with momentum q (all momenta below are measured in units of the electron mass) reduces to the replacement in the photon propagator,

$$\frac{1}{q^2} \to \frac{\alpha}{\pi} I_1(q)\,, \tag{8}$$

where $(\alpha/\pi)I_1(q)$ is the well known representation of the one-loop vacuum polarization,[47]

$$\frac{\alpha}{\pi} I_1(q) = \frac{\alpha}{\pi} \int_0^1 dv \frac{v^2(1-\frac{v^2}{3})}{1-v^2} \frac{1}{q^2 + \frac{4}{1-v^2}}\,. \tag{9}$$

The contribution of the diagrams in Fig. 2 is obtained by insertion of the one-loop photon polarization squared $(\alpha/\pi)^2 q^4 I_1^2(q)$ in the integrand in Eq. (5). In the general case of unequal masses the contribution to HFS has the form

$$\Delta E = 3\frac{\alpha^2(Z\alpha)}{\pi^3} E_F \frac{mM}{M^2-m^2} \int_0^\infty dq^2 f_\mu(q) q^4 I_1^2(q), \tag{10}$$

where the factor 3 before the integral has the combinatorial origin. We checked numerically that in the small mass ratio limit this integral reproduces the sum of all known analytically terms[15,49] of the expansion of this contribution in the small mass ratio.

In the case of positronium $M = m$ and we obtain the contribution to HFS of the diagrams with two one-loop polarization insertions in Fig. 2,

$$\Delta E_1 = \left(\frac{6\pi^2}{35} - \frac{8}{9}\right) \frac{\alpha^3}{\pi^3} E_F^{\rm Ps} = 0.803\,043\,294 \frac{\alpha^3}{\pi^3} E_F^{\rm Ps}. \tag{11}$$

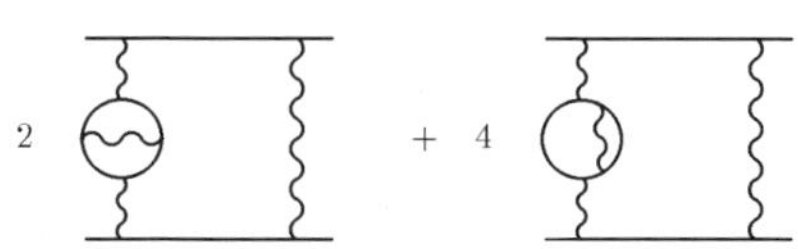

Fig. 3. Diagrams with two-loop polarization insertions

The contribution of the two-loop vacuum polarization in Fig. 3 can be obtained by the insertion of the two-loop photon polarization

$(\alpha^2/\pi^2)q^2I_2(q)$ [47,48] in the integrand in Eq. (5),

$$\left(\frac{\alpha}{\pi}\right)^2 I_2(q) = \frac{2}{3}\left(\frac{\alpha}{\pi}\right)^2 \int_0^1 dv \frac{v}{4+q^2(1-v^2)} \Bigg\{ (3-v^2)(1+v^2)\left[\mathrm{Li}_2\left(-\frac{1-v}{1+v}\right)\right.$$
$$+ 2\mathrm{Li}_2\left(\frac{1-v}{1+v}\right) + \frac{3}{2}\ln\frac{1+v}{1-v}\ln\frac{1+v}{2} - \ln\frac{1+v}{1-v}\ln v\Bigg]$$
$$+\left[\frac{11}{16}(3-v^2)(1+v^2)+\frac{v^4}{4}\right]\ln\frac{1+v}{1-v}$$
$$+\left[\frac{3}{2}v(3-v^2)\ln\frac{1-v^2}{4} - 2v(3-v^2)\ln v\right] + \frac{3}{8}v(5-3v^2)\Bigg\}, \tag{12}$$

where the dilogarithm is defined as $\mathrm{Li}_2(z) = -\int_0^1 dt \ln(1-zt)/t$.

In the case of unequal masses the integral for the contribution to HFS of the diagrams with the two-loop polarization in Fig. 3 has the form

$$\Delta E = 2\frac{\alpha^2(Z\alpha)}{\pi^3}E_F\frac{mM}{M^2-m^2}\int_0^\infty dq^2 f_\mu(q)q^2 I_2(q), \tag{13}$$

where the factor 2 before the integral is due to combinatorics. We checked numerically that in the small mass ratio case the integral in Eq. (13) coincides with the sum of the known terms [15,49] of the expansion of this contribution to HFS in the small mass ratio.

In the case of equal masses, for positronium, the contribution to HFS of the diagrams in Fig. 3 admits an analytic calculation, and we obtain

$$\begin{aligned}\Delta E_2 &= \left[-\frac{217}{30}\zeta(3) + \frac{28\pi^2}{15}\ln 2 + \frac{\pi^2}{675} + \frac{403}{360}\right]\frac{\alpha^3}{\pi^3}E_F^{\mathrm{Ps}} \\ &= 5.209\ 219\ 614\frac{\alpha^3}{\pi^3}E_F^{\mathrm{Ps}}.\end{aligned} \tag{14}$$

2.2. *One-Loop Electron Factor and One-Loop Polarization Insertion in the Exchanged Photon*

The diagrams in Fig. 4 are obtained from the skeleton diagrams in Fig. 1 by one-loop radiative insertions in one of the exchanged photons and one of

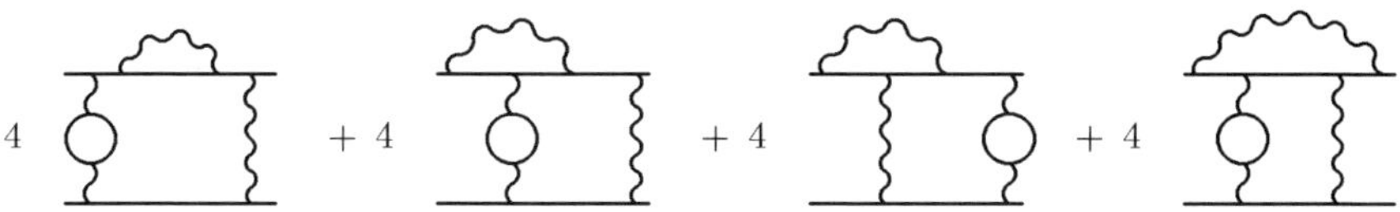

Fig. 4. Diagrams with one-loop polarization and radiative photon insertions

the fermion lines.† To describe these radiative insertions it is convenient to introduce the one-loop electron factor that is defined as a gauge invariant sum of the diagrams in Fig. 5 where the external electron lines are on-shell and carry zero spatial momenta (plus the diagrams with the exchanged external photon vertices). Physically the electron factor is a sum of one-loop corrections to the spin-dependent amplitude of the virtual forward Compton scattering.

We have derived explicit integral representations for these form factors long time ago calculating contributions to HFS in muonium.[50,51] Using this expression we calculated contribution to HFS in muonium numerically and checked that the integral with the electron factor reproduces the sum of the leading terms of the expansion[16,18,49] of this contribution to HFS in the small mass ratio

$$\Delta E = \frac{\alpha^2(Z\alpha)}{\pi^3}E_F \left[\pi^2\left(-\frac{4}{3}\ln^2\frac{1+\sqrt{5}}{2} - \frac{20}{9}\sqrt{5}\ln\frac{1+\sqrt{5}}{2} - \frac{64}{45}\ln 2 + \frac{\pi^2}{9} + \frac{3}{8} + \frac{1043}{675}\right) + \frac{m}{M}\left(\frac{5}{2}\ln^2\frac{M}{m} + \frac{10}{3}\ln\frac{M}{m} + 11.41788\right)\right]. \tag{15}$$

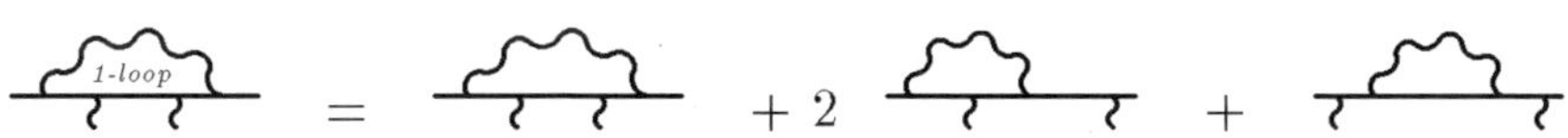

Fig. 5. One-loop fermion factor

In the case of equal masses there is an extra factor 2 before the diagrams

†Multiplicity factors in these diagrams correspond to the case of positronium, not muonium.

in Fig. 4. This factor arises because now there are two ways to insert the fermion factor in one of the lepton lines. Calculating the integral we obtain the contribution of the diagrams in Fig. 4 to HFS in positronium,

$$\Delta E_3 = -1.287\ 09\ (1)\ \frac{\alpha^3}{\pi^3} E_F^{Ps}. \tag{16}$$

2.3. *One-Loop Polarization Insertion in the Electron Factor*

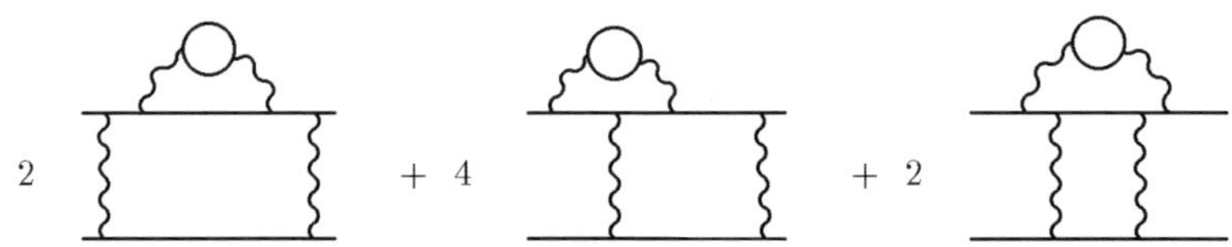

Fig. 6. Diagrams with one-loop polarization insertions in radiative photons

Consider now the diagrams in Fig. 6 with the one-loop polarization insertions in the radiative photon.‡ Effectively these diagrams contain a massive radiative photon, see Eq. (8) and Eq. (9). We consider this electron factor as a sum of the contributions corresponding to the separate diagrams in Fig. 6 with the self-energy, vertex and spanning photon insertions in the electron line. Each of these terms is calculated as a one-loop diagrams with a massive photon and then integrated over the auxiliary parameter v.

We have derived an integral representation for the contribution to HFS of the diagrams in Fig. 6. We have checked numerically that in the case of unequal masses this integral reproduces the expansion in the small mass ratio

$$\begin{aligned} \Delta E =& \frac{\alpha^2(Z\alpha)}{\pi^3} E_F \Bigg\{ -0.310\ 742\ \pi^2 \\ &+ \frac{m}{M}\left[\frac{3}{4}\ln^2\frac{M}{m} + \left(\pi^2 - \frac{53}{6}\right)\ln\frac{M}{m} + 7.08072\right]\Bigg\} \end{aligned} \tag{17}$$

derived earlier.

‡Multiplicity factors in these diagrams correspond to the case of positronium, not muonium.

In the case of equal masses the contribution of the diagrams in Fig. 6 to HFS is equal to

$$\Delta E_4 = -3.154\ 41\ (1)\ \frac{\alpha^3}{\pi^3} E_F^{Ps}\,. \tag{18}$$

2.4. *Light-by-Light Scattering Contribution*

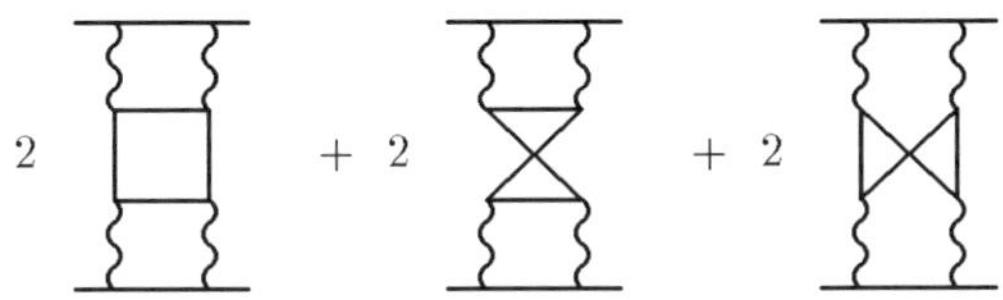

Fig. 7. Diagrams with light-by-light scattering insertions

Due to gauge invariance the light-by-light scattering block fast decreases with the momenta of the external (virtual) photons. Therefore, effectively all integration momenta in the diagrams in Fig. 7 are hard, of order of the electron mass (or in the interval from m to M in the case of unequal masses).

We derived an explicit integral representation for the the light-by-light scattering contribution to HFS in the general case of unequal masses.[24,26] We have checked numerically that in the small mass ratio case this contribution coincides with the sum of the first terms of the expansion that were calculated during the years[23,24,52,53]

$$\Delta E = \frac{\alpha^2(Z\alpha)}{\pi^3} E_F \Bigg\{ -0.472\ 514\ (1)\ \pi^2 + \frac{m}{M}\left[\frac{9}{4}\ln^2\frac{M}{m} + \left(-3\zeta(3) - \frac{2\pi^2}{3} + \frac{91}{8}\right)\ln\frac{M}{m} + 5.9949(1)\right]\Bigg\}. \tag{19}$$

In the case of equal masses we obtain

$$\Delta E_5 = -0.706\ 27\ (5)\ \frac{\alpha^3}{\pi^3} E_F^{Ps}, \tag{20}$$

which coincides with the result first obtained in Ref. 38.

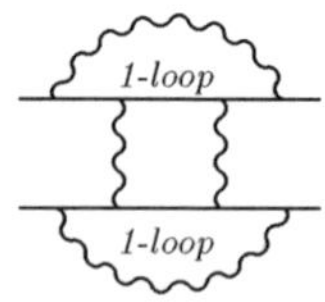

Fig. 8. Diagrams with one-loop radiative photon insertions in both fermion lines

2.5. *Two-One Loop Electron Factors*

The diagrams in Fig. 8 contain the one-loop fermion factors from Fig. 5 in both fermion lines. The first terms of the expansion of the contribution to HFS of the diagrams in Fig. 8 in the small mass ratio are already known for some time[17,54]

$$\begin{aligned}\Delta E =& \frac{\alpha(Z^2\alpha)(Z\alpha)}{\pi^3}E_F\left[\frac{\pi^2}{2}\left(\ln 2 - \frac{13}{4}\right)\right.\\ &+\frac{m}{M}\left(-\frac{9}{8}\ln\frac{M}{m} - \frac{15}{8}\zeta(3) + \frac{15\pi^2}{4}\ln 2 + \frac{37\pi^2}{24} - \frac{147}{32}\right)\\ &\left.+\frac{9}{16}\frac{Mm}{M^2-m^2}\ln\frac{M^2}{m^2}\right].\end{aligned} \tag{21}$$

We have derived an integral representation for the contribution to HFS of the diagrams in Fig. 8 and checked numerically that in the case of small mass ratio this integral representation coincides with sum above.

In the case of equal masses we obtain the contribution of the diagrams in Fig. 8 to HFS in positronium,

$$\Delta E_6 = -4.739\ 55\ (40)\ \frac{\alpha^3}{\pi^3}E_F^{Ps}. \tag{22}$$

3. Conclusions

We have derived explicit expressions for hard three-loop contributions to hyperfine splitting generated by the six gauge-invariant sets of diagrams with closed electron loops in Figs. 2–4 and in Figs. 6–8. In the case of unequal lepton masses we confirm numerically the already known results for muonium obtained earlier in the form of an expansion in the small mass ratio. We have calculated the contributions of these diagrams to HFS in the case of equal masses, for positronium. Collecting the results in Eqs. (11), (14), (16), (18), (20), and (22), we obtain the total hard contribution to HFS in positronium of order $m\alpha^7$ generated by all diagrams with closed electron loops in Figs. 2–4 and in Figs. 6–8,

$$\Delta E = -3.875\ 0\ (4) \left(\frac{\alpha}{\pi}\right)^3 E_F^{Ps} = -1.291\ 7\ (1) \frac{m\alpha^7}{\pi^3} = -5.672\ \text{kHz}. \quad (23)$$

Taking into account other recent theoretical results[37,38,41] we obtain the theoretical prediction for HFS in positronium

$$\Delta E_{theor} = 203\ 391.89\ (25)\ \text{MHz}. \quad (24)$$

This result does not contradict the experimental result in Eq. (4). Further reduction of both the experimental and theoretical uncertainties is warranted. Calculation of the remaining ultrasoft and hard nonlogarithmic contributions of order $m\alpha^7$ is the next task for the theory.

Acknowledgements

This work was supported by the NSF grant PHY-1402593. The work of V.S. was supported in part by the RFBR grant 14-02-00467.

References

1. W. Liu, M.G. Boshier, S. Dhawan *et al.*, *Phys. Rev. Lett.* **82**, 711 (1999).
2. K. Shimomura, *AIP Conf. Proc.* **1382**, 245 (2011).
3. M.I. Eides, H. Grotch, and V.A. Shelyuto, *Theory of Light Hydrogenic Bound States,* Springer, Berlin, Heidelberg, New York 2007.
4. P.J. Mohr, B.N. Taylor, and D.B. Newell, *Rev. Mod. Phys.* **84**, 1527 (2012).
5. D. Hanneke, S. Fogwell, and G. Gabrielse, *Phys. Rev. Lett.* **100**, 120801 (2008).
6. K.P. Jungmann, arXiv:nucl-ex/0404013 (2004).
7. M.A. Beg and G. Feinberg, *Phys. Rev. Lett.* **33**, 606 (1974); [Errata, *ibid.* **35**, 130(E) (1975)].
8. M.I. Eides, *Phys. Rev.* **A53**, 2953 (1996).
9. T. Kinoshita and M. Nio, *Phys. Rev.* **D53**, 4909 (1996).
10. J.R. Sapirstein, unpublished (1996).
11. K.P. Jungmann, arXiv:physics/9809020 (2008).
12. M.I. Eides and V.A. Shelyuto, *Phys. Rev.* **A68**, 042106 (2003).
13. M.I. Eides and V.A. Shelyuto, *Phys. Rev.* **A70**, 022506 (2004).
14. M.I. Eides and V.A. Shelyuto, *Can. J. Phys.* **85**, 509 (2007).
15. M.I. Eides, H. Grotch, and V.A. Shelyuto, *Phys. Rev.* **D65**, 013003 (2001).

16. M.I. Eides, H. Grotch, and V. A. Shelyuto, *Phys. Rev.* **D67**, 113003 (2003).
17. M.I. Eides, H. Grotch, and V.A. Shelyuto, *Phys. Rev.* **D70**, 073005 (2004); [*ibid.* **D88**, 119901(E) (2013)].
18. M.I. Eides and V.A. Shelyuto, *Can. J. Phys.* **83**, 363 (2005).
19. M.I. Eides and V.A. Shelyuto, *Phys. Rev. Lett.* **103**, 133003 (2009).
20. M.I. Eides and V.A. Shelyuto, *Zh. Eksp. Teor. Fiz.* **137**, 24 (2010) [*J. Exp. Theor. Phys.* **110**, 17 (2010)].
21. M.I. Eides and V.A. Shelyuto, *Phys. Rev.* **D80**, 053008 (2009).
22. M.I. Eides and V.A. Shelyuto, *Phys. Rev.* **A86**, 024501 (2012).
23. M.I. Eides and V.A. Shelyuto, *Phys. Rev.* **D87**, 013005 (2013).
24. M.I. Eides and V.A. Shelyuto, *Phys. Rev.* **D89**, 014034 (2014).
25. M.I. Eides and V.A. Shelyuto, *Phys. Rev. Lett.* **112**, 173004 (2014).
26. M.I. Eides and V.A. Shelyuto, *Phys. Rev.* **D90**, 113002 (2014).
27. M. Deutsch, *Phys. Rev.* **82**, 455 (1951).
28. M. Deutsch and E. Dulit, *Phys. Rev.* **84**, 601 (1951).
29. A.P. Mills, Jr. and G.H. Bearman, *Phys. Rev. Lett.* **34**, 246 (1975).
30. M.W. Ritter, P.O. Egan, V.W. Hughes, and K.A. Woodle, *Phys. Rev.* **A30**, 1331 (1984).
31. A.P. Mills, *Phys. Rev.* **A27**, 262 (1983).
32. A. Ishida, T. Namba, S. Asai, T. Kobayashi, H. Saito, M. Yoshida, K. Tanaka, and A. Yamamoto, arXiv:1310.6923 [hep-ex].
33. A. Miyazaki, T. Yamazaki, T. Suehara, T. Namba, S. Asai, T. Kobayashi, H. Saito, Y. Tatematsu, I. Ogawa, and T. Idehara, arXiv:1403.0312 [hep-ex].
34. J. Pirenne, *Arch. Sci. Phys. Nat.* **29**, 265 (1947).
35. V.B. Berestetskii, *Sov. Phys. JETP* **19**, 1130 (1949).
36. R.A. Ferrell, *Phys. Rev.* **84**, 858 (1951).
37. M. Baker, P. Marquard, A. A. Penin *et al.*, *Phys. Rev. Lett.* **112**, 120407 (2014).
38. G.S. Adkins and R.N. Fell, *Phys. Rev.* **A89**, 052518 (2014).
39. M.I. Eides and V.A. Shelyuto, *Phys. Rev.* **D89**, 111301 (2014).
40. M.I. Eides and V.A. Shelyuto, *Phys. Rev.* **D92**, 013010 (2015).
41. G.S. Adkins, C. Parsons, M.D. Salinger, R. Wang, and R.N. Fell, *Phys. Rev.* **A90**, 042502 (2014).
42. G.S. Adkins, C. Parsons, M.D. Salinger, and R. Wang, *Phys. Lett.* **B747**, 551 (2015).
43. G.S. Adkins, M. Kim, C. Parsons, and R. N. Fell, arXiv: 1507.07841 [hep-ph].

44. M.I. Eides, H. Grotch, and V.A. Shelyuto, *Phys. Rept.* **342**, 63 (2001).
45. K. Pachucki, *Phys. Rev.* **A54**, 1994 (1996).
46. T. Kinoshita and M. Nio, *Phys. Rev.* **D55**, 7267 (1997).
47. J. Schwinger, *Particles, Sources and Fields*, Addison-Wesley, Reading, MA, 1973, Vol. 2.
48. G. Källnand A. Sabry, *K. Dan. Vidensk. Selsk. Mat. Fys. Medd.* **29**, 17 (1955).
49. M.I. Eides, S.G. Karshenboim, and V.A. Shelyuto, *Phys. Lett.* **B229**, 285 (1989); *Pis'ma Zh. Eksp. Teor. Fiz.* **50**, 3 (1989) [*JETP Lett.* **50**, 1 (1989)]; *Yad. Fiz.* **50**, 1636 (1989) [*Sov. J. Nucl. Phys.* **50**, 1015 (1989)].
50. V.Yu. Brook, M.I. Eides, S.G. Karshenboim, and V.A. Shelyuto, *Phys. Lett. B* **216**, 401 (1989).
51. M.I. Eides, H. Grotch, and V.A. Shelyuto, *Phys. Rev.* **D58**, 013008 (1998).
52. M.I. Eides, S.G. Karshenboim, and V.A. Shelyuto, *Phys. Lett.* **B268**, 433 (1991); *ibid.* **316**, 631 (E) (1993); *ibid.* **319**, 545 (E) (1993); *Yad. Fiz.* **55**, 466 (1992); *ibid.* **57**, 1343 (E) (1994) [*Sov. J. Nucl. Phys.* **55**, 257 (1992); *ibid.* **57**, 1275 (E) (1994)].
53. M.I. Eides, S.G. Karshenboim, and V.A. Shelyuto, *Phys. Lett.* **B216**, 405 (1989); *Yad. Fiz.* **49**, 493 (1989) [*Sov. J. Nucl. Phys.* **49**, 309 (1989)].
54. N. Kroll and F. Pollock, *Phys. Rev.* **84**, 594 (1951); *ibid.* **86**, 876 (1952).
55. S.G. Karshenboim, M.I. Eides, and V.A. Shelyuto, *Yad. Fiz.* **48**, 1039 (1988) [*Sov. J. Nucl. Phys.* **48**, 661 (1988)].

M. Eides, Chernogolovka, 2015.

ELECTROMAGNETIC PRODUCTION OF POSITRONS AND ELECTRONS IN COLLISIONS OF HEAVY NUCLEI

I.B. KHRIPLOVICH

St. Petersburg State University, 198504 St. Petersburg, Russia,
E-mail: khios231@mail.ru

We consider the electromagnetic production of positrons and electrons in collisions of slow heavy nuclei. This process is dominated by emission of positrons, with the electrons captured by nucleus.

Positron production in collisions of heavy nuclei, as well as the production e^+e^- pairs, were addressed in numerous papers (see, for instance, Refs. 1, 2). Here we consider this problem with simple qualitative arguments.

We start with production of the pair e^+e^-, as presented in Fig. 1. Double lines therein describe the propagation of heavy nuclei, single and wavy lines refer to $e^{\pm}$ and virtual photons, respectively. The electromagnetic vertices on these double lines reduce to $(p' + p)/2M \approx v$. We neglect the difference between p' and p. For the velocity of nuclei we assume $v \simeq 0.1$. Thus, taking into account both nuclear lines in Fig. 1, we conclude that our amplitude is proportional to v^2.

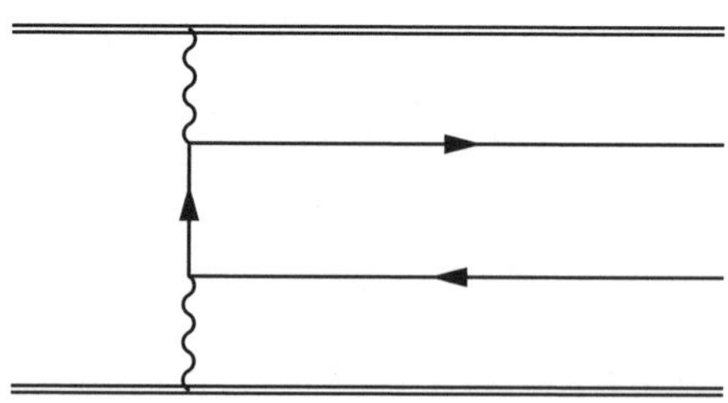

Fig. 1. Production of pair e^+e^-.

As to the discussed cross-section, it is obviously proportional to

$$v^4/v = v^3, \tag{1}$$

where in the denominator v originates from the particle flux. Here and below we use units such that $c = 1$.

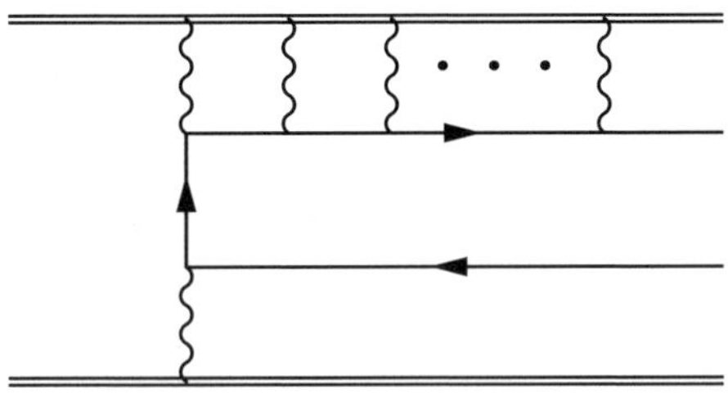

Fig. 2. Electron capture graph.

We address now the channel where the produced electron is captured by one of the nuclei, and the produced positron goes to infinity, as presented in Fig. 2. The captured electron here has no direct relation at all to the velocity discussed. Indeed, the attraction of this electron to nucleus is strong, and there is no reason to expect any extra v-dependence in the corresponding vertex. Therefore, the amplitude here is proportional to v, with the corresponding cross-section proportional to

$$v^2/v = v. \tag{2}$$

Thus, the positron production cross-section dominates the effect, since

$$v/v^3 \simeq 10^2. \tag{3}$$

In other words, the cross-section of the positron production, with the capture of the produced electron, is the dominating one, it is about hundred times more than that of e^+e^- production. The attraction of the produced electron to nucleus is strong, and there is no reason to expect any extra v-dependence in the corresponding vertex.

The electromagnetic cross-section of e^+e^- production was estimated in Ref. 3 as

$$\sigma_\pm \sim 10^{-26} - 10^{-25}\,\text{cm}^2\,. \tag{4}$$

Then, the reasonable estimate for the cross-section of positron production is

$$\sigma_+ \sim 10^{-24} - 10^{-23}\,\text{cm}^2\,. \tag{5}$$

Acknowledgements

I am grateful to V.M. Shabaev for useful discussions.

References

1. U. Mueller, T.D. Reus, J. Reinhardt, B. Mueller, W. Greiner, and G. Soff, *Phys. Rev.* **A37**, 1449 (1988).
2. I.A. Maltsev, V.M. Shabaev, I.I. Tupitsin, A.I. Bondarev, Y.S. Kozhedub, G. Plunien, and Th. Stolker, *Phys. Rev.* **A91**, 032708 (2015).
3. I.B. Khriplovich, *JETP Lett.* **100**, 552 (2014).

I.B. Khriplovich, Chernogolovka, 2015.

PRODUCTION OF QUARKONIA AT RHIC

RÓBERT VÉRTESI
Wigner Research Centre for Physics of the Hungarian Academy of Sciences RMI, H-1525 Budapest 114, P.O.Box 49, Hungary
and
Nuclear Physics Institute of the Academy of Sciences of the Czech Republic, Řež 130, 25068 Husinec-Řež, Czech Republic
E-mail: vertesi.robert@wigner.mta.hu

The production of different quarkonium states provides unique insight to the hot and cold nuclear matter effects in the strongly interacting medium that is formed in high energy heavy ion collisions. While LHC explores the energy frontier, RHIC has a broad physics program to explore the nuclear modification at different energies in a wide range of systems. Some of the most interesting recent results on J/ψ and Υ production in p+p, d+Au and A+A collisions from PHENIX and STAR are summarized in this work.

Keywords: Brookhaven RHIC Coll, quarkonium: heavy, quarkonium: production, quark gluon: plasma

1. Introduction

In the first decade of its operation, the Relativistic Heavy Ion Collider (RHIC) mostly concentrated on measurements of the more abundantly produced particles such as light mesons. These investigations led to the discovery of a strongly interacting quark-gluon plasma (sQGP), and the understanding of its basic properties.[1–5] In more recent years, however, vastly improved luminosity and newly installed detector subsystems allowed for rarer probes, such as comprehensive heavy flavor measurements. These probes provide us with tools to explore the thermal and dynamical properties of the sQGP from new aspects.

It has long been suggested that, due to the screening of the heavy quark potential, the production of heavy quarkonia (charmonium and bottomonium states) is suppressed in heavy ion collisions compared to expectations from p+p collisions. Charmonium suppression was, in fact, anticipated as a key signature of quark-gluon plasma (QGP) formation.[6] Moreover, different quarkonium states are expected to dissociate ("melt") in the QGP at different temperatures due to their different binding energies.[7] It has therefore been suggested that comprehensive quarkonium measurements can be used as a QGP thermometer.[8]

A nuclear modification that is chiefly governed by melting in the sQGP implies a monotonous increase in the suppression of a given quarkonium state with the beam energy. Recent measurements at RHIC and LHC, however, outline a more complicated situation: the beam energy dependence of J/ψ production is found to be rather weak, moreover, data at LHC energies show less suppression than at RHIC energies.[9] Our current understanding on this observation is that several nuclear matter effects influence the resulting yields besides melting.[10–12] In hot nuclear matter, uncorrelated quark-antiquark pairs can recombine in later stages, thus increasing the overall production. Cold nuclear matter effects can also be significant, such as nuclear shadowing and anti-shadowing, initial state energy loss or co-mover absorption of quarks. There is also a sizable but largely uncertain contribution to the yields from feed-down processes such as χ_c, ϕ and B decays to J/ψ, and χ_b, Υ(2S) and Υ(3S) to Υ(1S). Detailed experimental studies are required to disentangle these various effects.

2. Experimental approach

RHIC is an extremely versatile accelerator and detector complex that can be utilized to measure different quarkonium states produced in several collisional systems in a wide range of energy, centrality, momentum and rapidity. Symmetric and asymmetric collisions are studied in p+p, p+Au, d+Au, Cu+Cu, Cu+Au, Au+Au and U+U systems over energies ranging from $\sqrt{s_{\rm NN}}$=7.7 to 200 GeV for heavy ions, and up to 510 GeV in the case of (polarized or unpolarized) protons. Both the PHENIX[13] and STAR[14] experiments that currently take data at the RHIC ring have a rich heavy flavor program. STAR is capable of measurements in the dielectron channel in full azimuth at midrapidity ($|\eta|<0.5$) using the Time Projection Chamber for tracking, aided by the Barrel Electromagnetic Calorimeter for particle identification. From 2014 onwards, an outer Muon Telescope Detector (MTD) allows for measurements in the dimuon channel with significantly improved resolution.[15] Tracking in PHENIX is done using the Drift and Pad Chambers, while the Electromagnetic Calorimeter serves for mid-rapidity particle identification in the central arm (covering $|\eta|<0.35$ in pseudorapidity and $2 \times \pi/2$ in azimuth). The muons are tracked and identified in the forward arms ($1.2<|\eta|<2.2$).

The most common experimental approach is to measure the nuclear modification factor (R_{AA}) of a given state. It is computed by comparing the corrected number of particles measured in A+A collisions to the yield in p+p collisions scaled by the average number of binary nucleon-nucleon collisions, as

$$R_{\mathrm{AA}} = \frac{\sigma_{pp}^{inel}}{\sigma_{AA}^{inel}} \frac{1}{\langle N_{\mathrm{coll}} \rangle} \frac{d\sigma^{AA}/dy}{d\sigma^{pp}/dy}.$$

Here $\sigma_{AA(pp)}^{inel}$ is the total inelastic cross section of the A+A (p+p) collisions, $d\sigma^{AA(pp)}/dy$ denotes the production cross section of the observed state in A+A (p+p) collisions, and $\langle N_{\mathrm{coll}} \rangle$ is the average number of binary nucleon-nucleon collisions from the Glauber model. In the case of quarkonia, where the reconstruction is done via the dielectron (dimuon) decay channel, usually the $B_{ee(\mu\mu)} \times d\sigma^{AA(pp)}/dy$ is specified, where $B_{ee(\mu\mu)}$ is the branching ratio corresponding to the given decay process of the state to an electron-positron ($\mu^+\mu^-$) pair. A similarly defined R_{dAu} is used for describing the nuclear modification in d+Au collisions.

3. J/ψ measurements

Besides providing reference for nuclear modification in heavier systems, p+p data serve as a benchmark for QCD models. Fig. 1 shows the transverse momentum spectrum of inclusive J/ψ production at RHIC in p+p collisions at $\sqrt{s_{NN}}$ = 200 GeV,[16–18] compared to several theoretical calculations on prompt J/ψ production. The prompt color evaporation model[19] may provide an acceptable description of the spectrum over the whole transverse momentum range. However, the same models that describe the spectrum face difficulties understanding the J/ψ polarization.[20]

Collisions of d+Au (p+A) systems are generally considered adequate for CNM effect studies because they are too cold for the plasma to form.[3,4] Fig. 2 shows the nuclear modification of the J/ψ mesons in d+Au collisions at $\sqrt{s_{NN}}$=200 GeV, measured by PHENIX.[21] While there is a suppression in the whole rapidity range, the strongest effect is in the gold-going (backward) direction, while it gradually decreases through mid-rapidity and the R_{AA} is marginally consistent with unity in the deuteron-going (forward) region. This pattern is reasonably explained by the EPS09 model,[22] containing nuclear shadowing and final state breakup of the $c\bar{c}$ pair.

Fig. 5 shows the J/ψ nuclear modification factor at forward rapidities for $\sqrt{s_{NN}}$ =39, 62.4 and 200 GeV Au+Au collisions at PHENIX,[23] as a

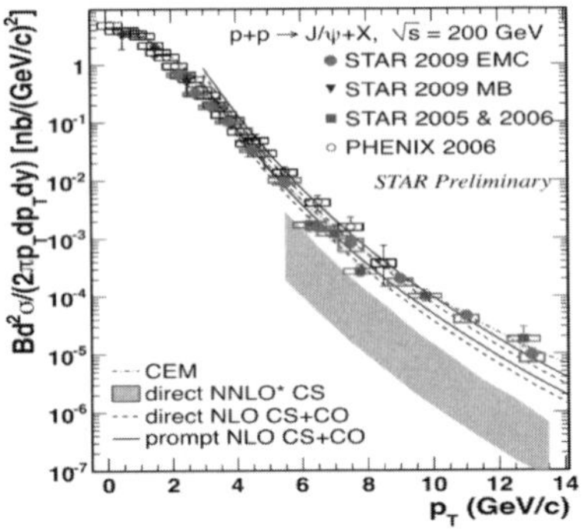

Fig. 1. p_T-spectrum of J/ψ mesons in p+p collisions at $\sqrt{s_{NN}}$=200 GeV at mid-rapidity measured by PHENIX (open symbols) and STAR (solid symbols), compared to calculations for prompt J/ψ production.

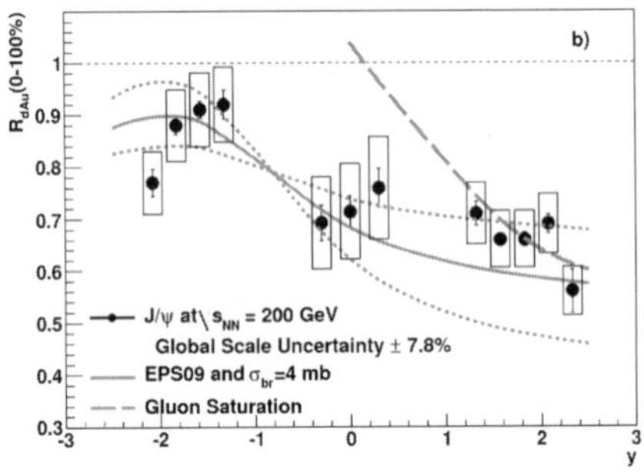

Fig. 2. Nuclear modification factor R_{dAu} versus rapidity in d+Au collisions at $\sqrt{s_{NN}}$=200 GeV measured by PHENIX.

function of the number of participant nucleons (N_part). A significant suppression is observed at all energies in mid-peripheral to central collisions, and all three datasets are consistent with each other within uncertainties. STAR data at mid-rapidity exhibits similar trends.[24] Model predictions that include in-medium dissociation of the J/ψ as well as later regeneration from $c\bar{c}$ pairs are consistent with data.[25] The weak dependence of R_AA on collision energy suggests that the different contributions to the J/ψ nuclear modification largely cancel each other. Fig. 6 shows PHENIX forward rapidity measurements of R_AA vs. N_part in U+U and Au+Au systems. The two datasets mostly overlap each other, similarly to other measurements of open and hidden heavy flavor at RHIC.[28] However, there is a significant discrepancy at higher N_part values that may be a hint that there is more coalescence in collisions of non-spherical Uranium nuclei.

We saw that it is difficult to disentangle the sources of nuclear modification, because the R_AA is rather insensitive to beam energy over a wide range and also to the number of nucleons in the chosen heavy ion system. At high momentum, however, the impact of CNM effects and regeneration decreases and hot nuclear modification becomes dominant.[25] Fig. 3 shows the nuclear modification factor for J/ψ mesons with a transverse momentum p_T>5 GeV/c, measured by STAR at mid-rapidity in Au+Au collisions at $\sqrt{s_{NN}}$=200 GeV, compared to J/ψ mesons from the same collisions without momentum selection.[16] The suppression of high-p_T J/ψ at higher N_part values is a clear sign of nuclear modification in the sQGP.

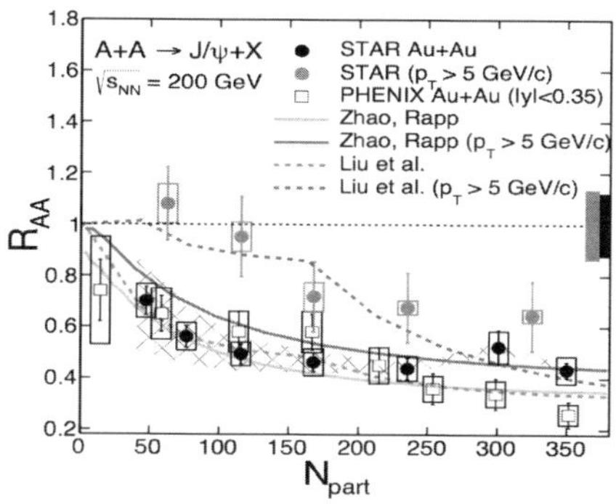

Fig. 3. STAR and PHENIX minimum bias (black circles and empty squares) J/ψ vs. $N_{\rm part}$, as well as STAR high-p_T J/ψ vs. $N_{\rm part}$ (shaded circles) in Au+Au collisions at $\sqrt{s_{NN}}$=200 GeV at mid-rapidity, compared to models.

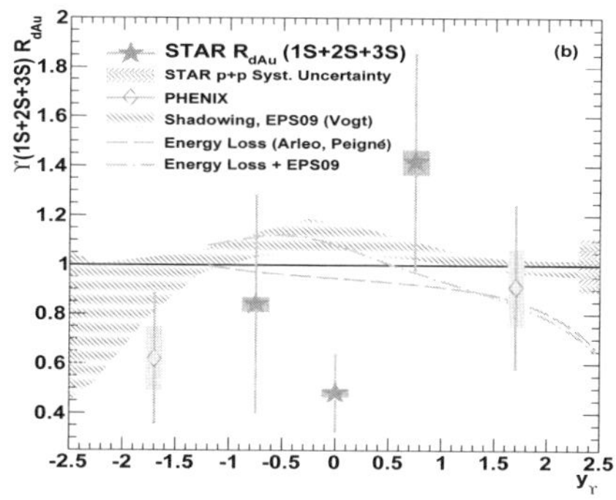

Fig. 4. Υ(1S+2S+3S) vs. rapidity in $\sqrt{s_{NN}}$=200 GeV d+Au collisions from STAR (stars) and PHENIX (diamonds), compared to theoretical calculations.

4. ϒ production

Theoretical calculations that include shadowing and/or parton energy loss[11,12] predict an $R_{\rm dAu} \approx 1$ at mid-rapidity. While the cross sections for Υ production in $\sqrt{s_{NN}}$=200 GeV p+p collisions[30] are consistent with NLO pQCD color evaporation model predictions,[11] the nuclear modification in d+Au, shown on Fig. 4, is stronger than what the above models can explain.[30] This suggests that CNM effects may not be well understood or that CNM effects alone may not be enough to explain the suppression in the d+Au mid-rapidity bin ($|y|$<0.5). However, the uncertainties on the measurement are still quite high. Evaluating the high-statistics p+Au data taken at RHIC in 2015 is essential to see whether there is an effect beyond model predictions or not.

While the most heavy-ion luminosity in RHIC is collected in Au+Au collisions, central U+U data at $\sqrt{s_{NN}}$ = 193 GeV are estimated to have a 20% higher average energy density than that of Au+Au,[29] thus allowing for the extension of measurements towards higher $N_{\rm part}$ numbers and further tests of the sequential suppression hypothesis. Fig. 7 shows the nuclear modification factor of Υ(1S+2S+3S) from Au+Au collisions at $\sqrt{s_{NN}}$ = 200 GeV as well as from U+U collisions at $\sqrt{s_{NN}}$=193 GeV, with respect to $N_{\rm part}$. Comparison to theoretical calculations shows the strong $q-\bar{q}$ binding potential scenarios[26,27] are favored by the data. $R_{\rm AA}$ of central Υ(1S) and the excited states Υ(2S+3S) at 0-60% centrality are compared

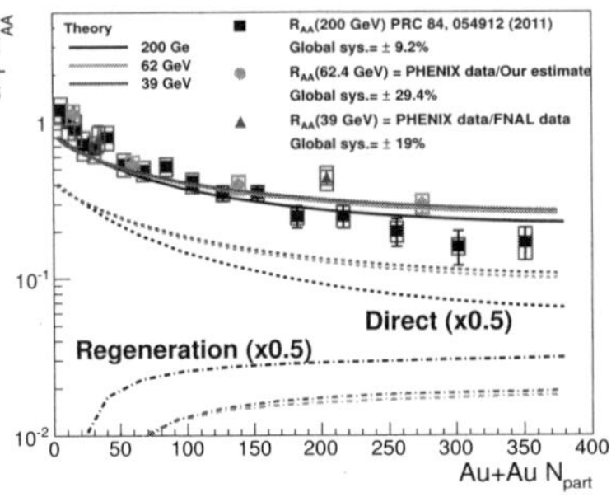

Fig. 5. PHENIX forward rapidity J/ψ R_{AA} vs. N_{part} in Au+Au collisions at $\sqrt{s_{NN}}$=39, 62.4 and 200 GeV compared to theoretical calculations.

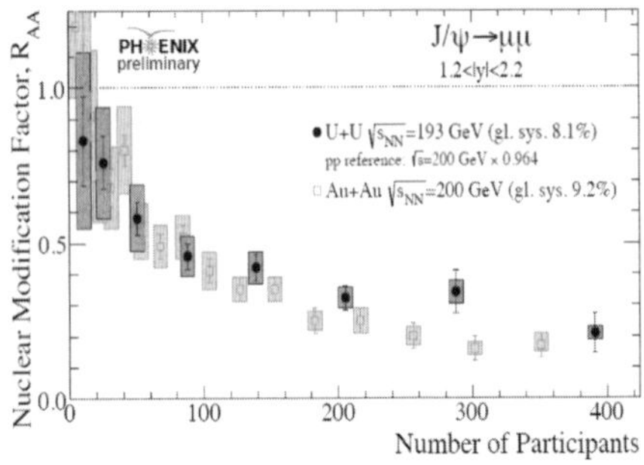

Fig. 6. PHENIX forward rapidity J/ψ R_{AA} vs. N_{part} in Au+Au collisions at $\sqrt{s_{NN}}$=200 GeV (squares) and U+U collisions at $\sqrt{s_{NN}}$=193 GeV (circles).

to central high-p_{T} J/ψ mesons in $\sqrt{s_{NN}}$ = 200 GeV Au+Au collisions[16,17] on Fig. 8. A significant suppression of Υ states is observed in RHIC central heavy ion collisions.[30,31] While the Υ(1S) is similarly suppressed to the high-p_{T} J/ψ mesons in central Au+Au collisions at $\sqrt{s_{NN}}$=200 GeV, there is a stronger suppression in the case of the excited states. This attests to the sequential melting picture in the presence of a deconfined medium. The suppression pattern seen in the case of the U+U collisions is consistent with the trend marked in Au+Au collisions and extends it towards higher N_{part} values.

5. Summary and outlook

Recent RHIC measurements show that the presence of a hot medium affects quarkonium production. The significant suppression of high-p_{T} J/ψ yields and Υ(1S) states in central heavy ion collisions, and an even stronger suppression of Υ(2S+3S) states in 0-60% Au+Au collisions is in accordance with the sequential melting picture. The trends in U+U measurements are generally an extension of those observed in Au+Au to higher number of participant nucleons, although there is some indication in the J/ψ data that there may be more coalescence in U+U than in Au+Au. There is no strong dependence of the J/ψ suppression on the beam energy between $\sqrt{s_{NN}}$=39 and 200 GeV, and the suppression in LHC 2.75 TeV collisions is even weaker than at RHIC, which attests to the role of J/ψ regeneration. Measurements in d+Au systems show that cold nuclear matter effects are significant in the case of J/ψ and also may not be negligible in the case of

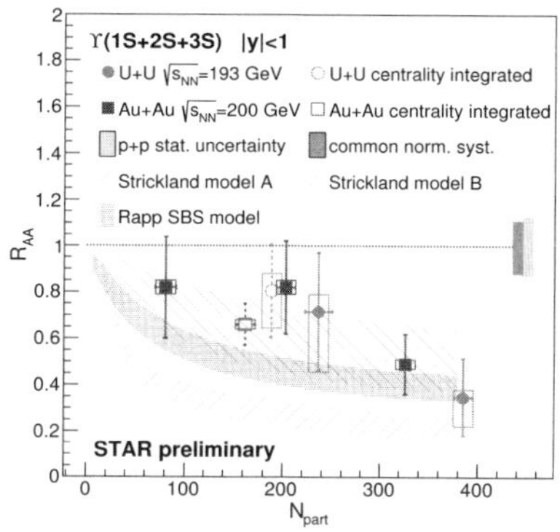

Fig. 7. ϒ(1S+2S+3S) vs. $N_{\rm part}$ in $\sqrt{s_{NN}}$=200 GeV Au+Au (squares) and $\sqrt{s_{NN}}$=193 GeV U+U collisions (circles, preliminary), compared to different popular theoretical calculations.

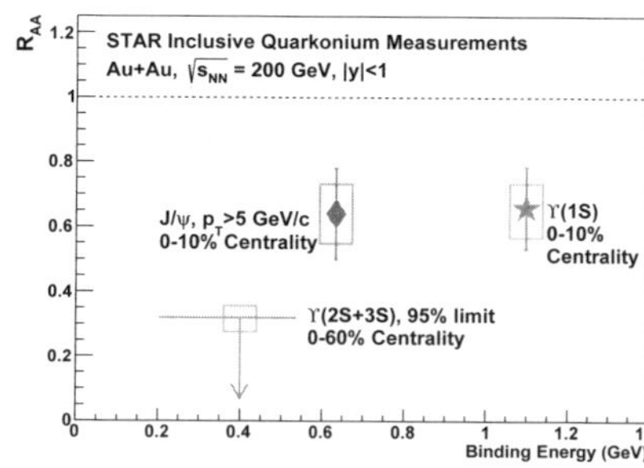

Fig. 8. $R_{\rm AA}$ of central ϒ(1S) and the excited states ϒ(2S+3S) at 0-60% centrality are compared to central high-$p_{\rm T}$ J/ψ mesons in $\sqrt{s_{NN}}$ = 200 GeV Au+Au collisions.

ϒ mesons.

The developments of PHENIX and STAR, paired with the increased luminosity delivered by RHIC in recent years, will highly increase precision of the measurements in the near future. The first results with STAR MTD have already been presented.[32] There are ongoing measurements with the inner silicon tracking systems of both experiments, where the prompt and non-prompt J/ψ contributions will be separated, thus providing means to estimate the fraction of J/ψ mesons coming from feed-down from B mesons. Besides heavy ion measurements, high luminosity p+Au data recorded in 2015 will help the understanding of cold nuclear matter effects.

Acknowledgments

The author would like to thank the Editors of the *Gribov-85 Memorial Volume — Exploring Quantum Field Theory*, especially Júlia Nyíri for the invitation of this contribution; and the organizers of the *Joint Wigner-CCNU High-Energy Heavy-Ion Balaton Workshop 2015, Tihany, Hungary* for the invitation to present this material. The work has been supported by the grant 13-20841S of the Czech Science Foundation (GAČR), as well as the Hungarian OTKA grants NK 101438, NK 106119 and the TET grant TET_12_CN-1-2012-0016.

References

1. I. Arsene *et al.* [BRAHMS Collaboration], *Nucl. Phys.* **A757**, 1 (2005).
2. B.B. Back *et al.* [PHOBOS Collaboration], *Nucl. Phys.* **A757**, 28 (2005).
3. J. Adams *et al.* [STAR Collaboration], *Nucl. Phys.* **A757**, 102 (2005).
4. K. Adcox *et al.* [PHENIX Collaboration], *Nucl. Phys.* **A757**, 184 (2005).
5. A. Adare *et al.* [PHENIX Collaboration], *Phys. Rev.* **C81**, 034911 (2010).
6. T. Matsui and H. Satz, *Phys. Lett.* **B178**, 416 (1986).
7. S. Digal, P. Petreczky, and H. Satz, *Phys. Lett.* **B514**, 57 (2001).
8. Á. Mócsy and P. Petreczky, *Phys. Rev. Lett.* **99**, 211602 (2007).
9. R. Arnaldi [ALICE Collaboration], *Nucl. Phys.* **A904**, 595c (2013).
10. L. Grandchamp, S. Lumpkins, D. Sun, H. van Hees, and R. Rapp, *Phys. Rev.* **C73**, 064906 (2006).
11. R. Vogt, R.E. Nelson, and A.D. Frawley, *PoS ConfinementX*, **203** (2012).
12. F. Arleo and S. Peigne, *JHEP* **1303**, 122 (2013).
13. K. Adcox *et al.* [PHENIX Collaboration], *Nucl. Instrum. Meth.* **A499**, 469 (2003).
14. K.H. Ackermann *et al.* [STAR Collaboration], *Nucl. Instrum. Meth.* **A499**, 624 (2003).
15. L. Ruan *et al.*, *J. Phys.* **G36**, 095001 (2009).
16. L. Adamczyk *et al.* [STAR Collaboration], *Phys. Lett.* **B722**, 55 (2013);
17. L. Kosarzewski [STAR Collaboration], *Acta Phys. Polon. Supp.* **5**, 543 (2012).
18. A. Adare *et al.* [PHENIX Collaboration], *Phys. Rev.* **D82**, 012001 (2010).
19. A. D. Frawley, T. Ullrich and R. Vogt, Phys. Rept. **462**, 125 (2008).
20. B. Trzeciak [STAR Collaboration], *Nucl. Phys.* **A904-905**, 607c (2013).
21. A. Adare *et al.* [PHENIX Collaboration], *Phys. Rev. Lett.* **107**, 142301 (2011).
22. K.J. Eskola, H. Paukkunen, and C.A. Salgado, *JHEP* **0904**, 065 (2009).
23. A. Adare *et al.* [PHENIX Collaboration], *Phys. Rev.* **C86**, 064901 (2012).
24. W. Zha [STAR Collaboration], *Nucl. Phys.* **A931**, 596 (2014).
25. X. Zhao and R. Rapp, *Phys. Rev.* **C82**, 064905 (2010).

26. M. Strickland and D. Bazow, *Nucl. Phys.* **A879**, 25 (2012).
27. A. Emerick, X. Zhao and R. Rapp, *Eur. Phys. J.* **A48**, 72 (2012).
28. R. Vértesi [STAR Collaboration], *Nucl. Part. Phys. B. Conf. Suppl.* **273-275**, 1588 (2016), arXiv:1410.3959 [nucl-ex].
29. D. Kikoła, G. Odyniec, and R. Vogt, *Phys. Rev.* **C84**, 054907 (2011).
30. L. Adamczyk *et al.* [STAR Collaboration], *Phys. Lett.* **B735**, 127 (2014) [*Phys. Lett.* **B743**, 537 (2015)].
31. R. Vértesi [STAR Collaboration], *Proceedins of Hard Probes 2015, Montreal, Canada*, to be published in *Nucl. Phys. B Conf. Suppl.*, arXiv:1509.05359 [hep-ex].
32. R. Ma [STAR Collaboration], *Proceedins of Hard Probes 2015, Montreal, Canada*, to be published in *Nucl. Phys. B Conf. Suppl.*, arXiv:1509.06440 [nucl-ex].

ULTRA-FAST NEUTRINOS: WHAT CAN WE LEARN FROM A FALSE DISCOVERY?

DEZSŐ HORVÁTH

MTA Wigner Research Centre for Physics, Budapest, Hungary
and
MTA Institute of Nuclear Research (Atomki), Debrecen, Hungary
E-mail: horvath.dezso@wigner.mta.hu

We overview the history of superluminal neutrinos, how they were first observed then refuted, and the emerged physical interpretations. It is shown that in certain cases such false experimental discoveries can help the development of science.

1. Neutrino mysteries

Neutrino physics was and still is full of mysteries. To start with, *Wolfgang Pauli* introduced an unobservable, weakly interacting neutral particle named *neutron* in order to restore energy conservation in nuclear decays, and when the real neutron appeared, *Enrico Fermi* added the Italian diminutive to make it *neutrino.* With neutrinos the parity violation of weak interactions is extreme: practically only left (opposite to momentum) polarized neutrinos and right-polarized antineutrinos are produced. One of the earliest and most important results of the high-energy electron-positron colliders was to establish from the measurement of width of the Z boson that there are just three flavours of light neutrinos belonging to charged leptons, electron, muon and tau.

The neutrinos could be Majorana particles, i.e., their own antiparticles: this is still an open question.

For a long time it was assumed that they were massless and they could not change flavor; the observation of neutrino oscillation led to the discovery of their nonzero masses. Neutrino oscillation was proposed by Bruno Pontecorvo[1] in 1957 and later used by *Vladimir Gribov* and *Bruno Pontecorvo*[2] to explain the solar neutrino deficit. According to the theory the phase of a neutrino is approximately $1.27\Delta m_{ij}^2 L/E$ where $\Delta m_{ij}^2 = |m_i^2 - m_j^2|$, the mass-squared difference of the two neutrino states in units of eV^2, L is the distance between source and detector in km and E is the neutrino energy in GeV. Those early papers by Gribov and Pontecorvo laying the foundation of neutrino oscillations are among the most cited ones in neutrino physics.

The discovery of neutrino oscillations[4] has solved several mysteries of neutrino physics while raising new ones. It clarified the partial disappearance of electron neutrinos from the Sun and of muon neutrinos from the interactions of cosmic rays in the atmosphere of Earth. Neutrinos of different flavors (electron, muon and tau) can convert into each other in flight: the atmospheric muon neutrinos in the diameter of Earth into tau neutrinos, the electron neutrinos of the Sun in the Sun–Earth distance into the other two kinds.

Neutrino physics is probably the most intensively cultivated field of particle physics: according to Wikipedia 35 neutrino experiments were taking data in 2015, 8 were under construction, 10 proposed and 20 closed off (see Fig. 1). Several of them study neutrino oscillations on Earth distances: beams of muon neutrinos are sent from proton accelerators to underground detectors located far away: in America Fermilab to MINOS in the Soudan mine (735 km), in Japan J-Parc to Super-Kamiokande in the Kamioka mine (T2K experiment, 295 km) and CERN to the detectors in LNGS, the Gran Sasso underground laboratory (CNGS beam, 732 km).

While neutrino oscillations seem to solve the problems of neutrino disappearance they raise other problems. Usually, oscillation between particle states happens when their mass eigenstates and weak interaction eigenstates are different due to different interaction eigenstates. However, according to the Standard Model neutrinos have just one interaction, the weak one, and thus in principle they should not have different flavor and mass eigenstates. On the other hand, we *artificially* put the fermion masses in the Standard Model Lagrangian as the mechanism of spontaneous symmetry breaking allows it, so formally we do not need a fifth force to take care of splitting neutrino states.

The MINOS Collaboration reported[5] in 2007, that while studying neutrino oscillations they have measured the traversing velocity of muon neutrinos. As compared to the velocity of light *in vacuo* the neutrinos have arrived 126 ± 32 (stat) ± 64 (syst) ns too early. The difference was not significant, less than twice the total uncertainty and the systematic error made further data collection senseless. The conclusion was that neutrinos travel approximately at the speed of light.

There was a similar velocity measurement 20 years before. Neutrinos from the SN1987A supernova[6] were detected a few hours before the flare-

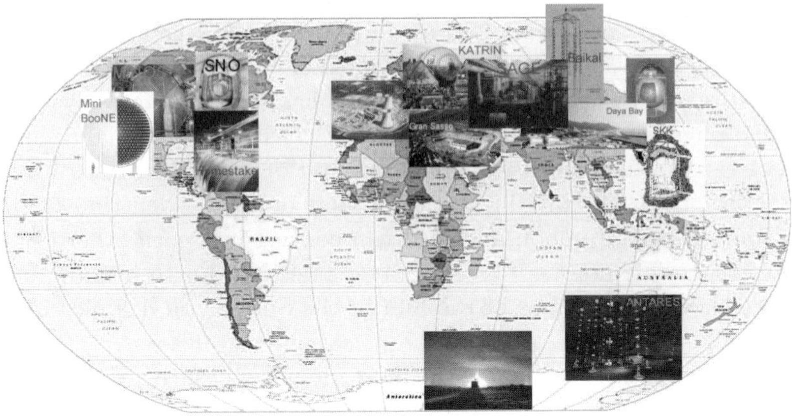

Fig. 1. Neutrino experiments in mines and tunnels, under water or ice and near nuclear reactors. Layers of earth or water shields the detector against charged particles from the Sun and the ice or water can be used as Cherenkov radiators.

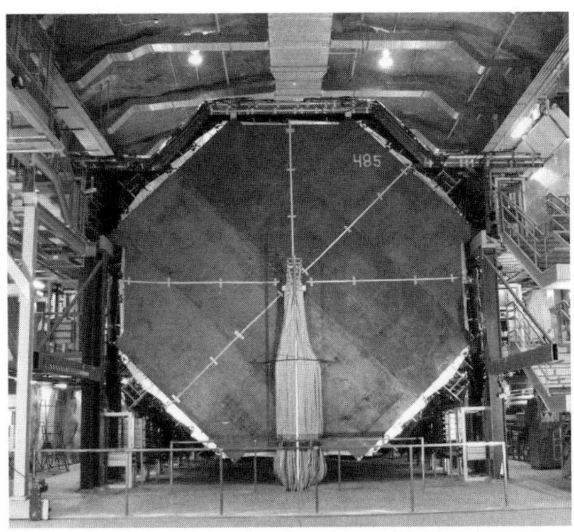

Fig. 2. The detection system of the MINOS (Main Injector Neutrino Oscillation Search) experiment in the Soudan mine.

up by the Japanese Kamiokande (Kamioka Nucleon Decay Experiment) and the American IMB (Irvine-Michigan-Brookhaven) experiments. Most of the supernova energy leaves in neutrinos in a few seconds whereas the heat has to get out of the star material: the light from SN1987A was seen

Fig. 3. The OPERA experiment in LNGS, the Gran Sasso underground laboratory.

for more than a year afterwards. Studying SN1987A led to the conclusion that neutrinos travel at speeds very close to that of the light *in vacuo*: $v_\nu = (1 + \delta \cdot 10^{-5})c$, where $|\delta| < 10^{-9}$.

2. The OPERA experiment

The OPERA (Oscillation Project with Emulsion tRacking Apparatus) experiment at LNGS announced on 22 Sep. 2011 that they measured neutrinos from CERN which traveled the distance of 732 km faster than light *in vacuo* by $\delta \approx 2.5 \times 10^{-5}$. They have submitted a long paper in arXiv[7] and gave a talk at CERN. OPERA was built to study the conversion of muon to tau neutrinos. It had two similar modules (Fig. 3) consisting of alternating layers of scintillators and photoemulsion plates. In its five years of work four tau neutrinos were identified coming from the CNGS beam of CERN.

The observation of superluminal neutrinos was received by the scientific community with extreme scepticism. We had several news of observations contradicting to previous experimental experience which subsequently were refuted (dibaryons, pentaquarks, various Higgs-bosons and other exotic particles). The OPERA result leaked out a few days before the announcement the same way as a year later Nature-online published two days before the official announcement the discovery of the 125 GeV Higgs boson. Rolf-Dieter Heuer, the Director-General of CERN issued a warning to all CERN users asking for moderation in discussing the result: he agreed to give the talk as the laboratory must allow the scientific community to discuss this result. Out of the 200 participants of OPERA 175 signed the paper[7]: they stated that they spent 6 months with checking the result and could not find anything wrong. The error was discovered half a year later only: apparently a faulty optical connection caused a 60 ns delay to the signal synchronizing the two clocks at CERN and at OPERA which pushed the apparent arrival time of the neutrinos to that much earlier.

OPERA measured the velocity of neutrinos the following way. The Super-proton Synchrotron (SPS) at CERN puts 450 GeV protons into a target and produces very high energy, relativistic pions. Those fly and part of them decay into muons and muon neutrinos, ($\pi^{\pm} \rightarrow \mu^{\pm}\nu_{\mu}$) in a 1 km long vacuum pipe. The rest of the pions are stopped via strong interaction in an absorber which lets the charged muons through; those hit a muon detector and then also stop. The very fast pion emits the muon and neutrino forward, but the muon slows down before decaying and the daughter particles of the decays $\mu^{+} \rightarrow e^{+}\overline{\nu}_{\mu}\nu_{e}$ and $\mu^{-} \rightarrow e^{-}\overline{\nu}_{e}\nu_{\mu}$ are emitted in three directions. Thus the beam sent to Gran Sasso through Earth's crust will consist of muon neutrinos only, although there it will have a diameter of almost 3 km. In the beginning the SPS injected the protons in 10 μs long pulses, but after OPERA announced its velocity measurement they also sent short pulses of 3 ns length.

The travel time of neutrinos and the flight distance were measured by Swiss and Italian geodesic experts with atomic clocks at CERN and at Gran Sasso synchronized by GPS. Timing was started by the proton kicker of SPS — the flight time of the relativistic pions in less than 1 km adds very little to the uncertainty of the travel time of neutrinos in 732 km, estimated to be about 10 ns. The main uncertainty component of timing was due to the fact that the GPS signals had to be sent from the surface to the laboratory 1400 m under ground.

OPERA first announced that the neutrinos arrived 57.8 ± 7.8 (stat) ± 7.4 (sys) ns (i.e., in total 58 ± 10 ns) earlier than expected by the speed of light. In Ref. 7 this difference was slightly modified to 60.7 ± 6.9 (stat) ± 7.4 (sys) ns.

It was easy to estimate that if the neutrinos have constant speed then at least part of the neutrinos from SN1987A located at 168000 light years from us should have arrived 4 years before the light pulse.[8] In the early 80's two large neutrino detectors worked, the Soviet BAKSAN and the American IMB and none of them detected higher counts than usual that time. Once again the superiority of event registration was demonstrated: one could check these events 30 years after the actual measurement.

However, there are large differences between experimental conditions of the supernova and accelerator measurements. SN1987A emitted electron neutrinos of MeV energies which then traveled *in vacuo* whereas the anomaly was measured with ν_μ traveling through earth at much higher, GeV energies.

3. Interpretation of the OPERA anomaly

The first reactions were quite characteristic for both the general and the physics communities. Newspapers reported "Relativity is ruled out!" whereas the physicists were very sceptic, all of us thought that the measurement failed somehow. We tried to find (1) mistakes in the measuring technique and (2) contradictions to previous measurements (SN1987A was one of those) or (3) proposed control measurements. Of course, behind this there was the thought: (4) what if it is really true, how can one put this into the present picture of physics? In a very short time after the OPERA announcement a tremendeous number of papers were published[17] on this topic, trying to understand, refute or interpret of the OPERA effect. It is not possible, not even that sensible any more to properly review them; here we present just a sample of the attempts.

3.1. *Fault in the experimental apparatus*

One of the first, trivial objections against the OPERA method was how it is possible to determine the arrival of 10000 ns long neutrino pulses with the precision of 10 ns. In order to check that at the end of 2011 CERN sent very short, 3 ns long particle pulses to Gran Sasso and OPERA confirmed the 60 ns anomaly.

The method of measurement was described[7] by OPERA in quite precise detail. They have detected a few cm shift of the laboratory at an earthquake, so the 20 cm distance uncertainty was quite plausible. All objections against the time determination were also well refuted: GPS coordinates, general relativity and temperature effects, etc. Half a year later, ICARUS, another Gran Sasso experiment reported[9] that the CERN neutrinos arrived to them in proper time, and then OPERA itself found the error allegedly in the time synchronization line and their final publication[10] reported a timing difference of 6.5 ± 11 ns, i.e., no anomaly.

3.2. *Theoretical and experimental objections*

3.2.1. *Cohen-Glashow effect*

Cohen and Glashow have published[11] almost immediately after the OPERA announcement a strong theoretical objection: a superluminal neutrino, similarly to the Cherenkov radiation, should lose energy via producing electron-positron pairs in weak interaction, distorting the energy distribution of the neutrino beam to the extent that muon neutrinos with energies above 12.5 GeV could not reach Gran Sasso. The ICARUS experiment, located in LNGS next to OPERA, announced[9] that they see neither the distortion of the neutrino energy spectrum, nor photons and electron-positron pairs from neutrino bremsstrahlung. Moreover, the underground neutrino detectors (like ICECUBE in the Antarctic ice) see neutrinos of very high energies, above 100 TeV that limits the relative difference between the speeds of light and neutrinos to $\delta = |v_\nu - c|/c < 1.7 \times 10^{-11}$, six orders of magnitude below the OPERA result.

The Cohen-Glashow paper got a lot of citations and had a serious influence on neutrino physics. There were several proposals to test it and using its effect to check the OPERA result.

3.2.2. *Neutrino oscillation*

Neutrino oscillation was observed among all three kinds of neutrinos, with very small mass differences, $|m_i^2 - m_j^2| \sim 10^{-3} - 10^{-4}$ eV^2. Not even the larger mass difference of the sterile neutrino allegedly observed (but never confirmed) by the LSND (Liquid Scintillator Neutrino Detector at Los Alamos) experiment[12] could be sufficient for superluminality. In any case, it would be very hard to explain an oscillation between superluminal neutrinos and even harder that between superluminal muon and subluminal electron neutrinos. Furthermore, the MINOS experiment[15] saw muon neutrinos converted in flight to electron ones with a probability about 34% without any significant change in flight time. Let us note, however, that if there had been some superluminal component in the electron neutrinos from SN1987A, due to oscillation they had to extend so much in timing by the time of arrival that it could have been impossible to detect them. Thus the fact that they were not observed does not refute a possible partial superluminality of the supernova neutrinos.

3.3. *Test experiments*

In particle physics new observations are never accepted as discoveries until confirmation by another independent measurement. A good example is the above mentioned extra neutrino observed by LSND[12]: as the Mini-BOONE[13] experiment built to check it could neither fully confirm nor disprove it, it is still in doubt.

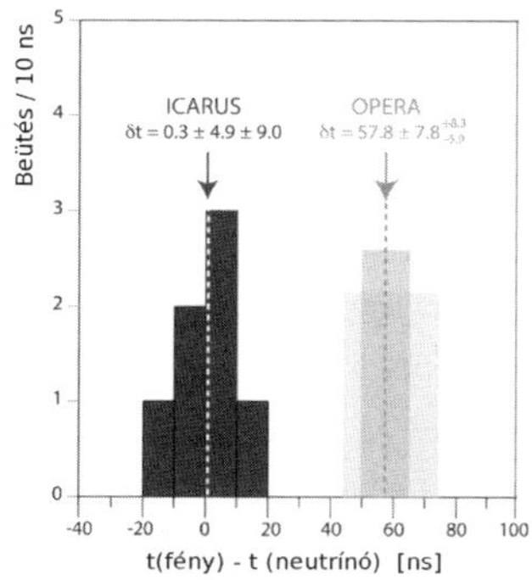

Fig. 4. Arrival times of CNGS neutrinos to the OPERA and ICARUS detectors as compared to that expected for light *in vacuo.*[9]

The J-PARC (Japan Proton Accelerator Research Complex) facility built in Tokai announced that the time resolution of the T2K (Tokai to Kamioka with 295 km flight path) experiment is not sufficiently good to check the OPERA effect, but MINOS was prepared to do that. There were numerous other proposals. For instance, the authors of Ref. 3 planned to propose a short-distance experiment with a 2 km long flight path in vacuum where the neutrinos could fly together with light, providing thereby a straight comparison between the flight times of neutrinos and light. When the OPERA anomaly disappeared it led to an idea how to measure the neutrino mass[14] using time of flight.

However, all this interest was lost when the ICARUS experiment located next to OPERA at LNGS announced[9] that they see no anomaly with the flight time of CERN neutrinos (Fig. 4), and then OPERA reported the experimental error.

Several technical control studies were summarized in a dedicated mini-workshop[16] organized at LNGS in March 2012. OPERA and the LVD (Large Volume Detector) detector in its neighborhood had a long collaboration detecting cosmic muons coming horizontally through the side of the Gran Sasso mountain and detected by both experiment. LVD noticed that the timing difference between the two detectors was different between August 2008 and December 2011 by $\Delta t = -73 \pm 9$ ns from that measured before and after, so OPERA must had measured in that interval flight times earlier by $|\Delta t|$. The OPERA collaboration confirmed LVD's finding: the problem was apparently fixed in 2012.

3.4. *Interpretation attempts*

Assuming the possible validity of the observation of superluminal neutrinos raises many questions. Is the effect global or local? Does it depend on flight distance, direction, time, medium or other conditions? There were an avalanche of papers on this topic, here we quote a few examples only, the reader is directed to the reference data base[17] for details. The aim of this review is to illustrate how positively can even a false discovery influence the development of science.

3.4.1. *Distance dependence?*

Maybe superluminal movement is a local feature, connected to some extra force field on Earth, which is coupled to neutrinos differently than to other particles? The interaction of the neutrinos with the new field could possibly modify the metrics of space-time and the neutrino could propagate faster than light without violating Lorentz invariance. The new force field could in principle be even the magnetic field of Earth; one should repeat the experiment in extreme electromagnetic noise. If the typical wave length of the new field is of the order of Earth's diameter, its effect in galactic sizes is negligible, and that explains its not observation for SN1987A.

Mixing with a fourth kind (possibly sterile) neutrino could also cause superluminality: if the extra neutrino could propagate in extra dimensions the ordinary neutrino states when detected could apparently be faster than light. In principle this mixing could be enhanced by the presence of baryonic matter. In order to check whether the velocity increase depends on the path made in earth it was proposed[18] that CERN send neutrino pulses to the NESTOR (Neutrino Extended Submarine Telescope with Oceanographic Research) detector placed on the bottom of Aegean sea. Its distance from CERN is 1676 km and the superluminal neutrinos should arrive there 140 ns earlier than light if it is distance dependent.

3.4.2. *Tachyons?*

Dozens of papers have shown that whichever way we assume the existence of superluminal neutrinos it leads to contradictions. The *tachyons*, those hypothetic faster-than-light particles, have higher speed at lower energies (their velocity scaling with E^2), they approach the speed of light at infinite energy.[19] Therefore if the anomalous muon neutrinos were tachyons, they should have arrived to the detector at various times depending on energy.

Another problem is that the neutrino speed measured by OPERA assumes a tachyon mass of about 100–130 MeV/c^2 whereas in pion decay only about 40 MeV is left for neutrinos. If the $\sim$ 10 GeV neutrinos from CNGS arrive to Gran Sasso at 730 km 60 ns earlier, the anomalous neutrinos from SN1987A should have arrived 137000 years earlier, with no chance of observation. Furthermore, if just one kind of neutrino is tachyonic, it cannot mix and oscillate with the ordinary ones.

3.4.3. *Fifth force?*

Let us assume that in addition to the known four interactions, strong, electromagnetic, weak and gravitational, there exists a fifth one which couples differently to neutrinos and other particles.[20] Then the neutrinos could move in different metrics, and sense different distances from other particles and apparently move at speeds faster than light. There were several attempts to build such a world using various force fields (scalar, vector or tensor fields) and reproduce the OPERA effect, even the magnetic field of Earth was used for that. Several authors tried to use extra space dimensions, shortening spatial distances. Using earlier measurements severe limits were found for the new couplings and dimension sizes. Even a new concept of *dark neutrino* was proposed[21] which would be free of the Cohen–Glashow effect. The gravitational field of Earth also could have an anomalous term which changes the metric for neutrinos, such a model could also describe all measured quantities.

In Ref. 22 the Finsler spacetime was proposed to account for the superluminal phenomena, it allows superluminal speeds proportional to energy per unit mass of the particle while keeping causality. Such a superluminal speed formula was shown to be consistent with data of OPERA, MINOS and Fermilab-1979 neutrino experiments as well as the observations on neutrinos from SN1987a.

3.4.4. *Lorentz violation?*

The dispersion relation which describe the interdependence of energy, mass and momentum can be modified[23] by the additional small term $p^2 = E^2 - \vec{p}^2 = m^2 + f(\vec{p}, E)$ where $p = [E, \vec{p}]$ is the energy-momentum 4-vector, m is the mass of the particle and $f(p)$ is an anomalous term violating Lorentz invariance. Although the Standard Model includes Lorentz invariance, i.e., the general spatial isotropy of the physical laws, several extension were built to the model with small enough Lorentz-violations to be still in agreement with the experimental data. The results of the neutrino experiments give limits for these theories, but they can be brought in agreement.[19] An anisotropic Lorentz violation would make neutrinos move with different velocities in various directions, and introducing one additional free parameter is sufficient to reproduce all experimental data.

4. Conclusion

The announcement of the OPERA collaboration of the observation of superluminal neutrinos invoked a feverish activity in the particle physics community. In the final count the experiment itself was faulty, at our present knowledge anomalous neutrinos do not exist. Nevertheless, the false discovery caused a brain storming: the inSPIRE data base lists more than 350 citations of the original paper, and many of them date after 2012 when the experimental error was announced. Although many physicists think that it was undoubtedly wrong to start with, the reaction of the community shows that even such false discoveries can promote the development of physics.

5. Acknowledgments

The author is indebted to Zoltán Trócsányi, István Nándori and Sándor Nagy for the collaborative effort to collect and sort out the publication avalanche on superluminal neutrinos in 2011-12 . This work was supported by the Hungarian Scientific Research Project OTKA K103917.

References

1. B. Pontecorvo, *Neutrino Experiments and the Problem of Conservation of Leptonic Charge*, *Sov. Phys. JETP* **26**, 984 (1968) [*Zh. Eksp. Teor. Fiz.* **53**, 1717 (1967)].
2. V.N. Gribov and B. Pontecorvo, *Neutrino astronomy and lepton charge*, *Phys. Lett.* **B28**, 943 (1969).
3. D. Horváth, S. Nagy, I. Nándori, Z. Trócsányi: *Superluminal Neutrinos: the first six months*, unpublished; *The rise and fall of ultrafast neutrinos*, *Fizikai Szemle* (Budapest) **2012/5**, 145-152 (2012) (In Hungarian).
4. Y. Fukuda *et al.* [Super-Kamiokande Collaboration], *Evidence for oscillation of atmospheric neutrinos*, *Phys. Rev. Lett.* **81**, 1562 (1998).
5. P. Adamson *et al.* [MINOS Collaboration]: *Measurement of neutrino velocity with the MINOS detectors and NuMI neutrino beam*, *Phys. Rev.* **D76**, 072005 (2007).

6. D.N. Schramm, J.W. Truran: *New physics from Supernova SN1987A*, *Phys. Rept.* **189**, 89-126 (1990).
7. T. Adam *et al.* [OPERA Collaboration], *Measurement of the neutrino velocity with the OPERA detector in the CNGS beam.*, arXiv:1109.4897v1 [hep-ex] (2011).
8. D. Fargion and D. D'Armiento, *Inconsistence in super-luminal Cern-Opera neutrino speed with observed SN1987A burst and neutrino mixing for any imaginary neutrino mass*, *J. Phys.* **G39**, 085002 (2012).
9. M. Antonello *et al.* [ICARUS Collaboration], *Measurement of the neutrino velocity with the ICARUS detector at the CNGS beam*, *Phys. Lett.* **B713**, 17 (2012).
10. T. Adam *et al.* [OPERA Collaboration], *Measurement of the neutrino velocity with the OPERA detector in the CNGS beam.*, arXiv:1109.4897v4, JHEP **1210**, 093 (2012).
11. A.G. Cohen and S.L. Glashow: *Pair Creation Constrains Superluminal Neutrino Propagation*, *Phys. Rev. Lett.* **107**, 181803 (2011).
12. A. Aguilar-Arevalo *et al.* [LSND Collaboration], *Evidence for neutrino oscillations from the observation of anti-neutrino(electron) appearance in a anti-neutrino(muon) beam*, *Phys. Rev.* **D64**, 112007 (2001).
13. R. Dharmapalan *et al.* [MiniBooNE+ Collaboration], *A new investigation of electron neutrino appearance oscillations with improved sensitivity in the MiniBooNE+ experiment*, arXiv:1310.0076 [hep-ex].
14. U.D. Jentschura, D. Horvath, S. Nagy, I. Nandori, Z. Trocsanyi, and B. Ujvari, *Weighing the Neutrino*, *Int. J. Mod. Phys.* **E23** (2014) 1450004
15. P. Adamson *et al.* [MINOS Collaboration], *Measurement of Neutrino Oscillations with the MINOS Detectors in the NuMI Beam*, *Phys. Rev. Lett.* **101**, 131802 (2008).
16. *LNGS results on the neutrino velocity topic*, mini-workshop, Gran Sasso, 28th march 2012, `http://agenda.infn.it/conferenceDisplay.py?confId=4896`
17. Most of the literature can be found in inSPIRE by `http://inspirehep.net/search?ln=en&p=refersto:recid:928153` and a detailed description is in Wikipedia, under the title "Faster than light neutrino anomaly".
18. A. Nicolaidis, *Neutrino Shortcuts in Spacetime*, *Mod. Phys. Lett.* **A27**

(2012) 1250127
19. G. Amelino-Camelia, G. Gubitosi, N. Loret, F. Mercati, G. Rosati, and P. Lipari, *OPERA-reassessing data on the energy dependence of the speed of neutrinos*, *Int. J. Mod. Phys.* **D20** (2011) 2623
20. G. Dvali and A. Vikman, *Price for Environmental Neutrino-Superluminality*, *JHEP* **1202**, 134 (2012).
21. I.Y. Aref'eva and I.V. Volovich, *Superluminal Dark Neutrinos*, arXiv:1110.0456 [hep-ph].
22. Z. Chang, X. Li, and S. Wang, *OPERA superluminal neutrinos and kinematics in Finsler spacetime*, Mod. Phys. Lett. A **27** (2012) 1250058
23. S.R. Coleman and S.L. Glashow, *High-energy tests of Lorentz invariance*, *Phys. Rev.* **D59**, 116008 (1999).

QFT AND BEYOND

MINIMAL STRING THEORY AND THE DOUGLAS EQUATION

A.A. BELAVIN

L.D. Landau Institute for Theoretical Physics, 142432 Chernogolovka, Russia
and
Institute for Information Transmission Problems, 127994 Moscow, Russia
and
Moscow Institute of Physics and Technology, 141700 Dolgoprudny, Russia
E-mail: belavin@itp.ac.ru

V.A. BELAVIN

P.N. Lebedev Physical Institute, 119991 Moscow, Russia
and
Institute for Information Transmission Problems, 127994 Moscow, Russia
E-mail: belavin@lpi.ru

We use the connection between the Frobenius manifold and the Douglas string equation to further investigate Minimal Liouville gravity. We search for a solution of the Douglas string equation and simultaneously a proper transformation from the KdV to the Liouville frame which ensures the fulfilment of the conformal and fusion selection rules. We find that the desired solution of the string equation has an explicit and simple form in the flat coordinates on the Frobenius manifold in the general case of (p,q) Minimal Liouville gravity.

1. Introduction

The purpose of this talk[1] is to present the progress in the study of the Minimal Liouville Gravity (MLG)[2] using an approach based on the Douglas string equation.[3] This study is a continuation of earlier works.[4–7]

The Liouville Gravity represents a consistent example of the noncritical String theory. In the initial continuous approach the Liouville Gravity is formulated as a BRST invariant theory composed of the matter sector, the Liouville theory and the ghosts system. MLG represents the theory, where the matter sector is taken to be a (p, q) Minimal Model of CFT.[8] The main problem of MLG is to calculate correlation functions of BRST invariant observables, which are given by integrals over moduli of Riemannian surfaces. Usually they are called the correlation numbers. Numerous examples show that the solution of the problem is quite nontrivial within the framework of the continuous approach.

An alternative approach to MLG has grown up from the idea of triangulations of two-dimensional surfaces realized in terms of Matrix Models.[9–15] One of the most important points of the approach is the string equation which was derived by Douglas[3] in the Matrix Models approach to two dimensional gravity. The subject of the string equation is the generating function of the correlation numbers which depends on the parameters of the problem (the so called KdV times). In our work, following Refs. 5–7, we will conjecture that the Douglas equation is applicable to the Minimal Liouville gravity as well as to Matrix Models of 2D gravity.

This conjecture requires the following two questions to be answered: how to choose the desired solution of the Douglas string equation and an appropriate form of the so called resonance transformation[4] from the KdV times to the Liouville coupling constants. Once these two questions are answered, the generating function of the correlation functions in MLG is given explicitly as an integrated one-form defined uniquely for each (p, q) MLG model and coincides with a special choice of the tau-function of the dispersionless limit[16,17] of the generalized KdV hierarchy.

In this talk, using the connection[6] of the approach to MLG[5] based on the string equation with the Frobenius manifold structure, we find the necessary solution of the string equation. We also show that this solution together with the suitable chosen resonance transformation lead to the results which are consistent with the main requirements of (p, q) models of MLG (the so-called selection rules). It is remarkable that the needed solution of the Douglas equation has a very simple form in the flat coordinates on the Frobenius manifold in the general case of (p,q) Minimal Liouville as well as it has been found recently in the case of unitary models of MLG.[7]

2. Frobenius manifolds

In this and the two next sections we give the definition and a short review of the main properties of the Frobenius manifolds needed for our purposes. Here we follow the paper by B. Dubrovin,[17] see also Ref. 6.

By definition a commutative associative algebra A with unity equipped with a nondegenerate invariant bilinear form $(\, ,)$ is called Frobenius algebra. The invariance of the bilinear form means that for any three vectors a, b, c in A:

$$(a \cdot b, c) = (a, b \cdot c). \tag{1}$$

Let M be n-dimensional manifold with a flat metric $\eta_{\alpha\beta}dv^\alpha dv^\beta$ which is constant in the flat coordinates v^α.

We introduce in the tangent space $T_{\mathbf{v}}M$ the structure of the Frobenius algebra by the following identification of the bases

$$\frac{\partial}{\partial v^\alpha} \to e_\alpha, \tag{2}$$

Thus, we can multiply tangent vectors at any point of M

$$e_\alpha e_\beta = C^\gamma_{\alpha\beta} e_\gamma. \tag{3}$$

The structure constants $C^\gamma_{\alpha\beta}$ may depend on v^α. Such a manifold M can be called quasi-Frobenius manifold.

Definition 2.1. *The manifold M is called Frobenius manifold if these two structures are adjusted with each other in such a way that*

(1) the invariant bilinear form $(\frac{\partial}{\partial v^\alpha}, \frac{\partial}{\partial v^\beta})$ is identical to the metric $\eta_{\alpha\beta}$;

(2) the structure of the Frobenius algebra at each point of M and the metric on M are constrained by the following relation

$$\nabla_\rho C_{\alpha\beta\gamma} = \nabla_\alpha C_{\rho\beta\gamma}. \tag{4}$$

The last requirement is equivalent to the requirement that there exists a function F on M which is connected with the structure constants of the Frobenius algebra as

$$C_{\alpha\beta\gamma} = \frac{\partial^3 F}{\partial v^\alpha \partial v^\beta \partial v^\gamma}, \tag{5}$$

where

$$C_{\alpha\beta\gamma} = \eta_{\alpha\rho} C^\rho_{\beta\gamma}. \tag{6}$$

The function F is called Frobenius potential. The consistency of this property with the associativity of the Frobenius algebra is known as WDVV condition[18]

$$\frac{\partial^3 F}{\partial v^\alpha \partial v^\beta \partial v^\rho} \eta^{\rho\lambda} \frac{\partial^3 F}{\partial v^\lambda \partial v^\mu \partial v^\nu} = \frac{\partial^3 F}{\partial v^\nu \partial v^\beta \partial v^\rho} \eta^{\rho\lambda} \frac{\partial^3 F}{\partial v^\lambda \partial v^\mu \partial v^\alpha}. \tag{7}$$

The following statement[17] follows from these properties of the Frobenius manifold M. There exist an one-parametric flat deformation $\widetilde{\nabla}_\alpha$ of the connection ∇_α

$$\widetilde{\nabla}_\alpha x^\gamma = \nabla_\alpha x^\gamma - z C^\gamma_{\alpha\beta} x^\beta, \tag{8}$$

or, equivalently,

$$[\widetilde{\nabla}_\alpha(z), \widetilde{\nabla}_\beta(z)] = 0. \tag{9}$$

The proof is based on the associativity of the Frobenius algebra and the equation (4). As a consequence of Eq. (9), there exist n linear independent solutions

$$\theta^{\alpha}(v,z) = \sum_{k=0}^{\infty} \theta_k^{\alpha}(v) z^k, \tag{10}$$

of the equation $\widetilde{\nabla}_{\alpha} d\theta^{\lambda}(v,z) = 0$, which is equivalent to

$$\frac{\partial^2 \theta^{\lambda}}{\partial v^{\alpha} \partial v^{\beta}}(v,z) = z C_{\alpha\beta}^{\gamma} \frac{\partial \theta^{\lambda}}{\partial v^{\gamma}}(v,z), \tag{11}$$

or

$$\frac{\partial^2 \theta_{k+1}^{\lambda}}{\partial v^{\alpha} \partial v^{\beta}}(v) = C_{\alpha\beta}^{\gamma} \frac{\partial \theta_k^{\lambda}}{\partial v^{\gamma}}(v). \tag{12}$$

The functions $\theta^{\alpha}(v,z)$ can be considered as the flat coordinates of the deformed connection $\widetilde{\nabla}_{\alpha}(z)$. We choose $\theta^{\lambda}(v,z)$ so that $\theta^{\lambda}(v,0) = \theta_0^{\lambda}(v) = v^{\lambda}$. From Eq. (12) it follows, that

$$\nabla(\nabla\theta^{\alpha}(v,z_1), \nabla\theta^{\beta}(v,z_2)) = (z_1 + z_2)\nabla\theta^{\alpha}(v,z_1) \cdot \nabla\theta^{\beta}(v,z_2), \tag{13}$$

and, hence, the scalar product $(\nabla\theta^{\alpha}(v,z), \nabla\theta^{\beta}(v,-z)) = Const(z)$ does not depend on v^{α}. For $z = 0$ we find $Const(0) = \eta^{\alpha\beta}$. Equation (12) is invariant with respect to the transformation

$$\theta^{\mu}(v,z) \to A_{\nu}^{\mu}(z)\theta^{\nu}(v,z), \tag{14}$$

where $A_{\nu}^{\mu}(0) = \delta_{\nu}^{\mu}$. Using these transformations one can fix the normalization in such a way that

$$(\nabla\theta^{\alpha}(v,z), \nabla\theta^{\beta}(v,-z)) = \eta^{\alpha\beta}. \tag{15}$$

3. Main example: Frobenius manifold of A_{q-1}-type

Our main example is A_{q-1} Frobenius manifold.[18] Let $Q(y)$ be a polynomial of y

$$Q(y) = y^q + u_1 y^{q-2} + \ldots + u_{q-1}, \tag{16}$$

and $\{u_{\alpha}\}$ represent some coordinates on M. We call $\{u_{\alpha}\}$ the canonical coordinates.

Definition 3.1. *A_{q-1} Frobenius algebra is the space of polynomials modulo polynomial $\frac{dQ}{dy}$:*

$$A_{q-1}(u) = \mathbb{C}[y] / \frac{dQ}{dy}. \tag{17}$$

The corresponding manifold M is called the Frobenius manifold of A_{q-1} type

The polynomials

$$P_\alpha(y) = \frac{\partial Q}{\partial u_\alpha}, \tag{18}$$

form a basis in the tangent space $T_{\mathbf{v}}M$. An invariant bilinear form (which is equivalent to the metric) is defined by

$$(P_\alpha, P_\beta) = \operatorname*{res}_{y=\infty}\left(\frac{P_\alpha(y)P_\beta(y)}{\frac{dQ}{dy}(y)}\right). \tag{19}$$

With this definition one can verify that the corresponding metric is flat and

$$C_{\alpha\beta\gamma} = \nabla_\alpha\nabla_\beta\nabla_\gamma F(u). \tag{20}$$

To this end we perform the transformation from the canonical coordinates $\{u_\alpha\}$ to the new coordinates $\{v^\alpha\}$ by means of the following relation

$$y = z - \frac{1}{q}\left(\frac{v^{q-1}}{z} + \frac{v^{q-2}}{z^2} + \cdots + \frac{v^1}{z^{q-1}}\right) + \mathcal{O}\left(\frac{1}{z^{q+1}}\right), \tag{21}$$

where $z^q = Q(y)$.

Some useful properties of the new coordinates are formulated in the following

Theorem 3.1. *From the transformation* (21) *it follows that*

(1) v^α *form flat coordinates, i.e., the metric in this coordinates is constant and*

$$\eta_{\alpha\beta} = -q\left(\frac{\partial Q}{\partial v^\alpha}, \frac{\partial Q}{\partial v^\beta}\right) = \delta_{\alpha+\beta,q}, \tag{22}$$

(2)

$$C_{\alpha\beta\gamma} = -q\operatorname*{res}_{y=\infty}\left(\frac{\frac{\partial Q}{\partial v^\alpha}\frac{\partial Q}{\partial v^\beta}\frac{\partial Q}{\partial v^\gamma}}{\frac{dQ}{dy}}\right) = \frac{\partial^3 F}{\partial v^\alpha\partial v^\beta\partial v^\gamma}. \tag{23}$$

(3)

$$\theta_{\alpha,k} = -c_{\alpha,k}\operatorname*{res}_{y=\infty} Q^{k+\frac{\alpha}{q}}(y), \tag{24}$$

where

$$c_{\alpha,k} = \frac{\Gamma(\frac{\alpha}{q})}{\Gamma(\frac{\alpha}{q}+k+1)}. \tag{25}$$

To prove these statements it is convenient to use the basis elements of A_{q-1} in flat coordinates defined by $\Phi_\alpha(y) = \frac{\partial Q(y)}{\partial v^\alpha}$ which possess the following property

$$\Phi_\alpha(y) = \frac{1}{\alpha}\frac{d}{dy}\left(Q^{\frac{\alpha}{q}}\right)_+. \tag{26}$$

In what follows we use the following convention

$$\theta_{\mu,k} = \theta_{\mu-q\lfloor\mu/q\rfloor,k+\lfloor\mu/q\rfloor}, \tag{27}$$

where $\lfloor\mu/q\rfloor$ is the integer part of μ/q. It is clear that (27) agrees with the definition (24).

4. Frobenius manifolds and Douglas string equation

4.1. *Integrable hierarchies*

Let $\mathcal{M}$ be a space of functions of x taking values in M. Let I and J be functionals on $\mathcal{M}$. We define the Poisson bracket on $\mathcal{M}$ as

$$\{I,J\} = \int \frac{\delta I}{\delta v^\alpha(x)}\eta^{\alpha\beta}\frac{d}{dx}\frac{\delta I}{\delta v^\beta(x)}dx, \tag{28}$$

or

$$\{v^\alpha(x), v^\beta(y)\} = \eta^{\alpha\beta}\delta'(x-y), \tag{29}$$

where, as usual in the calculus of variations, the integrand is defined modulo total derivatives. The functionals

$$H_{\alpha,k} = \int \theta_{\alpha,k+1}(\vec{v}(x))dx, \qquad \alpha = 1,...,n, \quad k \geq 0, \tag{30}$$

mutually commute among themselves

$$\{H_{\alpha,k}, H_{\beta,l}\} = 0. \tag{31}$$

As a result, the Hamiltonian flows

$$\frac{\partial v^\mu}{\partial t_k^\alpha} = \{v^\mu, H_{\alpha,k}\} = \eta^{\mu\nu}\frac{\partial}{\partial x}\frac{\partial\theta_{\alpha,k+1}}{\partial v^\nu} = C^{\mu\rho}_\lambda\frac{\partial\theta_{\alpha,k}}{\partial v^\rho}\frac{\partial v^\lambda}{\partial x}. \tag{32}$$

commute, i.e.,

$$\frac{\partial}{\partial t_l^\beta}\frac{\partial\vec{v}}{\partial t_k^\alpha} = \frac{\partial}{\partial t_k^\alpha}\frac{\partial\vec{v}}{\partial t_l^\beta}. \tag{33}$$

It follows from Eq. (32) that $t_0^1 = x$.

4.2. *Douglas string equation*

Let us define a function $S(v,t)$ on M which depends on the additional parameters $\{t_k^\alpha\}$

$$S(v,t_k^\alpha)=\sum_{\alpha=1}^{n}\sum_{k\geq 0} t_k^\alpha \theta_{\alpha,k}(v). \tag{34}$$

The equation

$$\frac{\partial S}{\partial v^\alpha}=0, \tag{35}$$

is called a string equation. In the case of Frobenius manifold of A_{q-1} type it is nothing but the Douglas string equation written in the form of the principle of least string action.[20] It can be shown that solutions $\vec{v}(t_k^\alpha)$ of the string equation (35) satisfy also Eq. (32).

4.3. *Equation for Tau-function*

We define the function $Z[t]=\log\tau(t)$, where

$$Z[t]=\frac{1}{2}\int_0^{v=v^*(t)}\Omega, \tag{36}$$

and

$$\Omega=C_\alpha^{\beta\gamma}(v)\frac{\partial S(v,t)}{\partial v^\beta}\frac{\partial S(v,t)}{\partial v^\gamma}dv^\alpha, \tag{37}$$

is the differential form and $v^*(t)$ is one of the solutions of the string equation (35). From the associativity of the algebra A_{q-1} and the equations (12) it follows that Ω is a closed one-form.

Lemma 4.1. *On the solution of the string equation* $Z(t)$ satisfies*

$$\frac{\partial^2 Z(t)}{\partial t_k^\alpha \partial t_0^1}=\theta_{\alpha,k}(v(t)). \tag{38}$$

In particular,

$$v^\alpha(t)=\eta^{\alpha\beta}\frac{\partial^2 Z}{\partial t_0^\beta \partial t_0^1}, \tag{39}$$

*To simplify our expressions we write $v(t)$ instead of $v^*(t)$ when it is clear from the context.

and for $v^{q-1}(t) = u_1(t)$

$$\frac{\partial^2 Z}{\partial x^2} = u_1(t). \tag{40}$$

Proof. Differentiating with respect to t_k^α and t_0^1 and taking into account the string equation, we find

$$\frac{\partial^2 Z}{\partial t_k^\alpha \partial t_0^1} = \int_0^{v^*(t)} C_\lambda^{\beta\gamma} \frac{\partial \theta_{\alpha,k}}{\partial v^\beta} \frac{\partial \theta_{1,0}}{\partial v^\gamma} dv^\lambda = \int_0^{v^*(t)} \frac{\partial \theta_{\alpha,k}}{\partial v^\lambda} dv^\lambda = \theta_{\alpha,k}. \tag{41}$$

Here we used that $\theta_{1,0} = v_1 = v^{q-1}$, $C_\lambda^{\beta,q-1} = \delta_\lambda^\beta$. □

Taking into account $\theta_{\alpha,0} = v_\alpha$ we obtain from Lemma 4.1 that

$$\frac{\partial^2 Z(t)}{\partial t_0^\alpha \partial t_0^1} = v_\alpha(t). \tag{42}$$

Since Z satisfy equations (35) and (40), it is a tau-function of the integrable hierarchy connected with the corresponding Frobenius manifold.

4.4. *A_{q-1} FM and dispersionless limit of Gelfand-Dikij Hierarchy*

The dispersionless limit of the Gelfand-Dikij equations is formulated as follows:

$$\frac{\partial Q}{\partial t_k^\alpha} = [A_{\alpha,k}, Q] = \frac{\partial A_{\alpha,k}}{\partial x} \frac{\partial Q}{\partial y} - \frac{\partial A_{\alpha,k}}{\partial y} \frac{\partial Q}{\partial x}, \tag{43}$$

where

$$Q = y^q + u_1(x) y^{q-2} + ... + u_{q-1}, \tag{44}$$

and

$$A_{\alpha,k} = \frac{1}{q} c_{\alpha,k} \left(Q^{k+\frac{\alpha}{q}} \right)_+. \tag{45}$$

One can show that these equations are equivalent to the Hamiltonian equation

$$\frac{\partial v^\mu}{\partial t_k^\alpha} = \eta^{\mu\nu} \frac{\partial}{\partial x} \frac{\partial \theta_{\alpha,k+1}}{\partial v^\nu}. \tag{46}$$

4.5. *Formula for tau-function*

As it was derived above the logarithm of the tau-function $Z[\{t_k^\alpha\}]$ is given by

$$Z[\{t_k^\alpha\}] = \frac{1}{2}\int_0^{v^*(t)} C_\alpha^{\beta\gamma}\frac{\partial S}{\partial v^\beta}\frac{\partial S}{\partial v^\gamma}dv^\alpha, \tag{47}$$

where

$$S = \sum_{\alpha=1}^{q-1}\sum_k t_k^\alpha \theta_{\alpha,k}\,. \tag{48}$$

5. Resonance problem in (p, q) MLG

5.1. *Homogeneity property of string equation. Spectrum for (p, q) case*

Let now only the finite number of the parameters $\{t_k^\alpha\}$ be non-zero. One of them we take equal to one and others be enumerated by two integers (m, n). Here $1 \leq m \leq q-1$, $1 \leq n \leq p-1$, where p, q are two coprime integers, $p > q$ and q is a degree of the polynomial Q defined in Eq. (44). Hence, the set of the parameters $\{t_k^\alpha\}$ is replaced by the set $\{t_{mn}\}$. Let us take the action in the form

$$S = \operatorname*{res}_{y=\infty}[Q^{\frac{p+q}{q}} + \sum_{m,n}^{pm-qn>0} t_{mn}Q^{\frac{pm-qn}{q}}], \tag{49}$$

It is easy to check that $Q[y, u_\alpha]$ and $S[u_\alpha, t_{mn}]$ are quasi-homogeneous functions

$$Q[\rho y, \rho^{r_\alpha}u_\alpha] = \rho^q Q[y, u_\alpha], \qquad S[\rho^{r_\alpha}u_\alpha, \rho^{\sigma_{mn}}t_{mn}] = \rho^{p+q}S[u_\alpha, t_{mn}]. \tag{50}$$

Here we denote

$$r_\alpha = q - \alpha - 1, \qquad \sigma_{mn} = p + q - |pm - qn|. \tag{51}$$

We call $\{\sigma_{mn}\}$ the set of the scaling indices of the set $\{t_{mn}\}$. As it was found by Douglas,[3] the numbers $\delta_{mn} = \frac{\sigma_{mn}}{2q}$ coincide with the gravitational dimensions of the physical fields in (p, q) Minimal Liouville gravity.[19]

The function $Z[t_{mn}]$ is a quasi-homogeneous function

$$Z[\rho^{2q\delta_{mn}}t_{mn}] = \rho^{2(p+q)}Z[t_{mn}]. \tag{52}$$

5.2. *The group of the resonance transformations*

Since the scaling indices are integer, the following relation can take place

$$\sigma_{mn} = \sigma_{k_1 l_1} + \sigma_{k_2 l_2} + \dots + \sigma_{k_N l_N}. \tag{53}$$

This is known as a resonance condition. The number of possible resonances grows when p and q increase. A transformation $t_{mn} \to \lambda_{mn}$ of the form

$$\begin{aligned} t_{mn} = \lambda_{mn} + &\sum_{k_1,l_1,k_2,l_2} A_{mn}^{k_1 l_1; k_2, l_2} \lambda_{k_1,l_1} \lambda_{k_2,l_2} \\ + &\sum_{k_1,l_1,k_2,l_2,k_3,l_3} A_{mn}^{k_1 l_1; k_2, l_2; k_3, l_3} \lambda_{k_1,l_1} \lambda_{k_2,l_2} \lambda_{k_3,l_3} + \dots, \end{aligned} \tag{54}$$

is called resonance transformation if Eq. (53) is satisfied for each term. Besides, by definition, we suggest that the scaling index of λ_{mn} equals to the one of t_{mn}.

It is obvious that

$$t_{mn}(\{\rho^{\sigma_{kl}} \lambda_{kl}\}) = \rho^{\sigma_{mn}} t_{mn}(\{\lambda_{kl}\}), \tag{55}$$

and that the resonance transformation does not change the homogeneity property of the partition function $Z[t_{mn}(\{\lambda_{kl}\})] = \widetilde{Z}[\lambda_{mn}]$

$$\widetilde{Z}[\{\rho^{\sigma_{mn}} \lambda_{mn}\}] = \rho^{p+q} \widetilde{Z}[\{\lambda_{mn}\}]. \tag{56}$$

Hence, if we find some solution of the string equation (35) and construct $Z[t_{mn}]$, then we get a family of the solutions $\widetilde{Z}[\{\lambda_{mn}\}] = Z[\{t_{mn}(\{\lambda_{kl})\}\}]$ having the same homogeneity properties with respect to the resonance transformations.

5.3. *The choice of the resonance transformation and of the solution of the string equation*

Now we are in the position to formulate the following problem. We are looking for solutions of the string equation and resonance transformations which gives function $\widetilde{Z}[\{\lambda_{mn}\}]$ satisfying infinite number of constraints known as fusion rules for observables of minimal CFT models $M(p,q)$ and their for the correlators. In what follows we restrict ourself by considering spherical topology. Then these rules can be formulated as follows.

We denote by Φ_{mn}, where $1 \leq m \leq p$ and $1 \leq n \leq q$, the primary fields in the minimal model $M(p,q)$ of Conformal field theory. The fields $\Phi_{m,n}$ and $\Phi_{q-m,p-n}$ correspond to the same primary field.

The following graphical representation allows to formulate these restrictions

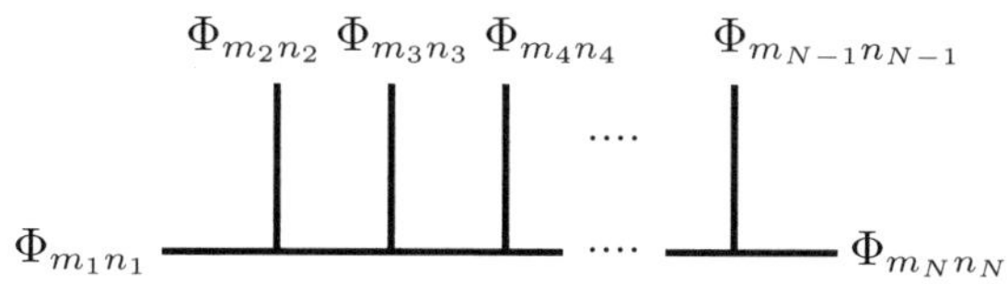

Here the external lines represent the (arbitrary arranged) primary fields in the correlator $\langle \Phi_{m_1n_1}\Phi_{m_2n_2}...\Phi_{m_Nn_N}\rangle$ (here we assume $N \geq 3$). The fusion rules result to the requirement that the correlation function must be equal to zero if there are no sets of pairs (k_i, l_i) assigned to the internal lines, for which in any vertex of the graph the following condition on the three pairs (m_i, n_i) $(i = 1, 2, 3)$ corresponding to the lines connected to this vertex

$$|m_1 - m_2| + 1 \leq m_3 \leq \min\{m_1 + m_2 - 1, 2q + 1 - m_1 - m_2\} \quad \text{step } 2, \tag{57}$$

$$|n_1 - n_2| + 1 \leq n_3 \leq \min\{n_1 + n_2 - 1, 2p + 1 - n_1 - n_2\} \quad \text{step } 2, \tag{58}$$

can not be satisfied any permutation of the pairs.

In addition, from the conformal selection rules for $N = 1$ it follows

$$\langle \Phi_{mn}\rangle = 0, \tag{59}$$

for $(m, n) \neq (1, 1)$ and for $N = 2$

$$\langle \Phi_{m_1n_1}\Phi_{m_2n_2}\rangle = 0, \tag{60}$$

for (m_1, n_1) not equal to (m_2, n_2) or $(q - m_2, p - n_2)$.

Now we are going to give a more precise formulation of our main conjecture.

Conjecture 5.1. *There exist a solution of the string equation and a choice of the resonance transformation described above, such that the function*

$$\begin{aligned}\widetilde{Z}[\{\lambda_{mn}\}] &= \langle \exp \sum_{m,n} \lambda_{m,n} O_{m,n}\rangle \\ &= \sum_{N=0}^{\infty} \sum_{m_i,n_i} \frac{\lambda_{m_1n_1}...\lambda_{m_Nn_N}}{N!} \langle O_{m_1n_1}...O_{m_Nn_N}\rangle,\end{aligned} \tag{61}$$

appears the generating function of the correlators in the Minimal Liouville Gravity.

In particular, all correlators $\langle O_{m_1n_1}...O_{m_Nn_N}\rangle$ forbidden by the conformal fusion rules vanish.

6. The plan of the solution of the problem

To solve the formulated above problem we write the action $S(v_\alpha, t_{mn})$ and the generating function $Z[\{t_{mn}\}]$ in terms of new variables $\{\lambda_{mn}\}$ using the resonance change of variables

$$\begin{aligned} t_{mn} &= \lambda_{mn} + A_{mn}\mu^{\delta_{mn}} + \sum_{m_1,n_1}^{\delta_{m_1n_1}\le\delta_{mn}} A_{mn}^{m_1n_1}\mu^{\delta_{mn}-\delta_{m_1n_1}}\lambda_{m_1n_1} + \\ &+ \sum_{m_1,n_1,m_2,n_2}^{\delta_{m_1n_1}+\delta_{m_2n_2}\le\delta_{mn}} A_{mn}^{m_1n_1,m_2n_2}\mu^{\delta_{mn}-\delta_{m_1n_1}-\delta_{m_2n_2}}\lambda_{m_1n_1}\lambda_{m_2n_2} + \dots, \end{aligned} \tag{62}$$

where $\mu = \lambda_{11}$ is called the cosmological constant in the continuum approach to MLG.

After performing this transform the action takes the form

$$\begin{aligned} \widetilde{S}[v_\alpha, \{\lambda_{mn}\}] =& S^{(0)}(v_\alpha) + \sum_{m,n}\lambda_{mn}\, S^{(mn)}(v_\alpha) + \\ &+ \sum_{m_1,n_1,m_2,n_2}\lambda_{m_1n_1}\lambda_{m_2n_2}\, S^{(m_1n_1,m_2n_2)}(v_\alpha) + \dots . \end{aligned} \tag{63}$$

The information about the form of the resonance transformation is encoded in the coefficients of $S^{(0)}$, $S^{(mn)}$, etc. From Eqs. (24) and (12.1) we find

$$S^{(0)} = \operatorname*{res}_{y=\infty}\left[Q^{\frac{p+q}{q}} + \sum_{l=1}^{s} A_{1l}\,\mu^{\frac{l+1}{2}}Q^{\frac{p-ql}{q}}\right], \tag{64}$$

where we introduced the new positive integer numbers s and p_0 such that $p = sq + p_0$ and $0 < p_0 < q$.

$$S^{(mn)} = \operatorname*{res}_{y=\infty}\left[Q^{\frac{pm-qn}{q}} + \sum_{l=n+2}^{sm+\lfloor\frac{p_0m}{q}\rfloor} A_{ml}^{mn}\,\mu^{\frac{l-n}{2}}\,Q^{\frac{pm-ql}{q}}\right], \tag{65}$$

where A_{kl}^{mn} are the coefficients of the resonance relations and $(l-n)$ is even. The higher coefficients can also be easily written in terms of the coefficients $A_{kl}^{\{m_in_i\}}$.

The generating function is given by

$$\widetilde{Z}[\{\lambda_{mn}\}] = \frac{1}{2}\int_0^{\mathbf{v}^*} C_\alpha^{\beta\gamma}(v)\frac{\partial\widetilde{S}}{\partial v^\beta}\frac{\partial\widetilde{S}}{\partial v^\gamma}dv^\alpha, \tag{66}$$

where $\mathbf{v}^*$ is defined as a function of the parameters $\{\lambda_{mn}\}$ of the Douglas string equation (35).

From now on we will skip the tilde over the functions $\widetilde{S}(\{u_\alpha\},\{\lambda_{mn}\})$ and $\widetilde{Z}(\{\lambda_{mn}\})$.

7. Appropriate solution

To compute the one-point function which is given by the integral

$$\langle O_{mn}\rangle = \int_0^{v_\alpha^0} C^\alpha_{\beta\gamma}\frac{\partial S^{(0)}}{\partial v_\beta}\frac{\partial S^{(mn)}}{\partial v_\gamma}dv_\alpha, \tag{67}$$

we need to know the upper limit in this integral v_α^0 which is the solution of the string equation for all couplings (except $\lambda_{11}=\mu$) equal to zero

$$v_\alpha^0 = v_\alpha^*(\lambda_{mn})\Big|_{\lambda_{mn}=0,\lambda_{11}=\mu}. \tag{68}$$

Explicitly, v_α^0 satisfies

$$\frac{\partial S^{(0)}}{\partial v_\mu}\Big|_{v_\alpha=v_\alpha^0} = 0. \tag{69}$$

Using Eqs. (64), (65) and (24), $S^{(0)}$ and $S^{(mn)}$ can be written as

$$S^{(0)} = -\frac{\theta_{p_0,s+1}}{c_{p_0,s+1}} - \sum_{l=1}^{s} A_{1l}\mu^{\frac{l+1}{2}}\frac{\theta_{p_0,s-l}}{c_{p_0,s-l}}, \tag{70}$$

$$S^{(mn)} = -\frac{\theta_{p_0m,sm-n}}{c_{p_0m,sm-n}} - \sum_{l=n+2}^{sm+\lfloor\frac{p_0m}{q}\rfloor} A^{mn}_{ml}\,\mu^{\frac{l+1}{2}}\frac{\theta_{p_0m,sm-l}}{c_{p_0m,sm-l}}. \tag{71}$$

We will use the following proposition from Ref. 7:

Proposition 7.1. *On the line* $v_{i>1}=0$,

$$\begin{cases} k \text{ even}: & \frac{\partial\theta_{\lambda,k}}{\partial v_\alpha} = \delta_{\lambda,\alpha}\,x_{\lambda,k}\left(-\frac{v_1}{q}\right)^{\frac{k}{2}q}, \\ k \text{ odd}: & \frac{\partial\theta_{\lambda,k}}{\partial v_\alpha} = \delta_{\lambda,q-\alpha}\,y_{\lambda,k}\left(-\frac{v_1}{q}\right)^{\frac{k-1}{2}q+\lambda}, \end{cases} \tag{72}$$

where

$$x_{\alpha,k} = \frac{\Gamma(\frac{\alpha}{q})}{\Gamma(\frac{\alpha}{q}+\frac{k}{2})(\frac{k}{2})!} \quad \textit{and} \quad y_{\lambda,k} = -\frac{\Gamma(\frac{\alpha}{q})}{\Gamma(\frac{\alpha}{q}+\frac{k+1}{2})(\frac{k-1}{2})!}. \tag{73}$$

Using this statement together with Eq.(70) it is not difficult to see that the string equation (69) has the solutions of the form $v^0_\alpha = 0$ for $\alpha \neq 1$ and the coordinate v^0_1 is a root of the equation

$$\frac{\partial S^{(0)}}{\partial v_{p_0}} = 0, \quad \text{if} \qquad s - \text{odd}, \tag{74}$$

or

$$\frac{\partial S^{(0)}}{\partial v_{q-p_0}} = 0, \quad \text{if} \qquad s - \text{even}. \tag{75}$$

Here we assume that after taking derivative we set all v_α for $\alpha \neq 1$ to zero. More explicitly these equations can be written as

$$\sum_{k=-1:2:s} B^{\text{odd}}_{p_0,k}\left(-\frac{v_1}{q}\right)^{\frac{s-k}{2}q} = 0, \quad \text{if} \qquad s - \text{odd}, \tag{76}$$

or

$$\sum_{k=-1:2:s} B^{\text{odd}}_{p_0,k}\left(-\frac{v_1}{q}\right)^{\frac{s-k-1}{2}q} = 0, \quad \text{if} \qquad s - \text{even}, \tag{77}$$

where

$$B^{\text{odd}}_{p_0,k} = \frac{x_{p_0,s-k}}{c_{p_0,s-k}} A_{1,k}\mu^{\frac{k+1}{2}}, \tag{78}$$

and

$$B^{\text{even}}_{p_0,k} = \frac{y_{p_0,s-k}}{c_{p_0,s-k}} A_{1,k}\mu^{\frac{k+1}{2}}, \tag{79}$$

where $A_{1,-1} = 1$.

8. One-point functions

As it was shown in Ref. 7, the structure constant in the flat coordinates on the line $v_{\alpha>0} = 0$, for $\alpha \geq \beta \geq \gamma$

$$C_{\alpha\beta\gamma} = \left(-\tfrac{v_1}{q}\right)^{\frac{\alpha+\beta+\gamma-q-1}{2}} \tag{80}$$

$$\text{if} \quad \tfrac{\alpha+\beta+\gamma-q-1}{2} \in \mathbb{N}_0 \quad \text{and} \quad \alpha+\beta-\gamma \in [1, q-1], \quad \text{otherwise } 0,$$

where $\mathbb{N}_0$ is the set of non-negative integers. Using Eq. (72) we find for s odd and $(sm - n)$ even

$$\langle O_{mn}\rangle = \int_0^{v_1^0} C_{q-1,p_0,p_0m} \frac{\partial S^{(0)}}{\partial v_{p_0}} \frac{\partial S^{(mn)}}{\partial v_{p_0m}} dv_1. \tag{81}$$

Taking into account Eq. (80) we conclude that the correlation function is zero for $m \neq 1$. Hence, in this case from the selection rules we obtain

$$\langle O_{1n}\rangle = \int_0^{v_1^0} \big(-\frac{v_1}{q}\big)^{p_0-1}\frac{\partial S^{(0)}}{\partial v_{p_0}}\frac{\partial S^{(1n)}}{\partial v_{p_0}}dv_1 = 0. \tag{82}$$

For s odd and $(sm-n)$ odd,

$$\langle O_{mn}\rangle = \int_0^{v_1^0} C_{q-1,p_0,q-p_0m}\frac{\partial S^{(0)}}{\partial v_{p_0}}\frac{\partial S^{(mn)}}{\partial v_{q-p_0m}}dv_1, \tag{83}$$

and the structure constant here is not equal to zero only if $q-p_0m = p_0$ as it follows from Eq. (80). Therefore the gravitational dimension

$$[\langle O_{mn}\rangle] = \frac{p+q}{q} - \delta_{mn} = \frac{sm-n}{2} + \frac{s+1}{2} + \frac{p_0m+p_0}{2q}, \tag{84}$$

is integer, the correlation function is analytic and we shell not consider it.[6]

Similarly, for s even and $(sm-n)$ even, we obtain the following consequence of the selection rules

$$\langle O_{1n}\rangle = \int_0^{v_1^0} \big(-\frac{v_1}{q}\big)^{q-p_0-1}\frac{\partial S^{(0)}}{\partial v_{q-p_0}}\frac{\partial S^{(1n)}}{\partial v_{q-p_0}}dv_1 = 0. \tag{85}$$

Finally, if s even and $(sm-n)$ odd, we find again that the expressions for the one point correlation functions are analytic.

A simple analysis shows that the number of these equations is equal to the number of the coefficients arising in the first order in the resonance relation. Hence the requirement of absence of the one point functions fixes uniquely unknown coefficients $B_{p_0,k}$ in the expressions (76) and (77).

Thus we arrive to the conclusion that the special solution of the string equation considered above ensure the requirements of the selection rules in agreement with the general prescription described in the previous section.

We note also that the variety of (p,q) models of minimal Liouville Gravity is split in two subclasses according to the condition that $\lfloor p/q \rfloor$ be either even or odd. In each case we find distinct sets of requirements formulated above leading to zero valued one point functions.

9. Two-point functions

We are now going to consider the two-point function. From Eq. (66) we find

$$\langle O_{m_1n_1}O_{m_2n_2}\rangle = \sum_{\gamma=1}^{q-1}\int_0^{v_1^0} dv_1\,\big(-\frac{v_1}{q}\big)^{\gamma-1}\frac{\partial S^{(m_1n_1)}}{\partial v_\gamma}\frac{\partial S^{(m_2n_2)}}{\partial v_\gamma}. \tag{86}$$

It follows from Eq. (72) that $\frac{\partial S^{(mn)}}{\partial v^\gamma} \neq 0$ if one of the following two conditions is satisfied

$$\begin{aligned} &1) \quad \gamma = mp_0 \bmod q \quad \text{and} \quad (sm-n) - \text{even}, \\ &2) \quad \gamma = q - mp_0 \bmod q \quad \text{and} \quad (sm-n) - \text{odd}. \end{aligned} \tag{87}$$

Similarly to the consideration in the previous section we find four cases where the two point function can be non-zero. In two cases: where the first pair (m_1, n_1) satisfies first condition while the second pair (m_2, n_2) is subject of the second condition and vice versa, we find the regular expression for the two point function. Thus, we are left with the two options where both pairs satisfy either the first or the second condition in Eq. (87).

Explicitly, in the case when both $(sm - n_1)$ and $(sm - n_2)$ are even we get the following requirement

$$\langle O_{mn_1} O_{mn_2} \rangle = \int_0^{v_1^0} dv_1 \left(-\frac{v_1}{q}\right)^{mp_0 - 1} \frac{\partial S^{(mn_1)}}{\partial v_{mp_0}} \frac{\partial S^{(mn_2)}}{\partial v_{mp_0}} = 0 \quad \text{if} \quad n_1 \neq n_2. \tag{88}$$

Making the substitution

$$t = 2\left(\frac{v_1}{v_1^0}\right)^q - 1, \tag{89}$$

and denoting

$$\frac{\partial S^{(mn)}}{\partial v_{mp_0}} = L_{\frac{sm-n}{2}}(t), \tag{90}$$

we find the following consequence of the diagonality condition

$$\langle O_{mn_1} O_{mn_2} \rangle = \int_{-1}^{1} dt\, (1+t)^{\frac{mp_0 - q}{q}} L_{\frac{sm-n_1}{2}}(t) L_{\frac{sm-n_2}{2}}(t) = 0 \quad \text{if} \quad n_1 \neq n_2. \tag{91}$$

Hence, the selection rules for the two-point correlation numbers requires that the polynomials $L_{\frac{sm-n}{2}}$ form an orthogonal set of Jacobi polynomials

$$\frac{\partial S^{(mn)}}{\partial v_{mp_0}} = \frac{pm - qn}{q} P_{\frac{sm-n}{2}}^{(0, \frac{mp_0 - q}{q})}(t), \qquad \text{for} \quad (sm-n) - \text{ even.} \tag{92}$$

In the second case, where both $(sm - n_1)$ and $(sm - n_2)$ are odd, we have

$$\langle O_{mn_1} O_{mn_2} \rangle = \int_0^{v_1^0} dv_1 \left(-\frac{v_1}{q}\right)^{q - mp_0 - 1} \frac{\partial S^{(mn_1)}}{\partial v_{q - mp_0}} \frac{\partial S^{(mn_2)}}{\partial v_{q - mp_0}} = 0 \quad \text{if} \quad n_1 \neq n_2. \tag{93}$$

Denoting

$$\frac{\partial S^{(mn)}}{\partial v_{q-mp_0}} = (1+t)^{\frac{mp_0}{q}} L_{\frac{sm-n-1}{2}}(t), \tag{94}$$

we find the following consequence of the diagonality condition for the two-point correlation function in this case

$$\langle O_{mn_1} O_{mn_2} \rangle = \int_{-1}^{1} dt\,(1+t)^{\frac{mp_0}{q}} L_{\frac{sm-n_1-1}{2}}(t) L_{\frac{sm-n_2-1}{2}}(t) = 0 \quad \text{if} \quad n_1 \neq n_2. \tag{95}$$

It means that

$$\frac{\partial S^{(mn)}}{\partial v_{q-mp_0}} = \frac{pm-qn}{q}(1+t)^{\frac{mp_0}{q}} P^{(0,\frac{mp_0}{q})}_{\frac{sm-n-1}{2}}(t) \qquad \text{for} \quad (sm-n)- \text{ odd.} \tag{96}$$

At last, inserting these explicit expressions for the derivatives of $S^{(mn)}$ to the equations (82) and (85) we arrive to the condition

$$\langle O_{1n} \rangle = \int_{-1}^{1} (1+t)^{\frac{p_0-q}{q}} L_{\frac{s+1}{2}}(t) P^{(0,\frac{p_0-q}{q})}_{\frac{s-n}{2}}(t) dt = 0, \tag{97}$$

in the case where s is odd and n is odd and greater than 1. And

$$\langle O_{1n} \rangle = \int_{-1}^{1} (1+t)^{\frac{p_0}{q}} L_{\frac{s}{2}}(t) P^{(0,\frac{p_0}{q})}_{\frac{s-n-1}{2}}(t) dt = 0, \tag{98}$$

in case where s is even and n is odd and greater than 1. Here we introduced the polynomial $L_n(t)$

$$\frac{\partial S^{(0)}}{\partial v_{p_0}}(t) = L_{\frac{s+1}{2}}(t), \tag{99}$$

for s odd,

$$\frac{\partial S^{(0)}}{\partial v_{q-p_0}}(t) = (1+t)^{\frac{p_0}{q}} L_{\frac{s}{2}}(t). \tag{100}$$

for s even.

Taking into account these equations, the order of the polynomials $\frac{\partial S^{(0)}}{\partial v_{p_0}}$ and $\frac{\partial S^{(0)}}{\partial v_{q-p_0}}$ and the string equations (74), (75) we obtain the following explicit expressions

$$\frac{\partial S^{(0)}}{\partial v_{p_0}} = \frac{p+q}{q}\left(P^{(0,\frac{p_0-q}{q})}_{\frac{s+1}{2}}(t) - P^{(0,\frac{p_0-q}{q})}_{\frac{s-1}{2}}(t)\right), \tag{101}$$

if s is odd and

$$\frac{\partial S^{(0)}}{\partial v_{q-p_0}} = \frac{p+q}{q}(1+t)^{\frac{p_0}{q}}\left(P^{(0,\frac{p_0}{q})}_{\frac{s}{2}}(t) - P^{(0,\frac{p_0}{q})}_{\frac{s-2}{2}}(t)\right), \tag{102}$$

if s is even.

10. Conclusions

In the present talk we have described the relation between the approach to (p, q) models of Minimal Liouville gravity based on the Douglas string equation, on one hand, and the Frobenius manifolds of A_{q-1} type on the other. As a result of this relation the generating function of correlation numbers in MLG is represented by the logarithm of the tau-function of the corresponding integrable hierarchy. All necessary information is encoded in the solution of the Douglas string equation and in the resonance relations between the parameters of the integrable hierarchy and the coupling constants of MLG. Using this relation and some special properties of the flat coordinates on the Frobenius manifold, we have found the appropriate solution of the Douglas string equation. This result generalizes analogues result found recently for Unitary models of Minimal Liouville gravity.[7] We have shown that the appropriate solution is consistent with the basic requirements of the conformal selection rules arising on the levels of one- and two-point correlation functions. Namely, the number of the parameters of the resonance transformations is exactly the number of the constraints following from the selection rules. Resolving these constraints we have found explicit form of the resonance transformations in terms of Jacoby polynomials. It would be interesting to investigate if this matching persists for multi-point correlation functions when the fusion rules of the underlying minimal models of CFT should be taken into account. This analysis requires also knowing the explicit form of the structure constants of the Frobenius algebra in the flat coordinates. We plan to study these questions in the near future.

References

1. A.A. Belavin, V.A. Belavin, *Frobenius manifolds, integrable hierarchies and minimal Liouville gravity*, *JHEP* **09**, 151 (2014).
2. A.M. Polyakov, *Quantum Geometry of Bosonic Strings*, *Phys. Lett.* **B103**, 207–210 (1981).
3. M.R. Douglas, *Strings in less than one-dimension and the generalized KdV hierarchies*, *Phys. Lett.* **B238**, 176 (1990).
4. G.W. Moore, N. Seiberg, and M. Staudacher, *From loops to states in 2-D quantum gravity*, *Nucl. Phys.* **B362**, 665–709 (1991).
5. A. Belavin and A. Zamolodchikov, *On Correlation Numbers in 2D Minimal Gravity and Matrix Models*, *J. Phys.* **A42**, 304004 (2009),

arXiv:0811.0450 [hep-th].

6. A. Belavin, B. Dubrovin, and B. Mukhametzhanov, *Minimal Liouville Gravity correlation numbers from Douglas string equation*, *JHEP* **1401**, 156 (2014), arXiv:1310.5659 [hep-th].
7. V. Belavin, *Unitary Minimal Liouville Gravity and Frobenius Manifolds*, arXiv:1405.4468 [hep-th].
8. A. Belavin, A.M. Polyakov, and A. Zamolodchikov, *Infinite Conformal Symmetry in Two-Dimensional Quantum Field Theory*, *Nucl.Phys.* **B241**, 333–380 (1984).
9. V. Kazakov, A.A. Migdal, and I. Kostov, *Critical Properties of Randomly Triangulated Planar Random Surfaces*, *Phys. Lett.* **B157**, 295–300 (1985).
10. V. Kazakov, *Ising model on a dynamical planar random lattice: Exact solution*, *Phys. Lett.* **A119**, 140–144 (1986).
11. V. Kazakov, *The Appearance of Matter Fields from Quantum Fluctuations of 2D Gravity*, *Mod. Phys. Lett.* **A4**, 2125 (1989).
12. M. Staudacher, *The Yang-Lee edge singularity on a dynamical planar random surface*, *Nucl.Phys.* **B336**, 349 (1990).
13. E. Brezin and V. Kazakov, *Exactly solvable field theories of closed strings*, *Phys. Lett.* **B236**, 144–150 (1990).
14. M.R. Douglas and S.H. Shenker, *Strings in Less Than One-Dimension*, *Nucl.Phys.* **B335**, 635 (1990).
15. D.J. Gross and A.A. Migdal, *Nonperturbative Two-Dimensional Quantum Gravity*, *Phys. Rev. Lett.* **64**, 127 (1990).
16. I. Krichever, *The Dispersionless Lax equations and topological minimal models*, *Commun. Math. Phys.* **143**, 415–429 (1992).
17. B. Dubrovin, *Integrable systems in topological field theory*, *Nucl.Phys.* **B379**, 627–689 (1992).
18. R. Dijkgraaf, H.L. Verlinde, and E.P. Verlinde, *Topological strings in d less than 1*, *Nucl. Phys.* **B352**, 59–86 (1991).
19. V. Knizhnik, A.M. Polyakov, and A. Zamolodchikov, *Fractal Structure of 2D Quantum Gravity*, *Mod. Phys. Lett.* **A3**, 819 (1988).
20. P.H. Ginsparg, M. Goulian, M. Plesser, and J. Zinn-Justin, *(p, q) String actions*, *Nucl. Phys.* **B342**, 539–563 (1990).
21. P. Di Francesco and D. Kutasov, *Unitary Minimal Models Coupled To 2-d Quantum Gravity* , *Nucl.Phys.* **B342**, 589–624 (1990).

A.A. Belavin, Chernogolovka, 2015.

THE ϵ-EXPANSION FOR QED

L. DI PIETRO, Z. KOMARGODSKI, I. SHAMIR, and E. STAMOU

Department of Particle Physics and Astrophysics
Weizmann Institute of Science, Rehovot 76100, Israel
E-mail: lorenzo.dipietro, zohar.komargodski, itamar.shamir, emmanuel.stamou@weizmann.ac.il

We study Quantum Electrodynamics in $d = 3$ (QED_3) coupled to N_f flavors of fermions. The theory flows to an IR fixed point for N_f larger than some critical number N_f^c. For $N_f \leq N_f^c$, chiral symmetry breaking is believed to take place. In analogy with the Wilson-Fisher description of the critical $O(N)$ models in $d = 3$, we use the existence of a fixed point in $d = 4 - 2\epsilon$ to study the three-dimensional theory. We show how the ϵ-expansion can be used to study the anomalous dimension of 2- and 4-fermion operators. The latter leads to an estimate of the critical number N_f^c. An important novelty compared to the $O(N)$ models is that, because of the structure of spinors, the theory in $d = 3$ has an enhanced symmetry. We identify the operators in $d = 4 - 2\epsilon$ that correspond to the additional conserved currents at $d = 3$.

1. QED_3 and Chiral Symmetry Breaking

We consider QED in three dimensions with $2N_f$ complex two-component massless fermions of unit charge ψ^i ($i = 1, \ldots, 2N_f$). When N_f is sufficiently large the theory flows to an interacting fixed point that preserves the $SU(2N_f)$ global symmetry of the microscopic Lagrangian. This fixed point is stable in the sense that it has no relevant operators that respect all the symmetries. However, it is believed that a critical number N_f^c exists,[1,2] such that when $N_f \leq N_f^c$ the theory exhibits spontaneous symmetry breaking[3]

$$SU(2N_f) \to SU(N_f) \times SU(N_f) \times U(1). \tag{1}$$

This can be viewed as coming from a chiral condensate of the parity invariant operator

$$\sum_{a=1}^{N_f} (\bar{\psi}_a \psi^a - \bar{\psi}_{a+N_f} \psi^{a+N_f}). \tag{2}$$

In condensed matter physics this theory has been advocated as an effective description of various strongly correlated materials. QED_3 can arise as the continuum limit of spin systems with various values of N_f, e.g. $N_f = 2, 4$.[4–6] The theory with $N_f = 2$ is also applied to high-temperature superconductivity.[7–9]

A salient feature of QED in 3 dimensions is the positive mass dimension of the gauge coupling. The theory is strongly coupled in the IR, and is not amenable to ordinary perturbation theory. One possible scheme of approximation is to study the theory in the large N_f limit. This was considered by various authors.[9–14] However, as mentioned above, physically interesting models occur for values of N_f of order one, where this approach may not be reliable.

We propose to study QED in 3.99 dimensions. The theory admits an IR fixed point in $d = 4 - 2\epsilon$ that can be studied perturbatively in ϵ. This is analogous to the Wilson-Fisher fixed point in $O(N)$ models.[15] By plugging $\epsilon = 1/2$ we obtain an approximation to the observables of the theory in the physical dimension $d = 3$.

As an application we show how this can be used to estimate the critical number N_f^c. The idea is as follows. The theory has two independent $SU(2N_f)$ invariant 4-fermion operators

$$\mathcal{O}_1 = (\bar{\psi}_i \sigma^\mu \psi^i)^2, \qquad \mathcal{O}_2 = (\bar{\psi}_i \psi^i)^2. \tag{3}$$

Close to the UV Gaussian fixed point the dimensions of $\mathcal{O}_{1,2}$ is 4 and the operators are irrelevant. However, along the flow their dimensions receive large corrections and as a consequence they may become relevant in the IR. This relevant perturbation can start a flow to the chiral-symmetry breaking phase. The plan is therefore to extract N_f^c from the condition that the dimension of $\mathcal{O}_1$ or $\mathcal{O}_2$ is equal to 3.

2. The ϵ-expansion

Consider the QED Lagrangian in $d = 4 - 2\epsilon$

$$\mathcal{L} = -\frac{1}{4e^2} F_{\mu\nu} F^{\mu\nu} + i\bar{\Psi}_a \gamma^\mu D_\mu \Psi^a. \tag{4}$$

In 4 dimensions, Ψ^a are N_f four component Dirac spinors, which may be identified with the three dimensional two-component spinors via

$$\Psi^a = \begin{pmatrix} \psi^a \\ i\sigma^2 \psi^{a+N_f} \end{pmatrix}, \qquad a = 1, \ldots, N_f. \tag{5}$$

In $d = 4 - 2\epsilon$ the coupling constant of the theory e is of positive mass dimension. We choose a renormalization scale μ, and define a dimensionless coupling by $\hat{e} = e\mu^{-\epsilon}$. At one-loop the beta function takes the form

$$\beta(\hat{e}) = -\epsilon\hat{e} + \frac{N_f}{12\pi^2}\hat{e}^3. \tag{6}$$

The first term originates from the classical scaling of the coupling in $4 - 2\epsilon$ dimensions, while the second term is the leading quantum correction. Since the one-loop term is positive, below 4 dimensions there is a fixed point with coupling

$$\hat{e}^2 = \frac{12\pi^2\epsilon}{N_f}, \tag{7}$$

in which the classical and quantum terms balance each other. Close to four dimensions the theory is weakly coupled and perturbation theory is reliable.

Let us comment on the continuation of spinors to non-integer dimensions. Writing the 4-fermion operators (3) in terms of the spinors in non-integer dimensions requires a definition of γ^5. We will be using the 't Hooft–Veltman prescription[16–18] in which $\gamma^5 = i\gamma^0\gamma^1\gamma^2\gamma^3$ for any d. This breaks Lorentz invariance in d dimensions but reproduces the axial anomaly in $d = 4$. We will discuss below another interesting aspect of the ϵ-expansion for QED – the appearance of new symmetries in integer dimensions.*

As a first example, we consider scalar bilinears of the fermions. Their anomalous dimensions may be measured at the fixed point. They descend from two types of 4-dimensional operators. The first are operators that are scalars also in $d = 4$

$$(B_1)_a{}^b = \bar{\Psi}_a\Psi^b \quad , \quad \bar{\Psi}_a\gamma^5\Psi^b. \tag{8}$$

Operators in this class can preserve at most an $SU(N_f)$ diagonal subgroup of the $SU(N_f) \times SU(N_f)$ that exists in $d = 4$. The one-loop calculation gives

$$\Delta_{IR}(B_1) = 3 - 2\epsilon - \frac{9\epsilon}{2N_f}. \tag{9}$$

The second type descends from operators which are 3-forms in $d = 4$

$$(B_{2\mu\nu\rho})_a{}^b = \bar{\Psi}_a\gamma_{\mu\nu\rho}\Psi^b \quad , \quad \bar{\Psi}_a\gamma_{\mu\nu\rho}\gamma^5\Psi^b. \tag{10}$$

*For previous application of the ϵ-expansion to fermionic systems see Refs. 19, 20, 21.

In $d=3$ we can contract with $\epsilon_{\mu\nu\rho}$ to obtain scalars. The chiral condensate (2) and the $SU(2N_f)$ invariant $\sum_i \bar{\psi}_i\psi^i$ belong to this class of operators. For these operators the anomalous dimension vanishes at one-loop

$$\Delta_{IR}(B_2) = 3 - 2\epsilon. \tag{11}$$

The ϵ-expansion therefore suggests that these operators have an IR dimension close to the classical value $\Delta = 2$.

Let us now consider the 4-fermion operators. When continued to d dimensions they take the form

$$\mathcal{O}_1 = (\bar{\Psi}_a\gamma^\mu\Psi^a)^2, \qquad \mathcal{O}_2 = (\bar{\Psi}_a\gamma^\mu\gamma^5\Psi^a)^2. \tag{12}$$

The matrix of anomalous dimensions at one-loop is†

$$\gamma_{\mathcal{O}} = \frac{\hat{e}^2}{16\pi^2}\begin{pmatrix} \frac{8}{3}(2N_f+1) & 12 \\ \frac{44}{3} & 0 \end{pmatrix} \tag{14}$$

The operator corresponding to the smallest eigenvalue has IR dimension

$$\Delta_{IR} = 2d - 2 + \frac{\epsilon}{N_f}\left(2N_f + 1 - 2\sqrt{N_f^2 + N_f + 25}\right). \tag{15}$$

Here we used the fixed point value of $\hat{e}^2$. If the anomalous dimension is such that $\Delta_{IR} \leq d$ then the operator becomes relevant. For $\epsilon = \frac{1}{2}$ this can happen if

$$N_f \leq 2.25. \tag{16}$$

To one-loop approximation, this argument predicts that the Goldstone phase can exist only for $N_f = 1$ or 2. This result is consistent with lattice simulations[23–25] and with the bound obtained from using the F-theorem.[26]

3. Input from $d = 2$

For the estimate of N_f^c we can also use inputs from $d=2$. The $O(N)$ models show that such $d=2$ boundary conditions are often useful to improve the accuracy of the predictions of the ϵ-expansion. Let us try to understand the appropriate boundary condition. In $d=2$ the quadrilinear operator is marginal already at tree-level. Therefore, the criterion above implies

---

†To compute the anomalous dimensions of $\mathcal{O}_{1,2}$ it is crucial to also take into account

$$\mathcal{O}_3 = \bar{\Psi}_a\gamma^\mu\Psi^a\left(\tfrac{1}{e}\partial^\nu F_{\mu\nu} - \bar{\Psi}_b\gamma^\mu\Psi^b\right). \tag{13}$$

This operator classically vanishes on the equations of motion but its mixing with $\mathcal{O}_{1,2}$ cannot be neglected in loop computations.[22]

that in $d = 2$ the anomalous dimension must evaluate to 0 at N_f^c. This is satisfied for $N_f^c = \infty$. This value is consistent with the IR behavior of QED in $d = 2$: for every finite N_f the theory flows to an $SU(2N_f)$ Wess-Zumino-Witten (WZW) interacting CFT[27,28] namely a σ-model with coset target space deformed by a WZW term (exactly as one expects for the Lagrangian of Nambu-Goldstone bosons). Even though Nambu-Goldstone bosons do not exist in $d = 2$, it appears natural to interpret this theory as the continuation of the chirally broken phase to $d = 2$. (See also Refs. 29, 30.) Assuming that the anomalous dimension approaches 0 as $1/N_f$ for $N_f \to \infty$, the divergence of $N_f^c(d)$ for $d = 2$ is given by a simple pole. This suggests using the modified ansatz $N_f^c(d) = (d-2)^{-1} f(d)$ in the equation for N_f^c, which can then be solved for f perturbatively in ϵ. With this ansatz, the leading-order estimate based on our one-loop computation is $N_f^c \leq 4.5$. The difference with the previous estimate are higher-order terms in ϵ; it may be viewed as a measure of uncertainty. Improving on this requires computations beyond one-loop.[31]

4. Emergent symmetries in $d = 3$

As discussed above, when $N_f > N_f^c$ the 3 dimensional theory is characterized by an unbroken $SU(2N_f)$. However, it is not possible to preserve all this symmetry when the theory is extended to $d > 3$. This has to be contrasted with the $O(N)$ model, in which one has the same manifest global symmetries for any d.

For QED$_d$ it is expected that additional currents emerge at $d = 3$. This means that there are vector operators at the fixed point, whose dimension is precisely 2. It is possible to construct operators in d dimensions which are identified with the additional currents in $d = 3$.

Let us start from $d = 4$. The theory then preserves only an $SU(N_f) \times SU(N_f)$ chiral symmetry generated by the currents

$$\begin{aligned}(J_\mu)_a{}^b &= \bar{\Psi}_{\bar{a}}\gamma_\mu\Psi^{\bar{b}}\,(\delta_{\bar{b}}^b\delta_a^{\bar{a}} - N_f^{-1}\delta_a^b\delta_{\bar{b}}^{\bar{a}}),\\ (J_\mu^5)_a{}^b &= \bar{\Psi}_{\bar{a}}\gamma_\mu\gamma_5\Psi^{\bar{b}}\,(\delta_{\bar{b}}^b\delta_a^{\bar{a}} - N_f^{-1}\delta_a^b\delta_{\bar{b}}^{\bar{a}}).\end{aligned} \tag{17}$$

Since the anomalous dimension of conserved currents J_μ vanishes to all orders in perturbation theory, their dimensions at the fixed point is precisely $d-1$. In particular, in $d = 3$ they have dimensions 2 and are conserved by unitarity. The situation with J_μ^5 is more subtle since they are explicitly broken by the regulator for non-integer d.[32] For our purposes it suffices to say that at one-loop the anomalous dimension is vanishing.

The enhancement of $SU(N_f) \times SU(N_f)$ to $SU(2N_f)$ implies the existence of $2N_f^2 + 1$ emergent conserved currents. One of them corresponds to the singlet axial current $J_\mu^s = \sum_a \bar{\Psi}_a \gamma_\mu \gamma_5 \Psi^a$. Indeed the continuation of the anomaly operator $F \wedge F$ vanishes for $d = 3$. As for the remaining $2N_f^2$ currents, we note that in $d = 4 - 2\epsilon$ we can define the following antisymmetric tensor operators

$$(K_{\mu\nu})_a{}^b = \bar{\Psi}_a \gamma_{\mu\nu} \Psi^b \ , \ \bar{\Psi}_a \gamma_{\mu\nu} \gamma_5 \Psi^b \ . \tag{18}$$

They carry the correct flavor and Lorentz quantum numbers to be identified with the additional currents, because in $d = 3$ we can use the totally antisymmetric tensor $\epsilon_{\mu\nu\rho}$ and dualize them to spin-1 operators.

We are led to the expectation that the IR dimension of J_μ^s and $K_{\mu\nu}$ should evaluate to 2 for $\epsilon = 1/2$. The one-loop computation gives

$$\Delta_{IR}(J^s) = 3 - 2\epsilon$$
$$\Delta_{IR}(K) = 3 - 2\epsilon + \frac{3\epsilon}{2N_f} \ .$$

Note that the anomalous dimension of J^s only starts at two-loop order.[33,34] Plugging $N_f = 3$ and $\epsilon = 1/2$ in Eq. (19) we find $\Delta_{IR}(K) = 2.25$. This agrees to within 10% with the desired value, with precision improving for larger values of N_f. It is essential to go beyond one-loop to elucidate this matter. Indeed, preliminary higher-order computations show support for this claim.[31] Note that we can also study the anomalous dimension of the operator $F_{\mu\nu}$. The Bianchi identity is obeyed for all d, and one can verify that $\Delta_{IR}(F) = 2$ holds to all orders in ϵ.

5. Concluding remarks

We have presented here a proposal to study three-dimensional QED by using the ϵ-expansion. It was shown how it can be applied to obtain an approximation of the critical number N_f^c, which is consistent with known results. One naturally would like to improve the accuracy in the computation of the anomalous dimensions by going beyond one-loop. This requires using resummation methods and presents new challenges. We believe this merits further investigation.[‡]

[‡]The computation of F in QED_3 via the ϵ-expansion will be presented in Ref. 36.

Acknowledgments

We thank Ofer Aharony, Leon Balents, Jan de Boer, Simone Giombi, Igor Klebanov, Sung-Sik Lee, Prithvi Narayan, Yu Nakayama, Hugh Osborn, Silviu Pufu, Slava Rychkov, Subir Sachdev, Adam Schwimmer, Nathan Seiberg, Tarun Sharma, Philipp Strack and Grigory Tarnopolsky for useful discussions.

References

1. T. Appelquist, D. Nash, and L.C.R. Wijewardhana, *"Critical Behavior in (2+1)-Dimensional Qed,"* *Phys. Rev. Lett.* **60**, 2572 (1988).
2. T. Appelquist and L.C.R. Wijewardhana, *"Phase Structure of Non-compact QED3 and the Abelian Higgs Model,"* arXiv:hep-ph/0403250.
3. C. Vafa and E. Witten, *"Eigenvalue Inequalities for Fermions in Gauge Theories,"* *Commun. Math. Phys.* **95**, 257 (1984).
4. J.B. Marston and I. Affleck, *"Large-N Limit of the Hubbard-Heisenberg Model,"* *Phys. Rev.* **B39**, 11538 (1989).
5. Y. Ran, M. Hermele, P.A. Lee, and X.G. Wen, *"Projected-Wave-Function Study of the Spin-1/2 Heisenberg Model on the Kagom Lattice,"* *Phys. Rev. Lett.* **98**, no. 11, 117205 (2007).
6. M. Hermele, Y. Ran, P. Lee, and X.G. Wen, *"Properties of an algebraic spin liquid on the kagome lattice,"* *Phys. Rev.* **B77**, no. 22, 224413 (2008).
7. W. Rantner, and X.G. Wen, *"Electron Spectral Function and Algebraic Spin Liquid for the Normal State of Underdoped High T_c superconductors"*, *Phys. Rev. Lett.* **86**, 3871 (2001).
8. W. Rantner and X.G. Wen, *"Spin correlations in the algebraic spin liquid: Implications for high-Tc superconductors"*, *Phys. Rev.* **B66**, 144501 (2002).
9. M. Hermele, T. Senthil, and M.P.A. Fisher, *"Algebraic spin liquid as the mother of many competing orders,"* *Phys. Rev.* **B72**, 104404 (2005).
10. R.D. Pisarski, *"Chiral Symmetry Breaking in Three-Dimensional Electrodynamics,"* *Phys. Rev.* **D29**, 2423 (1984).
11. T.W. Appelquist, M.J. Bowick, D. Karabali and L.C.R. Wijewardhana, *"Spontaneous Chiral Symmetry Breaking in Three-Dimensional QED,"* *Phys. Rev.* **D33**, 3704 (1986).
12. J.A. Gracey, *"Electron Mass Anomalous Dimension at O(1/(Nf(2)) in Quantum Electrodynamics,"* *Phys. Lett.* **B317**, 415 (1993) arXiv:hep-th/9309092.

13. J.A. Gracey, *"Computation of Critical Exponent Eta at O(1/N_f ^{2}) in Quantum Electrodynamics in Arbitrary Dimensions,"* *Nucl. Phys.* **B414**, 614 (1994), arXiv:hep-th/9312055.
14. C. Xu, *"Renormalization group studies on four-fermion interaction instabilities on algebraic spin liquids,"* *Phys. Rev.* **B78**, 054432 (2008).
15. K.G. Wilson and M.E. Fisher, *"Critical exponents in 3.99 dimensions,"* *Phys. Rev. Lett.* **28**, 240 (1972).
16. G. 't Hooft and M.J.G. Veltman, *"Regularization and Renormalization of Gauge Fields,"* *Nucl. Phys.* **B44**, 189 (1972).
17. P. Breitenlohner and D. Maison, *"Dimensional Renormalization and the Action Principle,"* *Commun. Math. Phys.* **52**, 11 (1977).
18. D.A. Akyeampong and R. Delbourgo, *"Anomalies via dimensional regularization,"* *Nuovo Cim.* **A19**, 219 (1974).
19. P. Ponte and S.S. Lee, *"Emergence of supersymmetry on the surface of three dimensional topological insulators,"* *New J. Phys.* **16** 1, 013044 (2014), arXiv:1206.2340 [cond-mat.str-el].
20. D. Dalidovich and S.S. Lee, *"Perturbative non-Fermi liquids from dimensional regularization,"* *Phys. Rev.* **B88**, 245106 (2013), arXiv:1307.3170 [cond-mat.str-el].
21. A.A. Patel, P. Strack, and S. Sachdev, *"Hyperscaling at the Spin Density Wave Quantum Critical Point in Two Dimensional Metals,"* arXiv:1507.05962 [cond-mat.str-el].
22. M. Beneke and V.A. Smirnov, *"Ultraviolet Renormalons in Abelian Gauge Theories,"* *Nucl. Phys.* **B472**, 529 (1996), arXiv:hep-ph/9510437.
23. S.J. Hands, J.B. Kogut, and C.G. Strouthos, *"Noncompact QED(3) with N(f) greater than or equal to 2,"* *Nucl. Phys.* **B645**, 321 (2002), arXiv:hep-lat/0208030.
24. S.J. Hands, J.B. Kogut, L. Scorzato, and C.G. Strouthos, *"Noncompact QED(3) with N(f) = 1 and N(f) = 4,"* *Phys. Rev.* **V70**, 104501 (2004), arXiv:hep-lat/0404013.
25. C. Strouthos and J.B. Kogut, *"The Phases of Non-Compact QED(3),"* *PoS LAT* **2007**, 278 (2007), arXiv:0804.0300 [hep-lat].
26. T. Grover, *"Entanglement Monotonicity and the Stability of Gauge Theories in Three Spacetime Dimensions,"* *Phys. Rev. Lett.* **112**, no. 15, 151601 (2014), arXiv:1211.1392 [hep-th].
27. D. Gepner, *"Nonabelian Bosonization and Multiflavor QED and QCD in Two-dimensions,"* *Nucl. Phys.* **B252**, 481 (1985).
28. I. Affleck, *"On the Realization of Chiral Symmetry in (1+1)-dimensions,"* *Nucl. Phys.* **B265**, 448 (1986).

29. E. Witten, *"Nonabelian Bosonization in Two-Dimensions,"* *Commun. Math. Phys.* **92**, 455 (1984).
30. A.M. Polyakov, *"Supermagnets and sigma models,"* arXiv:hep-th/0512310.
31. L. Di Pietro, and E. Stamou, in progress.
32. J.C. Collins, *Renormalization*, Cambridge University Press, 1987.
33. J. Kodaira, *"QCD Higher Order Effects in Polarized Electroproduction: Flavor Singlet Coefficient Functions,"* *Nucl. Phys.* **B165**, 129 (1980).
34. S.A. Larin, *"The Renormalization of the Axial Anomaly in Dimensional Regularization,"* *Phys. Lett.* **B303**, 113 (1993), arXiv:hep-ph/9302240.
35. J.A. Gracey, *"Three Loop Ms-Bar Tensor Current Anomalous Dimension in QCD,"* *Phys. Lett.* **B488**, 175 (2000), arXiv:hep-ph/0007171.
36. S. Giombi, I.R. Klebanov, and G. Tarnopolsky, *"Conformal Qed_D, F-Theorem and the ϵ Expansion,"* arXiv:1508.06354 [hep-th].

BETA FUNCTIONS IN CHIRALLY DEFORMED SUPERSYMMETRIC SIGMA MODELS IN TWO DIMENSIONS

ARKADY VAINSHTEIN

Department of Physics, University of Minnesota, Minneapolis, MN 55455, USA
and
William I. Fine Theoretical Physics Institute, University of Minnesota, Minneapolis, MN 55455, USA

We study two-dimensional sigma models where the chiral deformation diminished the original $\mathcal{N}=(2,2)$ supersymmetry to the chiral one, $\mathcal{N}=(0,2)$. Such heterotic models were discovered previously on the world sheet of non-Abelian stringy solitons supported by certain four-dimensional $\mathcal{N}=1$ theories. We study geometric aspects and holomorphic properties of these models, and derive a number of exact expressions for the β functions in terms of the anomalous dimensions analogous to the NSVZ β function in four-dimensional Yang-Mills. Instanton calculus provides a straightforward method for the derivation.

1. Introduction

This presentation is based on Ref. 1, where I collaborated with Jin Chen, Xiaoyi Cui and Misha Shifman. Although so many years passed since Gribov's death, each time I do something new I am always thinking what would be B.H.'s opinion of this and how I would try to present it to him.

Chirally deformed $\mathcal{N}=(0,2)$ sigma models to be considered below emerged as low-energy world sheet theories on non-Abelian strings supported in some $\mathcal{N}=1$ four-dimensional Yang-Mills theories[2,3] (for a recent review see Ref. 4). Particularly, in this case the target space was CP$(N-1)$ but the heterotic modification could be considered for a wide class of the Kähler manifolds. The heterotic models, although remaining largely unexplored, appear to be important in various problems. Some previous results can be found in Refs. 5–9, for a general discussion of (0,2) models see Refs. 10–15. Here we report further results in the study of geometric structure, holomorphic anomaly and exact β functions in these models.

Heterotic $\mathcal{N}=(0,2)$ models have a rich mathematical structure. In perturbation theory two-dimensional $\mathcal{N}=(0,2)$ models were shown to share some features with $\mathcal{N}=1$ super-Yang-Mills theories in four dimensions (e.g., Ref. 7). They are asymptotically free in the ultraviolet (UV), have different phases in the infrared (IR), and admit large-N solution.[8,9] These facts can be interpreted within 4D–2D correspondence (e.g., Ref. 2, 3) and the Dijkraaf-Vafa type deformation. The same correspondence was noted in more general 4D–2D coupled systems, the study of which requires a thorough knowledge of the two-dimensional side. A number of insights were obtained from string theory, see Refs. 13, 14. However, considerations of (0,2) models in quantum field theory are scarce. In particular, beyond chiral operators, nothing was explored until quite recently.

Two-dimensional sigma models present a natural playground for geometric explorations. They encode the geometry of the target space, that of the worldsheet, and the geometry of various moduli spaces. Essentially everything is known for the undeformed $\mathcal{N}=(2,2)$ models. With $\mathcal{N}=(0,2)$ supersymmetry, one can test the robustness of the target space geometry — whether or not quantization provides us with some kind of geometrical deformation. In particular, the models we will consider have both, isometries and a global symmetry realized in a nontrivial way. The interplay between geometry and quantum effects could be enlightening on both sides.

To explain the nature of heterotic modifications let us start with reminding geometry of unmodified $\mathcal{N}=(2,2)$ sigma models. It was pointed out by Zumino[16] that the target space of these models should have the Kähler geometry. Moreover, to be characterized just by one coupling g, it should be a symmetric space which can be described as a homogeneous space G/H for a Lie group G and the stabilizer H. For the projective CP$(N-1)$ space $G=\mathrm{SU}(N)$ and $H=\mathrm{S}\,(\mathrm{U}(N-1)\times\mathrm{U}(1))$. It is a particular case of Grassmannian spaces $\mathrm{SU}(n+m)/\mathrm{SU}(n)\times\mathrm{SU}(m)\times\mathrm{U}(1)$ (see, e.g., Ref. 17 for a full list of the symmetric Kähler spaces).

In these homogeneous spaces the Ricci tensor $R_{i\bar{j}}$ is proportional to the metric $G_{i\bar{j}}$,

$$R_{i\bar{j}}=b\,\frac{g^2}{2}\,G_{i\bar{j}}\,. \tag{1}$$

This feature is a definition of the Einstein spaces. The constant b is equal to the dual Coxeter number T_G for the group G.,

$$b_{G/H}=T_G\,. \tag{2}$$

Correspondingly, for the CP$(N-1)$ space

$$b_{\mathrm{CP}(N-1)}=T_{\mathrm{SU}(N)}=N\,. \tag{3}$$

The same constant $b = T_G$ defines the β function

$$\beta(\mathcal{N}=(2,2)) = \mu \frac{dg^2}{d\mu} = -T_G \frac{g^4}{4\pi}, \tag{4}$$

which is exhausted by one loop in the (2,2) theories.[17,18]

To diminish the number of supercharges from 4 in $\mathcal{N} = (2,2)$ to 2 in $\mathcal{N}=(0,2)$ one needs to break partially a partnership between bosonic and fermionic fields. In the (2,2) case each bosonic field ϕ^i has two fermion partners, right- and left-movers, ψ^i_R and ψ^i_L. A simple way to diminish supersymmetry to (0,2) is just to discard all ψ^i_R. Such chiral models, which can be called minimal, generically suffer from internal diffeomorphism anomaly.[19] In particular, only CP(1) out of the entire CP($N-1$) series is anomaly free and presents a consistent minimal model.

Our study is focused on a different heterotic modification. Namely, instead of deleting right-moving fermions ψ^i_R, one extra right-mover ζ_R is added to the content of the (2,2) theory. A new coupling which mixes ζ_R and ψ^i_R in the background of bosonic field leads to breaking of the (2,2) supersymmetry to (0,2). In contrast to the minimal (0,2) models this heterotic coupling can be switched on perturbatively which is sufficient to show an absence of internal anomaly problem.

Thus, our task in the present paper is to analyse the heterotically deformed (nonminimal) $\mathcal{N}=(0,2)$ models. We present a more complete geometric formulation of the class of nonminimal models which will be studied in this paper. Holomorphic properties of such models are revealed. They have two coupling constants, the original g and the heterotic h. Correspondingly, there are two β functions. As in four-dimensional Yang-Mills,[20] we have to differentiate between the holomorphic coupling constants which are renormalized at most at one loop, and their nonholomorphic counterparts. The latter appear in conventional perturbation theory and are sometimes referred to as canonic.

We calculate both β functions in more than one way. In particular, we derive exact relations between the β functions and the anomalous dimensions γ, analogous to the NSVZ relations[21] in four-dimensional Yang-Mills using the instanton calculus. For instance, for β_g it will be shown that to all orders in perturbation theory

$$\beta_g = \mu \frac{dg^2}{d\mu} = -\frac{g^2}{4\pi} \frac{T_G\, g^2 \left(1 + \gamma_{\psi_R}/2\right) - h^2 \left(\gamma_{\psi_R} + \gamma_\zeta\right)}{1 - (h^2/4\pi)} \tag{5}$$

where γ_{ψ_R}, γ_ζ are the anomalous dimensions of the ψ_R, ζ_R fields.

In Ref. 1 we computed the anomalous dimensions up to two loops implying prediction for three loops in β_g . Our two-loop results for anomalous dimensions also confirm the fact that there exists a fixed point for the ratio $\rho \equiv h^2/g^2$. The critical value ρ_c depends on a manifold geometry and equals to 1/2 for CP($N-1$).[6] At this point three-loop β_g reduces to

$$\beta_g^{(3)} = -T_G \frac{g^4}{4\pi} \frac{1}{\left(1-(h^2/4\pi)\right)}, \tag{6}$$

i.e., to the one-loop expression up to the factor $1/(1-(h^2/4\pi))$.

2. Heterotic $\mathcal{N} = (0,2)$ models in 2D

2.1. *Formulation of the model*

We start to describe the model to be studied in this paper by introducing two types of chiral $\mathcal{N} = (0,2)$ superfields A and B. The first, bosonic superfield A describes a chiral supermultiplet which on mass shell consists of a complex bosonic field and a left-moving Weyl fermion,

$$A(x_R + 2i\theta^\dagger\theta, x_L, \theta) = \phi(x_R + 2i\theta^\dagger\theta,\, x_L) + \sqrt{2}\,\theta\,\psi_L(x_R + 2i\theta^\dagger\theta,\, x_L)\,. \tag{7}$$

Here $x_{R,L}$ are light-cone coordinates $x_{R,L} = t \pm x$ and θ is the one-component Grassmann variable corresponding to θ_R. The second, fermionic superfield B refers to the Fermi supermultiplet which on mass shell contains only the right-moving fermion (F is an auxiliary field),

$$B(x_R + 2i\theta^\dagger\theta, x_L, \theta) = \psi_R(x_R + 2i\theta^\dagger\theta,\, x_L) + \sqrt{2}\,\theta F(x_R + 2i\theta^\dagger\theta,\, x_L)\,. \tag{8}$$

Note that the $\mathcal{N} = (2,2)$ chiral field $\Phi(x_R + 2i\theta_R^\dagger\theta_R, x_L - 2i\theta_L^\dagger\theta_L, \theta_R, \theta_L)$ decomposes in the $\mathcal{N} = (0,2)$ superfields A and B as

$$\begin{aligned}&\Phi(x_R + 2i\theta_R^\dagger\theta_R, x_L - 2i\theta_L^\dagger\theta_L, \theta_R, \theta_L)\\ &= A(x_R + 2i\theta_R^\dagger\theta_R, x_L - 2i\theta_L^\dagger\theta_L, \theta_R) + \sqrt{2}\,\theta_L B(x_R + 2i\theta_R^\dagger\theta_R, x_L, \theta_R).\end{aligned} \tag{9}$$

The $\mathcal{N} = (0,2)$ supersymmetry transformations are as follows:

$$\begin{aligned}&\delta x_R = -2i\theta^\dagger\epsilon\,, \quad \delta x_L = 0\,, \quad \delta\theta = \epsilon\,, \quad \delta\theta^\dagger = \epsilon^\dagger\,,\\ &\delta\phi = \sqrt{2}\,\epsilon\,\psi_L\,, \qquad \delta\psi_L = -\sqrt{2}\,i\epsilon^\dagger\partial_L\phi\,,\\ &\delta\psi_R = \sqrt{2}\,\epsilon\,F\,, \qquad \delta F = -\sqrt{2}\,i\epsilon^\dagger\partial_L\psi_R\,,\end{aligned} \tag{10}$$

where $\partial_L = \partial_t + \partial_x = 2\,\partial_{x_R}$.

The undeformed $\mathcal{N}=(2,2)$ model in terms of the $\mathcal{N}=(0,2)$ superfields (7) and (8) contains equal number of bosonic A^i and fermionic B^i superfields. In the particular case of CP$(N-1)$ we have $i=1,2,...,N-1$.

The heterotic deformation to be considered below is induced by adding a singlet fermionic superfield $\mathcal{B}$,

$$\mathcal{B} = \zeta_R(x_R + 2i\theta^\dagger\theta,\, x_L) + \sqrt{2}\,\theta\mathcal{F}(x_R + 2i\theta^\dagger\theta,\, x_L)\,. \tag{11}$$

The Lagrangian can be written as

$$\begin{aligned}\mathcal{L} = &\frac{1}{4}\int d^2\theta \left[K_i(A,A^\dagger)\left(i\partial_R A^i - 2\kappa\,\mathcal{B}B^i\right) + \text{H.c.}\right] \\ &+\frac{1}{2}\int d^2\theta \left[Z\,G_{i\bar{j}}(A,A^\dagger)\,B^{\dagger\bar{j}}B^i + \mathcal{Z}\,\mathcal{B}^\dagger\mathcal{B}\right].\end{aligned} \tag{12}$$

Here K is the Kähler potential viewed as a function of the bosonic superfields. By definition

$$K_i(A,A^\dagger) \equiv \partial_{A^i}\,K(A,A^\dagger)\,. \tag{13}$$

Moreover, $G_{i\bar{j}}$ is the metric on the target space,

$$G_{i\bar{j}} = K_{i\bar{j}}(A,A^\dagger) \equiv \partial_{A^i}\partial_{A^{\dagger\bar{j}}}\,K(A,A^\dagger)\,. \tag{14}$$

Two Z factors (for the fields B^i and $\mathcal{B}$) are introduced in Eq. (12), in anticipation of their renormalization group (RG) evolution. In the CP$(N-1)$ model the Fubini-Study metric comes from the Kähler potential of the form

$$K(A,A^\dagger) = \frac{2}{g^2}\,\log\left(1 + \sum_i^{N-1} A^{\dagger\,i}\,A^i\right).$$

One can check that the above Lagrangian is target-space invariant. However, the target space invariance is implicit in Eq. (12) because $K_i(A,A^\dagger)$ is not explicitly target-space invariant. It becomes explicit upon passing to the integration over the Grassmann half-space in the first line of Eq. (12),

$$\begin{aligned}\mathcal{L} = &-\frac{1}{4}\int d\theta\, G_{i\bar{j}}(A,A^\dagger)(\bar{D}A^{\dagger\bar{j}})\left(i\partial_R A^i - 2\kappa\,\mathcal{B}B^i\right) + \text{H.c.} \\ &+\frac{1}{2}\int d^2\theta \left[Z\,G_{i\bar{j}}(A,A^\dagger)\,B^{\dagger\bar{j}}B^i + \mathcal{Z}\,\mathcal{B}^\dagger\mathcal{B}\right].\end{aligned} \tag{15}$$

The F-term structure $G_{i\bar{j}}(\bar{D}A^{\dagger\bar{j}})i\partial_R A^i$ in the first line is an analogue of that for the gauge term $W^\alpha W_\alpha$ in 4D gauge theories, while another F-structure,

$G_{i\bar{j}}(\bar{D}A^{\dagger\bar{j}})\,\mathcal{B}B^i$, is an analogue of superpotential in 4D. The chiral nature of these terms plays a crucial role in their renormalization.

The target space invariance is also transparent if one rewrites the Lagrangian in components,

$$\mathcal{L} = G_{i\bar{j}}\left[\partial_R\phi^{\dagger\bar{j}}\partial_L\phi^i + \psi_L^{\dagger\bar{j}}\, i\nabla_R\,\psi_L^i + Z\,\psi_R^{\dagger\bar{j}}\, i\nabla_L\psi_R^i\right] + Z\,R_{i\bar{j}k\bar{l}}\,\psi_L^{\dagger\bar{j}}\psi_L^i\,\psi_R^{\dagger\bar{l}}\psi_R^k$$

$$+\mathcal{Z}\,\zeta_R^\dagger\, i\partial_L\,\zeta_R + \left[\kappa\,\zeta_R\,G_{i\bar{j}}\big(i\,\partial_L\phi^{\dagger\bar{j}}\big)\psi_R^i + \text{H.c.}\right] + \frac{|\kappa|^2}{Z}\zeta_R^\dagger\,\zeta_R\big(G_{i\bar{j}}\,\psi_L^{\dagger\bar{j}}\psi_L^i\big) \tag{16}$$

$$-\,\frac{|\kappa|^2}{\mathcal{Z}}\big(G_{i\bar{j}}\psi_L^{\dagger\bar{j}}\psi_R^i\big)\big(G_{k\bar{l}}\psi_R^{\dagger\bar{l}}\psi_L^k\big)\,.$$

Here $\nabla_{L,R}$ are covariant derivatives, $\nabla_{L,R}\psi^i_{R,L} = \partial_{L,R}\psi^i_{R,L} + \Gamma^i_{kl}\partial_{L,R}\phi^k\,\psi^l_{R,L}$. The first line in this equation refers to the undeformed $\mathcal{N} = (2,2)$ theory, the subsequent terms bring in the (0,2) deformation.

2.2. *Geometry of heterotic deformation*

What is the geometrical meaning of the heterotic deformation? For the Kähler manifold M of the complex dimension d (for CP$(N{-}1)$ the dimension $d = N-1$) we have d right-moving fermions ψ_R^i, $i = 1,\ldots,d$, plus ζ_R. They can be viewed as defined on the tangent bundle $T\,(M\times C)$. Let us denote $\zeta_R = \psi_R^{d+1}$. Similarly, for superfields $\mathcal{B} = B^{d+1}$. Then, the Lagrangian for the right-moving fermions can be written as

$$\mathcal{L}_B = \frac{1}{2}\int d^2\theta\left\{G^{(B)}_{i\bar{j}}B^{\dagger\,\bar{j}}B^i + \left[T_{ik}B^iB^k + \text{H.c.}\right]\right\}, \quad (i,k,\bar{j} = 1,\ldots,d+1). \tag{17}$$

Here the metric $G^{(B)}_{i\bar{j}}$ and antisymmetric potential T_{ik} are functions of the bosonic superfields A^i, $A^{\dagger\bar{j}}$ with $i,\bar{j} = 1,\ldots,d$. Comparing with the previous definitions we see that nonvanishing components of $G^{(B)}_{i\bar{j}}$ and T_{ik} are

$$G^{(B)}_{i\bar{j}} = \begin{cases} ZG_{i\bar{j}}\,, & i,\bar{j} = 1,\ldots,d, \\ \mathcal{Z}\,, & i = d+1,\ \ \bar{j} = d+1\,, \end{cases} \tag{18}$$

$$T_{(d+1)\,i} = -T_{i\,(d+1)} = -\frac{\kappa}{2}\,K_i\,, \quad i = 1,\ldots,d\,. \tag{19}$$

The potential T_{ik} is not uniquely defined but its curvature

$$\mathcal{H}_{ik\bar{j}} = T_{ik,\bar{j}} = \frac{\partial T_{ik}}{\partial A^{\dagger\,\bar{j}}} \tag{20}$$

is a good object. This curvature defines the chiral form for the heterotic modification,

$$\mathcal{L}_B = \frac{1}{2}\int d^2\theta\, G^{(B)}_{i\bar{j}}\, B^{\dagger\,\bar{j}} B^i - \frac{1}{2}\int d\theta \Big[\mathcal{H}_{ik\bar{j}}(\bar{D}A^{\dagger\bar{j}})B^i B^k + \text{H.c.}\Big], \tag{21}$$

In the model at hand the nonvanishing components of $\mathcal{H}_{ik\bar{j}}$

$$\mathcal{H}_{(d+1)\,i\bar{j}} = -\mathcal{H}_{i\,(d+1)\,\bar{j}} = -\frac{\kappa}{2}\, G_{i\bar{j}} \qquad (i,\bar{j} = 1,\ldots,d) \tag{22}$$

are expressed via the metric tensor $G_{i\bar{j}}$. It looks even simpler for $\mathcal{H}^j_{ik} = \mathcal{H}_{ik\bar{j}}G^{j\bar{j}}$:

$$\mathcal{H}^j_{(d+1)\,i} = -\mathcal{H}^j_{i\,(d+1)} = -\frac{\kappa}{2}\,\delta^j_i \qquad (i,j = 1,\ldots,d)\,. \tag{23}$$

In differential geometry the heterotic construction presented above can be described as an action of the C^* one-dimensional algebra on the odd tangent vector bundle.* Note that there is a diagonal U(1) which rotates B^i and $\mathcal{B}$ in the opposite directions.

The consideration above is applicable to modifying any Kähler manifold, we are not limited to CP(N–1). What is less clear, whether or not it is possible to add more right-moving fermions so that the extra fermionic bundle is not just TC. To this end a three-form $\mathcal{H}$ consistent with the target-space invariance should exist. We are not aware of such generalizations.

2.3. *Holomorphy and its breaking*

The deformed (0,2) theory contains four bare parameters:

$$\frac{1}{g^2}\,,\ \frac{\kappa}{g^2}\,,\ Z\,,\ \mathcal{Z}\,. \tag{24}$$

The first two, $1/g^2$ and κ/g^2, enter as coefficients of the F terms in Eq. (15) and can be taken to be complex, while parameters Z and $\mathcal{Z}$ should be real. The imaginary part of $1/g^2$ defines the vacuum θ angle, $\text{Im}(1/g^2) = \theta/4\pi$, while the phase of (κ/g^2) produces an addition to this θ angle. These angles do not contribute to physical effects due to the presence of massless fermionic fields whose phase can be redefined.

Nonrenormalization of superpotential (i.e. the κ term in Eq. (15)) implies the absence of loop corrections to the holomorphic coupling κ/g^2, and, in particular, the absence of its running,

$$M_{\text{uv}}\frac{d}{dM_{\text{uv}}}\,\frac{\kappa}{g^2} = 0\,. \tag{25}$$

*We are indebted to Alexander Voronov for explanations.

This means that the curvature $\mathcal{H}_{ik\bar{j}}$ is the renormalization group invariant tensor with no higher-loop corrections. For a detailed derivation of the nonrenormalization theorem see Ref. 5.

The situation is more complicated for the "main" coupling $1/g^2$ appearing in the target space metric. The coupling in the bare Lagrangian (i.e. at $M_{\rm uv}$) is holomorphic. This means that it can receive only one-loop renormalization, implying one-loop β function. Much in the same way as in four-dimensional $\mathcal{N}=1$ Yang-Mills theory the holomorphic anomaly showing up in loops defies this theorem. The coupling constant the running of which is calculated in conventional perturbation theory is nonholomorphic. For the time being let us denote it by square brackets $1/[g^2]$, as in Ref. 20. In $\mathcal{N} = (2,2)$ sigma models there is no holomorphy violation, and $1/g^2$ and $1/[g^2]$ coincide.

In other words, in the undeformed (2,2) theory, i.e. at $\kappa = 0$, the holomorphicity of $1/g^2$ is maintained. It implies that only one-loop running of $1/g^2$ is allowed, higher loops are absent. In the 4D case this phenomenon is also known in $\mathcal{N}=2$ gauge theories. With less supersymmetry, i.e. in $\mathcal{N}=1$ gauge theories in 4D, holomorphicity is broken. It happens usually at two-loop level but in certain cases appears already at the level of the first loop, see Refs. 22, 23. Likewise, our κ term leads to breaking of holomorphicity for $1/g^2$. This happens at the level of the first loop. More specifically, the first loop provides a finite $|\kappa|^2$ correction to $1/g^2$ which then leads to the nonholomorphic running of g^2 in the second loop.

Iteration in κ involves integration over the quantum ζ_R and ψ_R fields in the form of a polarization operator, see Fig. 1. The polarization operator

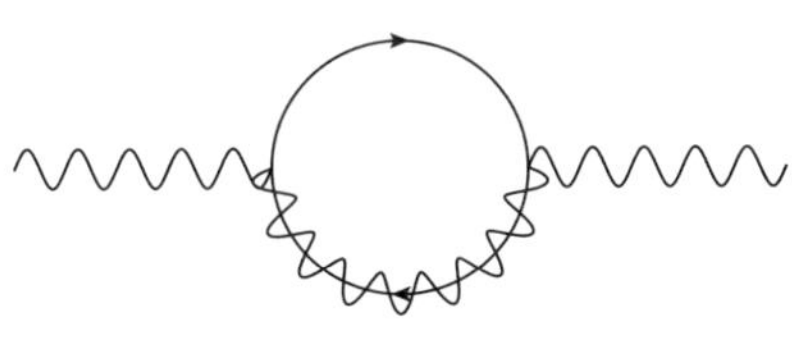

Fig. 1. One-loop finite correction to the canonical coupling g. The wave line denotes the background field A. The solid line denotes the propagator of B, while the solid line with a wavy line superposed denotes that of $\mathcal{B}$. We shall follow the same notation throughout this paper.

Π_{RR} is defined as

$$\Pi^{i\bar{j}}_{RR}(x,y) = i|\kappa|^2\Big\langle T\left\{\zeta_R(x)\psi^i_R(x)\,\psi^{\dagger\bar{j}}_R(y)\zeta^\dagger_R(y)\right\}\Big\rangle_{\rm bck} = i|\kappa|^2 S_\zeta\, S^{i\bar{j}}_{\psi_R}\,, \quad (26)$$

where S_ζ and S_{ψ_R} are the propagators of ζ and ψ_R fields in the background of the bosonic field ϕ. We give here the result for Π_{RR} ,

$$\Pi^{i\bar{j}}_{RR}(x,y) = -\frac{|\kappa|^2}{4\pi Z\mathcal{Z}}\,\langle x|\,G^{i\bar{j}}\,\nabla_R\,\frac{1}{\nabla_L\nabla_R}\,\nabla_R\,|y\rangle\ . \tag{27}$$

Let us emphasize that there is no ambiguity in the chiral fermion loop for Π_{RR} due to its nonzero Lorentz spin. Generally speaking, the polarization operator $\Pi_{\mu\nu}$ can contain local terms such as $g_{\mu\nu}$ or $\epsilon_{\mu\nu}$. These terms have zero Lorentz spin and do not contribute to Π_{RR}. Note also that in Eq. (27) the ordering of operators is not important because we neglect by commutator $[\nabla_R,\nabla_L]^i_k = R_k{}^i{}_{m\bar{n}}\big(\partial_R\phi^{\dagger\bar{n}}\partial_L\phi^m - \partial_L\phi^{\dagger\bar{n}}\partial_R\phi^m\big)$. Additional terms with this commutator are infrared ones and do not contribute to the running of couplings we are after.

The fermion loop of Fig. 1 then results in the following addition to the action,

$$\begin{aligned}\Delta_\kappa S &= \int d^2x d^2y\, G_{i\bar{k}}\partial_L\phi^{\dagger\bar{k}}(x)\,\Pi^{i\bar{j}}i(x,y)\,\partial_L\phi^l(y)G_{l\bar{j}}\\ &= -\frac{|\kappa|^2}{4\pi Z\mathcal{Z}}\int d^2x\, G_{i\bar{j}}\partial_L\phi^{\dagger\bar{j}}\partial_R\phi^i\ .\end{aligned} \tag{28}$$

Here we used the relation $\nabla_R\,\partial_L\phi = \nabla_L\,\partial_R\phi$, which makes the expression for the heterotic correction $\Delta_\kappa\mathcal{L}$ to the original bosonic Lagrangian local,

$$\Delta_\kappa\mathcal{L} = -\frac{|\kappa|^2}{4\pi Z\mathcal{Z}}\,G_{i\bar{j}}\partial_R\phi^{\dagger\bar{j}}\partial_L\phi^i\ . \tag{29}$$

The resulting correction to the metric $\Delta G_{i\bar{j}}$ can be rewritten in a more geometrical form in terms of the curvature $\mathcal{H}_{ik\bar{j}}$,

$$\Delta_\kappa G_{i\bar{j}} = -\frac{|\kappa|^2}{4\pi Z\mathcal{Z}}\,G_{i\bar{j}} = -\frac{1}{2\pi}\,\mathcal{H}_{lk\bar{j}}\bar{\mathcal{H}}_{\bar{l}\bar{k}i}G^{(B)l\bar{l}}G^{(B)k\bar{k}} = -\frac{1}{2\pi}\,\mathcal{H}_{lk\bar{j}}\bar{\mathcal{H}}_i^{\;lk}\,, \tag{30}$$

where the indices in $\bar{\mathcal{H}} = \mathcal{H}^*$ are raised by the $G^{(B)i\bar{j}}$ metric tensor (inverse to $G^{(B)}_{i\bar{j}}$ defined in Eq. (18)).

Equation (29) clearly demonstrates the breaking of holomorphicity by the fermion loop depicted in Fig. 1. This loop is also related to the axial anomaly in the fermionic current. Correspondingly we claim the absence of higher-loop corrections to Eq. (29).

From the above considerations we see that perturbation theory is governed by two real couplings, $[g^2]$ and a real nonholomorphic combination

$$h^2 = \frac{|\kappa|^2}{Z\mathcal{Z}}\,. \tag{31}$$

We will also use the ratio ρ of the couplings,

$$\rho \equiv \frac{h^2}{g^2} \,. \tag{32}$$

Then Eq. (29) implies

$$\frac{1}{g^2} - \frac{1}{4\pi}\,[\rho] = \frac{1}{[g^2]} \,. \tag{33}$$

It is convenient to rewrite Eq. (33) as

$$\frac{1}{g^2} = \frac{1}{[g^2]} + \frac{1}{4\pi}[\rho] \,, \tag{34}$$

where the holomorphic coupling (i.e. renormalized only at one loop) on the left-hand side is presented as a combination of two nonholomorphic terms.

3. Beta functions

3.1. *Generalities*

Considering two, introduced above, couplings, $[g^2]$ and h^2 as functions of the UV cut-off $M_{\rm uv}$ we define two β functions:

$$\beta_g \equiv \frac{d\,[g^2](M_{\rm uv})}{dL} \,, \qquad \beta_h \equiv \frac{d\,h^2(M_{\rm uv})}{dL} \,, \qquad L = \log M_{\rm uv} \,. \tag{35}$$

In what follows we will use also the β function for ρ, see Eq. (32),

$$\beta_\rho \equiv \frac{d\,[\rho](M_{\rm uv})}{dL} \,. \tag{36}$$

We will omit below the square brackets in $[g^2]$ dealing with the 1PI definition of couplings.

As was discussed above nonrenormalization of the superpotential in Eq. (15) implies that the ratio κ/g^2 does not depend on $M_{\rm uv}$, see Eq. (25). This equation can be rewritten as

$$\frac{d}{dL}\,\frac{|\kappa|^2}{g^4} = \frac{d}{dL}\,\frac{h^2 Z\mathcal{Z}}{g^4} = \frac{h^2 Z\mathcal{Z}}{g^4}\left[\frac{\beta_h}{h^2} - 2\,\frac{\beta_g}{g^2} - \gamma\right] = 0\,, \tag{37}$$

where the anomalous dimension γ is defined as

$$\gamma \equiv -\frac{d\,\log\,(Z\mathcal{Z})}{dL} = \gamma_{\psi_R} + \gamma_\zeta\,, \quad \gamma_{\psi_R} \equiv -\frac{d\,\log\,Z}{dL}\,, \quad \gamma_\zeta \equiv -\frac{d\,\log\,\mathcal{Z}}{dL}\,. \tag{38}$$

This fixes the β function for h^2 in terms of the β function for g^2 and the sum of the anomalous dimensions for B^i and $\mathcal{B}$ fields,

$$\beta_h = h^2\left[\frac{2}{g^2}\,\beta_g + \gamma\right] . \tag{39}$$

For β_ρ we get

$$\beta_\rho = \rho\left[\frac{1}{g^2}\,\beta_g + \gamma\right]. \tag{40}$$

3.2. *Beta functions at one loop*

The relations (39) and (40) are exact to all loops. At the one-loop level all β functions and anomalous dimensions have been calculated earlier:[5,6]

$$\beta_g^{(1)} = -T_G\,\frac{g^4}{4\pi}\,,\quad \gamma_\zeta^{(1)} = d\,\frac{h^2}{2\pi}\,,\quad \gamma_{\psi_R}^{(1)} = \frac{h^2}{2\pi}\,,\quad \gamma^{(1)} = (d+1)\,\frac{h^2}{2\pi}\,, \tag{41}$$

$$\beta_h^{(1)} = -\frac{h^2}{2\pi}\left[T_G g^2 - (d+1)h^2\right],\quad \beta_\rho^{(1)} = (d+1)\,\frac{h^2}{2\pi}\left[\rho - \frac{T_G}{2(d+1)}\right]. \tag{42}$$

The results are for the Kähler manifolds of the complex dimension d, which are homogeneous spaces G/H, and T_G is a dual Coxeter number of the group G. It is a straightforward generalization of calculations of Refs. 5, 6, where the CP($N-1$) sigma model was considered, for which $d = N-1$ and $T_G = T_{\mathrm{SU}(N)} = N$.

An interesting feature of these one-loop results is that they exhibit a fixed point at $\rho = \rho_c = T_G/2(d+1)$, which becomes $\rho_c = 1/2$ for CP($N-1$). At this point

$$\frac{\beta_g^{(1)}}{g^2}\bigg|_{\rho=\rho_c} = \frac{\beta_h^{(1)}}{h^2}\bigg|_{\rho=\rho_c} = -\gamma^{(1)}\Big|_{\rho=\rho_c} = -\frac{T_G g^2}{4\pi}\,,\qquad \rho_c = \frac{T_G}{2(d+1)}\,. \tag{43}$$

In terms of the geometrical interpretation we can present all one-loop results as corrections to the bosonic metric $G_{i\bar{j}}$, and to the right-moving fermion extended metric $G_{i\bar{j}}^{(B)}$. These one-loop corrections are

$$\begin{aligned}
\Delta G_{i\bar{j}}\Big|^{\text{one-loop}} &= -\frac{1}{2\pi}\left\{\mathcal{H}_{lk\bar{j}}\,\bar{\mathcal{H}}_{\bar{l}\bar{k}i}\,G^{(B)l\bar{l}}G^{(B)k\bar{k}} + R_{i\bar{j}}\log M_{\text{uv}}\right\},\\
\Delta G_{i\bar{j}}^{(B)}\Big|^{\text{one-loop}} &= -\frac{\log M_{\text{uv}}}{2\pi}\left\{4\mathcal{H}_{ik\bar{l}}\,\bar{\mathcal{H}}_{\bar{j}\bar{k}l}\,G^{(B)k\bar{k}}G^{(B)l\bar{l}} + R_{i\bar{j}}^{(B)}\right\}.
\end{aligned} \tag{44}$$

As mentioned above there are no loop corrections of any order to the heterotic curvature tensor $\mathcal{H}_{ik\bar{j}}$.

4. Exact beta functions from instanton calculus

In this section we will briefly outline the instanton derivation of the β functions along the lines of Refs. 5, 7, 20, 21. In particular, we will use the nonrenormalization theorem for the second and higher loops in the instanton background. Only zero modes and the one-loop contribution have to be considered. All parameters that will appear in the derivation below are those from the bare Lagrangian.

To warm up let us briefly review the instanton calculation in Ref. 21. Consider the instanton measure in four-dimensional $\mathcal{N}=1$ Yang-Mills theory with the SU(N) gauge group. In the quasiclassical approximation the renormalization group invariant (RGI) prefactor in the instanton measure is

$$\mu_{\rm inst}^{(1)} = M_{\rm uv}^{3N} \exp\left(-\frac{8\pi^2}{g^2}\right) . \tag{45}$$

The factor in the exponent is the classical instanton action, while the pre-exponential factor comes from the zero modes. There are $n_B = 4N$ bosonic zero modes and $n_F = 2N$ fermion ones to produce $M^{n_B}/M^{n_F/2} = M^{3N}$. In terms of perturbation theory Eq. (45) gives us the one-loop running of the holomorphic coupling. What happens in higher loops?

The only change (to all orders in the coupling constant) is the emergence of another pre-exponential factor $g^{n_F}/g^{n_B} = 1/g^{2N}$, due to normalization of the zero modes, namely,

$$\mu_{\rm inst}^{\rm exact} = M_{\rm uv}^{3N} \frac{1}{[g^2]^N} \exp\left(-\frac{8\pi^2}{[g^2]}\right) . \tag{46}$$

Simultaneously, $1/[g^2]$ in the exponent becomes nonholomorphic. The combination (46) of $M_{\rm uv}$ and $[g^2](M_{\rm uv})$ is renormalization group invariant. Differentiating over $\log M_{\rm uv}$ we arrive at the NSVZ β function for SU(N). Generalization to an arbitrary gauge group G is just a substitution of N in expressions above by the dual Coxeter number T_G.

If matter fields are added, the only further changes in $\mu_{\rm inst}^{\rm exact}$ are as follows: (i) the power of $M_{\rm uv}$ is changed appropriately; (ii) $Z^{-1/2}$ factor appears in the pre-exponent for each matter-sector fermion zero mode. The number of such fermion zero modes is given by $2T(R)$ where $T(R)$ is the Dynkin index of representation R. In this way one obtains the full exact NSVZ β function,

$$\beta_{\rm NSVZ}(g^2) = -\frac{g^4}{8\pi^2}\left[3\,T_G - \sum_{\rm matter} T(R_i)(1-\gamma_i)\right]\left(1-\frac{T_G\, g^2}{8\pi^2}\right)^{-1} . \tag{47}$$

Now, let us see how the same strategy can be implemented in the κ deformed CP(N–1) sigma model under consideration. Let us start first with the non-deformed (2,2) case. At the classical level the instanton-generated exponent is

$$\exp\left(-\frac{4\pi}{g^2}\right), \tag{48}$$

see, e.g., Ref. 24. At the one-loop level (and at higher loops as well) nonzero modes cancel out. In the CP($N-1$) model there are $n_B = 2N$ bosonic zero modes and $n_F = 2N$ fermion zero modes, This produces the $M^{n_B}/M^{n_F/2} = M^N$ pre-exponential factor. As for normalization of the zero modes the corresponding factors cancel out between bosonic and fermion modes, $g^{n_B}/g^{n_F} = 1$. Thus, we come to

$$\mu_{\text{inst}}(2,2) = M^N \exp\left(-\frac{4\pi}{g^2}\right) = \text{RGI}, \tag{49}$$

which leads to the one-loop exact β function and unbroken holomorphicity in the (2,2) theory.

Now let us switch on the κ modification. As we discussed in Sec. 2.3 the holomorphicity is broken in perturbation theory already in the order $|\kappa|^2$. Here comes a surprise: such breaking does not occur in the instanton background!

Indeed, the κ terms in the Lagrangian (16) has the form

$$i\kappa\, G_{i\bar{j}}\, \zeta_R\, \psi_R^i\, \partial_L \phi^{\dagger\bar{j}} - i\kappa^* G_{i\bar{j}}\, \psi_R^{\bar{j}}\, \zeta_R\, \partial_L \phi^i\,; \tag{50}$$

the product of these two terms enters in the fermion loop calculation. After Euclidean continuation the instanton (or anti-instanton) background leads to vanishing either $\partial_L \phi^{\dagger\bar{j}}$ or $\partial_L \phi^i$. Therefore, the $|\kappa|^2$ iteration is not possible. It means that holomorphicity is not broken in one loop for the instanton, and the instanton action stays $4\pi/g^2$ with the original $1/g^2$. In terms of the running $1/[g^2]$ and $[\rho]$ it means that the combination (34) enters into the instanton exponent,

$$\exp\left(-\frac{4\pi}{g^2}\right) = \exp\left(-\frac{4\pi}{[g^2]} - [\rho]\right). \tag{51}$$

Moreover, in the instanton background the only effect of an additional right-mover ζ_R is its admixture to ψ_R. This triangle mixing does not change the eigenvalues so the cancellation of non-zero modes stays the same as in the (2,2) case. Also, the counting of the zero modes does not change. What appears in the pre-exponential factor is an additional $Z^{-1/2}$ factor for each zero mode of ψ_R because of its normalization.[25] There are N such modes so we arrive at the following exact expression for the measure:

$$\mu_{\text{inst}}^{\text{exact}} = \frac{1}{Z^{N/2}}\, M_{\text{uv}}^N \exp\left(-\frac{4\pi}{[g^2]} - [\rho]\right). \tag{52}$$

The Z factor is defined in Eq. (15), see the first term in the second line. All effects due to two loops and higher, associated with nonzero modes, cancel[5,7] much in the same way as in the (2,2) case.

Equation (52) implies

$$T_G \log M_{\text{uv}} - \frac{T_G}{2} \log Z - \frac{4\pi}{[g^2]} - [\rho] = \text{RGI}\,, \tag{53}$$

where we substitute N by the Coxeter index T_G to consider a generic Kähler manifold G/H.

Differentiating over $\log M_{\text{uv}}$ and using Eqs. (38) and (40) for β_ρ we arrive at the full exact β function (5), relating β_g to the anomalous dimensions, much in the same way as the NSVZ formula.

Combining the full β function in (5) with Eqs. (39) and (40) we derive the "secondary" β functions,

$$\begin{aligned} \beta_h &= -\frac{h^2}{1-(h^2/4\pi)}\left[T_G\,\frac{g^2}{2\pi}\left(1+\frac{1}{2}\gamma_{\psi_R}\right) - \gamma\left(1+\frac{h^2}{4\pi}\right)\right], \\ \beta_\rho &= -\frac{\rho}{1-(h^2/4\pi)}\left[T_G\,\frac{g^2}{4\pi}\left(1+\frac{1}{2}\gamma_{\psi_R}\right) - \gamma\right]. \end{aligned} \tag{54}$$

5. Conclusion

In this paper, we analysed various quantum effects in the $\mathcal{N} = (0,2)$ deformed (2,2) two-dimensional sigma models. The target spaces we consider generalize CP(N–1) to the Kähler manifolds which are homogeneous spaces G/H. The $\mathcal{N}=(0,2)$ deformation studied in this paper goes under the name of the *nonminimal* models.

We showed that quantum effects will not deform geometry. Note that this is in drastic contradistinction with the chiral models in the minimal case, see Ref. 26. Another phenomenon closely related to isometries is integrability. One could ask whether the Lax relation holds, which would make these classes of models integrable, as their undeformed cousins. We do not have an answer to this question.

The models we studied are characterized by two independent coupling constants. We analysed them in perturbation theory. A crucial role belongs to the graph depicted in Fig. 1 which is associated with the anomaly in the current that mixes the right-moving fermions. It is also anomalous in the sense that it produces a nonholomorphic contribution proportional to $|\kappa|^2$ to the renormalization of $1/g^2$ at one loop. This effect then penetrates into higher orders.

Using nonrenormalization theorems,[5] analogous to those in four-dimensional Yang-Mills,[20,21] we derive a number of (perturbatively) exact relations between the β functions and the anomalous dimensions of the fields B and $\mathcal{B}$. These relations are verified in Ref. 1 by explicit two-loop calculations. It means that the certain β functions are known up to three loops.

In Ref. 1 we also study the relation between the perturbative β functions and the general analysis of Dumitrescu and Seiberg[27] of supermultiplets of currents. We find how the general structure of Ref. 27 is implemented in the nonminimal models under consideration and demonstrated that the analysis provides an alternative way for the β-function calculation.

The $\mathcal{N} = (0,2)$ models attracted attention in connection with developments in the studies of surface operators in four dimensions.[28] In the nonminimal models the protected quantities, i.e. the chiral ring, is preserved under the (0,2) deformation. It is interesting to pursue the calculation beyond the chiral sector exploring nonchiral sector of the world-sheet theory as a part of a 4D–2D coupled system. We hope that the results presented here can enlighten the very first step in pursuing such a goal.

Two-dimensional asymptotically free sigma models are long known to be excellent laboratories for modelling four-dimensional Yang-Mills theories. It was forty years ago that A. Polyakov emphasized[29] that asymptotically free two-dimensional sigma models could be the best laboratory for the four-dimensional Yang-Mills theories. His anticipation seems to be materializing. The nonminimal (0,2) sigma models we discuss present a close parallel to $\mathcal{N}=1$ super-Yang-Mills with matter in four dimensions.[7]

References

1. J. Chen, X. Cui, M. Shifman, and A. Vainshtein, *$\mathcal{N} = (0,2)$ deformation of (2, 2) sigma models: Geometric structure, holomorphic anomaly, and exact β functions*, *Phys. Rev.* **D90**, no. 4, 045014 (2014), arXiv:1404.4689 [hep-th].
2. M. Edalati and D. Tong, *Heterotic Vortex Strings*, *JHEP* **0705**, 005 (2007), arXiv:hep-th/0703045.
3. M. Shifman and A. Yung, *Heterotic Flux Tubes in $\mathcal{N}=2$ SQCD with $\mathcal{N}=1$ Preserving Deformations*, *Phys. Rev.* **D77**, 125016 (2008) [Erratum *ibid.* **D79**, 049901 (2009)], arXiv:0803.0158 [hep-th].
4. M. Shifman and A. Yung, *Two-Dimensional Sigma Models Related to Non-Abelian Strings in Super-Yang-Mills,* arXiv:1401.7067 [hep-th].

5. X. Cui and M. Shifman, $\mathcal{N}=(0,2)$ *Supersymmetry and a Nonrenormalization Theorem, Phys. Rev.* **D84**, 105016 (2011), arXiv:1105.5107 [hep-th].
6. X. Cui and M. Shifman, *Perturbative Aspects of Heterotically Deformed CP($N-1$) Sigma Model. I, Phys. Rev.* **D82**, 105022 (2010), arXiv:1009.4421 [hep-th].
7. X. Cui and M. Shifman, $\mathcal{N}$=$(0,2)$ *Deformation of CP(1) Model: Two-dimensional Analog of* $\mathcal{N}=1$ *Yang-Mills Theory in Four Dimensions, Phys. Rev.* **D85**, 045004 (2012), arXiv:1111.6350 [hep-th].
8. M. Shifman and A. Yung, *Large-N Solution of the Heterotic* $\mathcal{N}=(0,2)$ *Two-Dimensional CP($N-1$) Model, Phys. Rev.* **D77**, 125017 (2008) [Erratum *ibid.* **D81**, 089906 (2010)], arXiv:0803.0698 [hep-th].
9. P.A. Bolokhov, M. Shifman, and A. Yung, *Large-N Solution of the Heterotic CP(N-1) Model with Twisted Masses, Phys. Rev.* **D82**, 025011 (2010), arXiv:1001.1757 [hep-th].
10. E. Witten, *Two-dimensional models with (0,2) supersymmetry: Perturbative aspects, Adv. Theor. Math. Phys.* **11** (2007), arXiv:hep-th/0504078.
11. M.-C. Tan and J. Yagi, *Chiral Algebras of (0,2) Sigma Models: Beyond Perturbation Theory, Lett. Math. Phys.* **84**, 257 (2008), arXiv:0801.4782 [hep-th]; arXiv:0805.1410 [hep-th].
12. J. Yagi, *Chiral algebras of (0, 2) models, Adv. Theor. Math. Phys.* **16**, 1 (2012), arXiv:1001.0118 [hep-th].
13. A. Adams, A. Basu, and S. Sethi, *(0,2) duality, Adv. Theor. Math. Phys.* **7**, 865 (2004), arXiv:hep-th/0309226.
14. I. Melnikov, S. Sethi, and E. Sharpe, *Recent Developments in (0,2) Mirror Symmetry, SIGMA* **8**, 068 (2012), arXiv:1209.1134 [hep-th].
15. B. Jia, E. Sharpe, and R. Wu, *Notes on non-Abelian (0,2) theories and dualities,* arXiv:1401.1511 [hep-th].
16. B. Zumino, *Supersymmetry and Kahler Manifolds, Phys. Lett.* **B87**, 203 (1979).
17. A.Y. Morozov, A.M. Perelomov, and M.A. Shifman, *Exact Gell-mann-low Function Of Supersymmetric Kahler Sigma Models, Nucl. Phys.* **B248**, 279 (1984).
18. V.A. Novikov, M.A. Shifman, A.I. Vainshtein, and V.I. Zakharov, *Instantons And Exact Gell-mann-low Function Of Supersymmetric O(3) Sigma Model, Phys. Lett.* **B139**, 389 (1984).
19. G.W. Moore and P.C. Nelson, *The Etiology of σ Model Anomalies, Commun. Math. Phys.* **100**, 83 (1985).

20. M.A. Shifman and A.I. Vainshtein, *Solution of the Anomaly Puzzle in SUSY Gauge Theories and the Wilson Operator Expansion, Nucl. Phys.* **B277**, 456 (1986).
21. V.A. Novikov, M.A. Shifman, A.I. Vainshtein, and V.I. Zakharov, *Exact Gell-Mann-Low Function of Supersymmetric Yang-Mills Theories from Instanton Calculus, Nucl. Phys.* **B229**, 381 (1983);
Beta Function in Supersymmetric Gauge Theories: Instantons Versus Traditional Approach, Phys. Lett. **B166**, 329 (1986);
for a comprehensive review see M.A. Shifman and A.I. Vainshtein, *Instantons versus supersymmetry: Fifteen years later,* In M. Shifman, *ITEP Lectures on Particle Physics and Field Theory*, World Scientific, Singapore, 1999, Vol. 2, p. 485, arXiv:hep-th/9902018.
22. L.J. Dixon, V.Kaplunovsky, and J. Louis, *Moduli dependence of string loop corrections to gauge coupling constants, Nucl. Phys.* **B355**, 649 (1991).
23. M.A. Shifman and A.I. Vainshtein, *On holomorphic dependence and infrared effects in supersymmetric gauge theories, Nucl. Phys.* **B359**, 571 (1991).
24. V.A. Novikov, M.A. Shifman, A.I. Vainshtein, and V.I. Zakharov, *Two-Dimensional Sigma Models: Modeling Nonperturbative Effects of Quantum Chromodynamics, Phys. Rept.* **116**, 103 (1984).
25. J. Chen, *Instanton Fermionic Zero Modes of Heterotic CP(1) Sigma Model,* arXiv:0912.0571 [hep-th].
26. J. Chen, X. Cui, M. Shifman, and A. Vainshtein, *On Isometry Anomalies in Minimal N=(0,1) and N=(0,2) Sigma Models,* arXiv:1510.04324 [hep-th].
27. T.T. Dumitrescu and N. Seiberg, *Supercurrents and Brane Currents in Diverse Dimensions, JHEP* **1107**, 095 (2011), arXiv:1106.0031 [hep-th].
28. D. Gaiotto, S. Gukov, and N. Seiberg, *Surface Defects and Resolvents, JHEP* **1309**, 070 (2013), arXiv:1307.2578 [hep-th].
29. A.M. Polyakov, *Interaction of Goldstone Particles in Two-Dimensions. Applications to Ferromagnets and Massive Yang-Mills Fields, Phys. Lett.* **B59**, 79 (1975).

A NATURAL EXTENSION OF THE CONFORMAL LORENTZ GROUP IN A FIELD THEORY CONTEXT

ANDRÁS LÁSZLÓ

Wigner Research Centre for Physics of the Hungarian Academy of Sciences, RMI, Department of High Energy Physics
Budapest 1121, P.O.Box 49, H-1525, Hungary
E-mail: laszlo.andras@wigner.mta.hu

In this paper a finite dimensional unital associative algebra is presented, and its group of algebra automorphisms is detailed. The studied algebra can physically be understood as the creation operator algebra in a formal quantum field theory at fixed momentum for a spin 1/2 particle along with its antiparticle. It is shown that the essential part of the corresponding automorphism group can naturally be related to the conformal Lorentz group. In addition, the non-semisimple part of the automorphism group can be understood as "dressing" of the pure one-particle states. The studied mathematical structure may help in constructing quantum field theories in a non-perturbative manner. In addition, it provides a simple example of circumventing Coleman–Mandula theorem using non-semisimple groups, without SUSY.

Keywords: Algebra automorphism; Levi decomposition; conformal Lorentz group extension; quantum field theory

1. Introduction

It is a well known fact in quantum field theory (QFT) that any kind of unification of internal (gauge) symmetries with spacetime symmetries is not evident at all. The celebrated Coleman–Mandula theorem and its various versions[1] prohibit the most simple unification scenarios: under quite generic conditions one cannot find a larger symmetry group for a QFT model, which is composed as the semi-direct product of the Poincaré group and the internal symmetry group. One of the important assumptions of the pertinent no-go theorem is that the Lie group of internal symmetries is restricted to products of copies of U(1) group and compact semisimple Lie groups. This means that the total symmetries of matter fields at a fixed spacetime point —or equivalently, at a fixed point of momentum space— is a product of copies of U(1) group and a semisimple Lie group.

Our aim is to show a possible, physically natural mathematical example, when the total symmetry group of matter fields at a spacetime point —or equivalently, at a fixed point of momentum space— is some nontrivial extension of the (conformal) Lorentz group, or more precisely, its covering group. The pertinent extended group can be viewed as the automorphisms of the creation operator algebra for a spin 1/2 particle along with its antiparticle. The non-semisimpleness of the discussed group gives possibility to nontrivially extend the group of spacetime symmetries. The extended part becomes an idempotent normal subgroup, which can be regarded as "dressing transformations" of pure one-particle states in a formal QFT context at a fixed momentum. The presented algebraic construction may also help to improve mathematical formulation of QFTs: it shows a possibility to avoid building the theory based on first constructing the one-particle theory and then constructing the corresponding multi-particle model, *a posteriori*. That is because the "dressing transformations" mix the pure p-particle, q-antiparticle states with each-other to some extent, putting the possible particle combination states in a unified multiplet.

2. Preliminaries: automorphisms of a Grassmann algebra

First we review the properties of the automorphism group of finite dimensional Grassmann algebras as prototype problem.[2–5] Physically, an n-generator Grassmann algebra can be thought of as the algebra of creation operator polynomials in a formal QFT at fixed momentum of a fermion particle with n internal degrees of freedom.

Definition 1. *(Grassmann algebra, canonical generator system) A finite dimensional complex associative algebra G with unit is called a* ***Grassmann algebra*** *if there exists a minimal generating system $(e_1, \ldots, e_n)$ of G such that*

$$\begin{aligned}
&e_i e_j + e_j e_i = 0 \qquad\qquad (i, j \in \{1, \ldots, n\}) \\
\text{and}\quad &e_{i_1} e_{i_2} \ldots e_{i_k} \qquad (1 \le i_1 < i_2 < \cdots < i_k \le n,\ 0 \le k \le n) \\
&\text{are linearly independent.}
\end{aligned} \tag{1}$$

Such a minimal generating system shall be referred to as ***canonical generator system****.*

We will now explore the basic properties of the group of automorphisms $\mathrm{Aut}(G)$ of G. These are the $G \to G$ invertible complex-linear transformations, which preserve the algebraic product on G.

Remark 2. *The following basic properties are well known.*[2–6]

(i) Grassmann algebras do exist: the exterior algebra $\Lambda(V)$ of a finite dimensional complex vector space V is a Grassmann algebra, where a basis of V are canonical generators. In fact, all Grassmann algebras are (not naturally) isomorphic to some exterior algebra $\Lambda(V)$.

(ii) If $(e_1, \ldots, e_n)$ is a canonical generator system and $\alpha \in \mathrm{Aut}(G)$, then $(\alpha(e_1), \ldots, \alpha(e_n))$ are also. If, in addition, $(e'_1, \ldots, e'_n)$ is an other system of canonical generators, then a mapping $e_i \mapsto \alpha(e_i) := e'_i$ $(i \in \{1, \ldots, n\})$ uniquely determines an automorphism $\alpha \in \mathrm{Aut}(G)$. That is, automorphisms can uniquely be characterized by their action on an arbitrarily chosen canonical generator system.

(iii) Given a chosen system of canonical generators, the linear subspace of the pure k-th order polynomials of them are called the space of k-forms, and are denoted by Λ_k. The linear subspace of the pure even / odd polynomials of them are called the space of even / odd forms, and are denoted by Λ_{ev} / Λ_{od}. As such, one has $G = \bigoplus_{k=0}^{n} \Lambda_k$ and $G = \Lambda_{\mathrm{ev}} \oplus \Lambda_{\mathrm{od}}$. These splittings of G are referred to as $\mathbb{Z}$ and $\mathbb{Z}_2$-grading, respectively.

(iv) Let the unity be denoted by $\mathbb{1}$ and its complex linear span by B. Let $M := \bigoplus_{k=1}^{n} \Lambda_k$ be the linear subspace of at least 1-forms and observe that it is the maximal ideal of G, and therefore is an $\mathrm{Aut}(G)$-invariant subspace. Because of that, we have the $\mathrm{Aut}(G)$-invariant splitting $G = B \oplus M$ with corresponding $\mathrm{Aut}(G)$-invariant complementing projection operators $I - m$ and m. Because of $\mathrm{Aut}(G)$-invariance of unity, $I - m = \mathbb{1}b$ can be written with uniquely determined $b : G \to \mathbb{C}$, $\mathrm{Aut}(G)$-invariant map, picking out the scalar component.

(v) Since M is $\mathrm{Aut}(G)$-invariant, all its powers $M^l = \bigoplus_{k=l}^{n} \Lambda_k$ are $\mathrm{Aut}(G)$-invariant $(l \in \{1, \ldots, n\})$.

(vi) The center of G, denoted by $Z(G)$, consists of all elements commuting with G, and they form an $\mathrm{Aut}(G)$-invariant subspace.

Theorem 3. *(D. Z. Djokovic[3]) Let us define the following subgroups of $\mathrm{Aut}(G)$, given a canonical generator system $(e_1, \ldots, e_n)$ of G.*

(i) Let $\mathrm{Aut}_{\mathbb{Z}}(G)$ be the $\mathbb{Z}$-grading preserving automorphisms. These are of the form $e_i \mapsto \sum_{j=1}^{n} \alpha_{ij} e_j$ $(i \in \{1, \ldots, n\})$ with $(\alpha_{ij})_{i,j \in \{1, \ldots, n\}} \in \mathrm{GL}(\mathbb{C}^n)$.

(ii) Let N_{ev} be the $\mathbb{Z}_2$-grading preserving automorphisms acting on M^1/M^2 factor space as unity. These are of the form $e_i \mapsto e_i + b_i$ $(i \in \{1, \ldots, n\})$ with $b_i \in M^3 \cap \Lambda_{\mathrm{od}}$.

(iii) Let $\mathrm{InAut}(G)$ be the inner automorphisms, i.e. the ones of the form

$\exp(a)(\cdot)\exp(a)^{-1}$ *(with some* $a \in G$*). These are of the form* $e_i \mapsto e_i + [a, e_i]$ *(*$i \in \{1, \ldots, n\}$*) with some* $a \in G$*.*

With these, the semi-direct product splitting

$$\mathrm{Aut}(G) = \mathrm{InAut}(G) \rtimes N_{\mathrm{ev}} \rtimes \mathrm{Aut}_{\mathbb{Z}}(G) \tag{2}$$

holds.

Corollary 4. *As a consequence, we have that the* Λ_k *subspaces of* k*-forms are not* Aut(G)*-invariant. In fact, the list of indecomposable* Aut(G)*-invariant subspaces are the followings:* B*,* M^l *(*$l \in \{1, \ldots, n\}$*),* $M^l \cap Z(G)$ *(*$l \in \{2, \ldots, n\}$*).*

Physicswise, this means that if we think of G as the algebra of creation operator polynomials in a formal QFT of an n internal degrees of freedom fermion particle at fixed momentum, and we assume that the full Aut(G) acts on this algebra as symmetry group, then it becomes a unified multiplet. Particularly, the only Aut(G)-invariant decomposition is $G = B \oplus M$, i.e. the splitting to 0-particle and to at-least-1-particle states. The reason is that the normal subgroup $N := \mathrm{InAut}(G) \rtimes N_{\mathrm{ev}}$ mixes higher particle content to lower particle states. This motivates to call N the ***dressing transformations***, being an idempotent normal subgroup of Aut(G).

3. Spin algebra and its automorphisms

Motivated by the above findings, we define a physically more relevant setting.

Definition 5. *(*$^+$*-algebra) A finite dimensional complex associative algebra* A *with unit shall be called a* $^+$***-algebra*** *if it is equipped with a conjugate-linear involution satisfying* $(xy)^+ = x^+ y^+$ *for all* $x, y \in A$*.*

It is important that in the above definition the $^+$-adjoint does not reverse the order of products, i.e. it is slightly different than that of a usual *-algebra. This will, physically, model the charge conjugation in our construction.

Definition 6. *(spin algebra) A finite dimensional complex associative* $^+$*-algebra* A *with unit shall be called a* ***spin algebra*** *if there exists a minimal generating system* (e_1, e_2, e_3, e_4) *of* A *such that*

$$\begin{aligned} & e_i e_j + e_j e_i &= 0 \qquad & (i, j \in \{1, 2\} \text{ or } i, j \in \{3, 4\}) \\ \textit{and} \quad & e_i e_j - e_j e_i &= 0 \qquad & (i \in \{1, 2\} \text{ and } j \in \{3, 4\}) \end{aligned}$$

and $e_3 = e_1^+,\ e_4 = e_2^+$
and $e_{i_1} e_{i_2} \dots e_{i_k} \qquad (1 \le i_1 < i_2 < \dots < i_k \le 4,\ 0 \le k \le 4)$
are linearly independent. (3)

Such a minimal generating system shall be referred to as ***canonical generator system****, and the notation $n := 4$ will occasionally be used.*

Physicswise, a spin algebra can be thought of as the algebra of creation operator polynomials in a formal QFT of a 2 internal degree of freedom fermion particle along with its antiparticle, at fixed momentum. It is important to note that in this construction the annihilation operators of particles are not yet identified with the creation operator of antiparticles, and therefore a spin algebra is very different than that of a canonical anticommutation relation (CAR) algebra, also referred to as mixed exterior algebras.[7] Here, the $^+$-adjoining only models the charge conjugation operator, exchanging particle and antiparticle creation operators in a conjugate-linear way. We will now explore the basic properties of the group of automorphisms $\mathrm{Aut}(A)$ of A. These are the $A \to A$ invertible complex-linear transformations, which preserve the algebraic product on A along with the $^+$-adjoining.

Remark 7. *Similar properties hold as for a Grassmann algebra.*

(i) Spin algebras do exist: if S^* is a complex 2 dimensional vector space, called to be the cospinor space, and we take $\Lambda(\bar{S}^*) \otimes \Lambda(S^*)$, then it naturally becomes a spin algebra. In fact, all spin algebras are (not naturally) isomorphic to this. (The notation $\bar{(\cdot)}$ means complex conjugation.)

(ii) If $(e_1, \dots, e_n)$ is a canonical generator system and $\alpha \in \mathrm{Aut}(A)$, then $(\alpha(e_1), \dots, \alpha(e_n))$ are also. If, in addition, $(e'_1, \dots, e'_n)$ is an other system of canonical generators, then a mapping $e_i \mapsto \alpha(e_i) := e'_i$ $(i \in \{1, \dots, n\})$ uniquely determines an automorphism $\alpha \in \mathrm{Aut}(A)$. That is, automorphisms can uniquely be characterized by their action on an arbitrarily chosen canonical generator system.

(iii) Given a chosen system of canonical generators, the linear subspace of the pure polynomials of p pieces of $\{e_1, e_2\}$ and q pieces of of $\{e_1^+, e_2^+\}$ are called the space of p,q-forms, and are denoted by $\Lambda_{\bar{p},q}$. The subspace of those polynomials for which $p + q = k$ holds are called the space of k-forms. The linear subspace of the pure even / odd Λ_k-s are called the space of even / odd forms, and are denoted by Λ_{ev} / Λ_{od}. As such, one has $A = \bigoplus_{p,q=0}^{2} \Lambda_{\bar{p},q}$, $A = \bigoplus_{k=0}^{4} \Lambda_k$ and $A = \Lambda_{\mathrm{ev}} \oplus \Lambda_{\mathrm{od}}$. These splittings of A are referred to as the $\mathbb{Z} \times \mathbb{Z}$, $\mathbb{Z}$ and $\mathbb{Z}_2$-grading, respectively.

(iv) Let the unity be denoted by $\mathbb{1}$ *and its complex linear span by* B. *Let* $M := \bigoplus_{k=1}^{n} \Lambda_k$ *be the linear subspace of at least 1-forms and observe that it is the maximal ideal of* A, *and therefore is an* $\mathrm{Aut}(A)$*-invariant subspace. Because of that, we have the* $\mathrm{Aut}(A)$*-invariant splitting* $A = B \oplus M$ *with corresponding* $\mathrm{Aut}(A)$*-invariant complementing projection operators* $I - m$ *and* m. *Because of* $\mathrm{Aut}(A)$*-invariance of unity,* $I - m = \mathbb{1}b$ *can be written with uniquely determined* $b : A \to \mathbb{C}$, $\mathrm{Aut}(A)$*-invariant map.*

(v) Since M *is* $\mathrm{Aut}(A)$*-invariant, all its powers* $M^l = \bigoplus_{k=l}^{n} \Lambda_k$ *are* $\mathrm{Aut}(A)$*-invariant* ($l \in \{1, \ldots, n\}$).

(vi) The center of A, *denoted by* $Z(A)$, *consists of all elements commuting with* A, *and they form an* $\mathrm{Aut}(A)$*-invariant subspace. In fact,* $Z(A) = \Lambda_{\bar{0}0} \oplus \Lambda_{\bar{2}0} \oplus \Lambda_{\bar{0}2} \oplus \Lambda_{\bar{2}2}$.

Theorem 8. *(A. László[8]) Let us define the following subgroups of* $\mathrm{Aut}(A)$, *given a canonical generator system* (e_1, e_2, e_1^+, e_2^+) *of* A.

(i) Let $\mathrm{Aut}_{\mathbb{Z}\times\mathbb{Z}}(A)$ *be the* $\mathbb{Z} \times \mathbb{Z}$*-grading preserving automorphisms. These are of the form* $e_i \mapsto \sum_{j=1}^{2} \alpha_{ij} e_j$ *and* $e_i^+ \mapsto \sum_{j=1}^{2} \bar{\alpha}_{ij} e_j^+$ $(i \in \{1, 2\})$ *with* $(\alpha_{ij})_{i,j\in\{1,2\}} \in \mathrm{GL}(\mathbb{C}^2)$. *Here, as before,* $\overline{(\cdot)}$ *denotes complex conjugation.*

(ii) Let $\mathcal{J} := \{I, J\}$ *be the two element group of automorphisms,* I *being the identity and* J *being the involutive complex-linear operator of particle-antiparticle label exchanging:* $e_1 \mapsto e_3$, $e_2 \mapsto e_4$, $e_3 \mapsto e_1$, $e_4 \mapsto e_2$.

(iii) Let $\tilde{N}_{\mathrm{ev}}$ *be a subgroup of the* $\mathbb{Z}_2$*-grading preserving automorphisms defined by the relations* $e_i \mapsto e_i + b_i$ *and* $e_i^+ \mapsto e_i^+ + b_i^+$ $(i \in \{1, 2\})$ *with uniquely determined parameters* $b_i \in \Lambda_{\bar{1}2}$.

(iv) Let $\mathrm{InAut}(A)$ *be the inner automorphisms, i.e. the ones of the form* $\exp(a)(\cdot)\exp(a)^{-1}$ *(with some* $a \in \mathrm{Re}(A)$*). These are of the form* $e_i \mapsto e_i + [a, e_i] + \frac{1}{2}[a, [a, e_i]]$ $(i \in \{1, \ldots, n\})$ *with uniquely determined parameter* $a \in \mathrm{Re}(\Lambda_{\bar{1}0} \oplus \Lambda_{\bar{0}1} \oplus \Lambda_{\bar{1}1} \oplus \Lambda_{\bar{2}1} \oplus \Lambda_{\bar{1}2}) \subset \mathrm{Re}(A)$.

With these, the semi-direct product splitting

$$\mathrm{Aut}(A) = \mathrm{InAut}(A) \rtimes \tilde{N}_{\mathrm{ev}} \rtimes \mathrm{Aut}_{\mathbb{Z}\times\mathbb{Z}}(A) \rtimes \mathcal{J} \tag{4}$$

holds.

Proof The proof is based on the fact that $\mathrm{Aut}(A)$ elements are the linear transformations preserving the canonical generator relations. Although the proof is theoretically not complicated, it is a quite extended calculation, and will be published in a more detailed paper.[8] ∎

Corollary 9. *As a consequence, we have that the* $\Lambda_{\bar{p}q}$ *subspaces of* p, q*-forms are not* $\mathrm{Aut}(A)$*-invariant. In fact, the list of indecomposable* $\mathrm{Aut}(A)$*-invariant subspaces are the followings:* B*,* M^l *(*$l \in \{1, 2, 3, 4\}$*),* $M^2 \cap Z(A)$*,* $V := \Lambda_{\bar{1}0} \oplus \Lambda_{\bar{0}1} \oplus \Lambda_{\bar{2}0} \oplus \Lambda_{\bar{0}2} \oplus \Lambda_{\bar{2}1} \oplus \Lambda_{\bar{1}2} \oplus \Lambda_{\bar{2}2}$*,* $U := \Lambda_{\bar{2}0} \oplus \Lambda_{\bar{0}2} \oplus \Lambda_{\bar{2}1} \oplus \Lambda_{\bar{1}2} \oplus \Lambda_{\bar{2}2}$*,* $W := \Lambda_{\bar{1}1} \oplus \Lambda_{\bar{2}1} \oplus \Lambda_{\bar{1}2} \oplus \Lambda_{\bar{2}2}$*. These are illustrated in Fig. 1.*

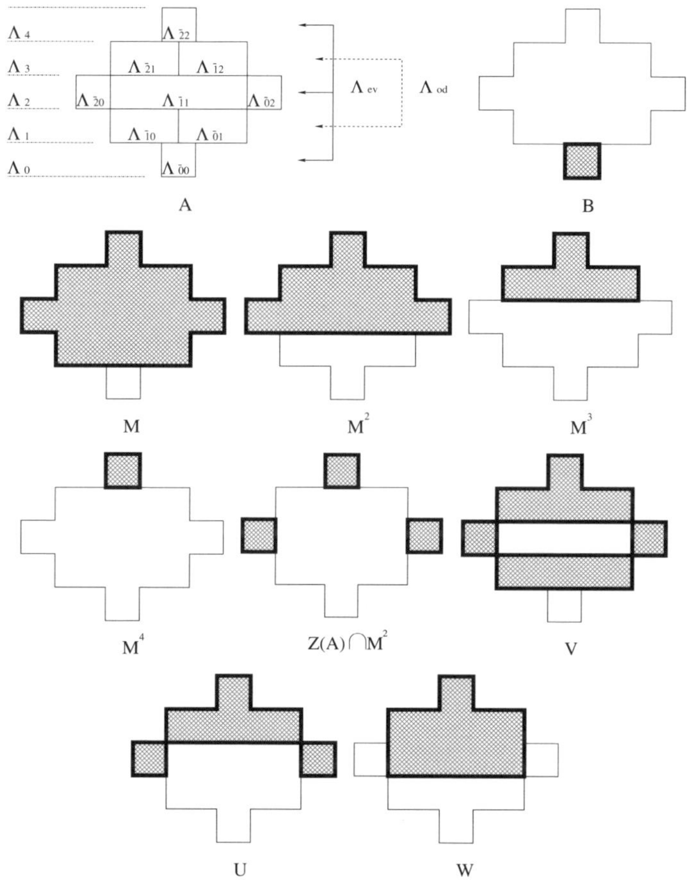

Fig. 1. Top left panel: illustration of $\mathbb{Z} \times \mathbb{Z}$, $\mathbb{Z}$ and $\mathbb{Z}_2$-graded structure of the spin algebra A. Other panels: illustration of $\mathrm{Aut}(A)$-invariant indecomposable subspaces of A. One unit box depicts one complex dimension on all panels, shaded framed regions depict the pertinent indecomposable $\mathrm{Aut}(A)$-invariant subspaces.

Physicswise, this result means that if we think of A as the algebra of creation operator polynomials in a formal QFT of a 2 internal degrees of freedom fermion particle along with its antiparticle, at fixed momentum, and we assume that the full $\mathrm{Aut}(A)$ acts on this algebra as symmetry group, then it becomes a unified multiplet. Particularly, the only $\mathrm{Aut}(A)$-invariant decomposition is $A = B \oplus M$, i.e. the splitting to 0-particle and to at-least-1-particle states. The reason is that the normal subgroup $N := \mathrm{InAut}(A) \rtimes \tilde{N}_{\mathrm{ev}}$ mixes higher particle content to lower particle states. This motivates to call N the ***dressing transformations***, being an idempotent normal subgroup of $\mathrm{Aut}(A)$.

When dealing with the $\mathbb{Z}\times\mathbb{Z}$-grading preserving part of $\mathrm{Aut}(A)$, i.e. the action of the subgroup $\mathrm{Aut}(A)_{\mathbb{Z}\times\mathbb{Z}} \equiv \mathrm{GL}(\mathbb{C}^2) \equiv \mathrm{D}(1) \times \mathrm{U}(1) \times \mathrm{SL}(\mathbb{C}^2)$, a fixed $\mathbb{Z}\times\mathbb{Z}$ grading may be used, and therefore the whole formalism can be thought of as an ordinary two-spinor calculus[9,10] on the tensor algebra of cospinors and complex conjugate cospinors. Using this, it is immediately seen, that $\mathrm{Aut}(A)$ acts on the maximal forms M^4 as a scaling by a positive real number, i.e. they are orientation preserving. Therefore, $\mathrm{Re}(M^4)$ may be split to the cones of positive and negative maximal forms $\mathrm{Re}_+(M^4)$ and $\mathrm{Re}_-(M^4)$, preserved by $\mathrm{Aut}(A)$.

Remark 10. *In order to proceed, we study the transpose action of* $\mathrm{Aut}(A)$ *on the dual vector space* A^* *of* A. *From simple linear algebra, it automatically follows, that the indecomposable* $\mathrm{Aut}(A)$*-invariant subspaces of* A^* *are:* $\mathrm{Ann}(M)$, $\mathrm{Ann}(B)$, $\mathrm{Ann}(B\oplus M)$ *(*$k \in \{2,4\}$*),* $\mathrm{Ann}(Z(A))$, $\mathrm{Ann}(B\oplus V)$, $\mathrm{Ann}(B \oplus W)$, *where* $\mathrm{Ann}(\cdot)$ *denotes the annulator subspace in* A^*. *These are illustrated in Fig. 2.*

We shall show that $\mathrm{Aut}(A)$ acts as the conformal Lorentz group on the four real dimensional $\mathrm{Aut}(A)$-invariant vector space $\mathrm{Re}(\mathrm{Ann}(B \oplus V)) \equiv \mathrm{Re}(\Lambda^*_{\bar{1}1})$. For this, we use the fact that a spin algebra may be equipped with a Hopf algebra structure, given a chosen $\mathbb{Z}\times\mathbb{Z}$-grading. As usual in the Hopf algebra context,[11,12] the unit is viewed as an $\eta : \mathbb{C} \to A$ linear map, and the algebraic product as an $\nabla : A\otimes A \to A$ linear map. The counit shall be an $\varepsilon : A \to \mathbb{C}$ linear map, and we set $\varepsilon := b$. The antipode shall be an $S : A \to A$ linear map, and we set $S(\cdot) := (\cdot)_{\mathrm{ev}} + (-1)(\cdot)_{\mathrm{od}}$. The swapping involution shall be an $\mathcal{I} : A\otimes A \to A\otimes A$ linear map, and we set to $x\otimes y \mapsto \mathcal{I}(x\otimes y) := \sum_{p,q=0}^{2} \sum_{r,s=0}^{2} (-1)^{pr+qs} x_{\bar{p}q} \otimes y_{\bar{r}s}$ (for all $x, y \in A$). The coproduct shall be a $\triangle : A \to A \otimes A$ linear map, defined by the requirements $\triangle(\mathbb{1}) := \mathbb{1} \otimes \mathbb{1}$, $\triangle(x) := \mathbb{1} \otimes x + x \otimes \mathbb{1}$ (for all $x \in \Lambda_{\bar{1}0} \oplus \Lambda_{\bar{0}1}$), and that $\triangle$ is an $A \to A \otimes A$ algebra homomorphism, where $A \otimes A$ is equipped with the skew-natural

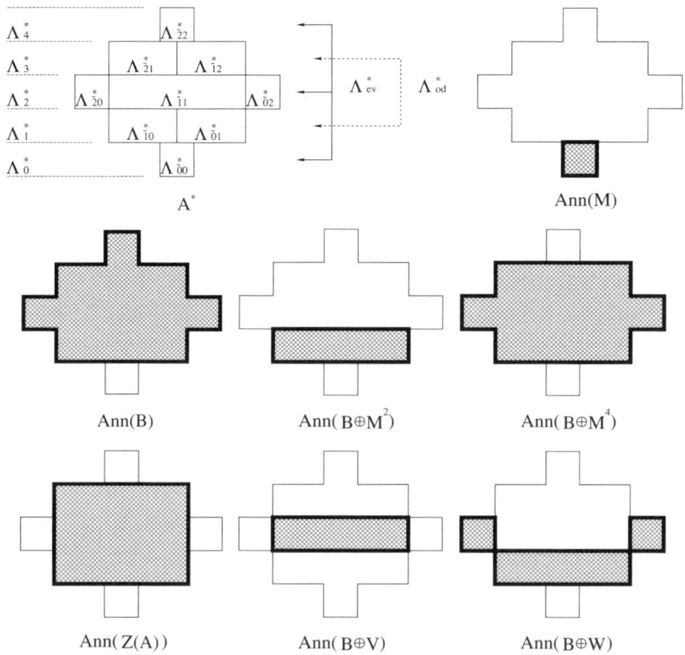

Fig. 2. Top left panel: illustration of $\mathbb{Z} \times \mathbb{Z}$, $\mathbb{Z}$ and $\mathbb{Z}_2$-graded structure of the dual A^* of a spin algebra A. Other panels: illustration of $\mathrm{Aut}(A)$-invariant indecomposable subspaces of A^*. One unit box depicts one complex dimension on all panels, shaded framed regions depict the pertinent $\mathrm{Aut}(A)$-invariant subspaces.

product $(\bigtriangledown \otimes \bigtriangledown) \circ (I \otimes (J \circ \mathcal{I}) \otimes I)$, I being the $A \to A$ identity map and J being the $A \otimes A \to A \otimes A$ swapping map. The holding of the Hopf relations on $(A, (\cdot)^+, \eta, \bigtriangledown, \varepsilon, \triangle, S, \mathcal{I})$ may be checked by direct calculations. Given the fixed spin algebra part $(A, (\cdot)^+, \eta, \bigtriangledown)$ it can be seen that the coalgebra part $(\varepsilon, \triangle, S, \mathcal{I})$ is not conserved by all $\mathrm{Aut}(A)$ transformations, only under the $\mathbb{Z}$-grading preserving transformations $\mathrm{Aut}_{\mathbb{Z}\times\mathbb{Z}}(A) \rtimes \mathcal{J} \subset \mathrm{Aut}(A)$: the dressing transformations N deform the part $(\triangle, S, \mathcal{I})$ to another compatible coalgebra structure.

Remark 11. *Equipped with the Hopf algebra notions, direct calculation shows that if we choose an* $\omega \in \mathrm{Re}_+(M^4) \setminus \{0\}$ *positive maximal form element of* A*, the bilinear map*

$$\begin{aligned} G(\omega) :\ & \mathrm{Re}(\mathrm{Ann}(B \oplus V)) \times \mathrm{Re}(\mathrm{Ann}(B \oplus V)) \to \mathbb{R}, \\ & (a, b) \mapsto G(\omega)(a, b) := (a \otimes b | \triangle(\omega)) \end{aligned} \tag{5}$$

is a Lorentz signature metric, where $(\cdot|\cdot)$ *denotes the duality pairing form. The action of* $\mathrm{Aut}(A)$ *preserves* $G(\omega)$ *up to a scaling factor, i.e. it acts on* $\mathrm{Re}(\mathrm{Ann}(B \oplus V))$ *as the conformal Lorentz group. The construction does not depend on the choice of the coproduct* $\triangle$, *i.e. does not depend on grading. The choice of* ω *is unique up to a scaling factor, and thus* $G(\omega)$ *is unique up to that as well.*

Next, we show that there are two charge-conjugate Dirac bispinor spaces embedded (not uniquely) in A, and thus the relation to the usual Clifford algebra formulation becomes clear. Given a coproduct $\triangle$, the four real dimensional $\mathrm{Aut}(A)$-invariant subspace $\mathrm{Re}(\mathrm{Ann}(B \oplus V)) \subset A^*$ may be embedded into $\mathrm{Re}(\mathrm{Lin}(A))$ using the rule $s \mapsto (s \otimes I) \circ \triangle$ (for each $s \in \mathrm{Re}(\mathrm{Ann}(B \oplus V))$), which we call a ***Pauli embedding***. Given a four real dimensional vector space T (modeling the spacetime tangent vectors at a spacetime point, or the momentum vectors of momentum space), a linear injection $T \to \mathrm{Re}(\mathrm{Ann}(B \oplus V))$ may be taken, which we call a ***Pauli injection***. The composition of a Pauli embedding and a Pauli injection shall be called ***Pauli map***, which is a $T \to \mathrm{Lin}(A)$ linear map. If we take such a Pauli map σ, in ordinary two-spinor calculus it corresponds to $\sigma_a^{A'A}$ in Penrose abstract indices,[9,10] where $_a$ corresponds to a T^* index, while $^{A'}$ and ^{A} are spinor indices. Keeping these in mind, one can see that the object $g(\sigma,\omega)_{ab} := b\,\sigma_a\sigma_b\,\omega$ is a Lorentz metric on T, because of Remark 11.

Definition 12. *(Dirac adjoint, Dirac gamma map) Let us fix a Pauli map* $\sigma : T \to \mathrm{Lin}(A)$ *and a positive maximal form* ω. *The conjugate-linear map*

$$\overline{(\cdot)} : A \to A^*,\ x \mapsto \overline{x} := \frac{1}{2} g(\sigma,\omega)^{ab}\, b\sigma_a \Big(x^+ \sigma_b(\cdot) + \sigma_b(x^+)(\cdot) \Big) \tag{6}$$

is called the ***Dirac adjoint***. *The linear map*

$$\gamma(\sigma,\omega) : T \to \mathrm{Lin}(A),\ u \mapsto u^a \gamma(\sigma,\omega)_a(\cdot) := u^a \sqrt{2} \Big(\sigma_a(\cdot) + \sigma_a(\omega)(\cdot) \Big) \tag{7}$$

is called the ***Dirac gamma map***.

In order to see the properties of the Dirac adjoint and the Dirac gamma map, the following important subspaces of A are introduced: let $D_+ := \Lambda_{\bar{1}0} \oplus \Lambda_{\bar{2}1}$ and $D_- := \Lambda_{\bar{0}1} \oplus \Lambda_{\bar{1}2}$, where the $\mathbb{Z} \times \mathbb{Z}$-grading is understood to be the one subordinate to the Pauli map σ, or equivalently, the one used in the underlying coproduct. Clearly, these are four complex dimensional subspaces of A with $D_+ \cap D_- = \{0\}$ and $(D_+)^+ = D_-$, $(D_-)^+ = D_+$. Direct calculations show that $\overline{(\cdot)}$ becomes non-degenerate and $\gamma(\sigma,\omega)$ satisfies the Clifford relations $\gamma(\sigma,\omega)_a\gamma(\sigma,\omega)_b + \gamma(\sigma,\omega)_b\gamma(\sigma,\omega)_a = 2\,I\,g(\sigma,\omega)_{ab}$

over D_+ and D_-. These subspaces, however, are not preserved by the dressing transformations N, but are transformed to other similarly behaving embedded Dirac bispinor spaces in A. This embedding is illustrated in Fig. 3.

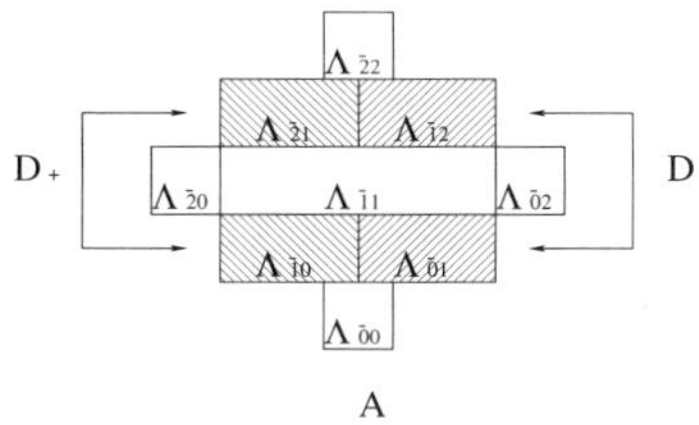

Fig. 3. Illustration of the embedding of the Dirac bispinor spaces D_+ and D_- into the spin algebra A, given a Pauli mapping σ. One unit box depicts one complex dimension. Shaded regions indicate the subspaces D_+ and D_-, respectively.

Finally, the relation of $(\cdot)^+$-adjoining to the usual Hilbert scalar product adjoining is given, which is used for the identification of antiparticle creation operators to particle annihilation operators in a traditional QFT setting. Let $u^a \in T$ be a future-directed timelike or null vector in terms of $g(\sigma,\omega)_{ab}$, modeling a momentum vector on mass shell. Then, the bilinear form $A \times A \to \mathbb{C}$, $(x,y) \mapsto b\,(x^+y) + u^a\,b\,\sigma_a\,((x - \mathbb{1}\,b\,x)^+(y - \mathbb{1}\,b\,y))$ is a positive semidefinite $\mathrm{Aut}(A)$-invariant inner product. Given a coproduct $\triangle$, this induces a scalar product on A, and the adjoining operator $(\cdot)^\dagger$ with respect to that is the usual Hilbert adjoining identifying antiparticle creation operators and particle annihilation operators at a fixed momentum $u^a \in T$. Since it is momentum dependent (has a parameter u^a), and is not $\mathrm{Aut}(A)$-invariant, it is not practical to consider $(\cdot)^\dagger$ as basic ingredient of the internal degrees of freedom, encoded in the mathematical structure of A: it is rather constructed as a derived quantity, as shown. Also in this approach, the problem of normal ordering does not appear: all polynomial expressions of A are automatically normally ordered, by construction.

4. Concluding remarks

A finite dimensional algebra was presented and its group of automorphisms was studied. The constructed algebra can physically be thought of as the creation operator algebra of a formal quantum field theory at fixed momentum of a spin 1/2 particle and its antiparticle. It was shown that the essential part of the pertinent automorphism group is basically the conformal Lorentz group. On the other hand, the remaining normal subgroup can be thought of as "dressing transformations" making "dressed" states from pure one-particle states. The proposed construction may be used in the construction of quantum field theories on a non-perturbative basis: our approach is different than that of usual approach when the one-particle theory is defined, and then a Fock space of multiparticle states is built on top of that. In addition, a similar approach could be used in GUT attempts, as it provides a possibility to circumvent Coleman–Mandula theorem: the non-semisimpleness of the pertinent symmetry group could allow to have the spacetime and internal symmetries to be connected via the non-semisimple part, without introducing SUSY. An interesting idea would be to study these results in the light of,[13] which provides an $\mathrm{U}(1) \times \mathrm{SU}(3)$ unification attempt, also using an algebra automorphism group.

Acknowledgements

I would like to thank Mátyás Domokos and Richárd Rimányi for valuable discussion and expert opinions on the algebra theory content of the paper, furthermore to Péter Vecsernyés for ideas related to the Hopf algebra context. I would like to thank Tamás Matolcsi, Péter Ván, Tamás Fülöp and Dezső Varga for motivations and discussion on the physical context of the presented idea. I would like to thank Júlia Nyíri for inviting the submission of this manuscript. This work was supported in part by the Momentum ("Lendület") program of the Hungarian Academy of Sciences, as well as the János Bolyai Research Scholarship of the Hungarian Academy of Sciences.

References

1. S. Weinberg, *The Quantum Theory of Fields III*, Cambridge University Press, Cambridge, 2000.
2. F. A. Berezin, *Mathematiceskie Zametki* **1**, 269 (1967).
3. D. Z. Djokovic, *Can. J. Math.* **30**, 1336 (1978).
4. V. V. Bavula, The Jacobian map, the Jacobian group and the group of automorphisms of the Grassmann algebra, *Lecture note for talk held on "Automorphisms and homology in noncommutative algebraic geometry"* CNRS Workshop, `arXiv:math/0703352`, 2007.
5. V. V. Bavula, *J. Alg. and its Appl.* **8**, 805 (2009).
6. N. Bourbaki, *Algebra I*, Springer-Verlag, Berlin, 1989.
7. J. R. Vanstone, *Can. J. Math.* **36**, 361 (1984).

8. A. László, The automorphism group of the spin algebra and extension of the conformal Lorentz group, Manuscript in preparation, 2015.
9. R. Penrose and W. Rindler, *Spinors and spacetime 1-2*, Cambridge University Press, Cambridge, 1984.
10. R. M. Wald, *General Relativity*, University of Chicago Press, Chicago, 1984.
11. L. A. Lambe and W. M. Seiler, *Georg. Math. J.* **9**, 723 (2002).
12. P. Cartier, *A Primer of Hopf Algebras*, IHES/M/06/40, Bures-sur-Yvette, IHES, 2006.
13. C. Furey, *Phys. Lett.* **B742**, 195 (2015).

RENORMALISATION GROUP DETERMINATION OF SCALAR MASS BOUNDS IN A SIMPLE YUKAWA-MODEL

A. JAKOVÁC

Institute of Physics, Eötvös University,
Budapest, H-1117, Hungary
E-mail: jakovac@caesar.elte.hu

I. KAPOSVÁRI

Eötvös University,
Budapest, H-1117, Hungary
E-mail: stevenkaposvari@gmail.com

A. PATKÓS

MTA-ELTE Research Group for Biological and Statistical Physics
Budapest, H-1117, Hungary,
E-mail: patkos@hector.elte.hu

The scalar mass is determined in the simplest cut-off regularized Yukawa-model in the whole range of stability of the scalar potential. Two versions of the Functional Renormalisation Group (FRG) equations are solved in the Local Potential Approximation (LPA), where also the possible existence of a composite fermionic background is taken into account. The close agreement of the results with previous studies taking into account exclusively the effect of the scalar condensate, supports a rather small systematic truncation error of FRG due to the omission of higher dimensional operators.

Keywords: ground state stability; Landau-singularity; mass bounds; renormalisation group

1. Motivation: from "Moscow-zero" to Higgs-mass bounds

The discovery of the "Moscow-zero" [1] (alias Landau-pole) appears from modern perspective as a true Archimedean "fixed point" on the historical path of renormalized quantum field theories. Its real importance has been recognized after desperate discussions concerning the very sense of quantum field theory. The present conference gives the opportunity to quote from the history notes of A.A. Anselm [2] that researchers of the Theory division of the Leningrad (at present Petersburg) Nuclear Physics Institute under the "natural scientific leadership" of V.N. Gribov contributed "many important papers during the 50's and the beginning of the 60's to the development of the "null-charge" problem." And the interest of Volodia Gribov did not diminish with time.[3]

At present two alternative attitudes are taken concerning the exact nature of the field theories which in the framework of (resummed) perturbation theory appear to possess Landau-poles. The concept of asymptotic safety[4] envisages the existence of an UV-attractive fixed point in the renormalisation flow of the model under study. The energy scale of this fixed point should be lower than the scale of the Landau-pole. Assuming the existence of such a fixed point in the joint theory of gravity and the Standard Model at the scale of the Planck-mass, relatively a accurate prediction(!) of the mass of the Higgs particle was achieved.[5] Some recent papers[6] express very optimistic views concerning the existence of a UV-safe completion of the Standard Model on its own, but to date the existence of any interactive fixed point could not be established.

The other alternative, phrased in the form of the triviality conjecture,[7] accepts the strict existence of a Landau-type singularity in the scalar sector of the Standard Model. This view gains support in the case of the four-dimensional Φ^4 theory from non-perturbative numerical simulations. In order to avoid the manifestation of the effects of singularity one has to restrict the momentum range of the modes participating in the dynamics of the model by applying some sort of momentum cut-off chosen below the location of the Landau-singularity. The scale of the cutoff indicates where the presently known fields and interactions are expected to reach the edge of validity. The Lagrangian of these effective models[8] is also allowed to contain operators which are perturbatively (in the weak coupling regime) non-renormalizable. Their coupling strengths are proportional to some negative power of the cut-off. The detailed choice of the Lagrangean at the cut-off exerts influence on the renormalisation flow towards the infrared. Conversely, the solution of the model which is designed to a certain predetermined set of low energy observables can be extended only up to a certain maximal momentum scale. In this sense the value of the maximal momentum (the cutoff Λ) is part of the physical data set characterizing the effective model. This approach has been used extensively for deriving upper bounds for the Higgs-particle in the scalar sector of the Standard Model.[9–13]

After the discovery of the Higgs-particle with mass 125 GeV, the actual question one asks is somewhat reoriented. Now the lower bound arising from requiring the stability of the Higgs potential is more in the focus. In particular, the possibility of the higher dimensional operators playing an essential role in the stability of the model is now actively investigated.[14–18] The accuracy of the renormalisation group equations (RGE) connecting the low energy momentum range with the operator content of the theory defined at the cutoff is of outstanding importance in finding a reliable answer. Since all methods of RGE-construction involve some approximations (truncations) it is of great interest to have a clear estimate of the sources and sizes of the truncation errors.

At present the most widely used form of the RGE is the Wetterich-Morris equation,[19–21] which describes the rate of evolution of the effective quantum action Γ_k at scale $k(t = \ln k/\Lambda)$:

$$\partial_t \Gamma_k[\Phi] = \frac{1}{2}\mathrm{Str}\left\{(\partial_t R_k)\left[\Gamma_k^{(2)} + R_k\right]^{-1}\right\} = \frac{1}{2}\partial_t \mathrm{Str}\log\left[\Gamma_k^{(2)} + R_k\right]. \quad (1)$$

The operation $\hat{\partial}_k$ appearing in front of the expressions on the right hand side of Eq. (9) requires the computation of the k-derivatives of the whole expression behind it, but taking into account only the k-dependence of the regulator functions $R_k^F(q)$ and $R_k^B(q)$. The functional super-trace on the right hand side contains $\Gamma^{(2)}$, which is the second functional derivative of $\Gamma_k[\Phi]$ depending on a set of fields denoted by Φ, together with an appropriately chosen infrared regulator function R_k. The super-trace over fermionic loops involves an extra minus sign relative to the bosonic loops. The aim of the present contribution is to construct and solve the RGE of a simple scalar-Yukawa model in an improved approximation scheme to this equation which goes beyond the accuracy of the previous investigations, in particular Ref. 15, where approximate solutions of Eq. (1) were constructed with a more restricted background configuration. The bounds found for the scalar mass upon fixed values for the fermion mass and the scalar condensate will be compared to the results of Ref. 15.

2. The scalar-fermion Yukawa model and its RGE

The solution of Eq. (1) is attempted for the scalar-fermion Yukawa model in Euclidean space-time using the following effective action Ansatz defining the field theory at scale k:

$$\Gamma_k = \int_x \left[Z_{\psi k}\bar{\psi}\gamma_m\partial_m\psi + \frac{1}{2}Z_{\sigma k}(\partial_m\sigma)^2 + h_k\sigma\bar{\psi}\psi + U_k(\rho)\right], \qquad \rho = \frac{1}{2}\sigma^2. \quad (2)$$

A non-zero condensate at scale $k = 0$ given as $v_0 = Z_{\sigma 0}^{1/2} < \sigma >_0 = \sqrt{2Z_{\sigma 0}\rho_0}$ in the scalar field σ violates the symmetry of the action under the discrete γ_5 transformation of the fermionic field ψ:

$$\sigma(x) \to -\sigma(x), \qquad \psi(x) \to \gamma_5\psi(x), \quad \bar{\psi}(x) \to -\bar{\psi}(x)\gamma_5. \tag{3}$$

The condensate generates a mass for the Fermi field: $m_\psi = h_0 v_0$. The index '0' in $h_0, \rho_0, Z_{\sigma 0}$ and v_0 emphasizes that the values to be used for the computation of these quantities should be evaluated from the action $\Gamma_{k=0}$, where $\Gamma_{k=\Lambda}$ arrives after running with help of Eq. (1) in form (2) from the cutoff Λ to $k = 0$. In order to maintain maximal analogy between our model and the top-Higgs sector of the Standard Model, in the discussion of the solution below we shall fix the $k = 0$ values of the vacuum condensate to $v_0 = 246$ GeV and the fermion mass to $m_\psi = 173$ GeV. In this preliminary report we shall neglect the running of the field renormalisations and set $Z_{\phi k} = Z_{\psi k} = 1$ in (2), which is the so-called Local Potential Approximation (LPA) to Eq. (1).

The second functional derivative of the effective action is represented as a 3x3 matrix:

$$\begin{pmatrix} \Gamma^{(2)}_{\sigma\sigma} & \Gamma^{(2)}_{\sigma\psi} & \Gamma^{(2)}_{\sigma\bar{\psi}^T} \\ \Gamma^{(2)}_{\psi^T\sigma} & \Gamma^{(2)}_{\psi^T\psi} & \Gamma^{(2)}_{\psi^T\bar{\psi}^T} \\ \Gamma^{(2)}_{\bar{\psi}\sigma} & \Gamma^{(2)}_{\bar{\psi}\psi} & \Gamma^{(2)}_{\bar{\psi}\bar{\psi}^T} \end{pmatrix}, \tag{4}$$

where for the representation of the fields we use the column-vector $(\sigma(x), \psi(x), \bar{\psi}^T(x))$ which corresponds to the Gor'kov-Nambu representation of the Fermi-field. As a first operation one can transform this matrix into a separate block-diagonal form in the scalar and the fermion-sector, which leads for the super-trace to

$$\begin{aligned} \frac{1}{2}\mathrm{Str}\log(\Gamma^{(2)} + R_k) &= -\frac{1}{2}\mathrm{Tr}\log(\Gamma^{(2)}_{\Psi^T\Psi} + R^F_k) \\ &\quad + \frac{1}{2}\mathrm{Tr}\log(\Gamma^{(2)}_{\sigma\sigma} - \Gamma^{(2)}_{\sigma\Psi}\Gamma^{(2)-1}_{\Psi^T\Psi}\Gamma^{(2)}_{\Psi^T\sigma}) \\ &= -\frac{1}{2}\mathrm{Tr}\log(\Gamma^{(2)}_{\Psi^T\Psi} + R^F_k) + \frac{1}{2}\mathrm{Tr}\log(\Gamma^{(2)}_{\sigma\sigma} + R^B_k) \\ &\quad + \frac{1}{2}\mathrm{Tr}\log\left[1 - \left(\Gamma^{(2)}_{\sigma\sigma} + R^B_k\right)^{-1}\Gamma^{(2)}_{\sigma\Psi}\left(\Gamma^{(2)}_{\Psi^T\Psi} + R^F_k\right)^{-1}\Gamma^{(2)}_{\Psi^T\sigma}\right]. \end{aligned} \tag{5}$$

Here $\Gamma^{(2)}_{\Psi^T\Psi}$ is a compact notation for a hypermatrix in the doubled bispinor field basis.

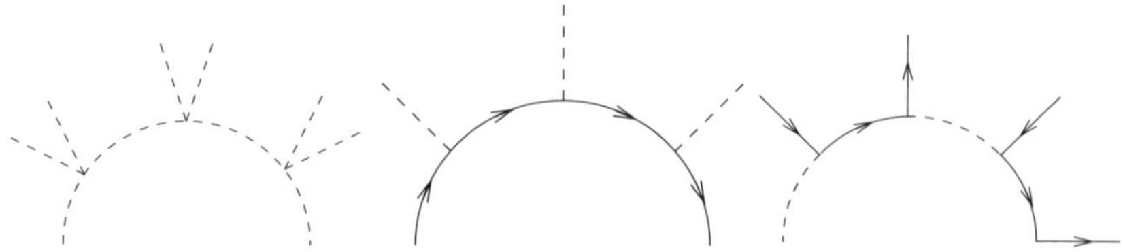

Fig. 1. Propagator sequences building the one-loop contributions to the tracelog of Eq. (5). On the left a piece of a scalar loop (dashed curves) immersed in a scalar condensate, in the middle a fermion loop (continuous curve) immersed in a scalar condensate, on the right a loop constructed of alternating sequence of fermion and scalar propagators immersed in a composite fermionic condensate is represented.

Each term has a diagrammatic representation. The first two terms correspond to the free fermion and free boson contributions to the super-tracelog computed in the background v_k. The last term is the sum of the infinite series of 1-loop contributions containing an increasing number of alternating boson and fermion propagators. At each vertex of fermion-to-boson transformation an external fermion leg joins the loop, due to the assumed presence of a composite fermionic background (see Fig. 1).

When one explicitly displays the matrix operations in the bispinor indices in the doubled fermionic sector, one finds for the last term of the previous equation

$$\begin{aligned} &\mathrm{Tr}\log\left[1-\left(\Gamma^{(2)}_{\sigma\sigma}+R^B_k\right)^{-1}\Gamma^{(2)}_{\sigma\Psi}\left(\Gamma^{(2)}_{\Psi^T\Psi}+R^F_k\right)^{-1}\Gamma^{(2)}_{\Psi^T\sigma}\right] \\ &=\mathrm{Tr}\log\left[1-\left(\Gamma^{(2)}_{\sigma\sigma}+R^B_k\right)^{-1}\left[\Gamma^{(2)}_{\sigma\psi}\left(\Gamma^{(2)}_{\bar\psi\psi}+R^F_k\right)^{-1}\Gamma^{(2)}_{\bar\psi\sigma}\right.\right. \\ &\qquad\left.\left.+\Gamma^{(2)}_{\sigma\bar\psi^T}\left(\Gamma^{(2)}_{\psi^T\bar\psi^T}+R^F_k\right)^{-1}\Gamma^{(2)}_{\psi^T\sigma}\right]\right]. \qquad (6) \end{aligned}$$

In earlier papers the super-trace was evaluated on a background exclusively consisting of a nonzero σ-condensate.[15] On such a background $\Gamma^{(2)}_{\sigma\Psi}=\Gamma^{(2)}_{\Psi^T\sigma}=0$ and the right hand side of the RGE simply reduces to the combination of a pure fermion- and a pure boson-loop. However, this choice is incompatible for a general choice of the background ρ_k with the field equation

$$\frac{\delta\Gamma_k}{\delta\sigma}\Big|_{\sigma=v_k,\psi=\psi_k}=h_k(\bar\psi\psi)_k+v_kU'_k(\rho_k)=0. \qquad (7)$$

In our recent publications clear arguments were put forward[22,23] how one can reconcile the Grassmannian nature of the Fermi fields with a nonzero coarse grained $[(\bar{\psi}\psi)_k]^n$, $n > 1$ background. Using the explicit expressions of the scalar-fermion (Yukawa) vertices and of the fermion propagators one quickly evaluates the contribution of the last term on the right hand side of Eq. (6) on a space-independent v_k and $(\bar{\psi}\psi)_k$ background:

$$\begin{aligned}\mathrm{Tr}\log\Bigg[1-\left(\Gamma^{(2)}_{\sigma\sigma}+R^B_k\right)^{-1}\Bigg[\Gamma^{(2)}_{\sigma\psi}\left(\Gamma^{(2)}_{\bar{\psi}\psi}+R^F_k\right)^{-1}\Gamma^{(2)}_{\bar{\psi}\sigma}\\ +\Gamma^{(2)}_{\sigma\bar{\psi}^T}\left(\Gamma^{(2)}_{\psi^T\bar{\psi}^T}+R^F_k\right)^{-1}\Gamma^{(2)}_{\psi^T\sigma}\Bigg]\Bigg]\\ =V_d\frac{1}{2}\int_q\log\left\{1-h_k^2\frac{1}{q_R^2+m_{B0}^2}\frac{2m_\psi(\bar{\psi}\psi)_k}{q_R^2+m_\psi^2}\right\}.\end{aligned} \tag{8}$$

Here V_d is the d-dimensional quantisation volume, $m^2_{B0} = U'_k(\rho_k) + 2\rho_k U''_k(\rho_k), m^2_\psi = 2h_k^2\rho_k$ and q_R^2 stands for the infrared regularized momentum-dependent part of the inverse propagator: $q^2 + R_k(q)$ (one chooses different regulators for bosons and fermions).

Within the LPA Ansatz (2) one finds the following RGE for the potential energy of the model after combining the three contributions on its right hand side:

$$\begin{aligned}\partial_t[U_k(\rho_k)+h_kv_k(\bar{\psi}\psi)_k]=\frac{1}{2}\hat{\partial}_t\int_q\Bigg[-4\log(q_R^2+m_\psi^2)+\log(q_R^2+m_{B0}^2)\\ +\log\left\{1-h_k^2\frac{1}{q_R^2+m_{B0}^2}\frac{2m_\psi(\bar{\psi}\psi)_k}{q_R^2+m_\psi^2}\right\}\Bigg].\end{aligned} \tag{9}$$

There are (at least) two strategies available how to deal with this equation. The first is simply to substitute Eq. (7) on the right hand side of Eq. (9) and arrive there at an expression which depends exclusively on ρ_k. This approach will be called *version A* and has the following RGE:

$$\partial_k h_k = 0, \tag{10}$$

$$\begin{aligned}\partial_t U_k(\rho_k)=\frac{1}{2}\hat{\partial}_t\int_q\Big\{-5\log(q_R^2+2h_k^2\rho_k)\\ +\log\left[(q_R^2+m_{B0}^2(q_R^2+2h_k^2\rho_k)+4h_k^2\rho_kU'(\rho_k)\right]\Big\}.\end{aligned} \tag{11}$$

This substitution step will not influence the change in the other couplings since the change of the equation of state is higher order in an infinitesimal step of scale change. The other reasonable attitude is to expand the right hand side of Eq. (9) in power series of $(\bar{\psi}\psi)_k$, equate the coefficients of the first power of the expansion on the two sides and substitute Eq. (7) only for the higher powers. This leads to the following coupled set of RGE:

$$\begin{aligned}\partial_t U_k(\rho_k) &= \frac{1}{2}\hat{\partial}_t \int_q \Big\{-5\log(q_R^2 + 2h_k^2\rho_k) \\ &\quad + \log\left[(q_R^2 + m_{B0}^2(q_R^2 + 2h^2\rho_k) + 4h_k^2\rho_k U'(\rho_k)\right]\Big\} \\ &\quad - \frac{1}{2}\hat{\partial}_t \int_q \frac{2h_k^3}{(q_R^2 + m_\psi^2)(q_R^2 + m_{B0}^2)} \times \frac{2\rho_k U_k'(\rho_k)}{h_k}, \\ \partial_t(h_k I_k) &= -\frac{1}{2}\hat{\partial}_t \int_q \frac{2h_k^3 I_k}{(q_R^2 + m_\psi^2)(q_R^2 + m_{B0}^2)}, \qquad I_k = v_k(\bar{\psi}\psi)_k. \end{aligned} \tag{12}$$

Note that the equation of the Yukawa-coupling is extracted directly from the potential energy density, since one has introduced also a composite fermion background in addition to ρ_k. This way of extracting can be seen to be equivalent to the projection realized with help of an appropriate third functional derivative of Γ_k (see Ref. 15).

In the actual computations the so-called Litim regulators were used,[24] which allow a quick evaluation of the integrals on the right hand side of RGE. The equations are rewritten next in terms of dimensionless rescaled quantities using the index 'r' for them:

$$\begin{aligned} v_k &= v_r k^{d/2-1}, \qquad \psi_k = \psi_r k^{(d-1)/2}, \qquad I_k = I_r k^{3d/2-1}, \\ h_k &= h_r k^{2-d/2}, \qquad U_k(\rho_k) = k^d u_r(\rho_r = \rho_k k^{-d+2}). \end{aligned} \tag{13}$$

One arrives at a single equation in *version A*, where $h_k = h_\Lambda$:

$$\partial_t u_r + du_r + (2-d)\rho_r u_r' = v_d\left(-\frac{5}{1+\mu_\psi^2} + \frac{2+\mu_\psi^2+\mu_\sigma^2}{(1+\mu_\psi^2)(1+\mu_\sigma^2) + 4h_r^2\rho_r u_r'}\right), \tag{14}$$

and a set of two coupled equations in *version B*:

$$\begin{aligned}\partial_t u_r + du_r + (2-d)\rho_r u_r' = v_d\bigg(&-\frac{5}{1+\mu_\psi^2} + \frac{2+\mu_\psi^2+\mu_\sigma^2}{(1+\mu_\psi^2)(1+\mu_\sigma^2) + 4h_r^2\rho_r u_r'} \\ &+ 4h_r^2\rho_r u_r' \frac{2+\mu_\psi^2+\mu_\sigma^2}{(1+\mu_\psi^2)^2(1+\mu_\sigma^2)^2}\bigg)\end{aligned} \tag{15}$$

and

$$\partial_t h_r^2 + (4-d)h_r^2 = 4h_r^4 v_d \frac{2+\mu_\psi^2+\mu_\sigma^2}{(1+\mu_\psi^2)^2(1+\mu_\sigma^2)^2}. \tag{16}$$

In these equations the following notations were introduced:

$$m_\sigma^2 = k^2\mu_\sigma^2, \qquad m_\psi^2 = k^2\mu_\psi^2, \qquad v_d = \frac{S_d}{d(2\pi)^d}, \tag{17}$$

(S_d is the surface of the d-dimensional unit sphere).

3. Estimating $m_\sigma(\lambda, \Lambda)$ with help of the solution of RGE

Whatever parametrisation is used at $k = \Lambda$ the couplings characterizing the Ansatz (2) should be tuned to arrive at $k = 0$ in the broken symmetry phase where the minimum of the potential U_0 is taken at $\rho_{0,min} = (246GeV)^2/2$ and the Yukawa coupling hits the value $h_0 = 173/246$. Then one determines the second derivative of the potential at the minimum, which in turn determines the value of the scalar mass:

$$m_\sigma^2 = 2\rho_{0,min}U_0''(\rho_{0,min}). \tag{18}$$

The minimal Ansatz which is flexible enough for the required tuning of the parameters at $k = \Lambda$ is a quartic potential in ϕ. It has different parametrisations in the symmetric (SYM) and the broken symmetry (SB) regimes:

$$u_r^{SYM}(\rho_r) = \lambda_{1k}\rho_r + \frac{1}{2}\lambda_{2k}\rho_r^2, \qquad u_r^{SB}(\rho_r) = \frac{1}{2}\lambda_{2k}(\rho_r - \kappa_k)^2. \tag{19}$$

The stability of the potential requires in both phases $\lambda_{2k} \geq 0$. In the symmetric phase $\lambda_{1k} > 0$. In the broken symmetry phase for the physical (dimensionful) quantities one has

$$v_{0,min}^2 = \lim_{k\to 0} 2k^2\kappa_k, \qquad m_\sigma^2 = \lambda_{2,0}v_{0,min}^2. \tag{20}$$

In *version A* one chooses the fixed value $h_0 = h_k = h_\Lambda$ and by tuning $\lambda_{1,\Lambda}$ and subsequently κ_Λ one arrives at $k = 0$ to the physical value of the condensate. When $\lambda_{2,\Lambda}$ is not too large, one crosses over at some intermediate scale corresponding to $\lambda_{1,k_{crit}} = 0$ to the symmetry broken (SB) phase and one has to continue with the symmetry breaking parametrisation of the potential. For large enough $\lambda_{2,\Lambda}$ one starts straight in the SB phase. The result of m_σ depends then on $\lambda_{2,\Lambda}$ and Λ itself. This dependence was carefully studied in Ref. 15 without any fermionic background and we shall compare their results with those obtained with the two variants of RGE introduced above.

In *version A* the explicit RGE is written as a set of first order coupled non-linear equations for the coefficients of the quartic potential. They look like in the symmetric phase as:

$$\partial_t \lambda_1 + 2\lambda_1 = v_d \left[10h_r^2 - \frac{2+\lambda_1}{(1+\lambda_1)^2}(3\lambda_2 + 2h_r^2 + 6\lambda_1 h_r^2) + \frac{2h_r^2 + 3\lambda_2}{1+\lambda_1} \right],$$

$$\begin{aligned}\partial_t \lambda_2 + (4-d)\lambda_2 = v_d \Bigg[& -40h_r^4 - 20\frac{\lambda_2 h_r^2 (2+\lambda_1)}{(1+\lambda_1)^2} \\ & + \frac{2(2+\lambda_1)}{(1+\lambda_1)^3}(3\lambda_2 + 2h_r^2 + 6h_r^2\lambda_1)^2 \\ & - \frac{2(2h_r^2 + 3\lambda_2)}{(1+\lambda_1)^2}(3\lambda_2 + 2h_r^2 + 6h_r^2\lambda_1) \Bigg]. \end{aligned} \tag{21}$$

The denominator $(1+\lambda_1)$ represents the symmetric phase denominator of the scalar propagator, while in the symmetric phase the fermion is massless. In the SB phase the following equations are projected out from the equation valid for a general potential:

$$\begin{aligned}\lambda_2(-\partial_t \kappa_r + (2-d)\kappa_r) = v_d \Bigg[& \frac{10h_r^2}{(1+2h_r^2\kappa_r)^2} + \frac{3\lambda_2 + 2h_r^2}{(1+2h_r^2\kappa_r)(1+2\lambda_2\kappa_r)} \\ & - \frac{2(1+\kappa_r(\lambda_2 + h_r^2))}{(1+2h_r^2\kappa_r)^2(1+2\lambda_2\kappa_r)^2}(3\lambda_2 + 14h_r^2\lambda_2\kappa_r + 2h_r^2) \Bigg]. \end{aligned} \tag{22}$$

$$\begin{aligned}\partial_t \lambda_2 + (4-d)\lambda_2 = v_d \Bigg[& -\frac{40h_r^4}{(1+2h_r\kappa_r)^3} - \frac{40h_r^2\lambda_2(1+\kappa_r(\lambda_2 + h_r^2))}{(1+2h_r^2\kappa_r)^2(1+2\lambda_2\kappa_r)^2} \\ & + \frac{4(1+\kappa_r(h_r^2+\lambda_2))(2h_r^2 + 3\lambda_2 + 14h_r^2\lambda_2\kappa_r)^2}{(1+2h_r^2\kappa_r)^3(1+2\lambda_2\kappa_r)^3} \\ & - \frac{2(2h_r^2+3\lambda_2)(2h_r^2 + 3\lambda_2 + 14h_r^2\lambda_2\kappa_r)}{(1+2h_r^2\kappa_r)^2(1+2\lambda_2\kappa_r)^2} \Bigg]. \end{aligned} \tag{23}$$

For *version B* the projected equations of the potential are rather similar to those appearing above, just taking into account the subtraction performed in Eq. (12). We save space by not showing them explicitly. The equations for the Yukawa coupling are also rather simple:

$$\begin{aligned} \partial_t h_r^2 + (4-d)h_r^2 &= 4h_r^4 v_d \frac{2+\lambda_1}{(1+\lambda_1)^2} \qquad \text{(SYM)}, \\ \partial_t h_r^2 + (4-d)h_r^2 &= 8h_r^4 v_d \frac{1+\kappa_r(h_r^2+\lambda_2)}{(1+2\lambda_2\kappa_r)^2(1+2h_r^2\kappa_r)^2} \qquad \text{(SB)}. \end{aligned} \tag{24}$$

The resulting scalar mass values m_σ depend monotonically on $\lambda_{2,\Lambda}$ at any fixed value of Λ. With increasing $\lambda_{2,\Lambda}$ its value meets a limiting value

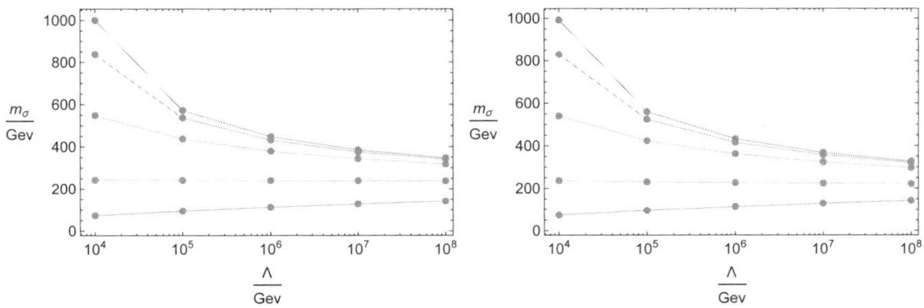

Fig. 2. The cutoff-dependence of m_σ for $\lambda_{2,\Lambda} = 0.001, 1., 10., 50., 100.$, from below upwards, obtained with *version A* (left) and *version B* (right) of RGE, respectively.

and apparently a limiting curve $m_\sigma(\lambda_{2,\Lambda} \to \infty, \Lambda)$ develops. It represents the triviality upper bound on the scalar mass. This is presented in Fig. 2. The left hand figure corresponds to *version A*, the right hand figure to *version B*. The figures are very similar not only qualitatively but also quantitatively, which argues for the robustness of the results against the different but consistent truncation schemes in LPA.

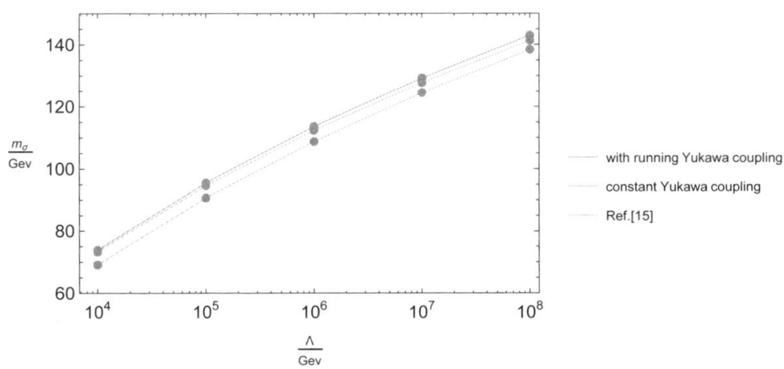

Fig. 3. The cutoff-dependence of m_σ for $\lambda_{2,\Lambda} = 0.001$ as obtained in three variants of LPA.

The stability of the results of Ref. 15 against including additional contributions due to a non-vanishing $(\bar{\psi}\psi)_k$ background is within 5% as one can see from Figures 3 and 4. Fig. 3 displays the variation of the scalar mass estimate for the smallest value $\lambda_{2,\Lambda} = 0.001$ which provides an estimate of the lower m_σ bound. One notices that all three estimates lie rather close to each other and follow the same functional dependence with Λ.

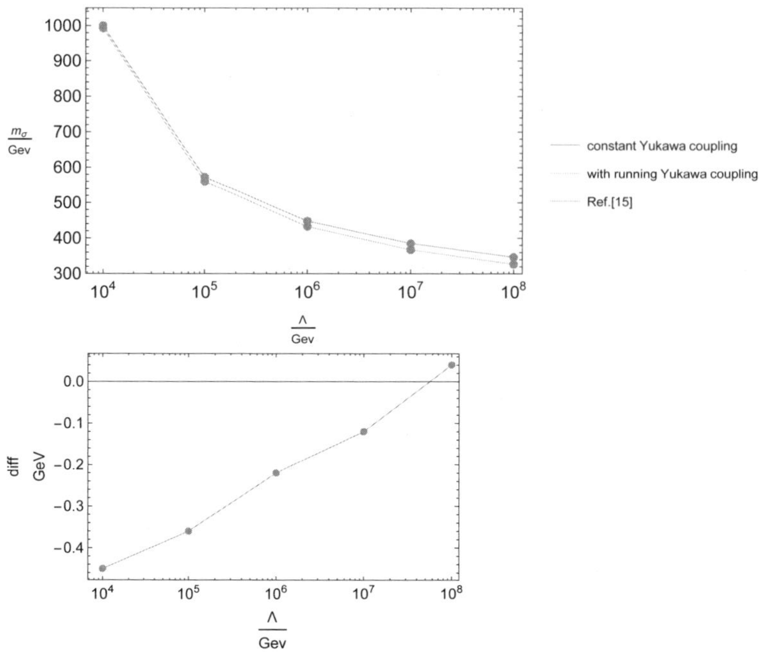

Fig. 4. The cutoff-dependence of m_σ for $\lambda_{2,\Lambda} = 100$ as obtained in three variants of LPA (top figure). The absolute difference of the estimates arising from our *version B* and the RGE used in Ref. 15 are shown in the lower figure

In case of the largest scalar self-coupling $\lambda_{2,\Lambda} = 100$ the points from Ref. 15 and our *version B* where the Yukawa-coupling non-trivially runs are optically indistinguishable, though the Λ-dependence of the difference shows some systematic tendency. These statements are illustrated in Fig. 4.

4. Conclusions

The non-perturbative mass estimates obtained in the lowest non-trivial truncation of the scalar potential, taking into account the effect of a non-trivial composite fermion background, coincides extremely closely with the results of a recent study where only scalar field background was included. This could have been expected with the present truncation, since it can explore only the close neighbourhood of the symmetry breaking minimum, where by Eq. (7) the additional diagrams included into our treatment in addition to those of Ref. 15 give only a minor contribution due to the nearly vanishing value of $(\bar{\psi}\psi)_k$. Certainly, this study should be continued in several aspects, before the conclusion drawn from this first experience could be confirmed. First, higher dimensional operators should be included into the scalar potential and check whether they influence the lower bound of m_σ the same way as was signalled in Refs. 15 and 18. Second, also the effect of the wavefunction renormalisation should be included into the RGE. Results on both questions will be published in a forthcoming communication.[25]

Acknowledgement

This research has been supported by the Hungarian Research Fund under the contract OTKA-K104292.

References

1. L.D. Landau, A.A. Abrikosov, and I.M. Khalatnikov, *Dokl. Akad. Nauk SSSR* **95**, 497, 773, 1177 (1954).
2. A.A. Anselm, http://thd.pnpi.spb.ru/history
3. V.N. Gribov, *Phys. Lett.* **B336**, 243 (1994).
4. S. Weinberg, in *"General Relativity: An Einstein centenary survey"*, ed. S.W. Hawking and W. Israel. Cambridge University Press, 1979. p. 790.
5. M. Shaposhnikov and C. Wetterich, *Phys. Lett.* **B683**, 196 (2010).
6. D. Litim and F. Sannino, *JHEP* **1412**, 178 (2014).
7. M. Lüscher and P. Weisz, *Nucl. Phys.* **B290**, 25 (1987); *ibid.* **B295**, 65 (1988).
8. S. Weinberg, *Physica* **96A**, 327 (1979).
9. L. Maiani, G. Parisi, and R. Petronzio, *Nucl. Phys.* **B136**, 115 (1978).
10. R. Dashen and H. Neuberger, *Phys. Rev. Lett.* **50**, 1897 (1983).
11. M. Lindner, *Z. Phys.* **C31**, 295 (1986).
12. J. Kuti,L. Lin and Y. Shen, *Phys. Rev. Lett.* **61**, 678 (1988).
13. T. Hambye and K. Riesselmann, *Phys. Rev.* **D55**, 7255 (1997).
14. Z. Fodor, K. Holland, J. Kuti, D. Nogradi, and C. Schroeder, *PoS* **LAT2007**, 056 (2007).
15. H. Gies, C. Gneiting, and R. Sondenheimer, *Phys. Rev.* **D89**,045012 (2014).
16. H. Gies and R. Sondenheimer, *Eur. Phys. J.* **C75**, 68 (2015).
17. A. Eichhorn, H. Gies, J. Jaeckel, T. Plehn, M.M. Scherer, and R. Sondenheimer, *JHEP* **1504**, 022 (2015).
18. D.Y.-J. Chu, K. Jansen, B. Knippschild, C.-J.D. Lin, and A. Nagy, arXiv:1501.05440 [hep-lat].

19. C. Wetterich, *Nucl. Phys.* **B352**, 529 (1991).
20. C. Wetterich, *Phys. Lett.* **B352**, 90 (1993).
21. T.R. Morris, *Int. J. Mod. Phys.* **A9**, 2411 (1994).
22. A. Jakovác and A. Patkós, *Phys. Rev.* **D88**,065008 (2013).
23. A. Jakovác, P. Pósfay, and A. Patkós, *Eur. Phys. J.* **C75**, 2 (2015).
24. D. Litim, *Phys. Rev.* **D64**, 105004 (2001).
25. A. Jakovác, I. Kaposvári, and A. Patkós, (work in progress).

A. Patkós, Chernogolovka, 2015.

MESON AND DIQUARK CONDENSATES IN THE NJL MODEL AND THE FERMI MOMENTUM

A. BHATTACHARYA

Department of Physics, Jadavpur University, Calcutta 700032, India, E-mail: phys.jdvu.ac.in

R. GHOSH

Department of Physics, Jogamaya Devi College, Kolkata, India, E-mail: rismita.ghosh@gmail.com

B. CHAKRABARTI

Department of Physics, Jogamaya Devi College, Kolkata, India, E-mail: ballari_chakrabarti@yahoo.co.in

Quark–anti quark and diquark condensation have been investigated in the framework of the NJL model. The Pauli–Villars regularization scheme have been used for the investigation of meson condensation in three dimension whereas the four dimensional case has been studied using the Schwinger–Dyson equation considering the Hartree approximation. Diquark condensation in three and four dimension have also been studied considering the Pauli–Villars regularization scheme. Using the Fermi momentum (p_f) of the particle as cut-off parameter, the gap energy/coherence length for meson condensation $\langle q\bar{q}\rangle$ have been investigated whereas for diquark qq the critical gap energy/critical coherence length (the distance over which there would be no diquark condensation) have been extracted. The variation of the coherence length/gap energy with p_f have also been investigated. The results are compared with exciting data. Some interesting observations are made.

1. Introduction

The NJL model suggested by Nambu and Jona-Lasinio[1] is based on an effective Lagrangian of relativistic fermions (quarks) which interact through a local current assuming that the gluon degrees of freedom can be frozen into a point like an effective interaction between quarks. BCS theory of superconductivity has a close analogy with properties of Dirac particles leading to NJL model description of elementary particles. The basic idea is that a diquark pair at the Fermi surface in quark matter becomes bound and condensed in the relativistic analogue of the BCS mechanism for electronic superconductivity. The NJL model has been widely used to study mesons[2–4] and diquark condensate. It is known that if quark–quark interaction is strong enough, two quark Cooper pair may exist and results in Bose-Einstein condensation. An excellent review on the NJL model and it's implication to hadrons and nuclei has been done by Vogl *et al.*[5] whereas Ebert *et al.*[6] have investigated $q\bar{q}$ condensate in the context of NJL model considering the spontaneous breakdown of chiral symmetry. Hands *et al.*[7] have investigated diquark color superconductivity and have observed that due to the strong attractive qq force in QCD, the estimate of gap energy which opens up at the Fermi surface may be as large as 100 MeV. Hellstern *et al.*[8] have studied the diquark confinement when the NJL model is extended beyond the rainbow ladder approximation. They have pointed out that a diquark has many appealing phenomenological aspects and are useful tools to parametrize unknown and complicated structure. They are color anti-triplet state to build a colorless baryons when interacting with a third quark. Mineo *et al.*[9] have used the NJL model to investigate anti-diquark or diquark-diquark interaction with Bethe–Salpeter–Faddeev equation. Diquark colour condensate has been studied by Molodtsov *et al.*[10] in the context of instanton liquid model. They have observed that such an interaction increases the critical value of chemical potential which causes an increase of quark matter density and yields onset of superconducting phase. Kawashima[11] has investigated attractive diquark channel $\langle qq \rangle$ in the infrared region in the context of QCD and observed that the behavior is not decided by decoupling dynamics as conventionally considered.

In the present work we have investigated the meson and diquark condensation in the framework of NJL model. Using the gap equation we have estimated the gap energy with Fermi momentum as cutoff parameter. We have used the Pauli–Villars regularization scheme and the Schwinger–Dyson equation to evaluate the properties of physical interest like gap energy, coherence length for mesons and diquark correlation. Possible Fermi momentum values for the heavy and light diquarks and mesons have been studied and an estimate of the Fermi momentums for the particles have been extracted.

2. NJL Model

The effective Lagrangian describing the local quark–anti quark interaction in two flavor NJL model has been described as[6]:

$$L = L_0 + L_{int} \tag{1}$$

where

$$L_0 = \bar{q}(i\gamma - m_0)q \tag{2}$$

$$L_{int} = G[(q\bar{q})^2 + (\bar{q}i\gamma_5\overline{\tau}q)^2] \tag{3}$$

Here 'G' is a dimensional strong coupling constant with the dimension GeV^{-2}. The point-like character of this coupling can be explained by the assumption that the gluon degrees of freedom is absorbed in the effective quark–anti quark coupling constant and such a reduction is justified in the low energy limit of QCD.[11] m_0 is the current quark mass, q is the Dirac field for the quark.

2.1. *Meson Condensation in Three Dimension*

For meson condensation in three dimensions, the gap equation runs as[12]:

$$\frac{\sigma}{2G} = \frac{i}{\Omega}\,\text{Tr}(i\partial - M)^{-1} = 4iN_cN_fM\int\frac{d^4k}{(2\pi)^4}\frac{1}{k^2 - M^2} \tag{4}$$

where σ the gap energy is M-m_0, M is the constituent quark mass, Ω is the volume of space time.

The Pauli–Villars regularization is characterized by the fact that the integrations are not constrained so that the integration variables can be freely shifted and all the results are mostly covariant. The effective action for the Pauli–Villars regularization can be expressed as[12]:

$$\Gamma^E_{eff}[\sigma,\pi] = \frac{1}{4G}\int d\tau d^3x(\sigma^2 + \pi^2) - \frac{1}{2}\sum_j c_j\,\text{Tr}\log(D^\dagger D + j\Lambda^2) \tag{5}$$

where D is the Dirac operator. The coefficient c_j is chosen in order to cancel the divergences in the effective action. With Λ as a cut-off parameter the gap equation for the Pauli–Villars regularization in three dimensions can be expressed as[12]:

$$\sigma = \frac{G}{2\pi^2}N_cN_fM\sum_j c_jM_j^2\log M_j^2 \tag{6}$$

The sums may be explicitly expressed as[12]:

$$\sum_0^3 c_j\log M_j^2 = \log\frac{M^2}{M^2 + 3\Lambda^2} + 3\log(1 + \frac{\Lambda^2}{M^2 + \Lambda^2}) \tag{7}$$

with $c_{0,1,2,3}$ = 1, -3,3,-1 and $M_j^2 = M^2 + j\Lambda^2$.

$$\sum_0^3 c_j M_j^2 \log M_j^2 = M^2 \left[\log \frac{M^2}{M^2+3\Lambda^2} + 3\log(1+\frac{\Lambda^2}{M^2+\Lambda^2}) \right] - 3\Lambda^2 \log\left[1 - \frac{\Lambda^4}{(M^2+2\Lambda^2)^4}\right] \tag{8}$$

The equation relating the coherence length (ξ) and the gap energy (σ) runs as,

$$\xi = \frac{1}{\sigma} \tag{9}$$

With the input of m_u = 360 Mev, m_s = 540 MeV, m_c = 1.71 GeV , m_b = 4.2 GeV, $N_c = 3$, $N_f = 6$, G = 1.166 GeV^{-2} and Fermi momentum as a cut-off parameter for the gap equation we have investigated the coherence length ξ ($\xi = 1/\sigma$) as a function of Fermi momentum p_F for light mesons $\langle u\bar{u}\rangle$, $\langle s\bar{s}\rangle$ and heavy mesons $\langle c\bar{c}\rangle$,$\langle b\bar{b}\rangle$ by using the relations (6) and (9). The results are displayed in the Figs. 1, 2 and 3.

2.2. *Meson Condensation in Four Dimension*

The Lagrangian describes the local $q\bar{q}$ interaction. Using the quark propagator $G(p)^{-1}$ as,

$$G(p)^{-1} = \gamma p - \Sigma - m_0 + i\varepsilon \tag{10}$$

we arrive at the Schwinger–Dyson equation in the Hartree approximation as

$$\Sigma = 2iG \int \frac{d^4p}{(2\pi)^4} [\mathrm{tr}\, G(p) + i\gamma_5\, \mathrm{tr}\{i\gamma_5 G(p)\}] \tag{11}$$

Defining the $\langle \bar{q}q\rangle$ condensate in momentum space as

$$\langle \bar{q}q\rangle = -12i \int \frac{d^4p(\Sigma + m_0)}{(2\pi)^4(p^2 - (\Sigma + m_0)^2 + i\epsilon)} \tag{12}$$

With $\Sigma + m_0 = m$, the constituent quark mass, we arrive at the gap equation:

$$m = m_0 - 2G\bar{q}q \tag{13}$$

Equation (13) is referred as the gap equation of the NJL model in an analogy to the corresponding equation that determines the energy gap in a superconductor. To evaluate the expression in the one loop approximation we come across a quadratically divergent integral. The integral becomes finite with the cut-off Λ and can be expressed as $\int \frac{d^4p\theta(\Lambda^2-p^2)}{(m^2-p^2)}$. It has been suggested by Blaschke *et al.*[13] that the quark condensate can be obtained in a domain between UV and IF region by spontaneous breakdown of chiral symmetry where quark loops has no imaginary part. The gap equation then can be expressed with cut-off Λ/m with no contribution from IF part. Assuming the cut-off parameter Λ as the Fermi momentum p_f of the corresponding condensate we arrive at the expression of the gap energy as:

$$\Delta = m - m_0 = \frac{3Gmp_f^2}{\pi^2}\left(\sqrt{1+\frac{m^2}{p_f^2}} - \frac{m^2}{p_f^2}\ln\frac{p_f+\sqrt{p_f^2+m^2}}{m}\right) \tag{14}$$

The Δ must be a positive quantity for a meaningful interpretation of gap energy. With x=$\frac{p_f}{m}$<1 and using $ln[x+\sqrt{x^2+1}] = x - \frac{x^3}{6} + \frac{3x^5}{40} - ....$, the expression (14) may be expanded to obtain Δ as:

$$\Delta = \frac{3Gm}{\pi^2}\left(\frac{2}{3}m^2x^3 - \frac{1}{5}m^2x^5 + ...\right) \tag{15}$$

Thus we have come across an expression relating the Fermi momentum and gap energy, with g^2= GΛ'^2 (see Ref. 5) where 'g' is a dimensionless coupling constant. We have considered the strong coupling limit $g^2/4\pi \geq 1$ (see Ref. 5) and used 'g' $\sim = \sqrt{4\pi} = 3.54$. We have plotted Fermi momentum p_f correlation length ξ ($\frac{1}{\Delta}$)graph for light and heavy condensates and have shown in Fig. 1 with Λ'=1.25 GeV for the light sector[14] and 1 GeV for the heavy sector.[15]

2.3. *Diquark Condensation under Three and Four Dimensional Pauli-Villars Regularization Scheme*

The Lagrangian for scalar diquark can be expressed as[12]:

$$L_0 = \bar{q}(i\gamma - m_0)q + G[(\bar{q}q)^2 + (\bar{q}i\gamma_5\overline{\tau}q)^2] + H[\overline{q}_{bi}^C\gamma_5\epsilon_{abc}\epsilon_{ij}q_{cj}]^2 \tag{16}$$

where a, b, c, and i, j are color and flavor indices respectively. The critical coupling for diquark binding, when the corresponding poles enter the two particle continuum (at p^2= $4M^2$) may be recast as[12]:

$$H_c = \frac{1}{4[\frac{\sigma}{2GN_cM} + 8N_fM^2I(4M^2)]}. \tag{17}$$

Rearranging the above equation the gap equation for diquark can be expressed as:

$$\sigma = 2GN_cM\left[\frac{1}{4H_c} - 8N_fM^2I(4M^2)\right], \tag{18}$$

where

$$I(4M^2) = \int^{\Lambda/M} \frac{d^3k}{(2\pi)^3} \frac{1}{\epsilon_k(4\epsilon_k^2 - 4M^2)} \,. \tag{19}$$

For three dimensions we computed the above integral as:

$$I_{3D}(4M^2) = \frac{\Lambda/M}{16\pi^3} \tag{20}$$

and for four dimension as in Ref. 12,

$$I_{4D}(4M^2) = \frac{1}{16\pi^2}\left[\log(1+\frac{\Lambda^2}{M^2}) + \frac{\Lambda}{M}\,\text{arctg}\,\frac{M}{\Lambda}\right] . \tag{21}$$

With the input of $I_{3D}(4M^2)$, $I_{4D}(4M^2)$ and the suggested value of the critical fields $H_c = 3.67$ GeV^{-2} (see Ref. 9) and Fermi momentum as cut-off, we have studied the variation of gap energy with Fermi momentum p_F for diquark condensation using Eq. (18). Results are displayed in the Figs. 5 and 6 or three dimensional regularization whereas Figs. 7 and 8 represent the four dimensional regularization.

3. Results and Discussions

In the present work we have investigated the superconducting order parameter or excitation gap for meson and diquark condensate in the framework of NJL model. Using the Pauli-Villars regularization scheme, the gap energy has been estimated with Fermi momentum as a cut-off parameter. It is well known that at low energy the hadronic phase is realized with dynamically broken chiral symmetry by condensation of quark–anti quark pairs and quark–quark pairs. The variation of coherence length $(1/\xi)$ of mesons in three dimensional regularization with moment cut-off p_F have been estimated for the light quark sector and displayed in Fig. 1 whereas for $c\bar{c}$ and $b\bar{b}$ the variations are displayed in Figs. 2 and 3. Chernodub[16] has made a detailed comparison between the properties of vacuum superconductivity and normal superconductivity. Schwinger pair production has been discussed in detail and it has been observed that the vacuum pairs are exotic in nature and have a typical coherence length $\xi \sim 1$ fm. In the present work it has been observed that the coherence length decreases with increasing Fermi momentum and for coherence length to be ~ 1 fm the corresponding Fermi momentum should be 0.66 GeV for $u\bar{u}$ condensate whereas for $s\bar{s}$ the p_F is 0.685 GeV. p_F is obtained as 0.942 GeV which corresponds to the coherence length ~ 1 fm for heavy quark condensate $c\bar{c}$. The quark Cooper pair has been investigated by a number of authors.[8,17] Bali *et al.*[19] have investigated the dual Maxwell equations for color field and monopole current distribution around static quark–antiquark pair $q\overline{q}$ in vacua. They have estimated the coherence length ξ= 0.25(3) fm using dual superconductor picture with Ginzburg-Landau type analysis.

Using the Schwinger-Dyson equation the gap energy of the meson condensation in four dimension has been estimated. In the present work superconducting order parameter or excitation gap has been estimated with Fermi momentum as cut-off. The variation of ξ with Fermi momentum have been displayed in Fig. 4 for light and heavy sector of condensates. It has been observed that the coherence length decreases with increasing Fermi momentum. It may be mentioned that superfluidity in fermion condensate has been studied by Khodel *et al.*[20] in Landau-Luttinger theory of Fermi liquid of quasi particles with momentum dependent distribution function $n(p)$. They have argued that the properties of fermion condensate systems are sharply different from ordinary properties and have obtained a linear relationship between Δ with pairing constant Λ indicating the fact that the value of Δ is not exponentially small in such systems. They have estimated the coherence length ξ 0.25(3) fm using a dual superconductor picture with a Ginzburg-Landau type analysis. We have observed that the p_f = 0.272 GeV yields the coherence length 0.25 fm for $\langle s\overline{s}\rangle$ condensate whereas p_f=0.254 GeV yields the same value of coherence length (0.25 fm) for heavy condensate. In the present work it also has been observed that if the coherence length to be ~1 fm, the Fermi momentum should be 0.176 GeV and 0.172 GeV for $\langle u\overline{u}\rangle$ and $\langle s\overline{s}\rangle$ respectively, whereas for the heavy meson condensate p_f= 0.142 GeV yields the coherence length of ~1 fm. Kiuchi[21] has investigated the coherence length of quark Cooper pairs for two flavors numerically within the framework of a modified QCD-like gauge field theory with the lattice-QCD-based gluon propagator, which is derived from the lattice QCD data. Matsuzaki[17] has investigated special structure of quark Cooper pairs in color superconductor particularly with density, and have suggested that the quark Cooper pairs are more bosonic at higher density from the study of coherence length and gap energy, whereas Abuki *et al.*[18] have investigated the problem in two flavour color superconductivity for a wide range of baryonic density. BEC-BCS crossover in the frame work of NJL model has been investigated by a number of authors[22–25] where it has been suggested that diquark condensation in two color QCD and pion condensation in real QCD would undergo a BEC-BCS crossover as the chemical potential increases.

The variation of diquark condensation gap energy with Fermi momentum has been displayed in Figs. 5, 6, 7 and 8. Kerbikov[26] has investigated the color superconducting phase in the framework of QCD and has observed that the color superconducting phase occurs in the BCS-BE crossover region. The gap energy has been estimated to be 0.1 GeV with coherence length $\xi = 0.6\,\text{fm}$. They have asserted that the diquark pair in colour superconducting phase is compact. In color anti-triplet $\overline{3}$ state the one gluon exchange leads to the quark quark potential which is one or two times weaker than the quark–antiquark one. The NJL model also results in strong q-q attraction. It has been observed that the q-q interaction is stronger than photon mediated electron-electron interaction. Considering critical value of gap energy to be 100 MeV as in Ref. 26, we have found that the Fermi momentum is in the range $0.80 \sim 0.81\,\text{GeV}$ for light quark sector for three dimensional regularization whereas for four dimensional regularization p_F is found to be in the range $0.21 \sim 0.18\,\text{GeV}$. Kiuchi[21] has investigated coherence length of quark Cooper pairs in modified QCD like gauge theory with latice QCD based gluon propagator. He has observed that $\frac{\xi}{d} < 1$ at $\mu < 0.65\,\text{GeV}$ where μ is the chemical potential and d is the inter particle distance. Schafer[27] has investigated diquark condensation in high density baryon matter and has argued that cold quark matter is diquark Bose condensate. They have suggested that two light flavor instanton effects lead to a gap of order $\Delta = 50\,\text{MeV}$.

In the present work we have studied the quark–antiquark and quark–quark condensates formation from vacuum superconductivity and extracted the values of Fermi momentum studying the coherence length for both light and heavy mesons. We have also studied the critical coherence length for diquark condensates using dimensional regularization. Fermi momentum of a particle is a very important quantity, particularly for the study of CKM matrix elements. p_F for meson has been studied by a number of authors and a wide range of values have been suggested.[25] We have extracted values of Fermi momentum from vacuum superconductivity for wide range of particles from heavy to light mesons and diquarks in the present investigation.

References

1. Y. Nambu and G. Jona-Lasinio, *Phys. Rev.* **124**, 246 (1961).
2. X.Y. Guo, X.L. Chen, and W. Z. Deng, *Chin. Phys.* **C37**, 033102 (2012).

3. Y. Jiang, K. Ren, and P. Zhuang, *Eur. Phys. J.* **C71**, 1822 (2011).
4. T. Inagaki, D. Kimura, H. Kohyama, and A. Kvinikhidze, *Phys. Rev.* **D83**, 034005 (2011).
5. U. Vogl and W. Weise, *Prog. Part. Nucl. Phys.* **27**, 195 (1991).
6. D. Ebert, A.N. Ivanov, and M.K. Volkov, *Fortsch. der Physik* **37**, 487 (1989).
7. S. Hands and D.N. Walters, *Phys. Lett.* **B548**, 196 (2002).
8. G. Hellstern, R. Alkofer, and H. Reinhardt , *Nucl. Phys.* **A625**, 697-712 (1997).
9. H. Mineo, J.A. Tjon, K. Tsushima, and S.N. Yang, *Phys. Rev.* **C77**, 055203 (2008).
10. S.V. Molodtsov and G.M. Zinavjev, *Phys. Atom. Nucl.* **66**, 1349-1356 (2001).
11. K. Kawashima, *Int. J. Mod. Phys.* **A25**, 2023-2035 (2010).
12. T. Brauner, https://sites.google.com/site/braunercz/notes.
13. D. Blaschke *et al.*, arXiv:hep-ph/9812503.
14. D. Ebert *et al.*, arXiv:hep-ph/9406220.
15. S.I. Kruglov, *Act. Phys. Pol.* **B21**, 985 (1990).
16. M.N. Chernodub, *Int. J. Mod. Phys.* **A27**, 1260003 (2012).
17. M. Matsuzaki, *Phys. Rev.* **D62**, 074014 (2000).
18. H. Abuki, T. Hatsuda, and K. Itakura, *Phys. Rev.* **D65**, 074014 (2002).
19. G.S. Bali *et al.*, *Nucl. Phys. Proc. Suppl.* **73**, 638 (1999).
20. V.A. Khodel and V.R. Shaginyan, *JETP Lett.* **51**, 553 (1990).
21. H. Kiuchi, *J. Phys. Studs.* **11**, No. 4, 390-395 (2007).
22. D.S. Hwang, C.S. Kim, and W. Namgung, *Z. Phys.* **C69**, 107 (1995).
23. A.K. Rai *et al.*, *J. Phys. G: Nucl. Part. Phys.* **28**, 2275 (2002).
24. A. Bhattacharya *et al.*, *Eur. Phys. J.* **C2**, 671 (1998).
25. S. Pepin, M.C. Birse, J.A. McGovern, and N.R. Walet, *Phys. Rev.* **C61**, 055209 (2000).
26. B.O. Kerbikov, *Phys. Atom. Nucl.* **65**, issue 10, 1918-1920 (2002).
27. T. Schafer, *Nucl. Phys.* **A642**, 45c-50c (1998).
28. D.S. Hwang, C.S. Kim, and N. Wuk, *Phys. Rev.* **D53**, 4951-4956 (1996).

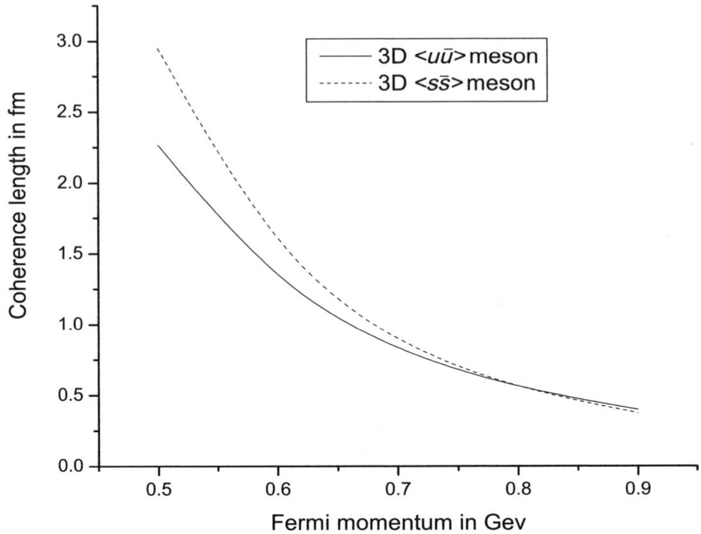

Fig. 1. Variation of coherence length (ξ) with Fermi momentum (p_F) for 3D light quark mesons.

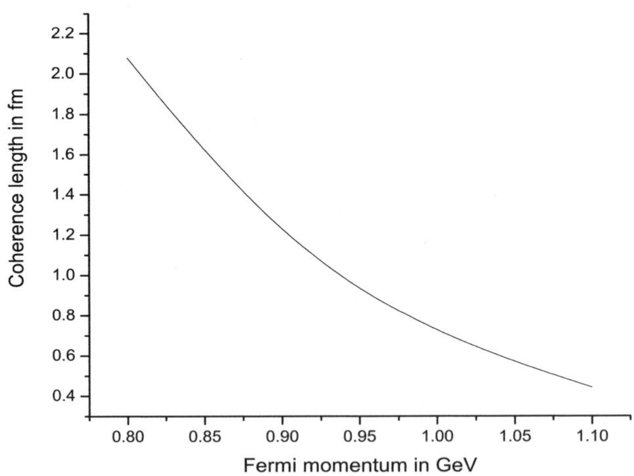

Fig. 2. Variation of coherence length (ξ) with Fermi momentum (p_F) for 3D charm quark meson.

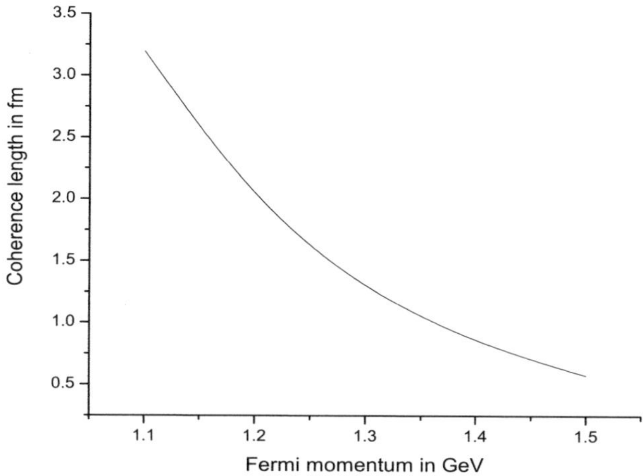

Fig. 3. Variation of coherence length (ξ) with Fermi momentum (p_F) for 3D bottom quark meson.

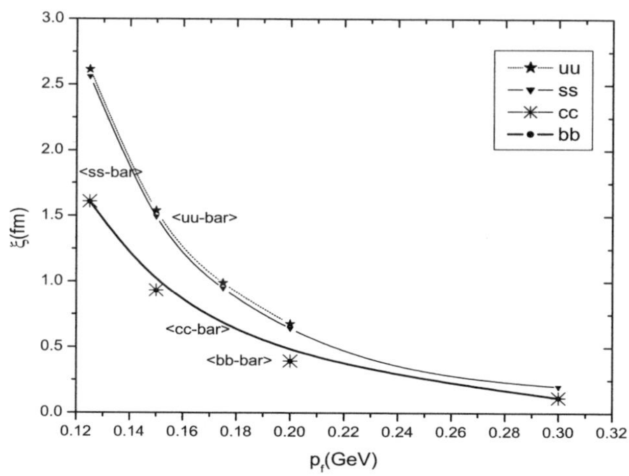

Fig. 4. Variation of coherence length (ξ) with Fermi momentum (p_F) for 4D meson.

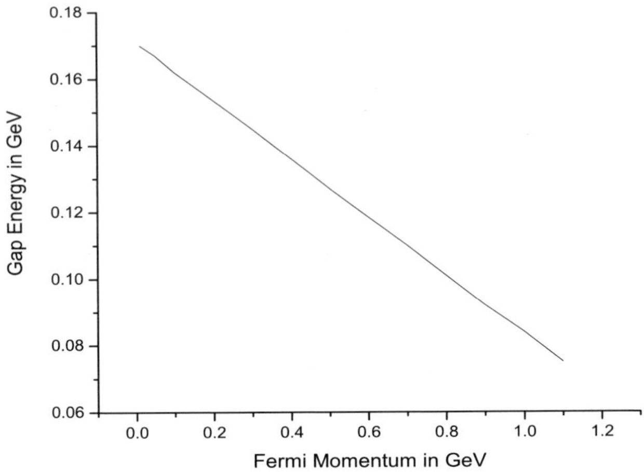

Fig. 5. Variation of Gap Energy (σ) with Fermi momentum (p_F) for 3D [uu] diquark.

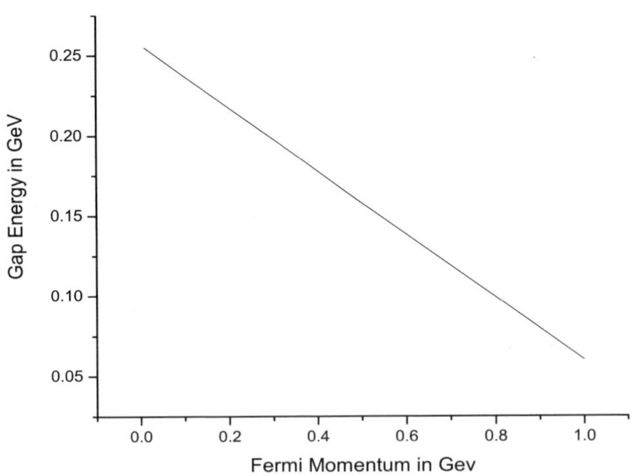

Fig. 6. Variation of Gap Energy (σ) with Fermi momentum (p_F) for 3D [ss] diquark.

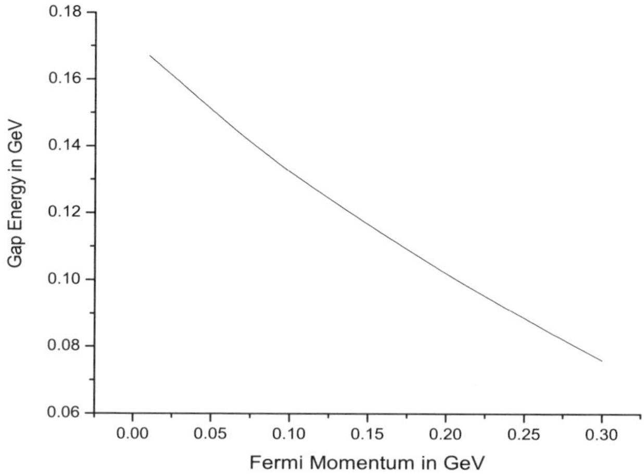

Fig. 7. Variation of Gap Energy (σ) with Fermi momentum (p_F) for 4D [uu] diquark.

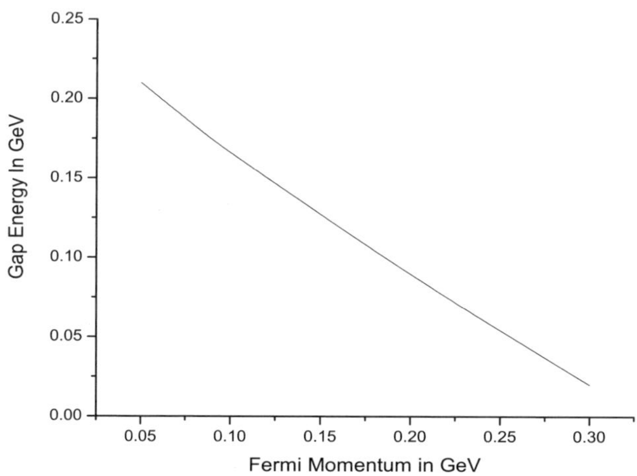

Fig. 8. Variation of Gap Energy (σ) with Fermi momentum (p_F) for 4D [ss] diquark.

SPIN TQFTS AND FERMIONIC PHASES OF MATTER

DAVIDE GAIOTTO

Perimeter Institute for Theoretical Physics,
Waterloo, Ontario, Canada N2L 2Y5

ANTON KAPUSTIN

Simons Center for Geometry and Physics,
Stony Brook, NY 11790

We study lattice constructions of gapped fermionic phases of matter. We show that the construction of fermionic Symmetry Protected Topological orders by Gu and Wen has a hidden dependence on a discrete spin structure on the Euclidean space-time. The spin structure is needed to resolve ambiguities which are otherwise present. An identical ambiguity is shown to arise in the fermionic analog of the string-net construction of 2D topological orders. We argue that the need for a spin structure is a general feature of lattice models with local fermionic degrees of freedom and is a lattice analog of the spin-statistics relation.

1. Introduction and summary

1.1. *Bosonic and fermionic gapped phases*

In condensed matter physics, topological phases of matter are often defined as equivalence classes of local gapped bosonic Hamiltonians, usually defined on a lattice, which can be deformed into each other without ever becoming gapless.[1,2] The notion of topological phase can be enriched by imposing additional constraints on the theories, such as a choice of global symmetry preserved by all the Hamiltonians. On the other hand, topological quantum field theories* can be thought of as describing the far infrared behavior of gapped bosonic quantum field theories (see, e.g., Section 4 of Ref. 3 or the monograph[4]).

*This is a somewhat looser notion of TQFT compared to some formal definitions. For example, we consider Chern-Simons theory to be a TQFT, even though it has a partition function which depends on a choice of metric on space-time. In other words, we allow the stress tensor to be non-zero, but proportional to the identity operator.

There is a close relation between topological phases of matter and topological quantum field theories, which can be thought of as a map from a topological phase of matter to the TQFT which encodes the low energy continuum limit of the corresponding Hamiltonian. In principle, one may imagine the map being many-to-one: it is not obvious that two local gapped Hamiltonians which map to the same TQFT will always be deformable into each other. Still, in practice we do not know of any observable which can distinguish two phases of matter, but cannot be formulated in terms of the TQFT data.†

In condensed matter physics, one also encounters the notion of a fermionic topological phase of matter, defined as an equivalence class of local gapped Hamiltonians which can involve fermionic degrees of freedom.[5,6] Perhaps surprisingly, some fermionic phases of matter are not expected to admit a purely bosonic realization. This is expected to be due to the difference in the notion of locality for bosonic and fermionic systems. Intuitively, if we partition a bosonic system in two parts, the total Hilbert space factors uniquely in the tensor product of the Hilbert spaces for the two parts. If we partition a fermionic system, though, the factorization has an intrinsic ambiguity, as observables in the tensor product of the Hilbert spaces for the two parts are defined up to a sign in the sector where both factors have odd fermion number.

1.2. *Spin structure dependence*

In unitary quantum field theory, fermions are naturally spinors and thus the low energy physics of a gapped fermionic theory is a spin-TQFT: a topological field theory defined on manifolds which can be equipped with a spin structure, whose correlation functions possibly depend on the choice of spin structure. The purpose of this paper is to explore the relation between fermionic topological phases of matter and spin-TQFTs. It is not obvious that such a relation should exist, as a lattice Hamiltonian involving fermionic degrees of freedom is usually written down without any reference to a spin structure on the manifold which is discretized by the lattice. One also cannot appeal to the spin-statistics relation, because the lattice

†It is also conceivable, perhaps, that some topological phase of matter may not give rise to a TQFT at low energy, i.e., that some anomaly/obstruction may prevent the definition of TQFT amplitudes on general manifolds in terms of the Hamiltonian data. But in all cases known to us one circumvent such obstructions by postulating that the TQFT depends on some additional geometric data, such as metric or framing.

destroys Lorenz and even rotational invariance which are the conditions of the spin-statistics theorem.‡

The first step of our analysis is to look carefully at the fermionic SPT phases constructed by Gu and Wen in Ref. 7. We find that the prescription used to define the partition function of such theories runs into an obstruction if applied to space-time manifolds of general topology, unless the second Stiefel–Whitney class $[w_2]$ of the manifolds vanishes, i.e., the manifold admits a spin structure. If the manifolds admits a spin structure, the obstruction can be eliminated, but the final answer will depend on the choice of spin structure η. In other words, these fermionic SPT phases define (invertible) spin-TQFTs.

Next, we look at other known constructions of fermionic phases of matter which are expected to admit a state-sum-like definition of their partition function: the construction of fermionic toric code in Ref. 8 and the general fermionic Turaev-Viro construction in Ref. 9. These references focus on the construction of a fixed-point Hamiltonian and wave-function for these fermionic phases of matter, rather than a partition function. It is straightforward, though, to assemble the same ingredients into a partition sum, borrowing some ideas from the Gu–Wen fermionic SPT phase construction. Again, we find an obstruction to define the partition sum unless the space-time manifolds admits a spin structure, in which case one can remove the obstruction and define a well defined partition function which depends on the choice of spin structure η. Thus these fermionic phases of matter are associated to spin-TQFTs.

We can describe the obstruction schematically here, referring the reader to Sections 2 and 6 for further details. State-sum models assemble the partition function from a triangulation of the space-time manifold X: each simplex is associated to some tensor in the tensor product of vector spaces associated to the faces and the legs of these tensors are contracted together as the simplices are glued along the corresponding faces of the triangulation. In a fermionic model, the vector spaces may be Grassmann-odd and Koszul signs occur when re-organizing and contracting the factors of the tensor products.

‡Taking the continuum limit and then applying the spin-statistics relation does not ameliorate the problem. The TQFT itself is, of course, Lorentz invariant, but the spin-statistics relation is a property of Lorenz invariant particle excitations. The continuum limit from the lattice theory to the low-energy TQFT only concerns the ground states of the system. A priori, massive excitations above these ground states do not need to transform properly under the Lorentz group.

These Koszul signs, arising from the anti-commutation of fermionic variables, are of course a key element of the problem. The non-local nature of these signs is precisely what should allow these fermionic phases of matter to be distinct from any bosonic phase. In order for the partition sum to be invariant under local changes in the triangulation of the manifold, one needs to cancel the change in the Koszul signs against the change in the local data attached to the simplices. The obstruction arises precisely when this cancellation is not possible.

We can express the obstruction neatly by encoding the fermion number of the vector spaces attached to faces in a $\mathbb{Z}_2$-valued $(d-1)$-cochain β_{d-1}. The cochain β_{d-1} is actually a cocycle, as the total fermion number of the tensors attached to simplices is even. It is useful to decompose the partition sum into a sum of terms $Z[X, \beta_{d-1}]$, which contain the parts of the state sum due to states of fermion number β_{d-1}.

We can encode a general change of triangulation of X into a triangulation of the $(d+1)$-dimensional manifold $X \times [0,1]$. Intuitively, we are gluing a sequence of $(d+1)$-dimensional simplices on top of our initial triangulation to get the final triangulation. We find that the triangulation invariance of the partition function is obstructed by some irreducible sign mismatch, which can be written schematically as

$$(-1)^{\int_{X\times[0,1]} w_2 \cup \beta_{d-1}} . \tag{1}$$

Here w_2 is a 2-cocycle with values in $\mathbb{Z}_2$ representing the second Stiefel–Whitney class of $X \times [0,1]$ and β_{d-1} is a lift to $X \times [0,1]$ of the cocycle β_{d-1}. If the cohomology class of $[w_2]$ is non-trivial and the theory involves choices of fermion numbers β_{d-1} which are non-trivial in cohomology, this sign mismatch cannot be absorbed by a redefinition of the local part of the partition function. This prevents us from constructing a well-defined partition sum and ruins the state-sum construction.

If we restrict X to be a spin manifold then $w_2(X)$ is exact and we can write $w_2 = \delta\eta$ for some 1-cochain η, which represents a choice of spin structure. This allows us to thus cancel the obstruction 1 by the variation of a local term,

$$(-1)^{\int_X \eta \cup \beta_{d-1}} , \tag{2}$$

so that the improved state sum,

$$Z[X, \eta] = \sum_{\beta_{d-1}} Z[X, \beta_{d-1}](-1)^{\int_X \eta \cup \beta_{d-1}} , \tag{3}$$

is fully invariant under changes of triangulations and defines a good theory. This theory is a spin-TQFT: it can only be defined on a spin manifold and depends on a choice of spin structure.

In Sections 3 and 4 we will look in further detail at the properties of the Koszul signs which occur in the state sum. The definition the partition function requires specific choices of how to order the factors in the tensor product associated to each simplex, and the two factors in the contraction of vector spaces at each face. Given some ordering choices Π, the permutations of the vector spaces involved in the state sum will produce some overall Koszul sign $\sigma_\Pi(X, \beta_{d-1})$, which depends only on the triangulation, on Π and on β_{d-1}.

The choice of order Π can be given independently of the other data in the state sum. The combined sign,

$$z_\Pi[X, \eta, \beta_{d-1}] = \sigma_\Pi(X, \beta_{d-1})(-1)^{\int_X \eta \cup \beta_{d-1}}\,, \tag{4}$$

appears to be a very useful object, which captures the intrinsically fermionic part of the full partition function. From now on we will drop the subscript Π. Our formulae will refer to the specific choice of order used in the Gu–Wen definition of fermionic SPT phases.[7] We will comment briefly on other choices of order in Section 3.

We can think about $z[X, \eta, \beta_{d-1}]$ as defining an effective action for a $(d-1)$-form $\mathbb{Z}_2$ gauge field with a very specific anomaly, or a very simple invertible spin-TQFT K_d equipped with an anomalous $(d-2)$-form $\mathbb{Z}_2$ global symmetry.

Under changes of triangulation, $z[X, \eta, \beta_{d-1}]$ changes by another interesting cocycle, the Steenrod square of β_{d-1}:

$$(-1)^{\int_{X\times[0,1]} Sq^2[\beta_{d-1}]} \equiv (-1)^{\int_{X\times[0,1]} \beta_{d-1}\cup_{d-3}\beta_{d-1}}\,. \tag{5}$$

We refer to Appendix B for the explicit definition of the higher cup products $\cup_a$. Their basic property is

$$A \cup_a B + B \cup_a A = \delta(A \cup_{a+1} B) + \delta A \cup_{a+1} B + A \cup_{a+1} \delta B\,, \tag{6}$$

with $\cup_0 \equiv \cup$.

Under gauge transformations, we find the precise form of the 't Hooft anomaly,

$$z[X, \eta, \beta_{d-1} + \delta\lambda_{d-2}] = z[X, \eta, \beta_{d-1}](-1)^{\int_X \beta\cup_{d-3}\lambda + \lambda\cup_{d-3}\beta + \lambda\cup_{d-3}\delta\lambda + \lambda\cup_{d-4}\lambda}\,. \tag{7}$$

Although $z[X, \eta, \beta_{d-1}]$ does not appear to admit a d-dimensional bosonic description, we also find that it is a quadratic refinement of a bosonic pairing:

$$z[X,\eta,\beta_{d-1}+\beta'_{d-1}] = z[X,\eta,\beta_{d-1}]z[X,\eta,\beta'_{d-1}](-1)^{\int_X \beta_{d-1}\cup_{d-2}\beta'_{d-1}}\,. \quad (8)$$

Finally, if X is a boundary of a compact oriented $(d+1)$-manifold Y and $d>2$, we find an explicit WZW-like expression for $z[X,\eta,\beta_{d-1}]$:

$$z[X,\eta,\beta_{d-1}] = (-1)^{\int_X \eta\cup\beta_{b-1}+\int_Y Sq^2[\beta_{d-1}]+w_2\cup\beta_{d-1}}\,. \quad (9)$$

Here we use the fact that for $d>2$ the cocycle β_{d-1} can be extended to Y. The action is independent of the choice of Y or of the way β_{d-1} is extended from X too Y because the expression $Sq^2[\beta_{d-1}]+w_2\cup\beta_{d-1}$ is exact for closed oriented Y. This formula is particularly useful for $d=3$, since any closed oriented 3-manifold X is a boundary of a compact oriented 4-manifold Y.

With a bit of extra work, we can rewrite the partition function $Z[X,\eta]$ of our spin-TQFT as the partition function of a $(d-1)$-form $\mathbb{Z}_2$ gauge theory,

$$Z[X,\eta] = \sum_{\beta_{d-1}} \tilde{Z}[X,\beta_{d-1}]z[X,\eta,\beta_{d-1}]\,, \quad (10)$$

where the gauge fields are coupled to two two sets of degrees of freedom: a standard bosonic TQFT equipped with a $(d-2)$-form $\mathbb{Z}_2$ global symmetry and partition function $\tilde{Z}[X,\beta_{d-1}]$, and the spin-TQFT K_d. The bosonic theory associated to $\tilde{Z}[X,\beta_{d-1}]$ must have a 't Hooft anomaly which cancels the 't Hooft anomaly of K_d, controlled by $Sq^2[\beta_{d-1}]$.

In order to make contact with concepts which are more familiar in condensed matter physics, it is useful to replace the notion of a TQFT with an anomalous global symmetry with the notion of a gapped boundary condition for a $(d+1)$-dimensional SPT phase, protected by a $(d-2)$-form $\mathbb{Z}_2$ global symmetry, with partition function

$$(-1)^{\int_Y Sq^2[\beta_{d-1}]}\,. \quad (11)$$

Then $\tilde{Z}[X,\beta_{d-1}]$ defines a bosonic gapped boundary condition for the $(d+1)$-dimensional SPT phase, while $z[X,\eta,\beta_{d-1}]$ defines a fermionic gapped boundary condition. The original spin TQFT can be recovered by gauging the $(d-2)$-form $\mathbb{Z}_2$ global symmetry on a slab, with one of these boundary conditions at either end.

In Section 7 we will argue that this construction has a close relation to the notion of fermionic anyon condensation. The generators of a non-anomalous $(d-2)$-form global symmetry G are loop observables which can be thought as worldlines of bosonic quasi-particles which fuse accordingly to the group law of G. Gauging the $(d-2)$-form symmetry is equivalent to proliferating these quasi-particles in correlation functions and should correspond to the standard notion of anyon condensation, at least in three space-time dimensions.

We will argue that the generators of a $(d-2)$-form $\mathbb{Z}_2$ global symmetry with the 't Hooft anomaly described above, instead, behave as fermionic quasi-particles. The challenge to define a fermionic analogue of the standard anyon condensation is mapped into the problem of gauging such anomalous symmetry. The kernel spin-TQFT K_d offers a solution to the problem: given some TQFT with fermionic quasi-particles we want to condense, we can tensor it with K_d to cancel the anomaly and gauge the $(d-2)$-form $\mathbb{Z}_2$ global symmetry. Essentially, the kernel spin-TQFT K_d uses the spin structure information to provide some extra signs which make the "fermionic" anyon condensation meaningful. This approach to fermionic anyon condensation appears to be closely related to work in progress by K. Walker.[25]

State-sum constructions of spin-TQFTs in two dimensions have been recently discussed by other authors.[26,27] It would be interesting to establish the precise connection between all these constructions.

Finally, using the notion of fermionic anyon condensation we sketch a rough argument demonstrating how one could potentially "simulate" a generic fermionic lattice Hamiltonian in $2+1$ dimensions given a copy of K_3 and a sufficiently rich bosonic system. This argument supports the idea that K_d may fully capture the non-local properties of a generic fermionic system.

1.3. *Conclusions and future directions*

Although we have demonstrated the link between fermionic phases of matter and spin-TQFTs only in a restricted set of examples, we believe that the relation will hold in greater generality. There are two natural ways one may try to extend our results

- It should be possible to adapt our analysis to the general mathematical framework of extended topological field theory. The analogue of a fermionic phase of matter should be an extended topological field theory such that the vector spaces attached to $(d-1)$-dimensional manifolds have Grassmann grading and the tensor products are twisted by the Koszul sign rule. The same sign combinatorics as in the state sum model should lead to an obstruction proportional to $[w_2]$. Thus we expect one could prove a theorem relating "fermionic" extended topological theories and extended spin-TQFTs.

- It should be possible to give a purely Hamiltonian version of our analysis. The fermionic SPT phase K_d associated to the $z[X, \eta, \beta_{d-1}]$ partition function should give us a recipe to build a one-dimensional Hilbert space from a given cocycle β_{d-1}, which could be described as a choice of sign on local patches of the space manifold. Hopefully, this recipe will capture the same sign ambiguities as one encounters in the construction of a general fermionic Hilbert space as a tensor product of local fermionic Hilbert spaces associated to local patches of the space manifold. Ideally, this would show in full generality, without reference to TQFTs, that any fermionic phase of matter can be obtained by combining the fermionic SPT phase K_d with some appropriate bosonic degrees of freedom and that fermionic phases of matter should generally require the existence of a spin structure on space.

2. Fermionic SPT phases and the Gu–Wen Grassmann integral

2.1. *The Gu–Wen construction*

The standard discrete action for a bosonic SPT phase protected by some symmetry group G is built from a $U(1)$-valued d-cocycle $\nu_d^b(g_0, \cdots, g_d)$ on BG, i.e., a function of $(d+1)$ G-valued variables, invariant under the action of G on itself,

$$\nu_d^b(gg_0, \cdots, gg_d) = \nu_d^b(g_0, \cdots, g_d), \tag{12}$$

and closed under the action of an appropriate differential δ.

Concretely, given a triangulation of a d-dimensional manifold X with a flat G connection, one evaluates the action as a product over all d-dimensional simplices of $\nu_d^{\pm}$ evaluated on a local trivialization of the connection. The G-symmetry of the cocycle makes the answer independent of the local trivialization and the cocycle condition $\delta\nu_d^b = 1$ insures invariance under changes of triangulation. Essentially, the ratio between the actions for two triangulations which differ by an elementary move equals the action for the boundary of a $(d+1)$-simplex, which by definition is the same as $\delta\nu_d^b$. More generally, two triangulations can be related by a sequence of moves which can be visualized as a triangulation of a cobordism $X \times [0, 1]$ from the manifold X to itself.

The Gu–Wen construction of a discrete partition function for fermionic SPT phases[7] involves two basic pieces of input: a $U(1)$-valued d- cochain $\nu(g_0, \cdots, g_d)$ on BG and a $\mathbb{Z}_2$-valued $(d-1)$-cocycle $n_{d-1}(g_0, \cdots, g_{d-1})$, such that

$$\delta \nu_d = (-1)^{Sq^2[n_{d-1}]} . \tag{13}$$

In other words, ν_d satisfies the cocycle condition up to signs, which are determined from n_{d-2} through the Steenrod square operation.

The Gu–Wen partition function can be decomposed into the product of three terms, which are not separately invariant under changes of triangulation. The first term, which we could denote as Z_ν, or $Z_\nu[T]$ if we want to indicate the specific choice of triangulation T of the space-time manifold X, is simply the product over all d-dimensional simplices of $\nu_d^{\pm 1}$, just as for a bosonic SPT phase. Because ν_d is not a cocycle, the sign of this term will jump under re-triangulation by the Steenrod square of n_{d-1} integrated over the cobordism from X to itself:

$$Z_\nu[T] = Z_\nu[T'] \exp i\pi \int_{X\times[0,1]} Sq^2[n_{d-1}] . \tag{14}$$

The second term, which we could denote as Z_θ or $Z_\theta[T]$, contains the "fermionic" degrees of freedom. It is a sign, defined by a Grassmann integral whose structure is determined by n_{d-1}. Schematically, one associate a pair of Grassmann odd variables to the two sides of each $(d-1)$-simplex such that n_{d-1} is 1. The integrand is built as a product over d-simplices of the Grassmann variables associated to that simplex. We do not expect to be able to write Z_θ as a standard bosonic action, i.e., the integral of the pull-back of some class on BG to the manifold: if we could do that, we would have reduced the system to a bosonic SPT phase. The Grassmann integral Z_θ only depends on the group variables through the image β_{d-1} of the cochain n_{d-1} computed on the faces of the triangulation, which is a standard $\mathbb{Z}_2$ cocycle. We will see later in Section 6.2 that the Grassmann integral Z_θ coincides with the function $\sigma_\Pi(X, \beta_{d-1})$ described in the introduction.

The third term, which we could denote as Z_m or $Z_m[T]$, is somewhat problematic: it is written in terms of a function $m_{d-2}(g_0, \cdots, g_{d-2})$ which is not G invariant, but satisfies $\delta m_{d-2} = n_{d-1}$. The expression for Z_m involves a product of $(-1)^{m_{d-2}}$ evaluated over a certain subset S of $(d-2)$-simplices in T defined in Ref. 7 by some local rule:

$$Z_m[T] = \prod_{s\in S} (-1)^{m_{d-2}(s)} . \tag{15}$$

We review the precise definition of S for $d = 2, 3, 4$ in Section 2.2.

The product is invariant under re-definitions of m_{d-2}, but the lack of G-invariance of m_{d-2} makes it problematic to define the model on a manifold with a non-trivial G-bundle. The wavefunctions built from this model have some m_{d-2} dependence which is stripped off by hand by the authors of Ref. 7 in a non-canonical way, opening the possibility for subtle sign changes and ambiguities in the corresponding TQFT. Indeed, we will argue here that removing the m_{d-2} dependence introduces naturally a dependence on a choice of spin structure on the manifold, so that the Gu–Wen fermionic SPT phases are a class of invertible spin-TQFTs.

In order to remove the spurious m_{d-2} dependence, we can imagine replacing the collection S of $(d-2)$-simplices used in Z_m with a collection of $(d-1)$ simplices E such that $\partial E = S$, so that the product over S of $(-1)^{m_{d-2}}$ in Z_m can be reorganized to a product Z_n^E over E of $(-1)^{n_{d-1}}$:

$$Z_n^E[T] = \prod_{e \in E} (-1)^{n_{d-1}(e)} . \tag{16}$$

Clearly, this will be only possible if S is exact, and then the result will depend on the choice of E: any two different choices of E differ by a cycle, which will have in general a non-zero pairing with n_{d-1}, and thus will give inequivalent partition functions. For a solid torus, this integration by parts should mimic the way the authors of Ref. 7 strip off the m_{d-2} dependence of wavefunctions.

At this stage, we have shown that we can construct an improved, well-defined partition function if the homology class of S vanishes, and that the improved partition function depends on a choice of trivialization E of S. Although the definition of S depends on the choice of triangulation, invariance of the partition function under changes of triangulation suggests that the homology class of S should capture some intrinsic triangulation-independent property of the underlying space-time manifold X.

We propose that the homology class of S captures precisely the second Stiefel–Whitney class $[w_2]$, which vanishes on spin manifolds. We also propose that S itself provides a canonical chain representative for w_2, so that the choice of E actually encodes a choice of a spin structure on the manifold.

The proposal implies that the partition function can only be defined on a spin manifold, and depends on a choice of spin structure. The combination of signs $Z_\theta Z_n^E$ will coincide with the function $z[X, \eta, \beta_{d-1}]$ defined in the introduction. Conversely, if the cohomology class of w_2 is non-trivial, the partition sum will have an unavoidable dependence on re-definitions of m_{d-2} which map to a non-trivial $(d-2)$ cohomology class on the manifold.

We can offer two strong checks of our proposal: a general consistency check we describe momentarily, and a direct calculation for a special class of triangulations, which are obtained by refining a generic triangulation by a barycentric subdivision. These triangulations are endowed with a canonical representative chain for w_2, which turns out to coincide with S for $d = 2, 3, 4$.

The first consistency check follows from the observation that a change of Z_m under a change of triangulation is a linear expression $\sum_{s \in V} m_{d-2}[s]$, a sum of m evaluated on some collection V of $(d-2)$-simplices of the cobordism $Y = X \times [0, 1]$ from X to itself. The change of Z_m should only depend on n_{d-1}, as it is cancelled by the change in Z_ν and Z_θ. It should thus be possible to write it as a linear expression $\sum_{s \in W} n_{d-1}[s]$, a sum of n_{d-1} evaluated on some collection W of $(d-1)$ simplices of Y such that $\partial W = V$.

The linear function defined by W must be such that the sum,

$$\sum_{s \in W} n_{d-1}[s] + \int_Y Sq^2[n_{d-1}] \,, \tag{17}$$

of the variations of Z_m and Z_ν can be cancelled by the variation of Z_θ.

There is a natural candidate for such a linear function: the integral of $w_2 \cup n_{d-1}$ over Y, where w_2 is some specific cochain which represents the second Stiefel–Whitney class. A neat property of w_2 is that $Sq^2[x] + w_2 \cup x$ is exact for any $(d-1)$-cocycle x and thus can be integrated by parts and plausibly cancelled by the variation of Z_θ. If we write $n_{d-1} = \delta m_{d-2}$ and thus $w_2 \cup n_{d-1} = \delta(w_2 \cup m_{d-2})$, we find that S must be a representative for w_2 on X.

We can give a direct proof that S is a representative for w_2 for a special class of triangulations BT, which are a barycentric subdivision of some rougher triangulation T. Each d-dimensional simplex s of T is subdivided in $(d+1)!$ simplices s_σ whose vertices are the barycenters of subsets of the vertices of s. More precisely, given a permutation σ of the vertices of s, the i-th vertex of s_σ is the barycenter of the vertices $\sigma(0), \cdots, \sigma(i)$. Thus the vertices of s_σ have a natural order, starting from a vertex of s and ending with the barycenter of s.

In the Gu–Wen construction, the triangulation is endowed with a branching structure, i.e., an orientation of the edges with no closed loops, which in turn provides a order to the vertices of simplices: the vertex with all outgoing edges is number "0", the vertex with a single incoming edge is number "1", etcetera.

If we have a barycentric triangulation, we can simply order the 1-simplices from the barycenter of fewer vertices to the barycenter of more vertices. That gives a useful canonical choice of branching structure on BT, which induces the same order of the vertices of simplices as the natural one of the barycentric subdivision. This branching structure has the useful property that each vertex of BT has the same position in the order of vertices in all simplices which include that vertex: vertices of T are always at position "0" in the order, midpoint of segments in T are always at position "1", etcetera.

The barycentric subdivision BT has an important property: the set of all $(d-2)$-simplices provides a chain which is a canonical representative for w_2. The collection S of $(d-2)$ simplices used in Ref. 7 to define Z_m is described as the sum over all $(d-2)$-simplices, plus a correction term given explicitly in dimensions $d = 2, 3, 4$. We will now show by direct inspection that the extra correction term vanishes for a barycentric triangulation BT, so that S is precisely the canonical representative for w_2!

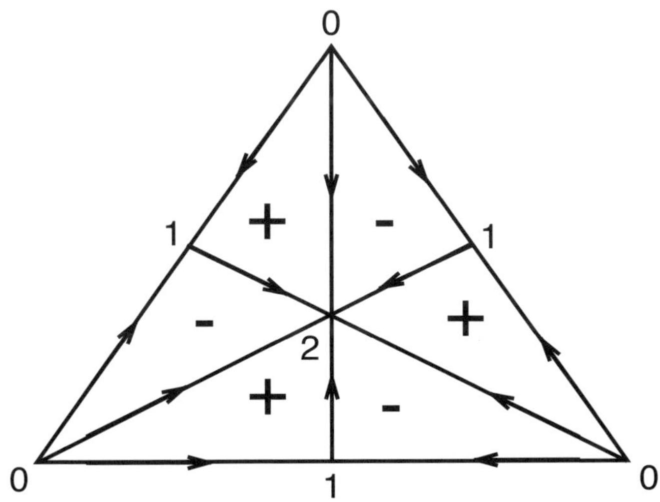

Fig. 1. The barycentric subdivision of a triangle

2.2. *Barycentric subdivisions in* $d = 2, 3, 4$

For a general triangulation, the branching structure gives an order to the vertices of each triangle: the n-th vertex has n incoming edges. The ordering of the vertices also gives an orientation of each simplex. This orientation may or not agree with the canonical counterclockwise orientation of the simplex. If it does, we have a "+" simplex, if not we have a "−" simplex. See Figure 2.

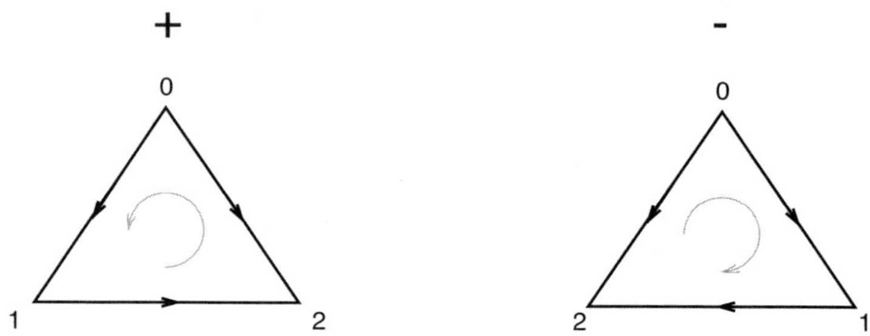

Fig. 2. The two possible orientations of a triangle and ordering of vertices induced by a branching structure on a 2d triangulation

In 2d, the partition function involves a factor of $(-1)^{m_0(g_v)}$ for each vertex v, and an extra factor of $(-1)^{m_0(g_v)}$ for each "−" triangle which has v as the vertex number 1.

If we pick our triangulation to be a barycentric subdivision BT, then each "1" vertex belongs to two "−" triangles, and we can disregard the second contribution. Thus $Z_m[BT]$ is the product over all vertices v of BT of $(-1)^{m_0(g_v)}$: This is simply the pairing of m_1 with the canonical 0-chain representative S for w_2.

Every 2d manifold admits a spin structure, so that w_2 is always exact. If we pick some 1-chain E on BT such that $\partial E = S$, we can replace $Z_m[BT]$ with the improved $Z_n^E[BT]$. We interpret the choice of E as a choice of spin structure on the discretized surface.

In 2d we can actually sketch a proof that S is a chain representative for w_2 even for a generic triangulation T. In 2d, we can build representatives for w_2 by taking a vector field and picking the points where the vector field vanishes, counting how many times the vector field winds around the origin in a neighbourhood of each point, modulo 2. That is the same as 1 plus the number of times the vector field is tangent in the counterclockwise direction to a small circle around the point. If we pick the vector field V in Figure A4, each vertex will contribute 1 plus the number of times the vertex appears at position "1" in a "$-$" triangle. That representative for w_2 coincides with S.

In 3d, the chain S of edges which appear in Z_m consists of all edges of the triangulation, together with the (02) edge for all "+" tetrahedra and the (13) edge for all "$-$" tetrahedra.

For a barycentric triangulation BT, the (02) edges join vertices of T and barycenters of triangles in T. They are shared by two "+" triangles. The (13) edges join midpoints of edges in T and the barycenters of tetrahedra in T. They are shared by two "$-$" triangles. Thus the extra contributions to S cancel out, and S coincides with the sum of all edges, i.e., canonical chain representative for w_2.

In 4d, the chain S consists of all triangles in the triangulation, together with the (013), (134), (123) triangles for a "+" 4-simplex, (024) for a "$-$" 4-simplex. In order to show these edges are shared by an even number of tetrahedra of the corresponding orientation, we can observe that they are fixed points of two reflections.

For example, a (013) triangle has vertices which are barycenters of the first, the first 2 and all vertices in a sequence of 4 (v_1, v_2, v_3, v_4). As it belongs to a face, it is shared by two tetrahedra in T. As is invariant under exchange of v_3 and v_4, it is fixed by a second reflection.

Similarly, a (134) triangle has vertices which are barycenters of the first 2, 4, 5 vertices of a sequence of 5 vertices. It is invariant under exchanging the first pair, or the second pair. A (123) triangle has vertices which are barycenters of the first 2, 3, 4 vertices of a sequence of 4 vertices. It belongs to a face and is invariant under exchange of the first pair of vertices. Finally, a (024) triangle has vertices which are barycenters of the first 1, 3, 5 vertices of a sequence of 5 vertices. It is invariant under exchange of the second and third vertices, and the fourth and fifth.

Thus all extra contributions to S cancel out, and S is given by the sum of all triangles, i.e., canonical chain representative for w_2.

In order to mimic in general dimension d the 2d analysis for a general triangulation T, we would need to build a w_2 representative in terms of $d-1$ vector fields on X. The representative would be concentrated at the locus where the vector fields fail to be linearly dependent, weighed by the number of times the vanishing linear combination of the vector fields winds around the other non-vanishing linear combinations in a neighbourhood of the locus. We expect it should be possible to define some canonical set of $d-1$ vector fields in each simplex, linearly dependent at $(d-2)$-simplices, which demonstrate the S is a representative for w_2.

3. The Gu–Wen Grassmann integral as a quadratic refinement

3.1. *Quadratic property of the Gu–Wen Grassmann integral*

Next, we will look at Z_θ. We aim to show that the combination

$$Z_n^E[BT]Z_\theta[BT] \equiv z[X, \eta, \beta_{d-1}] \tag{18}$$

is a quadratic function of the $\mathbb{Z}_2$ cocycle β_{d-1}, which refines the pairing on the space of $\mathbb{Z}_2$ $(d-1)$-dimensional cocycles defined by the higher cup product $\cup_{d-2}$.

The Grassmann integral only depends on the image $\beta_{d-1}(e)$ of the cochain n_{d-1} on the co-dimension one faces e of the triangulation. Thus we can consider some generic $(d-1)$-cochain β_{d-1}, assigning a $\mathbb{Z}_2$ element to each edge of the triangulation, and specialize it to the image of n_{d-1} later on. For notational clarity, we will usually omit the subscript $(d-1)$ from β.

In the following formulae we will often refer to a cochain A_p evaluated the simplex defined by vertices $a_0, \cdots a_p$ simply as $A_p(a_0 \cdots a_p)$. We will also denote the Grassmann variables associated to an edge with vertices a and b as θ_{ab}, $\bar{\theta}_{ab}$.

We will start with a 2d example, and then proceed to higher dimensions. If $\beta(e) = 1$ for an edge e, we assign to the edge Grassmann variables θ_e and $\bar{\theta}_e$: θ_e is associated to the side where the canonical orientation of the face agrees with the orientation of the edge, $\bar{\theta}$ on the side it disagrees.

To each triangle t we associate an even monomial $u(t)$, the product of the Grassmann variables attached to its edges, if present, in the order 12, 01, 02 according to the ordering of the vertices for a "+" triangle, opposite for a "−" triangle.

Thus a "+" triangle has monomial

$$\theta_{12}^{\beta(12)}\theta_{01}^{\beta(01)}\bar{\theta}_{02}^{\beta(02)} \tag{19}$$

and a "−" triangle has

$$\theta_{02}^{\beta(02)}\bar{\theta}_{01}^{\beta(01)}\bar{\theta}_{12}^{\beta(12)}\,. \tag{20}$$

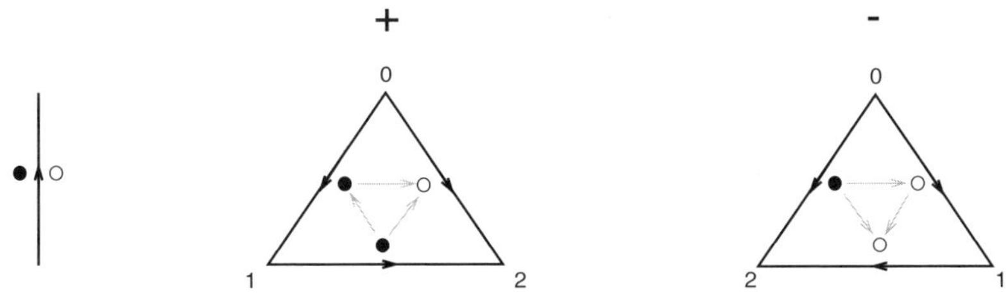

Fig. 3. Left: to each edge e such that $\beta(e) = 1$ we assign a θ_e Grassmann variable (black dot) and a $\bar{\theta}_e$ Grassmann variable (white dot). Each such edge contributes $d\theta_e d\bar{\theta}_e$ to the measure. Middle and right: as β is a cochain, each triangle is associated to two Grassmann variables, which are ordered in the integrand according to the grey arrows.

We are interested in the sign

$$\sigma(\beta) = \int \prod_{e|\beta(e)=1} d\theta_e d\bar{\theta}_e \prod_t u(t)\,. \tag{21}$$

We will show directly from the definition that

$$\sigma(\beta + \beta') = \sigma(\beta)\sigma(\beta')(-1)^{\int_X \beta \cup \beta'}\,, \tag{22}$$

i.e., $\sigma(\beta)$ is a quadratic function which refines the intersection pairing on 1-cocycles.

We can combine the two Grassmann integrals in $\sigma(\beta)\sigma(\beta')$ as

$$\sigma(\beta)\sigma(\beta') = \int \prod_{e|\beta(e)=1} d\theta_e d\bar{\theta}_e \prod_{e|\beta'(e)=1} d\theta'_e d\bar{\theta}'_e \prod_t u(t)[\beta, \theta, \bar{\theta}]u(t)[\beta', \theta', \bar{\theta}']\,. \tag{23}$$

There is a subset of "spurious" Grassmann variables which is associated to edges for which $\beta(e) = \beta'(e) = 1$. If we can systematically integrate them out, the remaining Grassmann variables will be associated to edges for which $\beta(e) + \beta'(e) = 1$ and coincide with the Grassmann variables in the formula for $\sigma(\beta + \beta')$.

Our strategy is to permute the variables until we form pairs $\theta_e\theta'_e$ and $\bar{\theta}_e\bar{\theta}'_e$ of spurious variables. These pairs are Grassmann even, and can be brought out of the integrand and eliminated against the $d\theta_e d\bar{\theta}_e d\theta'_e d\bar{\theta}'_e$ measure to give an overall $\prod_e (-1)^{\beta(e)\beta'(e)}$ sign.

The permutation of variables which brings the spurious pairs together is very simple: we interleave the variables in the product $u(t)[\beta, \theta, \bar{\theta}]u(t)[\beta', \theta', \bar{\theta}']$: for a "+" triangle we have

$$\left[\theta_{12}^{\beta(12)}\theta_{01}^{\beta(01)}\bar{\theta}_{02}^{\beta(02)}\right]\left[(\theta'_{12})^{\beta'(12)}(\theta'_{01})^{\beta'(01)}(\bar{\theta}'_{02})^{\beta'(02)}\right] =$$

$$= (-1)^{\beta(01)\beta'(12)+\beta(02)\beta'(12)+\beta(02)\beta'(01)} \cdot$$

$$\cdot \left[\theta_{12}^{\beta(12)}(\theta'_{12})^{\beta'(12)}\right]\left[\theta_{01}^{\beta(01)}(\theta'_{01})^{\beta'(01)}\right]\left[\bar{\theta}_{02}^{\beta(02)}(\bar{\theta}'_{02})^{\beta'(02)}\right], \quad (24)$$

and for a "−" triangle we have

$$\left[\theta_{02}^{\beta(02)}\bar{\theta}_{01}^{\beta(01)}\bar{\theta}_{12}^{\beta(12)}\right]\left[(\theta'_{02})^{\beta'(02)}(\bar{\theta}'_{01})^{\beta'(01)}(\bar{\theta}'_{12})^{\beta'(12)}\right] =$$

$$= (-1)^{\beta(01)\beta'(02)+\beta(12)\beta'(02)+\beta(12)\beta'(01)} \cdot$$

$$\cdot \left[\theta_{02}^{\beta(02)}(\theta'_{02})^{\beta'(02)}\right]\left[\bar{\theta}_{01}^{\beta(01)}(\bar{\theta}'_{01})^{\beta'(01)}\right]\left[\bar{\theta}_{12}^{\beta(12)}(\bar{\theta}'_{12})^{\beta'(12)}\right]. \quad (25)$$

Thus we have grouped together pairs of spurious variables, if present, and left the non-spurious variables in the correct order to be identified with $u(t)[\beta + \beta', \cdots]$.

It is useful to re-distribute the overall $\prod_e (-1)^{\beta(e)\beta'(e)}$ sign, by associating each factor of $(-1)^{\beta(e)\beta'(e)}$ with the triangle to the right of the oriented edge e. Thus each "+" triangle, which only sits to the right of its 02 edge, is associated to an overall factor

$$(-1)^{\beta(01)\beta'(12)+\beta(02)\beta'(12)+\beta(02)\beta'(01)+\beta(02)\beta'(02)} = (-1)^{\beta(01)\beta'(12)} \quad (26)$$

and to each "−" triangle, which only sits to the right of its 01 and 12 edges, is associated to an overall factor

$$(-1)^{\beta(01)\beta'(02)+\beta(12)\beta'(02)+\beta(12)\beta'(01)+\beta(01)\beta'(01)+\beta(12)\beta'(12)} = (-1)^{\beta(01)\beta'(12)}. \quad (27)$$

Thus the reorganization of the Grassmann integral $\sigma(\beta)\sigma(\beta')$ into the Grassmann integral $\sigma(\beta + \beta')$ produces a sign of $(-1)^{\beta(e_{01})\beta'(e_{12})}$ for each triangle. This is nothing else but the cup product of β and β'! Thus we have the desired quadratic property,

$$\sigma(\beta + \beta') = \sigma(\beta)\sigma(\beta')(-1)^{\int_X \beta \cup \beta'}. \quad (28)$$

We can readily extend the above argument to Z_θ in any dimension d. For general d, the Grassmann integral involves again pairs of variables associated to the two sides of each oriented $(d-1)$-simplex e such that $\beta(e) = 1$.

The integrand is a product of monomials $u[s]$ made out of all Grassmann variables associated to each d-simplex s. The order of the variables in the monomial is determined by a specific rule. We will describe the rule later, for now we only need to know that it gives a canonical order to the faces of each simplex.

If we denote as $\sigma(\beta)$ again the result of the Grassmann integral, the same interleaving operation as in the 2d case gives us immediately a proof that $\sigma(\beta)$ is quadratic:

$$\sigma(\beta+\beta') = \sigma(\beta)\sigma(\beta')\prod_s \epsilon[s,\beta,\beta'], \tag{29}$$

with

$$\epsilon[s,\beta,\beta'] = (-1)^{\sum_{e,e'\in s}^{e>e'}\beta(e)\beta'(e')+\sum_{e\in s}^{e>0}\beta(e)\beta(e')}, \tag{30}$$

with $e > e'$ in the order determined by $u[s]$ and $e > 0$ if $u[s,\beta]$ includes a $\bar{\theta}_e$ variable. We aim to identify the d-cochain in the exponent with the higher cup product $\beta \cup_{d-2} \beta'$.

The order of faces induced by $u[t]$ for a $+$ simplex places first the faces which omit the even vertices (first the one omitting 0, then the one omitting 2, etc.), which are labelled by θ_e variables, and then the faces which omit the odd vertices (first the one omitting 1, then the one omitting 3, etc.), which are labelled by $\bar{\theta}_e$ variables. The order for a $-$ simplex is the opposite (and the role of θ_e and $\bar{\theta}_e$).

We will now compute $\epsilon[s,\beta,\beta']$ in $d=3$ and $d=4$ for this choice of order and verify it is given by the standard higher cup products $\beta \cup_1 \beta'$ and $\beta \cup_2 \beta'$ respectively. A similar analysis in higher dimension should be straightforward.

Notice that if we were to pick a different choice of order Π for the factors in the measure, $\sigma_\Pi(\beta)$ would differ from the standard $\sigma(\beta)$ by a local linear term, which does not affect the quadratic refinement property. If we pick a different choice of order Π for the factors in $u[t]$, $\sigma_\Pi(\beta)$ will differ from the standard $\sigma(\beta)$ by a local quadratic term, some (β,β) pairing defined by a sum over simplices of the product of fermion numbers of the variables which have been permuted in $u[t]$ to get to the new order. That would change the quadratic refinement from $\beta \cup_{d-2} \beta'$ to some other

$$\beta \tilde{\cup}_{d-2} \beta' = \beta \cup_{d-2} \beta' + (\beta,\beta') + (\beta',\beta). \tag{31}$$

This is just a different choice of definition for the higher cup product: the basic relation in equation 6 remains valid with the re-definition

$$\beta \tilde{\cup}_{d-3} \beta' = \beta \cup_{d-2} \beta' + \delta(\beta,\beta') + (\delta\beta,\beta') + (\beta\delta\beta') \tag{32}$$

and all other products unchanged.

3.2. *Quadratic refinement in 3d*

At a + simplex, we have an order of faces (123), (013), (023), (012) and thus the signs computed from the quadratic pairing,

$$\begin{aligned}&\beta(012)\beta'(123) + \beta(012)\beta'(013) + \beta(012)\beta'(023) + \beta(023)\beta'(123)+\\&+ \beta(023)\beta'(013) + \beta(013)\beta'(123) + \beta(023)\beta'(023) + \beta(012)\beta'(012) =\\&\beta(023)\beta'(012) + \beta(013)\beta'(123)\,.\end{aligned} \tag{33}$$

At a − simplex, we have an order of faces (012), (023), (013), (123), and thus the signs computed from the quadratic pairing,

$$\begin{aligned}&\beta(123)\beta'(012) + \beta(123)\beta'(023) + \beta(123)\beta'(013) + \beta(013)\beta'(012)+\\&+ \beta(013)\beta'(023) + \beta(023)\beta'(012) + \beta(013)\beta'(013) + \beta(123)\beta'(123) =\\&\beta(023)\beta'(012) + \beta(013)\beta'(123)\,.\end{aligned} \tag{34}$$

Again, the dependence on the type of triangle drops out. We have a d-cochain $(\beta, \beta')_3$ given by the pairing $\beta(023)\beta'(012) + \beta(013)\beta'(123)$. This is precisely the definition of $\beta \cup_1 \beta'$. Thus we claim that in 3d we have

$$\sigma(\beta + \beta') = \sigma(\beta)\sigma(\beta')(-1)^{\int_X \beta\cup_1\beta'}\,. \tag{35}$$

3.3. *Quadratic refinement in 4d*

At a + simplex, we have an order of faces (1234), (0134), (0123),(0234), (0124) and thus the signs computed from the quadratic pairing,

$$\begin{aligned}&\beta(0124)\,[\beta'(1234) + \beta(0134) + \beta(0123) + \beta(0234)] +\\&+ \beta(0234)\,[\beta'(1234) + \beta'(0134) + \beta'(0123)] +\\&+ \beta(0123)\,[\beta'(1234) + \beta'(0134)] + \beta(0134)\beta'(1234)+\\&+ \beta(0124)\beta'(0124) + \beta(0234)\beta'(0234) =\\&\beta(0234)\beta'(0124) + \beta(0123)\beta'(1234)+\\&+ \beta(0123)\beta'(0134) + \beta(0134)\beta'(1234)\,.\end{aligned} \tag{36}$$

At a − simplex, we find the same. Thus have a d-cochain

$$\begin{aligned}(\beta, \beta')_4 =&\beta(0234)\beta'(0124) + \beta(0123)\beta'(1234)+\\&+ \beta(0123)\beta'(0134) + \beta(0134)\beta'(1234)\,,\end{aligned} \tag{37}$$

which is the standard expression for $\beta \cup_2 \beta'$.

4. The 't Hooft anomaly

4.1. *Gauge variation of the Gu–Wen Grassmann integral*

Next, we can look at the variation of the Grassmann integral under exact changes in the cocycle β_{d-1}. The calculation is greatly simplified by the quadratic refinement property.

$$\sigma(\beta+\delta\lambda)=\sigma(\beta)\sigma(\delta\lambda)(-1)^{\int_X \beta\cup_{d-2}\delta\lambda}=\sigma(\beta)\sigma(\delta\lambda)(-1)^{\int_X \beta\cup_{d-3}\lambda+\lambda\cup_{d-3}\beta}\,, \tag{38}$$

where we used the basic property of the higher cup product, i.e., Eq. 6: the violation of the Leibniz rule for $\cup_a$ equals the symmetrization of $\cup_{a-1}$.

We can specialize the quadratic refinement property to

$$\sigma(\delta\lambda+\delta\lambda')=\sigma(\delta\lambda)\sigma(\delta\lambda')(-1)^{\int_X \lambda\cup_{d-3}\delta\lambda'+\lambda'\cup_{d-3}\delta\lambda+\lambda\cup_{d-4}\lambda'+\lambda'\cup_{d-4}\lambda}\,, \tag{39}$$

which can be solved up to a linear ambiguity:

$$\sigma(\delta\lambda)=(-1)^{\sum_{s\in\tilde{S}}\lambda(s)+\int_X \lambda\cup_{d-3}\delta\lambda+\lambda\cup_{d-4}\lambda}\,, \tag{40}$$

In order to fix the ambiguity, it is useful to compute directly $\sigma(\delta\lambda)$ for the simplest possible case, where λ is non-zero only on a single $(d-2)$-simplex, so that $\delta\lambda$ equals 1 on all the $(d-1)$-simplices whose boundary include the selected $(d-2)$-simplex. Thus if we go around the $(d-2)$-simplex, we will encounter a sequence of Grassmann variables ϑ_i, with measure factors $\pm d\vartheta_{2i}d\vartheta_{2i+1}$ from the $(d-1)$-simplices and $u(t)$ factors $\pm\vartheta_{2i+2}\vartheta_{2i+1}$ from the d-simplices inserted around the $(d-2)$-simplex.

If all the signs above were $+$, the overall Grassmann integral would give a factor of -1. In general, the signs in the measure factors are determined by the orientation of the $(d-1)$-simplices and the signs in the integrand factors $u[t]$ are determined by the relative order of the two $(d-1)$-simplices Grassmann variables in $u[t]$.

Consider for simplicity a barycentric subdivision. We have several types of $(d-2)$ simplices, which can be labellet by two integers from 0 to d: they are simplices whose vertices do not include the barycenters of a vertices or of b vertices. They are associated to an alternating sequence of $(d-1)$ simplices which omit either barycenters of a vertices or of b vertices. As these simplices are also of alternating type $+$ or $-$, it is easy to see that we can pick a direction around the $(d-2)$-simplex such that the integrand factors will have all $+$ signs. Working through a few examples, it is easy to convince oneself that the signs in the measure factors multiply to 1. Thus if λ_s is non-zero only on a single $(d-2)$-simplex s in BT, we can write

$$\sigma(\delta\lambda_s)=-1\,. \tag{41}$$

As the quadratic part of equation 40 vanishes for λ_s, we find that $\tilde{S}$ for a barycentric subdivision BT coincides again with the canonical representative for w_2. We expect $\tilde{S}$ to essentially coincide with S and the canonical chain representative for w_2 for a general triangulation as well. In other words, we expect that

$$z[X, \eta, \delta\lambda_s] = 1\,. \tag{42}$$

Thus we can write

$$\sigma(\delta\lambda) = (-1)^{\int_X \lambda \cup_{d-3} \delta\lambda + \lambda \cup_{d-4} \lambda + w_2 \cup \lambda}\,, \tag{43}$$

and thus:

$$\sigma(\beta + \delta\lambda) = \sigma(\beta)(-1)^{\int_X \beta \cup_{d-3} \lambda + \lambda \cup_{d-3} \beta + \lambda \cup_{d-3} \delta\lambda + \lambda \cup_{d-4} \lambda + w_2 \cup \lambda}\,. \tag{44}$$

We can also write the 't Hooft anomaly of $z[X, \eta, \beta_{d-1}]$ under gauge transformations of β_{d-1}:

$$z[X, \eta, \beta_{d-1} + \delta\lambda_{d-2}] = z[X, \eta, \beta_{d-1}](-1)^{\int_X \beta \cup_{d-3} \lambda + \lambda \cup_{d-3} \beta + \lambda \cup_{d-3} \delta\lambda + \lambda \cup_{d-4} \lambda}\,. \tag{45}$$

4.2. *A WZW-like expression for a quadratic refinement*

For $d \geq 3$ one can construct a WZW-like expression for the quadratic function of the cocycle β as follows. Let us assume that X is a boundary of some compact oriented $(d+1)$-manifold Y. This is automatic if $d \leq 3$, since the oriented bordism group $\Omega_d^{SO}(pt)$ vanishes for $d = 2, 3$,[14] but in general it is a nontrivial constraint on X. If we are given a $(d-1)$-cocycle $\beta \in Z^{d-1}(X, \mathbb{Z}_2)$ on X, one can always choose Y so that β extends to a $(d-1)$-cocycle on Y. To see this, we regard β as map $\beta : X \to K(\mathbb{Z}_2, d-1)$ where $K(\mathbb{Z}_2, d-1)$ is an Eilenberg-MacLane space; then the statement we need is that the reduced oriented bordism group $\tilde{\Omega}_d^{SO}(K(\mathbb{Z}_2, d-1))$ vanishes. This follows from the Atiyah-Hirzebruch spectral sequence for (unreduced) bordism and the vanishing of the reduced homology of $K(\mathbb{Z}_2, d-1)$ in degree less than $d-1$. Moreover, if $d \geq 3$, the reduced oriented d-dimensional bordism of a product of several copies of $K(\mathbb{Z}_2, d-1)$ also vanishes, for the same reason.[§] This implies that one can choose Y to be independent of β. Such a Y is not unique, of course.

[§]This fails for $d = 2$, because of the Tor terms in the Künneth formula. In fact, it is easy to give an example of a closed oriented 2-manifold X and a pair of classes $\alpha_1, \alpha_2 \in H^1(X, \mathbb{Z}_2)$ such that there is no compact oriented 3-manifold Y such that $X = \partial Y$ and both α_1, α_2 arise from restriction of cohomology classes on Y. For example, one can take X to be a 2-torus, with α_1 and α_2 being the generators of $H^1(X, \mathbb{Z}_2)$.

Consider now the following WZW-like expression:

$$\tilde{\sigma}(\beta) = (-1)^{\int_Y \beta \cup_{d-3} \beta + w_2 \cup \beta}\,. \tag{46}$$

It is independent of the actual choice of Y, as $\beta \cup_{d-3} \beta + w_2 \cup \beta$ is known to be exact if β is a cocycle.

This expression is a quadratic refinement of $\cup_{d-2}$. Indeed, if β and β' are $(d-1)$-cocycles on X, we get

$$\tilde{\sigma}(\beta + \beta') = \tilde{\sigma}(\beta)\tilde{\sigma}(\beta')(-1)^{\int_X \beta \cup_{d-2} \beta'}\,. \tag{47}$$

It also transforms in the same way as σ under 1-form $\mathbb{Z}_2$ gauge symmetry:

$$\begin{aligned}\tilde{\sigma}(\delta\lambda) &= (-1)^{\int_Y \delta\lambda \cup_{d-3} \delta\lambda + w_2 \cup \delta\lambda} = \\ &= (-1)^{\int_Y \lambda \cup_{d-4} \delta\lambda + \delta\lambda \cup_{d-4} \lambda + \int_X \lambda \cup_{d-3} \delta\lambda + w_2 \cup \lambda} \\ &= (-1)^{\int_X \lambda \cup_{d-3} \delta\lambda + \lambda \cup_{d-4} \lambda + w_2 \cup \lambda}.\end{aligned} \tag{48}$$

This means that σ and $\tilde{\sigma}$ can only differ by a linear and gauge-invariant function of β.

Thus for $d \geq 3$ for all practical purposes we can write

$$z[X, \eta, \beta_{d-1}] = (-1)^{\int_X \eta \cup \beta_{d-1} + \int_Y \beta_{d-1} \cup_{d-3} \beta_{d-1} + w_2 \cup \beta_{d-1}}\,. \tag{49}$$

5. Fermionic SPT phases and spin cobordism

It was proposed in Ref. 11 that fermionic Short Range Entangled phases in d space-time dimensions with symmetry G and vanishing thermal Hall conductivity are classified by the Pontryagin dual of the torsion part of $\Omega_d^{Spin}(BG)$. Here $BG \simeq K(G,1)$ is the classifying space of G. Some checks of this were performed in Ref. 12. We can now compare with the Gu–Wen supercohomology proposal in low dimensions.

Let us begin with $d = 3$. In this dimension there are no nontrivial fermionic SRE phases in the absence of symmetry, so in the presence of symmetry there is no distinction between SRE and SPT phases. From the mathematical viewpoint, one has $\Omega_3^{Spin}(pt) = 0$, and thus $\Omega_3^{Spin}(BG)$ coincides with the reduced bordism group $\tilde{\Omega}_3^{Spin}(BG)$. The partition function of the model corrected by the spin-structure dependent term is

$$Z(X, A, \eta) = \exp\left(2\pi i \int_X A^* \nu_3\right) z(X, \eta, A^* \beta_2)\,, \tag{50}$$

where $\nu_3 \in C^3(BG, \mathbb{R}/\mathbb{Z})$, $\beta_2 \in Z^2(BG, \mathbb{Z}_2)$, and A is a gauge field on X regarded as a map $A : X \to BG$. The cochains μ_3 and β_2 satisfy

$$\delta\nu_3 = \frac{1}{2}\beta_2 \cup \beta_2 \,. \tag{51}$$

It is easy to see that this expression defines an element of the Pontryagin dual of $\Omega_3^{Spin}(BG)$. Indeed, it is clear that $Z(A, \eta)$ is multiplicative under disjoint union. Now, suppose there exists a compact spin 4-manifold Y with boundary (X, η) such that A extends to a map $A_Y : Y \to BG$. Then

$$\exp(2\pi i \int_X A^*\nu_3) = (-1)^{\int_Y A_Y^*\beta_2 \cup A_Y^*\beta_2} \,. \tag{52}$$

On the other hand, the WZW-like expression for $z(X, \eta, \beta_2)$ becomes

$$(-1)^{\int_X \eta \cup A^*\beta_2 + \int_Y w_2 \cup A_Y^*\beta_2 + \int_Y A_Y^*\beta_2 \cup A_Y^*\beta_2} \,, \tag{53}$$

which is clearly the same as Eq. (52). Since $Z(X, A, \eta)$ becomes 1 when evaluated on trivial bordism classes, it defines a homomorphism from $\Omega_3^{Spin}(BG)$ to $U(1)$.

Not all spin cobordism classes can be so obtained. For example, for $G = \mathbb{Z}_2$ it is known that $\Omega_3^{Spin}(B\mathbb{Z}_2) \simeq \mathbb{Z}_8$,[12] while the Gu–Wen construction only gives phases labeled by $\mathbb{Z}_4$. From the physics side, it is also known that 3d fermionic SPT phases with $\mathbb{Z}_2$ symmetry are classified by $\mathbb{Z}_8$.[13]

For $d > 3$ there may exist nontrivial fermionic SRE phases even in the absence of any symmetry. According to Ref. 11, they exist whenever $\Omega_d^{Spin}(pt)$ has torsion. If we want to focus on fermionic SPT phases, we can restrict to d-manifolds X which define a trivial class in $\Omega_d^{Spin}(pt)$. Then the same argument shows that if A extends to a compact spin $(d+1)$-manifold Y such that $\partial Y = X$, then the partition function of the Gu–Wen model is 1. Therefore, each Gu–Wen supercohomology class defines a homomorphism from the reduced bordism group $\tilde{\Omega}_d^{Spin}(BG)$ to $U(1)$. Again, in general we do not expect that all such homomorphisms can be obtained from the Gu–Wen construction.

In $d = 2$ the arguments are a bit different, since there is no WZW-like expression for $z(X, \eta, \beta_1)$. The proposal of Ref. 11 is that fermionic SPT phases are classified by the Pontryagin dual of the reduced bordism group $\tilde{\Omega}_2^{Spin}(BG)$. There are also fermionic SRE phases in the absence of any symmetry which are classified by the dual of $\Omega_2^{Spin}(pt) = \mathbb{Z}_2$. The Gu–Wen construction describes only the former. To describe the correspondence, recall that a spin structure on an oriented 2-manifold can be identified with a quadratic refinement of the intersection form on

$H_1(X, \mathbb{Z}_2) \simeq H^1(X, \mathbb{Z}_2)$.[15,16] This quadratic refinement is nothing but $z(X, \eta, \beta_1)$, see Appendix A for a detailed discussion. Moreover, it follows from the results of Ref. 15 that the value of $z(X, \eta, \beta_1)$ depends only on the bordism class of (X, η, β_1) in $\Omega_2^{Spin}(B\mathbb{Z}_2)$. Thus the Gu–Wen construction defines a map from $H^1(BG, \mathbb{Z}_2)$ to spin cobordism of BG. This map is not a homomorphism, because $z(X, \eta, \beta_1)$ is not linear but quadratic in β_1. But we should remember that every element $\nu_2 \in H^2(BG, \mathbb{R}/\mathbb{Z})$ also gives us an element in the spin cobordism of BG. Thus Gu–Wen SPT phases are described by pairs $(\nu_2, \beta_1) \in H^2(BG, \mathbb{R}/\mathbb{Z}) \times H^1(BG, \mathbb{Z}_2)$. The group structure is a nontrivial extension of $H^1(BG, \mathbb{Z}_2)$ by $H^2(BG, \mathbb{R}/\mathbb{Z})$:

$$(\nu_2, \beta_1) + (\nu_2', \beta_1') = (\nu_2 + \nu_2' + \frac{1}{2}\beta_1 \cup \beta_1', \beta_1 + \beta_1') \,. \tag{54}$$

It follows from the Atiyah-Hirzebruch spectral sequence that this extension is isomorphic to the dual of the reduced bordism $\tilde{\Omega}_2^{Spin}(BG)$, in agreement with the proposal of Ref. 11.

The nontrivial $d = 2$ fermionic SRE phase without any symmetry is realized by the Kitaev spin chain. The corresponding spin-TQFT has the Arf invariant as its partition function (the unique bordism invariant of spin structures in $d = 2$). We recall that the Arf invariant of (X, η) is essentially the average of the $z(X, \eta, \beta_1)$ over all $\beta_1 \in H^1(X, \mathbb{Z}_2)$.[16] This spin-TQFT can also be constructed using the Gu–Wen Grassmann integral, see Section 6.

6. Constructing spin-TQFTs in low dimensions

6.1. *State-sum constructions of 2d spin-TQFTs*

An oriented 2d TQFT can be defined axiomatically as a functor from a geometric category Cob_2 to the category of vector spaces. The category Cob_2 has closed 1-manifolds as objects and oriented bordisms between them as morphisms. It is well known that there is a 1-1 correspondence between oriented 2d TQFTs and commutative Frobenius algebras. The commutative Frobenius algebra corresponding to a given 2d TQFT encodes 2-point and 3-point functions on a sphere, and the rest of the correlators can be reconstructed from it.

An alternative approach to constructing 2d TQFTs is provided by state-sum models.[17,18] This approach is more natural from the statistical mechanics viewpoint and gives a manifestly local recipe for computing the TQFT partition function and correlators for a triangulated closed oriented d-manifold X. One starts with a not necessarily commutative semi-simple

Frobenius algebra A and defines the partition function as follows. Fix a basis e_i, $i \in I$, in A and denote by C^i_{jk} the structure constants of A in this basis. Let

$$g_{ij} = C^l_{ik} C^k_{jl}.$$

The matrix g_{ij} is non-degenerate if A is semi-simple. Let g^{ij} be its inverse. It is easy to see that

$$C_{ijk} = g_{il} C^l_{jk}$$

is cyclically symmetric. A coloring of a 2-simplex f of X is an assignment of an element of I to each boundary 1-simplex of f. A coloring of a triangulation is a coloring of each 2-simplex. If each 2-simplex is colored, each 1-simplex has two colors. The partition function of a triangulated manifold X is a sum over all colorings of the triangulation, with the weight of each coloring defined as product of weights of 1-simplices and 2-simplices. The weight of a 1-simplex colored by $i, j \in I$ is g^{ij}. The weight of a 2-simplex whose three edges are colored by i, j, k is C_{ijk}. Here we use the cycling ordering of the edges arising from the orientation of X. It is easy to show that the partition function thus defined is independent of the choice of a triangulation.[17,18]

The state-sum construction is somewhat redundant, as the partition function and the correlators depend only on the center $Z(A)$, which is a semi-simple commutative Frobenius algebra. The algebra A can be interpreted as the algebra of boundary operators for a particular boundary condition for the TQFT based on $Z(A)$. On the other hand, the state-sum construction is very explicit and can be easily extended to manifolds with boundaries.

Note that the state-sum construction always gives rise to a semi-simple $Z(A)$ and therefore does not produce the most general oriented 2d TQFT. For applications to condensed matter physics, this is not a serious drawback, since unitary TQFTs are automatically semi-simple.

It is easy to modify the state-sum construction to produce 2d spin-TQFTs. One starts with a $\mathbb{Z}_2$-graded semi-simple algebra A. Let e_i, $i \in I$, be its basis. Each basis vector is assumed to have a well-defined grading $\beta(i) \in \mathbb{Z}_2$. As in the bosonic case, we color each edge of each 2-simplex with an element of I, so that each 1-simplex is colored with a pair of elements of I. The weights assigned to 1-simplices and 2-simplices will be proportional to g^{ij} and C_{ijk}. Since the algebra A is $\mathbb{Z}_2$-graded, the matrix g^{ij} vanishes if $\beta(i) \neq \beta(j)$. Thus we may assume that if a 1-simplex is colored by i, j, then $\beta(i) = \beta(j)$. Thus each allowed coloring defines a $\mathbb{Z}_2$-valued 1-cochain β. Since C_{ijk} vanishes unless $\beta(i) + \beta(j) + \beta(k) = 0$, this 1-cochain is a cocycle.

In order to account for the Grassmann nature of elements of A, we correct the naive sum over the colour of edges by including an overall sign. This is the sign which arises from identifying C_{ijk} with an element of $A^* \otimes A^* \otimes A^*$ and g^{ij} with an element of $A \otimes A$: we pick some order of the factors in each individual weight and then re-order them to bring together the pairs of A and A^* spaces we want to contract.

The combinatorics of the Koszul signs is the same as in the Gu–Wen construction. The Grassmann variables θ_e, $\bar{\theta}_e$ act as placeholders for the Grassmann-odd generators of A^* in the weights of 2-simplices and the $d\theta_e$, $d\bar{\theta}_e$ as placeholders for the Grassmann-odd generators of A in the weights of 1-simplices. The contraction between generators of A and A^* is mimicked by the Grassmann integration. The result is just $\sigma(\beta)$.

Thus the naive weight of each allowed coloring will be the product of g^{ij} over 1-simplices, C_{ijk} over 2-simplices and $\sigma(\beta)$. We can rewrite the sum over colorings as a sum over colorings producing a particular 1-cocycle β followed by a sum over β. Let us denote by $Z[\beta]$ the result of the first summation. The discussion in Sec. 2 implies that $Z[\beta]$ is independent of the triangulation up to a sign. To make it completely independent, we need to choose a trivialization η of w_2 and multiply $Z[\beta]$ by a correction factor

$$(-1)^{\int_X \eta \cup \beta},$$

i.e., we take the correct weight to be the product of g^{ij} over 1-simplices, C_{ijk} over 2-simplices and $z(X, \eta, \beta)$. Then the partition partition function depends on the spin structure η on X, but not on a particular triangulation. Notice that in 2d $z(X, \eta, \beta)$ coincides with the well-known quadratic refinement $(-1)^{q_\eta(\beta)}$ of the intersection form, evaluated on β (see Appendix A).

The simplest example of a 2d spin-TQFT is obtained if we take A to be the Clifford algebra Cl(1). It is generated by 1 and an odd variable η satisfying $\eta^2 = 1$. The 2×2 matrix g_{ij}, $i \in \mathbb{Z}_2$, is given by $g_{ij} = 2\delta_{ij}$, while C_{ijk} is equal to either 2 or 0 depending on whether the sum of the indices is 0 or 1 modulo 2. In this case the coloring is completely determined by the 1-cochain β, and the weight is simply $z(X, \eta, \beta) = (-1)^{q_\eta(\beta)}$. More precisely, if we denote by E and F the number of 1-simplices and 2-simplices respectively, the partition function reduces to

$$2^{F-E}\sum_{\beta}(-1)^{q_\eta(\beta)} = 2^{1-g(X)}\mathrm{Arf}(X,\eta),$$

where $g(X)$ is the genus of X, and $\mathrm{Arf}(X,\eta) \in \{\pm 1\}$ is the Arf invariant of the spin structure η[15,16,23]:

$$\mathrm{Arf}(X,\eta) = 2^{-g(X)} \sum_{[\beta]\in H^1(X,\mathbb{Z}_2)} (-1)^{q_\eta(\beta)}.$$

The partition function of the state-sum is not ± 1, but it differs from it by an exponential of an integral of a local counter-term (the Euler density), and thus can be made ± 1 by a local redefinition. After such a redefinition, we get an invertible 2d spin-TQFT which describes the basic fermionic SRE phase in two space-time dimensions (the Majorana spin chain). This spin-TQFT has also been briefly discussed in Ref. 22.

6.2. *State-sum construction of 3d spin-TQFTs*

In 3d, the state-sum construction of bosonic TQFTs is known as the Turaev-Viro construction.[19,20] Again, it does not produce the most general oriented 3d TQFT[¶], but it has the advantage that it can be easily extended to manifolds with boundaries.

The starting point of the Turaev-Viro construction is a spherical fusion category $\mathcal{A}$. It can be thought of as a categorification of a finite-dimensional semi-simple algebra A. A fusion category is a semi-simple rigid monoidal category with a finite number of simple objects. Let I be the set of simple objects. The partition function is a sum over colorings of 1-simplices of the triangulation with elements of I. If we denote such a coloring by ϕ, we can write

$$Z = \sum_{\phi} w[\phi] Z_\phi\,. \tag{55}$$

The weights $w[\phi]$ are a product over 1-simplces of an appropriate function of their color, times a product over all 0-simplices of an appropriate $\mathcal{A}$-dependent constant.

Each individual term Z_ϕ in the partition function can be thought of as the evaluation of a tensor network: each face with edges of color i,j,k is

¶For example, Chern-Simons TQFTs for simple Lie groups do not arise from the Turaev-Viro construction.

associated to a pair of vector spaces V_{ijk}, V^*_{ijk}, each one associated to a side of the face, and each tetrahedron t to chosen element $F[\phi[t]]$ in the tensor product of the four vector spaces associated to its faces. The tensors are contracted together at the common faces. The partition function will be invariant under changes of triangulations if the $F[\phi[t]]$ tensors satisfy some basic axioms, such as the pentagon axiom, which guarantees invariance under a $2-3$ move.

The same construction applied to a manifold with a boundary produces wave-functions, which can be reproduced as ground states of a Hamiltonian built from the same data.[10]

The fermionic version of this construction is given in the literature in the "Hamiltonian" form, without a description of the corresponding partition partition function. In this section we will attempt to assemble the ingredients of the fermionic construction into a partition function.

The starting point is a spherical super-fusion category $\mathcal{A}$. As far as we know, this notion was first mentioned in Ref. 9 and appeared more recently in Ref. 25 in the context of fermion condensation (see below). We will now describe what this means in a somewhat informal way, relegating the details to Appendix C. A super-fusion category has a finite number of simple objects V_i, $i \in I$ representing independent quasi-particle excitations. There are no nonzero morphisms between V_i and V_j if $i \neq j$, while the space of morphisms from V_i to V_i is a $\mathbb{Z}_2$-graded division algebra (i.e., every nonzero element has an inverse). Unlike in the bosonic case, there are two possibilities for such a division algebra: $\mathbb{C}$ or a Clifford algebra with one generator Cl(1). Following Ref. 9 we will call the former kind of simple object a bosonic simple object, and the latter kind of object a Majorana simple object. Since $\mathrm{Cl}(1) \otimes \mathrm{Cl}(1) \simeq \mathrm{Cl}(2)$, the identity object is always bosonic.

This can be explained in a different way. One can tensor any object of $\mathcal{A}$ with a $\mathbb{Z}_2$-graded vector space. Tensoring with a bosonic vector space of dimension N is equivalent to taking N copies of the quasi-particle. Tensoring with $\mathbb{C}^{0|1}$ (a fermionic vector space of dimension 1) has the physical meaning of attaching a fermion to the quasi-particle. The latter operation maps an object to an isomorphic one, but the obvious isomorphism is odd (fermionic). There may or may not be an even isomorphism as well. Now, suppose V_i is a simple object. If V_i is even-isomorphic to $\mathbb{C}^{0|1} \otimes V_i$, then $\mathrm{Hom}(V_i, V_i) \simeq \mathrm{Cl}(1)$, i.e., V_i is Majorana. On the other hand, if V_i is not even-isomorphic to $\mathbb{C}^{0|1} \otimes V_i$, then $\mathrm{Hom}(V_i, V_i) \simeq \mathbb{C}$, i.e., V_i is bosonic.

As usual, one can fuse quasi-particles, and the fusion rules are described by the way tensor products of simple objects decompose into sums of simple objects:

$$V_i \otimes V_j \simeq \oplus_{k\in I} H_{ij}^k \otimes V_k \,. \tag{56}$$

Here the "coefficients" H_{ij}^k are finite-dimensional $\mathbb{Z}_2$-graded vector spaces. If some of the simple objects are Majorana, there is a subtlety: the absolute grading on some of H_{ij}^k is not well-defined. For simplicity, we will assume that all simple objects are bosonic, so that all H_{ij}^k have a well-defined grading.

The fusion of quasi-particles is associative, in the sense that for any $i, j, k \in I$ there is an even isomorphism,

$$\mathbf{a}(i,j,k) : (V_i \otimes V_j) \otimes V_k \to V_i \otimes (V_j \otimes V_k)\,.$$

Expanding both sides in terms of simple objects, we deduce that $\mathbf{a}(i,j,k)$ is determined by an even linear map of $\mathbb{Z}_2$-graded vector spaces,

$$F\begin{bmatrix} i & j & l \\ k & m & n \end{bmatrix} : H_{ij}^l \otimes H_{lk}^m \to H_{in}^m \otimes H_{jk}^n \,.$$

This map is known as a 6j symbol. The collection of all 6j symbols satisfy an associativity constraint which ensures that the two ways of constructing an isomorphism from $(((V_i \otimes V_j) \otimes V_k) \otimes V_l)$ to $(V_i \otimes (V_j \otimes (V_k \otimes V_l)))$ are the same. To write down a concrete form for it, we need to choose a basis $e_{\alpha_{ij}^k}$, $\alpha_{ij}^k \in J_{ij}^k$, in each space H_{ij}^k. Let $\epsilon(\alpha_{ij}^k)$ be the fermionic parity of the vector $e_{\alpha_{ij}^k}$. Then the associativity constraint (a.k.a. the fermionic pentagon equation) reads:

$$\sum_{t\in I} F\begin{bmatrix} i & j & m \\ k & n & t \end{bmatrix}_{\eta\phi}^{\alpha\beta} F\begin{bmatrix} i & t & n \\ l & p & s \end{bmatrix}_{\kappa\gamma}^{\phi\chi} F\begin{bmatrix} j & k & t \\ l & s & q \end{bmatrix}_{\delta\phi}^{\eta\kappa} =$$
$$= (-1)^{\epsilon(\alpha_{ij}^m)\epsilon(\delta_{kl}^q)} F\begin{bmatrix} m & k & n \\ l & p & q \end{bmatrix}_{\delta\epsilon}^{\beta\chi} F\begin{bmatrix} i & j & m \\ q & p & s \end{bmatrix}_{\phi\gamma}^{\alpha\epsilon} \,. \tag{57}$$

Note the sign on the r.h.s. It reflects the fact that composition of morphisms in the symmetric tensor category of $\mathbb{Z}_2$-graded vector spaces is defined using the Koszul sign rule.

A spherical structure on a super-fusion category assigns to every object V its dual V^*, so that $V^{**} \simeq V$ and that for every two objects U, V we have isomorphisms,

$$\mathrm{Hom}(U,V) \simeq \mathrm{Hom}(U \otimes V^*, 1) \simeq \mathrm{Hom}(V^* \otimes U, 1) \simeq$$
$$\simeq \mathrm{Hom}(1, U^* \otimes V) \simeq \mathrm{Hom}(1, V \otimes U^*)\,. \tag{58}$$

Let us denote by $V_{\bar{i}}$ the dual of a simple object V_i. The spherical structure ensures that the spaces $H_{ijk} = H_{ij}^{\bar{k}}$ are cyclically symmetric and that the space $H_{\bar{k}\bar{j}\bar{i}}$ is dual to H_{ijk} for all $i, j, k \in I$.

In the fermionic Turaev-Viro construction one considers colorings of 1-simplices of an oriented triangulated 3-manifold X with objects V_i. We can choose a branching structure on the triangulation, so that the vertices of each 3-simplex and each 2-simplex have an order. To each 2-simplex one can attach a $\mathbb{Z}_2$-graded vector spaces H_{ijk} and its dual, depending on the orientation. Next we would like to attach a 6j symbol, which is an element of a tensor product of four $\mathbb{Z}_2$-graded vector spaces, to a 3-simplex, and form the partition function by contracting these tensors along their shared 2-simplices. But there are two problems with this. First, the 6j symbol itself is an element of the tensor product with a particular ordering of $\mathbb{Z}_2$-graded factors H_{ij}^k and their duals. Changing this order may change the sign of some components of the 6j symbol, but there is no natural order on the faces of a 3-simplex. Second, the contraction of two odd elements of dual $\mathbb{Z}_2$-graded vector spaces changes sign if one exchanges their order, but for a given 2-simplex in an oriented triangulation there is no preferred order of two 3-simplices sharing it, unless the 2-simplex itself is given an orientation. Such an orientation is induced by a branching structure, but this means that the partition function might depend on the branching structure.

These choices are analogous to the choices made in the construction of fermionic SPT phases: the order of the factors in the tensor product defining 6j symbols is analogous to the order of the Grassmann variables in $u[t]$ and the choice of order of the two sides of a 2-simplex is analogous to the choice of $d\theta d\bar{\theta}$ or $d\bar{\theta} d\theta$ in the fermionic measure. One can make the analogy completely precise in the following way. First, we choose a basis

$$e_{\alpha_{ijk}}, \alpha_{ijk} \in J_{ijk} ,$$

in every space H_{ijk}.The weight of a particular coloring can be computed by summing over the variables α_{ijk} attached to every 2-simplex whose edges are colored by i, j, k. If we assume as before that the basis vectors have definite parity, each choice of α variables gives a $\mathbb{Z}_2$-valued 2-cochain. Further, since 6j symbols are even maps, this 2-cochain becomes a 2-cocycle β_2 when restricted to configurations of α variables for which all 6j symbols are nonzero. If we perform the summation over such variables first by fixing β_2, and then summing over all choices of β_2, then the Koszul sign is simply $\sigma_\Pi(\beta_2)$, where Π denotes the ordering choices made in the definition of the partition function.

Thus the tensor network amplitude restricted to some fermionic parity sector β_2 can be decomposed as

$$Z_\phi(\beta) = Z^b_\phi(\beta_2)\sigma_\Pi(\beta_2)\,, \tag{59}$$

where Z^b_ϕ is a "bosonized" amplitude computed without keeping track of Koszul signs in manipulating the vector spaces in the tensors, but rather pretends that all vector spaces are bosonic.

There is also a close analogy between the sign factor $(-1)^{\epsilon(\alpha_{ijm})\epsilon(\delta_{klq})}$ in the fermionic pentagon equation and the twisted cocycle condition satisfied by the 3-cochain ν_3 used in the construction of 3d fermionic SPT phases. Indeed, we can write the twisting factor as[||]

$$(-1)^{\epsilon(\alpha_{ijm})\epsilon(\delta_{klq})} \equiv (-1)^{\beta_2(012)\beta_2(234)} = (-1)^{[\beta_2\cup\beta_2](01234)}\,. \tag{60}$$

This is precisely the cup square of β_2 evaluated on the 4-simplex associated to the 2-3 move.[**] Notice that the fermionic pentagon relation involves a sum of terms with different fermionic parity in the internal faces, but the sign factor only involves the fermionic parity of the external faces.

The fermionic pentagon relation is written in terms of the components of the tensors, and thus does not include any Koszul signs. The sign factor describes a change in the sign of the bosonic weight under a 2-3 move. If we put the Koszul signs back in, the change of $\sigma_\Pi(\beta_2)$ under a 2-3 move will almost cancel the sign coming from the pentagon equation: it will trade the cup square of β_2 with the usual linear term $\int w_2\beta_2$ controlled by the canonical representative for w_2. Again, this linear term will provide an obstruction to assembling a well-defined partition function, unless the underlying manifold admits a spin structure.

It should be now clear how to assemble a well-defined spin-structure-dependent partition sum:

$$Z[\eta] = \sum_{\beta_2}\sum_\phi w[\phi]Z^b_\phi[\beta_2]z[X,\eta,\beta_2]\,, \tag{61}$$

[||]In order to write this formula, we have assigned a specific order 01234 to the vertices of 4-simplex, associated respectively to edges $imnp$, ijs, $kjmq$,knl, $lpqs$. This order is the one associated with the branching structure induced by the orientation of edges in the definition of the 6j symbols, such that a factor of H^k_{ij} is associated to a positively oriented face, with edges i and j positively oriented and k negatively oriented, and $(H^k_{ij})^*$ is associated to a negatively oriented face, with edges i and j negatively oriented and k positively oriented

[**]We would like to thank Zhengcheng Gu for pointing it out.

where we decomposed the partition sum over eigenspaces of $\hat{\beta}_2$, denoted the "bosonized" partition sum in each sector as $Z^b_\phi[\beta_2]$ and combined the Koszul signs and the spin-structure-dependent correction into $z[X,\eta,\beta_2]$. The change of $z[X,\eta,\beta_2]$ under 2-3 moves will cancel against the sign factor in the pentagon relation, giving a well-defined object.

Notice that $Z[\eta]$ is almost written as the partition function of a $\mathbb{Z}_2$ gauge theory with a 2-form gauge field β_2, except that $Z^b_\phi[\beta_2]$ is not well-defined on its own, as the $2-3$ move involves a summation over both bosonic and fermionic vector spaces when gluing the tetrahedra. Inspection of the $2-3$ move, though shows that the cocycle condition for β_2 almost fixes the Grassmann parity on internal faces, up to a binary choice, corresponding to a shift of β_2 by the coboundary of a cochain concentrated on the the "t" edge of the tetrahedra on the 3 side of the $2-3$ move.[††]

That means that we can define a sensible bosonic partition function by summing only over β_2 which differ by an exact cocycle,

$$\tilde{Z}^b[\beta_2] = \sum_{\lambda_1}\sum_{\phi} 2^{-\#(\text{edges})} w[\phi] Z^b_\phi[\beta_2+\delta\lambda_1](-1)^{\int_X \beta_2\cup\lambda_1+\lambda_1\cup\beta_2+\lambda_1\cup\delta\lambda_1}\,, \tag{62}$$

and re-write the fermionic partition function as

$$Z[\eta] = \sum_{\beta_2} \tilde{Z}^b[\beta_2] z[X,\eta,\beta_2]\,. \tag{63}$$

Thus the fermionic Turaev-Viro theories can be obtained by gauging the diagonal 1-form $\mathbb{Z}_2$ global symmetry of a product of appropriate bosonic theories and the spin-TFT K_3 defined by the partition function $z[X,\eta,\beta_2]$.

7. Fermionic anyon condensation

7.1. *Self-fermions and anomalous 1-form $\mathbb{Z}_2$ global symmetry*

In this section we would like to argue that in $2+1$ dimensions, a bosonic TQFT equipped with a 1-form $\mathbb{Z}_2$ global symmetry with the same 't Hooft

[††]Notice that because of that, the correction $\int_X \eta\cup\beta$ may add an extra relative sign between terms of a $2-3$ move, essentially controlled by the cup product between w_2 and a cochain concentrated on the "t" edge of the 4-symplex. That relative sign does not to affect the $2-3$ move with the standard branching structure, where the "t" edge has vertices (13) and does not contribute to a cup product. We expect that relative sign to be present in the pentagon relation for other choices of branching structure.

anomaly as K_3 is the same as a TQFT which has in the spectrum of quasi-particles a self-fermion ϵ which fuses with itself to the identity.

Our first observation is that a TQFT which has in the spectrum of quasi-particles a self-boson b which fuses with itself to the identity is automatically equipped with a non-anomalous 1-form $\mathbb{Z}_2$ global symmetry, whose generators are simply b worldlines. The operation of gauging the 1-form symmetry is the same as the operation of anyon condensation applied to b.

If we are given any two TQFTs with self-fermions ϵ and ϵ' respectively which fuse with itself to the identity, the product of the two theories has a self-boson $\epsilon\epsilon'$. This shows that if we take f and f' worldlines to be generators of 1-form $\mathbb{Z}_2$ global symmetries, the two $\mathbb{Z}_2$ global symmetries will have the same $\mathbb{Z}_2$-valued 't Hooft anomaly.

That means we only need to compute the 't Hooft anomaly for some representative TQFT with a self fermion ϵ which fuses with itself to the identity. All other TQFTs with such a quasi-particle will have the same 't Hooft anomaly.

A simple choice of such a TQFT is the toric code, aka $\mathbb{Z}_2$ gauge theory. We can describe it by the discrete partition function,

$$(-1)^{\int_X a_1 \cup \delta b_1}\,, \tag{64}$$

with a_1 and b_1 being discrete cochains.

The e,m and ϵ quasi-particles correspond to Wilson loops for a_1, b_1 and $a_1 + b_1$ respectively. The former two quasi-particles are generators of two non-anomalous 1-form $\mathbb{Z}_2$ symmetries.‡‡ For example, we can couple the theory to a $\mathbb{Z}_2$ 2-cocycle β_2^e as

$$(-1)^{\int_X a_1 \cup \delta b_1 + a_1 \cup \beta_2^e}\,. \tag{65}$$

The action is invariant under gauge transformations $\beta_2^e + \delta\lambda_1^e$, accompanied by $b_1 \to b_1 + \lambda_1^e$. Similarly, we can couple the theory to a $\mathbb{Z}_2$ 2-cocycle β_2^m as

$$(-1)^{\int_X a_1 \cup \delta b_1 + \beta_2^m \cup b_1}\,. \tag{66}$$

The action is invariant under gauge transformations $\beta_2^m + \delta\lambda_1^m$, accompanied by $a_1 \to a_1 + \lambda_1^m$.

On the other hand, the 1-form $\mathbb{Z}_2$ symmetry associated to ϵ is encoded in the action,

$$(-1)^{\int_X a_1 \cup \delta b_1 + a_1 \cup \beta_2 + \beta_2 \cup b_1}\,. \tag{67}$$

‡‡The two symmetries have a mixed anomaly.

Under gauge transformations $\beta_2 + \delta\lambda_1$, accompanied by $b_1 \to b_1 + \lambda_1$ and $a_1 \to a_1 + \lambda_1$, the action varies by

$$(-1)^{\int_X \lambda_1 \cup \delta\lambda_1 + \lambda_1 \cup \beta_2 + \beta_2 \cup \lambda_1} . \tag{68}$$

This is precisely the anomalous variation for K_3! This verifies our claim.

7.2. *Examples of fermionic anyon condensation*

We will come back to the toric code momentarily. Before that, we should give a simple examples of bosonic and fermionic theories related by fermionic anyon condensation.

Consider first a $U(1)_{4k}$ Chern-Simons theory. The Wilson loop of charge $2k$ fuses with itself to the identity. If k is even it is a self-boson b, while if k is odd it is a self-fermion ϵ. For even k, condensing b leads to a $U(1)_k$ CS theory. We expect the same to occur for odd k, but then the resulting $U(1)_k$ CS theory is a spin-TQFT. This is consistent with the fact that this is a fermionic anyon condensation.

It is interesting to observe that the toric code has a $\mathbb{Z}_2$ global symmetry which exchange the e and m particles. This symmetry is rather hard to make explicit in a concrete description of the model. For example, the action we wrote above is not invariant under a $\mathbb{Z}_2$ symmetry transformation exchanging a_1 and b_1:

$$(-1)^{\int_X a_1 \cup \delta b_1 + b_1 \cup \delta a_1} = (-1)^{\int_X \delta a_1 \cup_1 \delta b_1} . \tag{69}$$

Assuming that the $\mathbb{Z}_2$ global symmetry is not broken by the fermionic condensation of ϵ, the resulting spin-TQFT should still enjoy the $\mathbb{Z}_2$ global symmetry. Indeed, there are good reasons to believe this is a fermionic SPT phase, the root fermionic SPT phase which is not captured by the super group cohomology classification. For example, we can consider the Hilbert space of the fermionic theory on a general Riemann surface (equipped with spin structure). Sectors with a non-trivial 2-form gauge field flux are described by a Riemann surface with a single ϵ insertion, which do not contribute ground states. It is quite reasonable to assume that the 2^g ground states with no β_2 background flux connection will be projected down to 2^g one-dimensional Hilbert spaces labelled by the choice of spin structure.

A $\mathbb{Z}_2$-invariant boundary condition for the toric code has gapless modes: a non-chiral 2d Ising model coupled to the toric code in such a way that e ends on the spin operator $\sigma(z, \bar{z})$, m on the dual spin operator $\mu(z, \bar{z})$ and ϵ ends on chiral or anti-chiral fermion operators $\psi(z)$ and $\bar{\psi}(\bar{z})$, while the energy operator $\epsilon(z,\bar{z})$ can exist without a quasi-particle world line ending on it. The $\mathbb{Z}_2$ global symmetry is Kramer-Wannier duality.

After condensing the ϵ quasi-particle, the chiral or anti-chiral fermion operators $\psi(z)$ and $\bar{\psi}(\bar{z})$ in the boundary theory will be able to be inserted in correlation functions without a quasi-particle world line ending on it, Thus the boundary theory reduces to a theory of free Majorana fermions, one chiral and the other anti-chiral, with the $\mathbb{Z}_2$ global symmetry acting only on one of the two fermions. This is an the expected feature of the root fermionic SPT phase with $\mathbb{Z}_2$ global symmetry.

Another interesting example is the fermionic spin-TQFT "s-Ising" associated to the Ising 3d TQFT. The Ising TQFT has three quasi-particles: 1, σ and ϵ. In the presence of a boundary, these quasi-particles end on the corresponding operators of a chiral Ising model. We expect that upon condensing the ϵ quasi-particle, the corresponding chiral operator on the boundary, which is a free fermion operator, will be free to appear by itself. Thus the boundary theory should consist of a single chiral Majorana fermion, which can indeed be well-defined on a manifold equipped with a spin structure.

It is also interesting to look at the Hilbert space of s-Ising on Riemann surfaces of various genus, equipped with a spin structure. On a torus, the Ising TQFT has three states. Upon gauging the 1-form symmetry, we expect them to be projected to three one-dimensional Hilbert spaces labelled by the three even spin structures on the torus. As they have no β_2 flux on the Riemann surface, they should have even fermion number. Upon gauging the 1-form symmetry we also gain twisted sectors with a non-zero flux of β_2 on the Riemann surface and odd fermion number, i.e., a single insertion of ϵ. On the torus, that gives an extra state (labelled by a loop of σ with an ϵ insertion), which we expect to give a one-dimensional Hilbert spaces labelled by the single odd spin structure on the torus, of odd fermion number.

On a genus 2 Riemann surface the Ising TQFT has 10 untwisted sectors, which should map to 10 one-dimensional Hilbert spaces labelled by the 10 even spin structures on the torus. We can also find 6 states for the Ising TQFT on a genus two surface with an extra ϵ insertion, which should map to 6 one-dimensional Hilbert spaces labelled by the 6 odd spin structures on the torus.

In general, we expect the s-Ising spin-TQFT to have one-dimensional spaces of ground states for every choice of spin structure on a Riemann surface, with even fermion number if the spin structure is even and odd if the spin structure is odd. In other words, If we compactly the s-Ising TQFT on a circle with Ramond boundary conditions, we should be left with a 2d theory which computes the Arf invariant of Riemann surfaces.

It is somewhat surprising to observe that the s-Ising spin-TQFT has one-dimensional spaces of ground states, and yet supports chiral edge modes and would likely not be considered a fermionic SPT phase. Indeed, the standard classification of fermionic SPT phases in $2+1$ dimensions with fermion number symmetry only predicts no non-trivial theories.

7.3. *Bosonization in* $2+1$ *dimensions*

The operation of fermionic condensation on the toric code is somewhat intriguing from the perspective of defining a notion of bosonization in $2+1$ dimensions. Suppose that we were given a version of the toric code lattice Hamiltonian with the property that creating fermionic quasi-particle excitations could be created at much lower energy cost than the e and m quasi-particles. The Hilbert space of the low-energy theory will be the even part of a fermionic Fock space, but the interactions will not be completely local: fermion bilinear operators would still have to be represented by open string operators, corresponding to an ϵ worldline connecting the two fermionic creation/destruction operators.

Suppose also that we could tensor that system with K_3 and gauge the diagonal 1-form $\mathbb{Z}_2$ symmetry. Then the ϵ worldline string operators, dressed by the self-fermion quasiparticles of K_3, would become invisible in the gauged theory. This would liberate the endpoints of the open worldline string operators, which would behave as fermionic creation or destruction operators. The Hilbert space of the gauged low-energy theory should become a fermionic Fock space, which one could use to simulate some general fermionic Hamiltonian in a local way.

It would be interesting to make this recipe more precise. Ideally, one should find a lattice Hamiltonian description of the toric code which both allows the fermion quasiparticles to proliferate and which can be coupled appropriately to a β_2 cocycle.

Acknowledgments

We thank Z. Gu and D. Freed for helpful discussions and J. Morgan for educating one of us (AK) about spin structures and cobordism. The research of DG was supported by the Perimeter Institute for Theoretical Physics. Research at Perimeter Institute is supported by the Government of Canada through Industry Canada and by the Province of Ontario through the Ministry of Economic Development and Innovation. The work of AK was supported in part by the DOE grant DE-FG02-92ER40701 and by the Simons Foundation. Opinions and conclusions expressed here are those of the authors and do not necessarily reflect the views of funding agencies.

Appendix A. The Gu–Wen Grassmann integral in 2d as a quadratic refinement

Notice that we can depict β as a collection of closed, disjointed paths C_a in the dual graph to the triangulation, sequences of edges e_i^a with $\beta(e_i^a) = 1$ sharing a common triangle $t_{i\pm\frac{1}{2}}^a$ with the previous and next edges in the path. We can order the monomials in sequences along each path, and factorize the Grassmann integral as

$$\sigma(\beta) = \prod_a \left[\int \prod_i d\theta_{e_i^a} d\bar{\theta}_{e_i^a} \prod_i u(t_{i-\frac{1}{2}}^a) \right] , \tag{A.1}$$

where we ordered the integrand in one direction and the measure in the opposite direction. We assumed that $\epsilon[t, \beta]$ is 1 unless a triangle is crossed by some path C_a.

For each path C_a, denote as c_i^a the sequence of Grassmann variables encountered along the path. We have

$$\int \prod_i dc_{2i+1}^a dc_{2i}^a \prod_i c_{2i+1}^a c_{2i+2} = -1 \, , \tag{A.2}$$

where ℓ_a is the length of C_a.

The integration variables in $\prod_i d\theta_{e_i^a} d\bar{\theta}_{e_i^a}$ are encountered almost in the order $\prod_i dc_{2i+1}^a dc_{2i}^a$, up to permutations of two variables associated to the same edge. Similarly, the integration variables are encountered in the product $\prod_i u(t_{i-\frac{1}{2}}^a)$ almost in the same order as $\prod_i c_{2i+1}^a c_{2i+2}$, up to permutations of two variables associated to the same triangle.

Thus the overall sign of the integral can be computed by multiplying a sign for each edge and each triangle along C_a:

$$\sigma(\beta) = \prod_a \left[- \prod_i \epsilon_1[e_i^a] \prod_i \epsilon_2[t_{i-\frac{1}{2}}^a] \right] , \tag{A.3}$$

where the edge sign ϵ_1 is 1 if the edge orientation points to the right of C_a, -1 if it points to the left of C_a, while the triangle sign ϵ_2 is 1 if both edges of the triangle crossed by C_a point to the right of C_a, -1 if both point

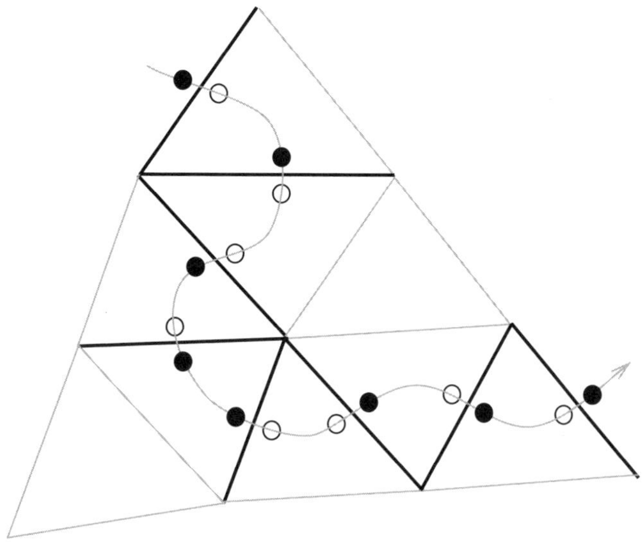

Fig. A1. A collection of edges e_i with $\beta(e_i) = 1$, organized into a path in the dual graph to the triangulation. The Grassmann variables are encountered along the path in an order which differs by simple local permutations from the order they appear with in the measure and integrand of the Grassmann integral.

to the left, while if they point in opposite directions the sign is 1 if C_a is turning left at the triangle, -1 if turning right.

It is useful to combine the sign associated with crossing an edge to the sign associated to transversing the subsequent triangle. The resulting sign is almost always $+1$, except when we cross a 01 edge and turn towards a 12 edge. See figures A2 and A3.

There is a simple interpretation of such sign. Consider a certain canonical vector field V inside each triangle, nonzero away from the vertices and continuous across the edges, which flows out of the vertex 0 and into the vertex 2, turning around near the vertex 1. This can be thought of as a continuation inside the triangle of the branching structure directions, as in Figure A4.

Given a smooth, non-self-intersecting path C in the triangulation, which does not pass through the vertices, we can measure how many times the vector field V restricted to C winds around compared to the tangent vector of C. We can measure such winding number $w(V, C)$, modulo 2, by count-

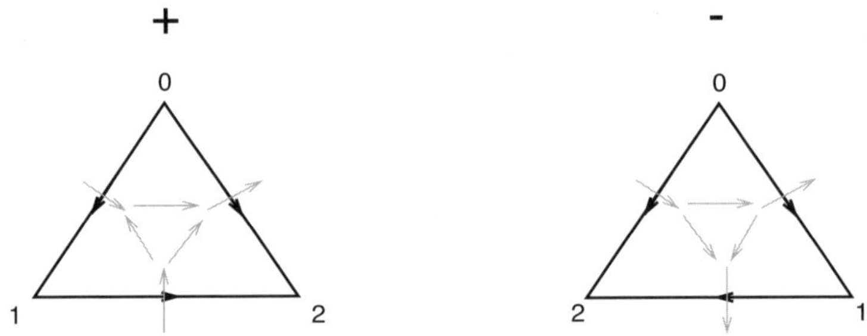

Fig. A2. As a path C proceeds along the triangulation, it accumulates a sign when crossing each edge, and a sign when passing through each triangle, depending on if the Grassmann variables it encounters are set in canonical order in the integral, or not. In this figure we indicate the directions along which a path receives a + sign. The opposite directions produce a − sign.

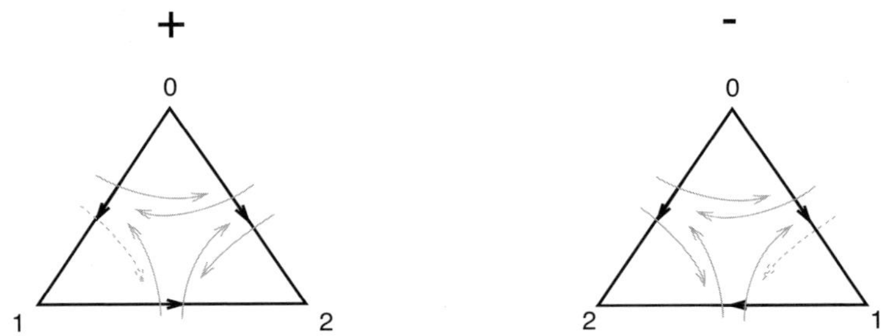

Fig. A3. As a path C proceeds along the triangulation from just before crossing an edge to just before crossing the next, it accumulates a sign as indicated in this figure: +1 in alms all cases, except the ones indicated by a dashed line.

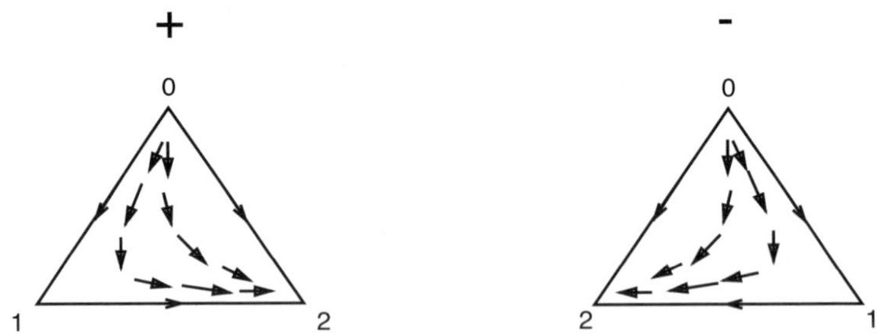

Fig. A4. A useful reference vector field V inside a triangle equipped with branching structure.

ing how many times the vector field V becomes tangent to the path C. Inspection of the sign rule of A3, shows that the overall sign accumulated by C is $(-1)^{w(V,C)+1}$.

Thus we conclude that

$$\sigma(\beta) = \prod_a (-1)^{w(V,C_a)+1}\,, \tag{A.4}$$

if β is represented by the collection of non-intersecting paths C_a.

The sign $\sigma(\beta)$ does depend on the choice of cochain β in a given cohomology class: if we sweep some path C across a vertex v of the triangulation, the winding number $w(V,C)$ will generically jump. It is natural to expect, and it should be easy to show, that the jump modulo 2 will be the same as the contribution of that vertex v to the set S of vertices which contribute to Z_m, i.e., 1 plus the number of times v is the vertex 1 of a "−" triangle.

Indeed, that would insure that

$$\sigma(\beta + \delta\lambda) = \sigma(\beta) \prod_{v\in S} (-1)^{\lambda(v)}\,, \tag{A.5}$$

so that the variation of $Z_\theta = \sigma(n)$ cancels against the variation of Z_m under $m_1 \to m_1 + \lambda$.

This fact is clearly true if the triangulation is a barycentric subdivision BT: the 0 vertices are sources, the 1 vertices simple saddles and the 2 vertices are sinks. All these configurations are associated to a jump of the winding number by one unit.

The combination,

$$(-1)^{q_E(\beta)} = \sigma(\beta) \prod_{e\in E} (-1)^{\beta(e)}\,, \tag{A.6}$$

is a known expression for the quadratic refinement of the intersection form on $\mathbb{Z}_2$ 1-cocycles on a surface.[23] Besides transforming in the obvious way under changes of spin structure E, it satisfies

$$q_E(\beta + \beta') = q_E(\beta) + q_E(\beta') + \int \beta \cup \beta'\,. \tag{A.7}$$

Thus

$$Z_n^E[BT] Z_\theta[BT] = (-1)^{q_E(n_1)}\,. \tag{A.8}$$

If we have a generic triangulation, we can still try to assemble a smooth vector field V and look at the zeroes of V, with appropriate multiplicity, as a representative chain S for w_2. In general, we expect solutions of $\partial E = S$ to still describe a choice of spin structure on the discretized surface. One may interprete measuring the winding numbers $w(V,C)$ modulo 2, corrected by the pairing with E, as a way to build a spin bundle on the triangulated surface.

Appendix B. Higher cup products

As described in the introduction, higher cup products satisfy the recursive property

$$A \cup_a B + B \cup_a A = \delta(A \cup_{a+1} B) + \delta A \cup_{a+1} B + A \cup_{a+1} \delta B\,, \qquad \text{(B.1)}$$

with $\cup_0 \equiv \cup$.

There is a rather explicit canonical formula for the higher cup products[24]:

$$[A_p \cup_a B_q](0,\cdots,p+q) = \sum_{i_0<\cdots<i_a} A(0,\cdots i_0, i_1, \cdots i_2, i_3, \cdots) B(i_0,\cdots,i_1,i_2,\cdots,i_3,i_4,\cdots) \qquad \text{(B.2)}$$

where the sequences of arguments of A_p and B_q must agree with the degree p,q of A_p and B_q.

For example, we recover the standard cup product:

$$[A_p \cup B_q](0,\cdots,p+q) = A(0,\cdots p)B(p,\cdots,p+q)\,. \qquad \text{(B.3)}$$

The next cup products are

$$[A_p \cup_1 B_q](0,\cdots,p+q-1) = \sum_{i_0} A(0,\cdots i_0, i_0+q, p+q-1) B(i_0,\cdots,i_0+q) \qquad \text{(B.4)}$$

$$\begin{aligned}[A_p \cup_2 B_q]&(0,\cdots,p+q-2) = \\ &= \sum_{i_0<i_1} A(0,\cdots i_0, i_1, \cdots p+i_1-i_0-1)\cdot \\ &\qquad \cdot B(i_0,\cdots,i_1,p+i_1-i_0-1,\cdots,p+q-2)\,. \end{aligned} \qquad \text{(B.5)}$$

and so on.

Appendix C. Super-fusion categories

A category $\mathcal{C}$ enriched over $\mathrm{Vect}_{\mathbf{k}}$ (the category of finite-dimensional vector spaces over a field $\mathbf{k}$) is a category whose morphism sets $\mathrm{Hom}(A,B)$ are vector spaces over $\mathbf{k}$, and the composition of morphisms is bilinear. It can be thought of as a generalization of the notion of an algebra over a field ("algebra with many objects"). Similarly, a category $\mathcal{C}$ enriched over $\mathrm{sVect}_{\mathbf{k}}$

(the category of $\mathbb{Z}_2$-graded vector spaces over a field $\mathbf{k}$) is a category whose morphism sets $\mathrm{Hom}(A, B)$ are $\mathbb{Z}_2$-graded vector spaces over $\mathbf{k}$, and the composition of morphisms is bilinear and is compatible with the grading. Such a catregory $\mathcal{C}$ can be thought of a generalization of the notion of a $\mathbb{Z}_2$-graded algebra over a field. In what follows we will suppress $\mathbf{k}$ (in physical applications one typically has $\mathbf{k} = \mathbb{C}$).

An important difference between ordinary algebras and $\mathbb{Z}_2$-graded algebras is the way the tensor product of algebras is defined. For ordinary algebras, the product of algebras C and D is an algebra whose underlying set is $C \otimes D$ and the multiplication is defined by

$$(c \otimes d) \cdot (c' \otimes d') = cc' \otimes dd', \quad c, c' \in C, \ d, d' \in D.$$

On the other hand, for $\mathbb{Z}_2$-graded algebras one defines the multiplication by

$$(c \otimes d) \cdot (c' \otimes d') = (-1)^{\epsilon(c')\epsilon(d)}(c \cdot c') \otimes (d \cdot d'), \quad c, c' \in C, \ d, d' \in D,$$

where $\epsilon(a) \in \mathbb{Z}_2$ is the fermionic parity of $a \in C$. This is the Koszul sign rule. This has an analog for categories: the Cartesian product of categories over sVect is defined differently from the Cartesian product of categories over Vect. In the former case, if C, C', C'' and D, D', D'' are objects of $\mathcal{C}$ and $\mathcal{D}$ respectively, the composition of morphisms in $\mathcal{C} \times \mathcal{D}$ is defined as follows:

$$\begin{aligned}(\alpha \otimes \beta) \cdot (\gamma \otimes \delta) = (-1)^{\epsilon(\gamma)\epsilon(\beta)}&(\alpha \cdot \gamma) \otimes (\beta \cdot \delta), \\ \alpha \in \mathrm{Hom}_{\mathcal{C}}&(C', C''), \gamma \in \mathrm{Hom}_{\mathcal{C}}(C, C'), \\ &\beta \in \mathrm{Hom}_{\mathcal{D}}(D', D''), \delta \in \mathrm{Hom}_{\mathcal{D}}(D, D'). \end{aligned} \tag{C.1}$$

while in the latter case the sign factor on the r.h.s. is absent.†

This in turn has consequences for how a tensor product on a category is defined. Both in the ordinary and the $\mathbb{Z}_2$-graded case, a tensor product on $\mathcal{C}$ is a functor m from $\mathcal{C} \times \mathcal{C}$ to $\mathcal{C}$, $m : (X, Y) \mapsto X \otimes Y$, $\forall X, Y \in \mathrm{Ob}(\mathcal{C})$, together with an associator. The associator is a natural isomorphism between two functors from $\mathcal{C} \times \mathcal{C} \times \mathcal{C}$ to $\mathcal{C}$ built from m. The first one is $m(m(-,-),-)$ and the second one is $m(-, m(-,-))$. The associator must satisfy the pentagon equation. Since the definition of the Cartesian product

†More generally, one can consider categories enriched over a symmetric tensor category. The definition of Cartesian product for such categories depends on the symmetric tensor category.

depends on whether we consider categories over Vect or sVect, so does the definition of the tensor structure.

Let us specialize to the case of (multi)-fusion categories. Multi-fusion categories are rigid (and in particular semi-simple) monoidal categories (categories with a tensor product and a unit object) over Vect or sVect which have finitely many isomorphism classes of simple objects. To emphasize the difference between the two cases, we will refer to multi-fusion categories over sVect as super-multi-fusion categories, while multi-fusion categories over Vect will be simply called multi-fusion categories. While multi-fusion categories are module categories over Vect, super-multi-fusion categories are module categories over sVect.

Let $\mathcal{C}$ be a super-multi-fusion category. For every object $X \in \mathrm{Ob}(\mathcal{C})$ we have a $\mathbb{Z}_2$-graded algebra $\mathrm{Hom}_{\mathcal{C}}(X, X)$. If X is simple, this algebra must be a division algebra, therefore it is isomorphic either to $\mathbb{C}$ or $\mathrm{Cl}(1)$. In the former case we will say that X is a bosonic simple object, while in the latter case we will say that X is a Majorana simple object.

A super-fusion category is a super-multi-fusion category whose unit object $\mathbf{1}$ is simple. Since $\mathbf{1} \otimes \mathbf{1} \simeq \mathbf{1}$, this implies that $\mathrm{Hom}(\mathbf{1}, \mathbf{1}) = \mathbb{C}$, i.e., $\mathbf{1}$ is a bosonic object.

Since a super-fusion category is rigid, for every object V we have a dual object V^*, and even morphisms $\mathrm{ev}_L : V^* \otimes V \rightarrow \mathbf{1}$, $\mathrm{coev}_L : \mathbf{1} \rightarrow V \otimes V^*$, $\mathrm{ev}_R : V \otimes V^* \rightarrow \mathbf{1}$ and $\mathrm{coev}_R : \mathbf{1} \rightarrow V^* \otimes V$ satisfying the usual identities. One can show[21] that V^{**} is isomorphic to V. A pivotal structure on a super-fusion category is a choice of such isomorphisms for all V in a way compatible with the tensor product. Given a pivotal structure, one can define the left and right dimensions of V by composing coev_L and ev_R or coev_R and ev_L. The left and right dimensions are complex numbers. If the left and right dimensions are equal for all V, one says that the pivotal structure is spherical. A spherical super-fusion category is a super-fusion category equipped with a spherical pivotal structure.

References

1. X. Chen, Z.-C. Gu, and X.-G. Wen, "Local unitary transformations, long-range quantum entanglement, wave function renormalization, and topological order," *Phys. Rev.* **B82**, 155138 (2010).
2. A. Kitaev, talk at the IPAM workshop "Symmetry and Topology in Quantum Matter," Jan 26-30, 2015.
3. E. Witten, "Supersymmetric index of three-dimensional gauge theory,"

In *Shifman, M.A. (ed.): The many faces of the superworld* 156-184, arXiv:hep-th/9903005.
4. X.G. Wen, "Quantum field theory of many-body systems: From the origin of sound to an origin of light and electrons," Oxford University Press, 2004.
5. X. Chen, Z.-C. Gu, and X.-G. Wen, "Classification of gapped symmetric phases in 1D spin systems," *Phys. Rev.* **83**, 035107 (2011).
6. L. Fidkowski, A. Kitaev, "Topological phases of fermions in one dimension," *Phys. Rev.* **B83**, 075103 (2011).
7. Z-C. Gu and X-G. Wen, "Symmetry-protected topological orders for interacting fermions: Fermionic topological nonlinear σ-models and a special group supercohomology theory," *Phys. Rev.* **B90**, 115141 (2014), arXiv:1201.2648 [cond-mat].
8. Z-C. Gu, Z. Wang, and X-G. Wen, "Lattice model for fermionic toric code," *Phys. Rev.* **B 90**, 085140 (2014), arXiv:1309.7032 [cond-mat].
9. Z-C. Gu, Z. Wang, and X-G. Wen, "A classification of 2D fermionic and bosonic topological orders," *Phys. Rev.* **B91**, 125149 (2015), arXiv:1010.1517 [cond-mat].
10. M.A. Levin and X.G. Wen, "String net condensation: A physical mechanism for topological phases," *Phys. Rev.* **B71**, 045110 (2005), arXiv:cond-mat/0404617.
11. A. Kapustin, "Symmetry protected topological phases, anomalies, and cobordisms: Beyond group cohomology," arXiv:1403.1467 [cond-mat.str-el].
12. A. Kapustin, R. Thorngren, A. Turzillo, and Z. Wang, "Fermionic symmetry protected topological phases and cobordisms," arXiv:1406.7329 [cond-mat.str-el].
13. Z-C. Gu and M. Levin, "The effect of interactions on 2D fermionic symmetry-protected topological phases with Z_2 symmetry," *Phys. Rev.* **B89**, 201113 (2014), arXiv:1304.4569 [cond-mat].
14. R.E. Stong, "Notes on cobordism theory," Princeton University Press, 1968.
15. M.F. Atiyah, "Riemann surfaces and spin structures," *Ann. Sci. Ecole Norm. Sup.* **4**, 47-62 (1971).
16. D. Johnson, "Spin structures and quadratic forms on surfaces," *J. London Math. Soc.* **22**, 365-373 (1980).
17. M. Fukuma, S. Hosono, and H. Kawai, "Lattice topological field theory in two-dimensions," *Commun. Math. Phys.* **161**, 157 (1994), arXiv:hep-th/9212154.

18. C. Bachas and P.M.S. Petropoulos, "Topological models on the lattice and a remark on string theory cloning," *Commun. Math. Phys.* **152**, 191 (1993), arXiv:hep-th/9205031.
19. V. Turaev and O.Ya. Viro, "State sum invariants of 3-manifolds and quantum 6j symbols," *Topology.* **31**, no. 4, 865-902 (1992).
20. J. Barrett and B. Westbury, "Invariants of piecewise-linear 3-manifolds," *Trans. Amer. Math. Soc.* **348**, no. 10, 3997-4022 (1996).
21. V. Drinfeld, S. Gelaki, D. Nikshych, and V. Ostrik, "On braided fusion categories," *Selecta Mathematica.* **16**, no. 1, 1-119 (2010).
22. G.W. Moore and G. Segal, "D-branes and K-theory in 2D topological field theory," arXiv:hep-th/0609042.
23. D. Cimasoni and N. Reshetikhin, "Dimers on surface graphs and spin structures," *Commun. Math. Phys.* **275**, no. 1, 187-208 (2007), arXiv:math-ph/0608070.
24. N. Steenrod, "Products of cocycles and extensions of mappings," *Ann. of Math.* **(2) 48**, 290 (1947).
25. K. Walker, talk at the IPAM workshop: "Symmetry and topology in quantum matter," Jan. 26-30, 2015.
26. J.W. Barrett and S.O.G. Tavares, "Two-dimensional state sum models and spin structures," arXiv:1312.7561 [math-ph].
27. S. Novak and I. Runkel, "State-sum constructions of two-dimensional topological quantum field theories on spin surfaces," arXiv:1402.2839 [math.QA].

FRAGMENTATION IN THE ϕ^3 THEORY AND THE LPHD HYPOTHESIS

K. URMOSSY

MTA Wigner RCP RMI, Budapest POB 49, H-1525 Hungary
karoly.uermoessy@cern.ch

J. RAK

University of Jyvaskyla, 9 Survontie Str., Jyvaskyla, Finland, FI-40014

We present analytic solution of the Dokshitzer–Gribov–Lipatov–Altarelli–Parisi (DGLAP) equation at leading order (LO) in the ϕ^3 theory in 6 space-time dimensions. If the ϕ^3 model was the theory of strong interactions, the obtained solution would describe the distribution of partons in a jet. We point out that the local parton-hadron duality (LPHD) conjecture does not work in this hypothetical situation. That is, treatment of hadronisation of shower partons is essential for the description of hadron distributions in jets stemming from proton-proton (pp) collisions at $\sqrt{s} = 7$ TeV and from electron-positron (e^+e^-) annihilations at various collision energies. We use a statistical model for the description of hadronisation.

1. Introduction

Recently, momentum fraction distributions of hadrons in jets stemming from electron-positron (e^+e^-) annihilations and proton-proton (pp) collisions have been described by simple analytic formulas obtained from statisitcal hadronisation models.[1–3] The obtained fragmentation functions (FF) have succesfully been used in a perturbative quantum-chromodynamics (pQCD) improved parton model calculation to obtain the transverse momentum (p_T) spectrum of charged pions stemming from pp collisions at $\sqrt{s} = 7$ TeV,[4] assuming $\log\log Q$ type scale evolution of the parameters of the FFs. This Q scale dependence was conjectured based on fits of the newly proposed FFs to AKK-type[5] light-quark and gluon FFs. However, a solution of the Dokshitzer–Gribov–Lipatov–Altarelli–Parisi (DGLAP)[6–8] equations and a global fit to measured data is still missing in the case of these new FFs. Before addressing the full QCD problem, we examine the situation in the simplest asymptotically free field theory, the ϕ^3 model in 6 spacetime dimensions, where there is only one type of parton.

As we shall see, the *local parton-hadron duality* (LPHD) hypothesis is not sufficient in the ϕ^3 theory when trying to model strong interactions. In QCD, the distribution of partons produced in the branching process inside a jet describes the energy distribution of hadrons stemming from e^+e^- annihilations (Sec. 7 in Ref. 9). However, in the ϕ^3 model, parton branching does not produce enough soft 'gluons' for the description of the low energy regime of hadron distributions at parton level. At least that is what the solution of the DGLAP equation with leading-order (LO) splitting function (presented in Sec. 2) suggests. To solve this problem, we introduce a statistical 'parton-hadronisation' function, and describe hadronic momentum fraction distributions in jets produced in e^+e^- and pp collisions in Sec. 3.

2. Distribution of Partons in a Jet in the ϕ^3_6 Theory

In this section, we conjecture that a jet is initiated by an on-shell parton of momentum $P_{init} = (Q, 0, 0, Q)$, and obtain the longitudinal momentum fraction distribution $D(z, Q^2)$ of partons of momenta $p = zP_{init}$ in the jet from the DGLAP equation in the ϕ^3 theory:

$$\frac{d}{d\log Q^2} D\left(z, Q^2\right) = g^2 \int_z^1 \frac{dy}{y} P(y)\, D\left(\frac{z}{y}, Q^2\right) . \tag{1}$$

At LO, the coupling $g^2(Q^2) = 1/\beta_0 \ln(Q^2/\Lambda^2)$, and the splitting function [10] is

$$P(z) = z(1-z) - \frac{1}{12}\delta(1-z) . \tag{2}$$

Eq. (1) factorizes in Mellin space:

$$\frac{d}{d\log Q^2} \tilde{D}\left(s, Q^2\right) = g^2 \tilde{P}(s)\, \tilde{D}\left(s, Q^2\right) , \tag{3}$$

where for a function f,

$$\tilde{f}(s) = \int_0^1 dz\, z^{s-1} f(z) , \qquad f(z) = \frac{1}{2\pi} \int_{-\infty}^{\infty} ds\, z^{-is} \tilde{f}(is) . \tag{4}$$

The sollution of Eq. (3) with $t = \ln(Q^2/\Lambda^2)$ is

$$\tilde{D}(s,t) = \tilde{D}(s,t_0)\, e^{\tilde{P}(s) \int_{t_0}^{t} dt' g^2(t')} = \tilde{D}(s,t_0)\, e^{\tilde{P}(s)\, b(t)} , \tag{5}$$

with

$$\tilde{P}(s) = \frac{1}{(s+2)(s+1)} - \frac{1}{12}, \qquad b(Q^2) = \frac{1}{\beta_0} \ln\left[\frac{\ln(Q^2/\Lambda^2)}{\ln(Q_0^2/\Lambda^2)}\right]. \tag{6}$$

As the initial parton had all of its own momentum, the initial parton distribution $D(z, Q_0^2) = \delta(z-1)$. Thus, from Eqs. (4–6), the distribution of partons in the jet becomes

$$D(z,Q^2) \sim \delta(z-1) + \sum_{k=1}^{\infty} \frac{b^k(Q^2)}{k!(k-1)!} \sum_{j=0}^{k-1} \frac{(k-1+j)!}{j!(k-1-j)!} \times z \ln^{k-1-j}\left[\frac{1}{z}\right]\left[(-1)^j + (-1)^k z\right]. \tag{7}$$

As can be seen in Fig. 1 (top panels), the distribution of partons in the jet Eq. (7) describes hadron momentum fraction distributions only in an intermediate range in the case of jets stemming from e^+e^- and pp collisions. This effect might have been predicted from the terms in the gluon-to-gluon splitting function in QCD

$$P_{gg}(z) \sim \frac{1-z}{z} + \frac{z}{1-z} + z(1-z) + c\,\delta(1-z), \tag{8}$$

not present in Eq. (2), and which enhance the production of very soft and very high-z gluons.

3. Introduction of a Hadronisation Function

In this section, we introduce a hadronisation function $d(z)$ to describe the probability of a parton stemming from the branching proccess to produce some hadrons. This case, the hadron distribution in a jet becomes

$$\frac{dN}{dz} = \int_z^1 \frac{dy}{y} D(y, Q^2)\, d\left(\frac{z}{y}\right). \tag{9}$$

Eq. (5) with initial condition $D(z, Q_0^2) = \delta(z-1)$ (that is, $\tilde{D}(s, Q_0^2) = 1$) shows that the number of branchings in the parton evolution process has Poissonian distribution, as $\tilde{D} \sim \sum(\tilde{P}b)^k/k!$. In z-space, products of $\tilde{P}$-s are convolutions, thus Eq. (9) can be written as

$$\frac{dN}{dz} = \sum_{k=0}^{\infty} \frac{b^k(Q^2)}{k!} \prod_{j=1}^{k} \int dy_j P(y_j) d(y_{k+1}) \delta(y_1 \cdots y_{k+1} - z). \tag{10}$$

If the splitting function $P(y)$ had a single peak at some y^*, Eq. (10) could be approximated by

$$\frac{dN}{dz} \approx \sum_{k=0}^{\infty} \frac{\left[b(Q^2)\, P(y^*)\right]^k}{k!y^{*k}} d\left(\frac{z}{y^{*k}}\right) . \tag{11}$$

To choose a simple model for the hadronisation function $d(z)$, we conjecture that this process is dominantly determined by the phase space of the produced hadrons. Arguments supporting such a conjecture can be found in Refs. 11 to 15. Furtheremore, we assume that hadrons are collinear to their parent parton, and we neglect masses. This way, the momentum fraction distribution of one hadron out of n hadrons stemming from the same parton becomes a one-dimensional microcanonical distribution[1–3]:

$$d_n(z) \sim (1-z)^{n-2} . \tag{12}$$

Though the maximum of the splitting function at LO in the ϕ^3 theory (Eq. (2)) at $y^* = 1/2$ is not at all peak-like, we try out the approximation following from Eqs. (11,12):

$$\frac{dN}{dz} \approx \sum_{k=0}^{\infty} \frac{b^k(Q^2)}{k!2^k} \left(1 - 2^k z\right)^{n-2} . \tag{13}$$

Fig. 1 (central panels) shows that Eq. (13) provides a reasonably good fit of dN/dz and dN/dx distributions of hadrons in jets of various energy, stemming from pp[16,17] and e^+e^-[18–24] collisions. As expected, the model fails to reproduce hadron spectra at very low z, due to the lack of soft 'gluon' radiation in the ϕ^3 theory. Besides, at $z \approx 1$, the curve of Eq. (13) becomes uneven, as an artefact of the replacement of the continuous integral in Eq. (9) by the integrand taken at $y* = 1/2$.

Throughout the fitting, parameter $n = 5$ has been used, while the obtained values of the $b(Q^2)$ are shown in Fig. 1 (bottom panel) with $Q = \sqrt{s}$ for e^+e^- and $Q = P_{jet}$ for pp collisions. The scale dependence of the b parameter was fitted by the LO ϕ_6^3 theory result Eq. (6) with $\Lambda = 0.2$ GeV (see Tab. 1). The values of β_0 and Q_0 coincide in case of jets stemming from e^+e^- and pp collisions with calorimetric jet reconstruction.[16] However, these values differ in the case of pp data with track-based jet reconstruction.[17]

Acknowledgement

This paper was also supported by the Hungarian OTKA Grant K104260.

Table 1. Scale dependence of the $b(Q^2)$ parameter of Eq. (13) obtained from fits shown in Fig. 1.

	β_0	Q_0
pp, Cal	0.168±0.008	1.484±0.171
pp, Trk	0.185±0.007	2.775±0.125
e^+e^-	0.188±0.007	1.403±0.100

References

1. K. Urmossy, G. G. Barnafoldi and T. S. Biro, Generalised Tsallis Statistics in Electron-Positron Collisions, *Phys. Lett.* **B701**, 111 (2011).
2. K. Urmossy, G. Barnafoldi and T. Biro, Generalised microcanonical statistics and fragmentation in electron-positron collisions, *Acta Phys. Polon. Supp.* **5**, 363 (2012).
3. K. Urmossy, G. Barnafoldi and T. Biro, Microcanonical Jet-Fragmentation in Proton-Proton Collisions at LHC Energy, *Phys. Lett.* **B718**, 125 (2012).
4. G. G. Barnafoldi, T. S. Biro, K. Urmossy and G. Kalmar, Tsallis-Pareto-like distributions in hadron-hadron collisions, in *Gribov-80*, eds. Y. L. Dokshitzer, P. Lévai and J. Nyiri (World Scientific, 2011).
5. S. Albino, B. Kniehl and G. Kramer, Fragmentation functions for light charged hadrons with complete quark flavor separation, *Nucl. Phys.* **B725**, 181 (2005).
6. Y. L. Dokshitzer, Calculation of the structure functions for deep inelastic scattering and e^+–e^- annihilation by perturbation theory in quantum chromodynamics., *Sov. Phys. JETP* **46**, 641 (1977).
7. V. Gribov and L. Lipatov, Deep inelastic e p scattering in perturbation theory, *Sov. J. Nucl. Phys.* **15**, 438 (1972).
8. G. Altarelli and G. Parisi, Asymptotic freedom in parton language, *Nucl. Phys.* **B126**, 298 (1977).
9. Y. L. Dokshitzer, V. A. Khoze, A. H. Mueller and S. I. Troian, *Basics of perturbative QCD* (Ed. Frontieres, 1991).
10. M. Grazzini, Semiinclusive DIS: An Explicit calculation in the target fragmentation region, *Nucl. Phys. Proc. Suppl.* **64**, 147 (1998).
11. V. V. Begun, M. Gazdzicki and M. I. Gorenstein, Particle spectra in statistical models with energy and momentum conservation, *Acta Phys. Polon.* **B43**, 1713 (2012).
12. V. V. Begun, M. Gazdzicki and M. I. Gorenstein, Power law in microcanonical ensemble with scaling volume fluctuations, *Phys. Rev.* **C78**, 024904 (2008).

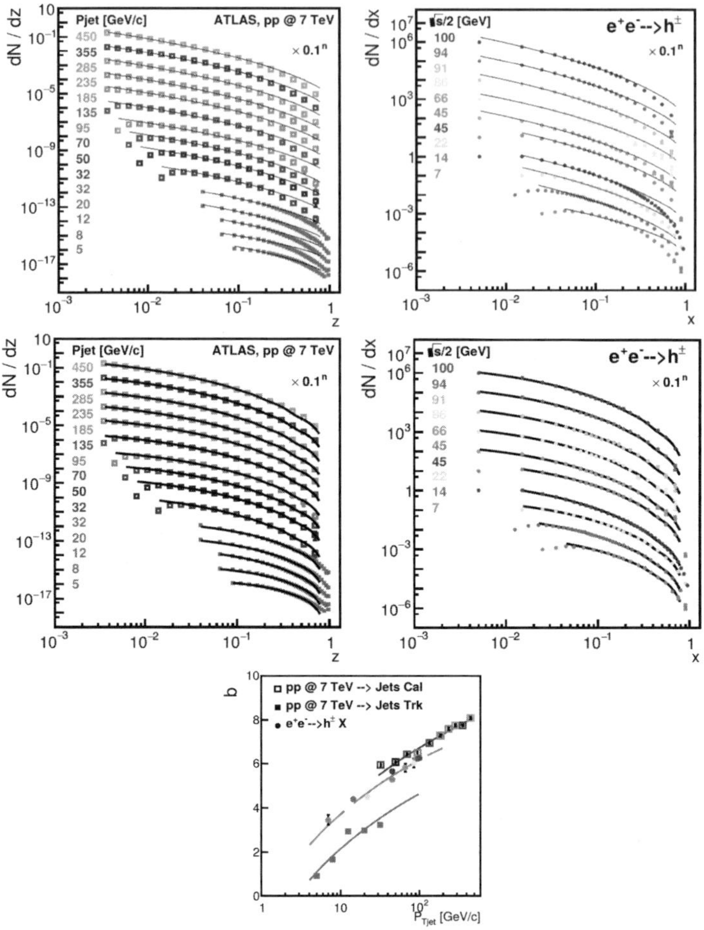

Fig. 1. Momentum fraction distributions of charged hadrons in jets stemming from pp (**left**), and e^+e^- (**right**) collisions. Data are compared to calculated results for partons Eq. (7) (**top**) and final state hadrons Eq. (13) (**center**). Obtained values of the $b(Q^2)$ parameter are fitted with Eq. (6) (**bottom**). Data obtained by calorimetric[16] and track-based[17] jet reconstruction are represented by open and full rectangles in the case of pp collisions.

13. C. Bignamini, F. Becattini and F. Piccinini, A Monte-Carlo generator for statistical hadronization in high energy e^+e^- collisions, *Eur. Phys. J.* **C72**, 2176 (2012).
14. F.-M. Liu and K. Werner, Microcanonical pentaquark production in e^+–e^- annihilations, *Phys. Rev.* **D74**, 034024 (2006).

15. M. Rybczyński and Z. Włodarczyk, Imprints of energy limitation in transverse momentum distributions of jets, *Eur. Phys. J.* **A51**, 80 (2015).
16. G. Aad *et al.*, Measurement of the jet fragmentation function and transverse profile in proton-proton collisions at a center-of-mass energy of 7 TeV with the ATLAS detector, *Eur. Phys. J.* **C71**, 1795 (2011).
17. G. Aad *et al.*, Properties of jets measured from tracks in proton-proton collisions at center-of-mass energy $\sqrt{s} = 7$ TeV with the ATLAS detector, *Phys. Rev.* **D84**, 054001 (2011).
18. W. Braunschweig *et al.*, Global jet properties at 14-GeV to 44-GeV center-of-mass energy in e^+e^- annihilation, *Z. Phys.* **C47**, 187 (1990).
19. H. Aihara, Charged-hadron inclusive cross sections and fractions in e^+e^- annihilation at $\sqrt{s} = 29$ GeV, *Phys. Rev. Lett.* **61**, 1263 (Sep 1988).
20. P. Abreu *et al.*, Determination of alpha-s from the scaling violation in the fragmentation functions in e^+–e^- annihilation, *Phys. Lett.* **B311**, 408 (1993).
21. R. Akers *et al.*, Measurement of the longitudinal, transverse and asymmetry fragmentation functions at LEP, *Z. Phys.* **C68**, 203 (1995).
22. G. Alexander *et al.*, QCD studies with e^+–e^- annihilation data at 130 GeV and 136 GeV, *Z. Phys.* **C72**, 191 (1996).
23. G. Abbiendi *et al.*, QCD studies with e^+–e^- annihilation data at 172-GeV - 189-GeV, *Eur. Phys. J.* **C16**, 185 (2000).
24. G. Abbiendi *et al.*, Charged particle momentum spectra in e^+–e^- annihilation at $\sqrt{s} = 192$ GeV to 209 GeV, *Eur. Phys. J.* **C27**, 467 (2003).

TWISTED NON-ABELIAN VORTICES

PÉTER FORGÁCS[a,b]

[a]*MTA Wigner RCP RMI, H1525 Budapest, POB 49*

[b]*LMPT CNRS UMR7350, Université de Tours, Parc de Grandmont, 37200 Tours, France*

ÁRPÁD LUKÁCS

MTA Wigner RCP RMI, H1525 Budapest, POB 49

FIDEL A. SCHAPOSNIK[c,d]

[c]*Departamento de Física, Universidad Nacional de La Plata,*

[d]*Instituto de Física La Plata C.C. 67, 1900 La Plata, Argentina*

Twisted non-Abelian flux-tube solutions are considered in the bosonic sector of a 4-dimensional $\mathcal{N} = 2$ super-symmetric gauge theory with $\mathrm{U}(2)_{\mathrm{local}} \times \mathrm{SU}(2)_{\mathrm{global}}$ symmetry, with two scalar doublets in the fundamental representation. Twist refers to a time-dependent matrix phase between the two doublets, and twisted strings have nonzero (global) charge, momentum, and in some cases even angular momentum per unit length. The planar cross section of a twisted string corresponds to a rotationally symmetric, charged non-Abelian vortex, satisfying 1st order Bogomolny-type equations and Gauss-type constraints. Quite unexpectedly some twisted strings lack cylindrical symmetry in $\mathbb{R}^3$.

1. Introduction

Vortex, resp. string-type solutions appear in many models, and they have many applications there is an enduring interest for them. In the plane, a vortex corresponds to a cross-section of a straight string in 3 spatial dimensions in the plane orthogonal to its direction. Non-Abelian vortices (some of them with an electric charge) have been first obtained in Ref. 4. The investigation of vortex-string solutions in supersymmetric non-Abelian gauge theories was started by the seminal papers[5] and the subject continues to attract attention. A simple model containing all essential features is an $U(N_c)$ gauge theory, coupled to N_f scalar fields in the fundamental representation, where $N_f \geq N_c$. Vortex solutions in such theories are usually referred to as non-Abelian vortices (NAV). NAV have attracted considerable interest, since they are at the heart of intriguing relationships between 2D sigma-models and 4D gauge theories. NAV in a $U(N)$ gauge theory possess "orientational moduli" whose low-energy dynamics is described by versions of $\mathbf{CP}^{N-1}$ sigma-models on the string worldsheet. Moreover, NAV also held the promise to be relevant to bring us closer to give a description of quark-confinement, for a review, see Ref. 6 and references therein. Static NAV are absolute minima of the energy functional in a fixed topological sector, characterized by the winding numbers associated to the Cartan subalgebra of the gauge group. Moreover NAV satisfy first order, Bogomolny-type equations admitting rather complicated, static multi-vortex solutions. Most remarkably a complete description of the NAV moduli space has been found.[7,8] An interesting application of non-Abelian vortices is illustrated by confined monopoles emerging as junctions of NAV with different moduli.[9,10] The dynamics of NAV based on the moduli approximation has been worked out in Refs. 11 to 13. In addition to NAV and related monopoles, domain walls have also received due attention,[6,14] and a number of intriguing relations between moduli spaces of monopoles and domain walls have been discovered.[15]

In this contribution based on our paper[16] we point out that by allowing for space-time dependent phases among the scalar fields, referred to as "twisting", new families of charged vortex-strings arise. More explicitly twisted configurations arise when the scalar fields, Φ, possess a $(t\,,z)$ dependent phase, i.e.

$$\Phi(x^\nu) = \Phi(x^i)e^{\frac{i}{2}M\omega_\alpha x^\alpha}\,, \quad i=1,2,\ \alpha=0,3\,,$$

where the flavours correspond to the columns of the (2×2) matrix Φ, M is a constant Hermitian matrix acting on the flavour indices, ω_α is vector in $(t\,,z)$ plane. Attention will be restricted to an U(2) gauge group broken by two scalar doublets with an appropriate scalar potential compatible with SUSY. It is known, that imposing the usual space-time symmetries on field configurations leading to vortex-type solutions, translational invariance in time and along the, say, x^3 or z direction, scalar fields may have a phase, with a linear dependence on $(t\,,z)$.[17–19] In general, the energy of *stationary* configurations with a nontrivial $(t\,,z)$ dependent phase is bigger than that of the static ones, however this does not make them necessarily unstable.[17] As is well known[17,18,20] when ω_α is a *light-like vector*, the field equations in the $(x^1\,,x^2)$ plane decouple completely from those in the $(t\,,z)$ directions. The chromo-electric components in the $(t\,, z)$ plane are determined by a set of linear, 2nd order equations (Gauss constraints), depending on the solution in the $(x^1\,, x^2)$ plane. Simple analysis of the Gauss constraints makes it very plausible that any NAV can be twisted by an arbitrary twisting matrix, although we have not attempted to formally prove this. There are global currents flowing in the z direction of an M-twisted vortex-string which has nonzero charge, momentum and, unless M is specially aligned in internal space, angular momentum. A twisted NAV has some genuine 3-dimensional structure, as there are global currents flowing along its axis, the fields have some nontrivial components orthogonal to its symmetry-plane, hence it is more conveniently thought of as a twisted string.

The simplest twisted NAV can be obtained from the rotationally symmetric "elementary" vortex solution of Ref. 5, in which case, for a general twisting matrix the problem reduces to a single 2nd order Gauss-constraint, which can be easily analysed. In fact our twisted NAV solution turns to be the same object as the "dyonic" vortex solution found in a "mass-deformed" SUSY gauge-field theory.[12] The bulk of our paper concerns the twisting of "composite-coincident" NAV,[7] corresponding to the rotationally-symmetric superimposition of elementary vortices. We characterize these superimposed NAV by a relative winding number and a moduli parameter. Interestingly we find that when the twisting matrix M, contains off-diagonal components the twisted string looses cylindrical symmetry in 3-dimensions.

The energy of an M-twisted string is higher than that of an untwisted one, it is given as a sum of the usual "magnetic" energy per unit length proportional to the magnetic flux, and of an "electric" contribution due to the rotating phase. The actual magnitude of the "electric" contribution depends essentially on the components of the twisting matrix, M. When M is diagonal the magnitude of the electric energy of a twisted NAV is typically much smaller than its magnetic one, in fact, it can be made arbitrarily small when the untwisted NAV is sufficiently close to a diagonal one. In the case when M contains off-diagonal components, the magnitude of the electric energy is comparable to the magnetic one.

2. Dimensional reduction

The 3+1 dimensional theory we consider is defined by the following Lagrangian:

$$\mathcal{L} = -\frac{1}{4g_1^2} F_{\mu\nu} F^{\mu\nu} - \frac{1}{4g_2^2} G^a_{\mu\nu} G^{\mu\nu a} + \mathrm{Tr}(D_\mu \Phi)^\dagger D^\mu \Phi - V\,, \qquad (1)$$

where $F_{\mu\nu} = \partial_\mu A_\nu - \partial_\nu A_\mu$; $G^a_{\mu\nu} = \partial_\mu C^a_\nu - \partial_\nu C^a_\mu + \epsilon^{abc} C^b_\mu C^c_\nu$ $(a = 1, 2, 3)$ denote the field strength tensors, while $D_\mu \Phi_A = (\partial_\mu - iA_\mu/2 - i\sigma^a C^a_\mu/2)\Phi_A$ stands for the covariant derivative. The two scalar doublets are encoded in the matrix $(\Phi)_{iA}$ with $i = 1\,,2$ being the gauge ("colour") and $A = 1\,,2$ the "flavour" index; the trace is taken over the flavour indices; the scalar potential, $V = V_1 + V_2$, can be written as:

$$V_1 = \frac{g_1^2}{8}\left(\Phi_A^\dagger \Phi_A - 2\xi\right)^2\,, \quad V_2 = \frac{g_2^2}{8}\left[(\Phi_1^\dagger\Phi_1 - \Phi_2^\dagger\Phi_2)^2 + 4|\Phi_2^\dagger\Phi_1|^2\right]\,, \quad (2)$$

where summation over repeated indices is understood. The Lagrangian in Eq. (1) is the bosonic sector of a 4-dimensional $\mathcal{N} = 2$ super-symmetric gauge theory. The fields transform under the $U(2)$ gauge and flavour symmetry as

$$\Phi \to \exp(i\Lambda)U\Phi V\,, \quad C_\mu \to UC_\mu U^\dagger + 2iU\partial_\mu U^\dagger\,, \quad A_\mu \to A_\mu + 2\partial_\mu\Lambda\,, \quad (3)$$

where $U(x)\,, V \in SU(2)$, with V being constant and $\Lambda(x)$ is a real function.

We now consider stationary and translationally symmetric fields in the x^3 direction. We decompose the coordinates as $x^\mu = (x^\alpha\,, x^i)$ with $\alpha = 0, 3$, $i = 1, 2$. Since the symmetries are generated by two commuting vector fields, there exist a gauge where the symmetric gauge fields are simply independent of the coordinates x^α as in Ref. 21. At the same time, however, the scalar doublets, Φ, may still depend linearly on x^α through an SU(2) phase. We shall consider the following Ansatz:

$$A_\mu(x^\nu) = (A_i(x^j), \omega_\alpha A(x^j))\,, \quad (4a)$$

$$C^a_\mu(x^\nu) = (C^a_i(x^j), \omega_\alpha C^a(x^j))\,, \quad (4b)$$

$$\Phi(x^\nu) = \Phi(x^i) e^{\frac{i}{2} M \omega_\alpha x^\alpha}\,, \quad (4c)$$

where $M = m^{\hat a}\sigma^{\hat a} = m^0 \mathbf{1} + m^a \sigma^a$, a constant Hermitian matrix, and ω_α is vector in $(t\,, z)$ plane.

The form of the Ansatz, Eq. (4) also restricts the symmetries of the model. Those symmetries, which preserve the Ansatz are flavour transformations which commute with the twist matrix M,

$$\Phi \to \Phi V\,, \quad V = \exp(iM\delta)\,, \quad (5)$$

and all gauge transformations where $U\partial_\alpha U^\dagger$ and $\partial_\alpha \Lambda$ only depends on x^i. The current generating these transformations is

$$K_\mu = m^{\hat a} K^{\hat a}_\mu\,, \quad K_\alpha = -\frac{\omega_\alpha}{2} Q\,, \quad Q = \mathrm{Tr}\left[\Phi^\dagger(\Phi M - \mathbb{C}\Phi)M\right]\,. \quad (6)$$

In the case when ω_α is *light-like*, the field equations in the (x^1, x^2) plane *decouple* from the field equations for the x^α components, and we shall concentrate on this case. Twisted non-Abelian strings of minimal energy are subject to a set of 1st order, Bogomolny-type equations in the $(x^1\,, x^2)$ plane and to Gauss-type constraints for the x^α components:

$$F_{ik} = \mp \frac{g_1^2}{2} \epsilon_{ik} (\mathrm{Tr}\,\Phi^\dagger \Phi - 2\xi)\,, \tag{7a}$$

$$G^a_{ik} = \mp \frac{g_2^2}{2} \epsilon_{ik}\, \mathrm{Tr}\,\Phi^\dagger \sigma^a \Phi\,, \tag{7b}$$

$$D_i \Phi = \mp i \epsilon_{ik} D_k \Phi\,, \tag{7c}$$

$$\frac{1}{g_1^2} \triangle A = \frac{1}{2}\,\mathrm{Tr}\left[(\Phi^\dagger \mathbb{C} - M\Phi^\dagger)\Phi\right]\,, \tag{7d}$$

$$\frac{1}{g_2^2} \hat{D}_i \hat{D}_i C^a = \frac{1}{2}\,\mathrm{Tr}\left[\left(C^a \Phi^\dagger + (A - M)\Phi^\dagger \sigma^a\right)\Phi\right]\,, \tag{7e}$$

Therefore the problem of finding twisted non-Abelian strings for $\omega^2 = 0$ reduces to solving Eqs. (7d)–(7e) in the *background* of a non-Abelian vortex in the (x^1, x^2) plane.

The total energy of an M-twisted string can be written as the sum of an "electric" $\mathcal{E}_0$ and of a "magnetic" $\mathcal{E}_1$ part:

$$E = \int d^2x T^0_{\ 0} = \int (\mathcal{E}_0 + \mathcal{E}_1)\, d^2x \equiv E_0 + E_1\,. \tag{8}$$

with

$$\mathcal{E}_0 = \frac{1}{4}(\omega_0^2 + \omega_3^2)(\mathbf{\Delta} + Q)\,, \tag{9}$$

where $\mathbf{\Delta} = \triangle\left(A^2/g_1^2 + C^a C^a/g_2^2\right)$ and Q is given by Eq. (6). The total "magnetic" energy, E_1, is given by the net Abelian flux through the (x^1, x^2) plane, which is given for configurations characterized by winding numbers $n_A\,, m_A$ as in Eq. (13) and subject to the 1st-order Eqs. (7):

$$E_1 = 2\pi\xi|n_1 + m_2| = 2\pi\xi|n_1 + m_1 + N|\,, \tag{10}$$

which of course also holds for the Ansatz (17).

An M-twisted string has momentum flowing along the z direction, which is easily obtained from the stress-energy tensor. The longitudinal momentum, $P = T_{03}$, carried by a twisted string, can be recast exploiting the Gauss constraints, Eqs. (7d)–(7e), as:

$$P = T_{03} = \frac{1}{2}\omega_0 \omega_3 (\mathbf{\Delta} + Q)\,. \tag{11}$$

M-twisted strings may also have angular momentum, J, as it can be seen from the angular momentum density

$$J = T_{0\vartheta} = \frac{1}{g_1^2} F_{0r} F_{\vartheta r} + \frac{1}{g_2^2} G^a_{0r} G^a_{\vartheta r} + \operatorname{Tr} D_0 \Phi^\dagger D_\vartheta \Phi + \operatorname{Tr} D_\vartheta \Phi^\dagger D_0 \Phi \,. \quad (12)$$

2.1. *The Ansatz; rotational symmetry and its loss*

Let us now impose rotational symmetry in the x^1, x^2 plane to the fields. Denoting the usual polar coordinates in the plane as $x^1 = r\cos\vartheta$, $x^2 = r\sin\vartheta$, rotational symmetry implies that by ϑ-dependent gauge transformations one can always achieve

$$\partial_\vartheta \{A_\mu, C^a_\mu\} = 0\,, \quad \Phi_A(x^i) = (\exp(in_A\vartheta)\Phi_{A\,1}(r), \exp(im_A\vartheta)\Phi_{A\,2}(r))\,. \quad (13)$$

In order to ensure consistency with the U(2) gauge and the global SU(2) flavour symmetry, the integers, n_A, m_A, satisfy the following relation: $n_2 - n_1 = m_2 - m_1 = N$, which can also be expressed on the 2 scalar doublets as

$$\Phi(x^i) = \begin{pmatrix} \phi_1(r)e^{in_1\vartheta} & \psi_1(r)e^{in_1\vartheta} \\ \phi_2(r)e^{im_1\vartheta} & \psi_2(r)e^{im_1\vartheta} \end{pmatrix} \begin{pmatrix} 1 & 0 \\ 0 & e^{iN\vartheta} \end{pmatrix} = \Phi_0(x^i)e^{i\mathbf{N}\vartheta}\,, \quad (14)$$

where $\mathbf{N} = \text{Diag}\{0, N\}$, which encodes the relative winding between the two flavours. Since in general the twisting matrix, M, mixes the 2 flavours, when $N \neq 0$ the right hand sides of Eqs. (7d)–(7e) depend explicitly on ϑ, breaking rotational symmetry in the x^α direction, i.e.

$$\partial_\vartheta \{A_\alpha, C^a_\alpha\} \neq 0\,. \quad (15)$$

To ensure rotational symmetry, the condition

$$[M, \mathbf{N}] = 0\,, \quad (16)$$

has to be imposed, implying that the twisting matrix be diagonal. In general the anisotropy generated by twisting matrices not commuting with $\mathbf{N}$, just rules out the possibility to consider rotationally symmetric configurations. Remarkably in the case of a light-like twist vector, due to the decoupling of the field Eqs. in the (x^1, x^2) plane decouple from Eqs. with $x^\alpha = (x^0, x^3)$ components, (7d)–(7e) one can have solutions which break rotational symmetry in the x^α direction. Whenever M does not commute with $\mathbf{N}$ the corresponding M-twisted strings have rotationally symmetric spatial sections in any plane orthogonal to the x^3-axis, however, the complete configuration is not rotationally symmetric in the whole space-time.

Our Ansatz in the radial gauge is given as

$$A_\alpha = A(r,\vartheta)\omega_\alpha\,, \quad A_r = 0\,, \quad A_\vartheta = a(r)\,, \tag{17a}$$

$$C^a_\alpha = C^a(r,\vartheta)\omega_\alpha\,, \quad C^a_r = 0\,, \quad C^a_\vartheta = \{c_1(r), 0\,, c_3(r)\}\,, \tag{17b}$$

$$\Phi(x^i) = \begin{pmatrix} \phi_1(r) & \psi_1(r)e^{iN\vartheta} \\ \phi_2(r) & \psi_2(r)e^{iN\vartheta} \end{pmatrix}, \tag{17c}$$

We can now display the angular momentum, J for our Ansatz (17) in a more explicit form. Exploiting Eqs. (7d)–(7e) we obtain:

$$J = \omega_0\left(\tilde{\mathbf{\Delta}} + \tilde{Q}\right), \quad \tilde{Q} = \mathrm{Tr}\left[\Phi^\dagger\Phi(\mathbf{N}M + M\mathbf{N}) - 2\Phi^\dagger\mathbb{C}\Phi\mathbf{N}\right] \tag{18}$$

where $\tilde{\mathbf{\Delta}} = \partial_i(a\partial_i A)/g_1^2 + \partial_i(C^a_\vartheta \hat{D}_i C^a)/g_2^2$.

3. Twisted vortices

3.1. *Twisted elementary vortices*

The simplest non-Abelian vortex solution is a 'diagonal' one, with just ϕ_1, ψ_2, a, c_3 being non-trivial, while ϕ_2, ψ_1, c_1 are all zero. Such solutions have been thoroughly investigated in Ref. 8. A larger family of 'elementary' vortex solutions, with ϕ_2, ψ_1, c_1 being also nontrivial, can be obtained by a 'colour-flavour' transformation from a "diagonal" one.[8] For diagonal vortices the relative winding between the 2 doublets is always trivial, $N = 0$, which remains so for the general elementary vortices. The general form of the elementary solution can be written as:

$$\Phi = \phi_+\mathbb{1} + \phi_- n^a\sigma^a\,, \quad A_\vartheta = a(r)\,, \quad C^a_\vartheta = n^a c_3\,, \quad \phi_\pm = (\phi_1 \pm \psi_2)/2\,, \tag{19}$$

where n^a is the 'orientational' unit vector of an elementary NAV, which can be parametrized by the 2 spherical angles as $n^a = (\sin\alpha\cos\beta\,, \sin\alpha\sin\beta\,, \cos\alpha)$. In this case the two moduli parameters are just the angles, α and β as in Refs. 6 and 8.

We note here that for the most general (with the 2 moduli) elementary vortex solution in (19) $C^2_\vartheta \neq 0$, hence for $\beta \neq 0$ it is not in the form of the minimal Ansatz (17), this has no influence, however, on our twisting of this solution. It is now simple to generalize, or deform the elementary vortex solution by M. A short computation shows that the Ansatz

$$A = m^0\,, \quad C^a = (m\cdot n)[1 - C(r)]n^a + C(r)m^a\,, \quad m\cdot n = m^a n^a\,, \tag{20}$$

containing just a single radial function, $C(r)$, reduces the Gauss constraints, Eqs. (7d)–(7e), to a single 2nd order inhomogeneous equation for $C(r)$:

$$\frac{1}{r}(rC')' - \frac{c_3^2}{r^2}C = g_2^2\left[\phi_+^2(C-1) + \phi_-^2(C+1)\right]. \tag{21}$$

Without going into more rigorous mathematics one can easily convince oneself that Eq. (21) admits a unique solution subject to boundary conditions guaranteeing regularity at $r = 0$ and at $r \to \infty$. Conversely, using maximum-principle-type arguments it is not difficult to show that all globally regular solutions are necessarily of the form of the Ansatz in Eq. (20).

The charge density Q, determining the energy resp. momentum, of the elementary string, is given as:

$$Q = \left(m{\cdot}m - (n{\cdot}m)^2\right)\left[\phi_1^2 + \psi_2^2 - 2C(r)\phi_1\psi_2\right] , \tag{22}$$

As one can easily see from Eq. (18) the angular momentum density of twisted elementary vortices vanishes. As already pointed out, this Subsection reproduces and extends previously obtained results of Ref. 12 using different methods, our starting point being rather different.

3.2. *Twisted coincident composite vortices*

Next we consider a vortex solution with nonzero relative winding between the 2 flavours. Assuming the minimal Ansatz, the vortex in the $(x^1\,, x^2)$ plane is rotationally symmetric, satisfying the 1st order Eqs. (7a-7c). Such solutions have already been analyzed in Ref. 7, where it has also been pointed out that they correspond to superimposed vortices on the top of each other. Therefore such vortices can be considered as composed of elementary ones. As we have already argued, deforming composite vortices with a general matrix, M induces a nontrivial angle-dependence in the $(x^0\,, x^3)$ plane.

Without losing generality one can parametrize the twisting matrix as

$$M = \frac{s}{2}\left(\mathbf{1} - \sigma^3\right) + \frac{m}{2}\left(\cos(\mu)\sigma^1 + \sin(\mu)\sigma^2\right) . \tag{23}$$

Taking into account the possible angle-dependence of the ω_α-components of the gauge fields, we introduce the following decomposition:

$$\begin{aligned} A &= sA_0 + mA_+e^{i(N\vartheta+\mu)} + mA_-e^{-i(N\vartheta+\mu)} , \\ C^a &= sC_0^a + mC_+^a e^{i(N\vartheta+\mu)} + mC_-^a e^{-i(N\vartheta+\mu)} , \end{aligned} \tag{24}$$

together with the conditions $A_+ = A_-^*$, $C_+^a = (C_-^a)^*$ ensuring reality of the fields, moreover $A_0, A_\pm$, and $C_0^a, C_\pm^a$ are functions of r.

Quite interestingly, the equations for the angle-dependent components, can be reduced to a quadrature, i.e. to solve a single 1st order, linear ordinary differential equation. The key observation is that $A_\pm = 0$, leading to an algebraic constraint between $C_\pm^1$ and $C_\pm^3$. As shown in Ref. 16 there is another algebraic relation among the functions $C^a_\pm$, and in terms of $C^1_\pm$ the remaining functions $C^2_\pm, C^3_\pm$ can be expressed algebraically.

The total electric energy, the longitudinal and the angular momenta can all be expressed in terms of the integral of the Noether charge Q, for details we refer to Ref. 16.

Acknowledgements

This work has been supported by the grants OTKA K101709 and TÉT 10-1-2011-0071 and MINCYT-Argentina HU/10/01. F.A.S. is partially financed by PIP-CONICET, PICT-ANPCyT and CICBA.

Bibliography

1. A.A. Abrikosov, *"On the Magnetic Properties of Superconductors of the Second Group"*, *Sov. Phys. JETP*, **5**, 1174 (1957); H.B. Nielsen and P. Olesen, *"Vortex-line models for dual strings"*, *Nucl. Phys.*, **B 61**, 45 (1973).
2. E. B. Bogomol'nyi, *"The stability of classical solutions"*, *Sov. J. Nucl. Phys.* **24**, 449 (1976).
3. H. J. de Vega and F. A. Schaposnik, *"Classical vortex solution of the Abelian Higgs model"*, *Phys. Rev.* **D14**, 1100 (1976).
4. H. J. de Vega and F. A. Schaposnik, *"Electrically charged vortices in non-Abelian gauge theories with Chern–Simons term"*, *Phys. Rev. Lett.* **56**, 2564 (1986); H. J. de Vega and F. A. Schaposnik, *"Vortices and electrically charged vortices in non-Abelian gauge theories"*, *Phys. Rev.* **D34**, 3206–3213, (1986).
5. A. Hanany and D. Tong, *"Vortices, instantons and branes"*, *JHEP* **0307**, 037 (2003); R. Auzzi, S. Bolognesi, J. Evslin, K. Konishi and A. Yung, *"Nonabelian superconductors: Vortices and confinement in N = 2 SQCD"*, *Nucl. Phys.* **B673**, 187 (2003) ; N. Markov, A. Marshakov, A. Yung, *"Non-Abelian vortices in $\mathcal{N} = 1*$ gauge theory"*, *Nucl. Phys.* **B709**, 267 (2005).
6. M. Shifman and A. Yung, *Supersymmetric Solitons*, Cambridge University Press, 2009.
7. R. Auzzi, M. Shifman, A. Yung, *"Composite non-Abelian flux tubes in $\mathcal{N} = 2$ sQCD"*, *Phys. Rev.* **D73**, 105012 (2006).
8. M. Eto, Y. Isozumi, M. Nitta, K. Ohashi, and N. Sakai, *"Solitons in the Higgs phase: the moduli matrix approach"*, *J. Phys. A: Math. Gen.* **39**, R315–R392 (2006);

M. Eto, Y. Isozumi, M. Nitta, K. Ohashi, and N. Sakai, *"Moduli space of non-Abelian vortices"*, *Phys. Rev. Lett.* **96**, 161601 (2006); M. Eto, K. Konishi, G. Marmorini, M. Nitta, K. Ohashi, W. Vinci, and N. Yokoi, *"Non-Abelian vortices of higher winding number"*, *Phys. Rev.* **D74**, 056021 (2006); M. Eto, T. Fujimori, S.B. Gudnason, K. Konishi, M. Nitta, K. Ohashi, and W. Vinci, *"Constructing non-Abelian vortices with arbitrary gauge groups"*, *Phys. Lett.* **B669**, 98–101 (2008).

9. D. Tong, *"Monopoles in the Higgs phase"*, *Phys. Rev.* **D69**, 065003 (2004); A. Hanany and D. Tong, *"Vortex strings and four-dimensional gauge dynamics"*, *JHEP* **0404**, 066 (2004); R. Auzzi, S. Bolognesi, J. Evslin, *"Monopoles can be confined by 0,1 or 2 vortices"*, *JHEP* **0502**, 046 (2002).
10. A. Gorsky, M. Shifman, and A. Yung, *"Non-Abelian Meissner effect in Yang-Mills theories at weak coupling"*, *Phys. Rev.* **D71**, 045010 (2005).
11. B. Collie, *"Dyonic non-Abelian vortices"*, *J. Phys.* **A42**, 085404 (2009).
12. M. Eto, T. Fujimori, M. Nitta, K. Ohashi, and N. Sakai, *"Dynamics of non-Abelian vortices"*, *Phys. Rev.* **D84**, 125030 (2011).
13. M. Eto, Y. Isozumi, M. Nitta, K. Ohashi, and N. Sakai, *"Manifestly supersymmetric effective Lagrangians on BPS solitons"*, *Phys. Rev.* **D73**, 125008 (2006).
14. D. Tong, *"The moduli space of BPS domain walls"*, *Phys. Rev.* **D66**, 025013 (2002).
15. A. Hanany and D. Tong, *"On monopoles and domain walls"*, *Commun. Math. Phys.* **266**, 647–663 (2006).
16. P. Forgács, Á. Lukács, F.A. Shaposnik, *"Non-Abelian vortices with a twist"*, *Phys. Rev.* **91**, 125001 (2015).
17. E. Abraham, *"Charged semilocal vortices"*, *Nucl. Phys.* **B399**, 197-210 (1993).
18. P. Forgács, S. Reuillon, and M.S. Volkov, *Phys. Rev. Lett.* **96**, 041601 (2006); *"Twisted superconducting semilocal strings"*, *Nucl. Phys.* **B751**, 390–418 (2006); Y. Brihaye, L. Honorez, *"Twisted semilocal strings in the MSSM"*, *Int. J. Mod. Phys.* **A23**, 581–597 (2008).
19. Y. Brihaye, Y. Verbin, *"Superconducting and spinning non-Abelian flux tubes"*, *Phys. Rev.* **D77**, 105019 (2008); J. Garaud and M.S. Volkov, *"Superconducting non-Abelian vortices in Weinberg-Salam theory – electroweak thunderbolts"*, *Nucl. Phys.* **B826**, 174–216 (2010).

20. L.A. Ferreira, *"Exact vortex solutions in an extended Skyrme-Faddeev model"*, *JHEP* **0905**, 001 (2009).
21. P. Forgács, N.S. Manton, *"Space-time symmetries in gauge theories"*, *Commun. Math. Phys.* **72**, 15 (1980).
22. J. Garaud and M.S. Volkov, *"Stability analysis of the twisted superconducting semilocal strings"*, *Nucl. Phys.* **B799**, 430–455 (2008); P. Forgács, Á. Lukács, *"Instabilities of twisted strings"*, *JHEP* **0912**, 064 (2009); B. Hartmann, P. Peter, *"Can type II semi-local cosmic strings form?"*, *Phys. Rev.* **D86**, 103516 (2012).
23. M. Hindmarsh, *"Existence and stability of semilocal strings"*, *Phys. Rev. Lett.* **68**, 1263–1266 (1992).

LIST OF PARTICIPANTS AND CONTRIBUTORS

Natalia V. ABRAMOVSKAYA
Novgorod State University,
ul. B. Peterburgskaya 41
173003 Veliky Novgorod, Russia
e-mail: nvrad@mail.ru

Victor ABRAMOVSKY
Novgorod State University,
ul. B. Peterburgskaya 41
173003 Veliky Novgorod, Russia
e-mail: Victor.Abramovsky@novsu.ru

Oleg D. ANDREEV
L.D. Landau Institute for Theoretical Physics
1A ul. Akad. Semenova
142432 Chernogolovka, Russia
e-mail: andre@itp.ac.ru

Vladimir V. ANISOVICH
Petersburg Nuclear Physics Institute
188300 Gatchina, St. Petersburg, Russia
e-mail: anisovich@thd.pnpi.spb.ru

Elena V. ARBUZOVA
Dubna University,
141980 Dubna, Russia

Yakov AZIMOV
Petersburg Nuclear Physics Institute
188300 Gatchina, St. Petersburg, Russia
e-mail: azimov@thd.pnpi.ru

Alexander BELAVIN
L.D. Landau Institute for Theoretical Physics
1A ul. Akad. Semenova
142432 Chernogolovka, Russia
e-mail: belavin@itp.ac.ru

Aparajita BHATTACHARYA
Department of Physics
Jadavpur University
Kolkata 700032, India
e-mail: aparajita_bh@yahoo.com.in

Ludmila N. BOGDANOVA
Institute of Theoretical and Experimental Physics
25 ul. B. Cheremushkinskaya
117218 Moscow, Russia

Konstantin BORESKOV
Institute of Theoretical and Experimental Physics
25 ul. B. Cheremushkinskaya
117218 Moscow, Russia
e-mail: boreskov@gmail.com

Máté CSANÁD
Department of Atomic Physics,
Eötvös University
1/A Pázmany Péter sétány
1117 Budapest, Hungary
e-mail: csanad@elte.hu

Yuri L. DOKSHITZER
Laboratoire de Physique Theorique et Hautes Energies (LPTHE)
Université Paris VI
Tour 24-25, 5eme etage, Boite 126 4 Place Jussieu
76252 Cedex 5, France
e-mail: yuri@lpthe.jussieu.fr

Alexandre D. DOLGOV
Polo Scientifico e Technologico - Edifico C
INFN, Dipartimento di Fisica, Universita di Ferrara
via Saragat 1
44100 Ferrara, Italy
e-mail: dolgov@fe.infn.it

Alexander DOROKHOV
Joint Institute for Nuclear Research
Dubna, Russia
e-mail: dorokhov@theor.jinr.ru

Igor M. DREMIN
RAS Lebedev Physics Institute
Moscow, Russia
e-mail: dremin@td.lpi.ru

Michael EIDES
Dept. of Physics and Astronomy,
University of Kentucky
177 Chemistry-Physics Bldg.
600 Rose Street
KY 40506-0055, Lexington, USA
e-mail: eides.ps.uky.edu

Gerasim ELIASHBERG
L.D. Landau Institute for Theoretical Physics
1A ul. Akad. Semenova
142432 Chernogolovka, Russia
e-mail: elias@itp.ac.ru

Victor S. FADIN
Budker Institute for Nuclear Physics and
Novosibirsk State University
Novosibirsk, Russia
e-mail: fadin@inp.nsk.ru

Leonid FRANKFURT
School of Physics and Astronomy
Tel-Aviv University
69978 Ramat Aviv, Tel-Aviv, Israel

Andor FRENKEL
MTA Wigner Research Centre for Physics
29-33 Konkoly-Thege Miklós street
1121 Budapest, Hungary
e-mail: frenkel.andor@wigner.mta.hu

Semen S. GERSHTEIN
Institute for High Energy Physics
142281 Protvino, Moscow Region, Russia
e-mail: Semen.Gerhtein@ihep.ru

Vahtang GOGOHIA
MTA Wigner Research Centre for Physics
29-33 Konkoly-Thege Miklós street
1121 Budapest, Hungary
e-mail: gogohia.vahtang@wigner.mta.hu

Alexander S. GORSKY
RAS Institute of Information Transmission Problems, Moscow, Russia
and
Moscow Institute of Physics and Technology, Dolgoprudny, Russia

Dezső HORVÁTH
Department of Experimental Particle Physics
MTA Wigner Research Centre for Physics
29-33 Konkoly-Thege Miklós street
1121 Budapest, Hungary
e-mail: horvath.dezso@wigner.mta.hu

Péter HRASKÓ
University of Pécs, Department of Theoretical Physics
Pécs, Hungary
e-mail: peter@hrasko.com

Boris Lazarevich IOFFE
Institute of Theoretical and Experimental Physics
25 ul. B. Cheremushkinskaya
117218 Moscow, Russia
e-mail: ioffe@ihep.ru

Laszlo JENKOVSZKY
Bogolubov Institute for Theoretical Physics,
National Academy of Sciences of Ukraine
14b Metrolohichna street
03680 Kiev, Ukraine
e-mail: jenk@bitp.kiev.ua

Dmitri KAZAKOV
Joint Institute for Nuclear Research
e-mail: kazakovd@theor.jinr.ru

Isaak M. KHALATNIKOV
L.D. Landau Institute for Theoretical Physics
1A ul. Akad. Semenova
142432 Chernogolovka, Russia

Dmitri KHARZEEV
Department of Physics
Brookhaven National Laboratory
Building 510, P.O.Box 5000
Upton, NY 11973-5000 USA
e-mail: kharzeev@bnl.gov

Iosif B. KHRIPLOVICH
St. Petersburg State University
Ulianovskaya 1,
198504 Petrodvorets, St. Petersburg, Russia
e-mail: khios231@mail.ru

Victor T. KIM
Petersburg Nuclear Physics Institute,
188300 Gatchina, St. Petersburg, Russia
e-mail: victor.t.kim@gmail.com

Zohar KOMARGODSKI
Department of Particle Physics and Astronomy,
Weizmann Institute of Science
Rehovot, Israel
e-mail: zohar.komargodski@weizmann.ac.il

Yury KUDENKO
RAS Institute for Nuclear Research
Moscow, Russia
e-mail: kudenko@inr.ru

Michael Yu. LASHKEVICH
L.D. Landau Institute for Theoretical Physics
1A ul. Akad. Semenova
142432 Chernogolovka, Russia
e-mail: lashkevi@landau.ac.ru

András LÁSZLÓ
MTA Wigner Research Centre for Phyics
29-33 Konkoly-Thege Miklós street
1121 Budapest, Hungary
e-mail: laszlo.andras@wigner.mta.hu

Vladimir V. LEBEDEV
L.D. Landau Institute for Theoretical Physics
1A ul. Akad. Semenova
142432 Chernogolovka, Russia
e-mail: lebede@itp.ac.ru

Péter LÉVAI
MTA Wigner Research Centre for Physics
29-33 Konkoly-Thege Miklós street
1121 Budapest, Hungary
e-mail: levai.peter@wigner.mta.hu

Dmitry LEVKOV
RAS Institute for Nuclear Research
Moscow, Russia
e-mail: levkov@ms2.inr.ac.ru

Lev N. LIPATOV
Petersburg Nuclear Physics Institute
188300 Gatchina, St. Petersburg, Russia
e-mail: lipatov@thd.pnpi.spb.ru

Árpád LUKÁCS
MTA Wigner Research Centre for Physics
29-33 Konkoly-Thege Miklós street
1121 Budapest, Hungary
e-mail: lukacs.arpad@wigner.mta.hu

Uri MAOR
Tel-Aviv University
School of Physics and Astronomy
69978 Ramat Aviv, Tel-Aviv, Israel
e-mail: maor@post.tau.ac.il

Larry McLERRAN
Department of Physics, Brookhaven National Laboratory,
Building 510, P.O.Box 5000
Upton NY 11973-5000 USA
e-mail: mclerran@mac.com

Aleksei V. NEFEDIEV
Institute of theoretical and Experimental Physics
25 ul. B. Cheremushkinskaya
117218 Moscow, Russia
e-mail: nefediev@itep.ru

Nikolai N. NIKOLAEV
L.D. Landau Institute for Theoretical Physics
1A ul. Akad. Semenova
142432 Chernogolovka, Russia
e-mail: nikolaev@itp.ac.ru

Victor A. NOVIKOV
Institute of Theoretical and Experimental Physics
25 ul. B. Cheremushkinskaya
117218 Moscow, Russia
e-mail: novikov@itep.ru

Júlia NYÍRI
MTA Wigner Research Centre for Physics
29-33 Konkoly-Thege Miklós street
1121 Budapest, Hungary
e-mail: nyiri.julia@wigner.mta.hu

Yuri F. ORLOV
Laboratory of Nuclear Studies,
Cornell University
Ithaca, New York, 14853 USA
e-mail: yfo1@cornell.edu

András PATKÓS
Department of Atomic Physics
Eötvös University
1/A Pázmány Péter sétány
1117 Budapest, Hungary
e-mail: patkos@galaxy.elte.hu

Grigorii B. PIVOVAROV
RAS Institute for Nuclear Research
Moscow, Russia
e-mail: gbpivo@gmail.com

Yaroslav P. PUGAI
L.D. Landau Institute for Theoretical Physics
1A ul. Akad. Semenova
142432 Chernogolovka, Russia
e-mail: slava@itp.ac.ru

Vladimir A. SALEEV
Samara State University,
and
Samara State Aerospace University
e-mail: saleev@samsu.ru

Wolfgang J. SCHAEFER
PAN Institute of Nuclear Physics
Krakow, Poland
e-mail: wolfgang.schafer@ifj.edu.pl

Dominique SCHIFF
Laboratoire de Physique Theorique d'Orsay
L'Universite Paris-Sud
Orsay, France
e-mail: Dominique.Schiff@th.u-psud.fr

Aleksandra V. SHIPILOVA
Samara State University,
and
Samara State Aerospace University
Samara, Russia
e-mail: alexshipilova@samsu.ru

Alexander J. SILENKO
Belarusian State University, Minsk, Belarus
and
Joint Institute for Nuclear Research, Dubna, Russia
e-mail: alsilenko@mail.ru

Noemi SMORODINSKAYA
Institute for Theoretical and Experimental Physics
25 ul. B. Cheremushkinskaya
117218 Moscow, Russia
e-mail: naya@itep.ru

Mark STRIKMAN
Department of Physics
Pennsylvania State University
104 Davey Laboratory
University Park, PA 16802-6300 USA
e-mail: strikman@phys.psu.edu

Antoni SZCZUREK
PAN Institute of Nuclear Physics
Krakow, Poland
e-mail: antoni.szczurek@ifj.pl

Oleg TERYAEV
Joint Institute for Nuclear Research
Dubna, Russia
e-mail: teryaev@theor.jinr.ru

Arkady VAINSHTEIN
William I. Fine Theoretical Physics Institute
School of Physics and Astronomy
University of Minnesota
116 Church Street S.E.
Minneapolis, MN 55455 USA e-mail: vainshte@umn.edu

Sergei N. VERGELES
L.D. Landau Institute for Theoretical Physics
1A ul. Akad. Semenova
142432 Chernogolovka, Russia
e-mail: vergeles@itp.ac.ru

Grigory E. VOLOVIK
Low Temperature Laboratory, Aalto University,
P.O.Box 15100, FI-00076 Aalto, Finland
and
L.D. Landau Institute for Theoretical Physics
1A ul. Akad. Semenova
142432 Chernogolovka, Russia
e-mail: volovik@boojum.hut.fi

Mikhail VYSOTSKY
Institute of Theoretical and Experimental Physics
25 ul. B. Cheremushkinskaya
117218 Moscow, Russia

Alan WHITE
Division of High Energy Physics,
Argonne National Laboratory
9700 Southcass Avenue
Argonne, IL 60439-4815 USA
e-mail: arw@hep.anl.gov

Bronislav G. ZAKHAROV
L.D. Landau Institute for Theoretical Physics
1A ul. Akad. Semenova
142432 Chernogolovka, Russia